Introductory Practical BIOCHEMISTRY

Introductory Practical
BIOCHEMISTRY

Editors
S.K. Sawhney
Randhir Singh

Alpha Science International Ltd.
Harrow, U.K.

Editors
S.K. Sawhney
Randhir Singh
Department of Biochemistry
CCS Haryana Agricultural University
Hisar, India

Alpha Science International Ltd.
Hygeia Building, 66 College Road
Harrow, Middlesex HA1 1BE, U.K.

ISBN 1-84265-245-1

Printed in India.

Preface

Biochemistry is now considered to be one of the fundamental subjects for students of biology and an increasing number of Indian Universities have begun to include it in their undergraduate programmes. With the exception of some of the Universities which have a well established and full fledged Department of Biochemistry, the main shortcoming in offering of this subject, particularly in affiliated degree colleges, is the inadequate laboratory training imparted to the students during the course. Unlike other basic science subjects like chemistry, botany and zoology, biochemistry practicals generally require rare chemicals and expensive equipments. Inability to garner adequate facilities for organizing meaningful practicals for large group of students happens to be the principal impediment in introduction and promotion of this discipline at undergraduate level. The present book includes a fairly large number of practicals taking fully into account the meagre resources at the disposal of the instructors. Various chapters dealing with different aspects of biochemistry, are written by the scientists who have long experience and are well versed with organizing and conducting practicals of the biochemistry courses offered to the undergraduate and postgraduate students.

Each chapter in this book begins with theoretical aspects of a group of techniques followed by suggested laboratory exercises which can be undertaken by students of different levels. Several of the experiments do not involve use of rare or expensive biochemicals or costly equipments and can effectively be conducted in departments having simple instruments such as colorimeter.

The later part of the book deals with fairly advanced techniques and will be suitable for students of postgraduate level. Hence, the book will prove equally useful for the students of undergraduate and postgraduate programmes and will provide a good coverage to practical component for teaching of biochemistry at college and university levels. Besides the book is also expected to meet the requirements of the students of other disciplines of life sciences such as botany, zoology, microbiology, molecular biology, biotechnology etc. In addition, it is hoped that the book would also serve as a useful companion for teachers engaged in teaching practical biochemistry to the above category of students.

The editors are thankful to all the contributors for their cooperation. Finally the authors solicit suggestions for improvement and enlargement of this book from the biochemists, instructors, students and any other person who may be interested in promoting this discipline.

S.K. Sawhney
Randhir Singh

Contributors

1. Dharam Singh Chopra (Chapter 13)
Department of Biochemistry
CCS Haryana Agricultural University
Hisar-125 004

2. L.K. Chugh (Chapter 1)
Regional Research Station
College of Agriculture
Kaul, Haryana

3. Santosh Dhillon (Chapters 5, 10 and 14)
Department of Biochemistry
CCS Haryana Agricultural University
Hisar-125 004

4. Kaushalya Gupta (Chapters 2 and 4)
Department of Biochemistry
CCS Haryana Agricultural university
Hisar-125 004

5. Sudhir Kumar (Chapter 11)
Department of Chemistry and Physics
CCS Haryana Agricultural University
Hisar-125 004

6. Shashi Madan (Appendix)
Dapartment of Plant Breeding
CCS Haryana Agricultural University
Hisar-125 004

7. Sarla Malhotra (Chapters 7, 9 and 12)
Department of Biochemistry
CCS Haryana Agricultural University
Hisar-125 004

8. M.S. Malik (Chapter 11)
Department of Chemistry and Physics
CCS Haryana Agricultural University
Hisar-125 004

9. Gaya Parsad (Chapter 13)
Department of Veterinary Microbiology
CCS Haryana Agricultural University
Hisar-125 004

10. S.K. Sawhney (Chapters 4, 8 and Appendix)
Department of Biochemistry
CCS Haryana Agricultural University
Hisar-125 004

11. Veena Sawhney (Chapter 3)
Directorate of Research
NARP Project, College of Agriculture
CCS Haryana Agricultural University
Hisar-125 004

12. H.R. Singal (Chapter 8)
Department of Biochemistry
CCS Haryana Agricultural University
Hisar-125 004

13. Randhir Singh (Chapters 7, 9 and 12)
Department of Biochemistry
CCS Haryana Agricultural University
Hisar-125 004

14. D.R. Sood (Chapter 6)
Department of Biochemistry
CCS Haryana Agricultural University
Hisar-125 004

Abbreviations

A_λ	Absorbance with wavelength (λ) denoted by the subscript
Å	Angstrom (1 Å = 0.1 nm)
bp	Base pairs
BSA	Bovine serum albumin
Ci	Curie
cpm	Counts per min
Conc	Concentrated
dpm	Disintegrations per min
DTT	Dithiothreitol
ε_λ	Molar extinction coefficient at the wavelength (λ) denoted by subscript
EDTA	Ethylenediamine tetraacetic acid
g	Gram
HEPES	N-2-hydroxyethylpiperazine-N′-2-ethanesulphonic acid
h	Hours
Kb	Kilobase
kD	Kilodalton
K_i	Dissociation constant for EI complex
K_m	Michaelis-Menten constant
L	Litres
M	Molar
min	Minutes
MOPS	[3–(N–morpholino) propane sulphonic acid]
N	Normal
NAD^+	Nicotinamide adenine dinucleotide (oxidised form)
NADH	Nicotinamide adenine dinucleotide (reduced form)
$NADP^+$	Nicotinamide adenine dinucleotide phosphate (oxidised form)
NADPH	Nicotinamide adenine dinucleotide phosphate (reduced form)
nm	Nanometer
O.D.	Optical density

P_i	Inorganic phosphate
POPOP	1,4-Bis-2-(5-phenyloxazol-2-yl)-benzene
PPO	2,5-Diphenyloxazole
psi	Pounds per square inch
R_f	Relative front
S	Sedimentation coefficient in Svedberg units
SDS	Sodium dodecyl sulphate
sec	Seconds
Tricine	[N-*tris*(hydroxymethyl) methyl glycine]
Tris	[*Tris*-(hydroxymethyl) aminomethane]
UV	Ultra-violet
v	Velocity
V_e	Elution volume
V_{max}	Maximum velocity
V_o	Void volume
vs	Versus
v/v	Volume by volume
w/v	Weight by volume

Contents

1. Solutions and Buffers

1.1 Introduction

Almost all the experimental techniques in biochemistry involve the use of solutions. These have to be used for extraction of biomolecules from tissues, their separation, quantitative estimation, purification and even for their physico-chemical characterization. Therefore, it is important that a student of biochemistry must be fully familiar with the proper preparation of solutions and buffers.

1.2 Solutions

A solution is homogeneous mixture of two or more non-reacting substances and has uniform properties such as chemical composition, density, refractive index etc. However, its composition can be varied within certain fixed limits.

A solution which is made up of two components is called binary solution. The dissolved substance in solution is called the solute and the medium in which it is dissolved is known as the solvent. The solute in a solution is always present in a smaller quantity than the solvent. The most commonly used solutions in biochemical work are of solid-liquid and liquid-liquid type. A few examples of such solutions are given below:

Solute	*Solvent*	*Example*
Solid	Liquid	Sugar in water, AMP in buffer, Tris in water
Liquid	Liquid	Glycerol in water, alcohol in water, 2-mercaptoethanol in buffer

Composition of a solution can be expressed in two ways: quantity and concentration. Quantity is the amount of any substance (solute) present in a solution/solvent irrespective of the amount/volume of the solvent/solution. In contrast, concentration refers to the quantity of the solute present in an exact or a specific amount of solvent or that of solution. For example, a solution containing 4 g NaOH in a volume of 100 ml or even 200 ml of a solution has a total quantity of 4 g NaOH. When this amount is expressed per unit volume such as 4 g/100 ml of the solution, then it is termed as 4% (concentration) solution. Thus in this example, per cent is a unit of concentration where as the amount is the unit of quantity.

The term mole, which is defined as the molecular weight of a compound in grams, is frequently used while indicating the quantity of a compound and for preparation of solutions. Thus,

1 mole = g molecular weight of the substance, and

1 mole of glucose = 180 g (Mol. wt. = 180)

1 mole of albumin = 68,000 g (Mol. wt. = 68,000)

1.2.1 Mode of expressing concentration of a solution

Following are some of the common ways of expressing the concentration of solutions:

1.2.1.1 Molarity (M)

This is the most common method for expressing the concentration of a solution in biochemical studies. The molarity of a solution is the number of moles of the solute dissolved per L of the solution. A solution which contains 1 mole of the solute in one L of the solution is called a molar solution. Molarity of a solution can be calculated as follows:

$$\text{Molarity} = \frac{\text{Weight of a solute in g/L of solution}}{\text{Mol. wt. of solute}}$$

It may be noted that in case of molar solutions, the combined total volume of the solute and solvent is one L. Thus for preparing 0.1 M NaOH, one may proceed as follows:

$$\text{Mol. wt. of NaOH} = 40$$

$$\text{Required molarity of solution} = 0.1 \text{ M}$$

$$\therefore \text{Amount (in g) of NaOH per L of solution} = \text{Mol. wt. of NaOH} \times \text{molarity}$$

$$= 40 \times 0.1 = 4 \text{ g}$$

Thus, weigh 4 g of NaOH, dissolve it in a small volume of solvent (water) and make the final volume to 1 L with the solvent.

Sometimes it is desirable to know number of moles of a substance in a reaction mixture. This can be calculated using a simple relationship:

1 M solution = 1 mole of the substance/L of solution.

= 1 mmole/ml of solution

= 1 μmole/μl of solution

1 mM solution = 1 mmole/L of solution

= 1 μmole/ml of solution

1.2.1.2 Molality (m)

A solution which contains 1 mole of the solute dissolved in 1 kg of the solvent is called a molal solution. Hence,

$$\text{Molality} = \frac{\text{Weight of a solute in g/kg of solvent}}{\text{Mol. wt. of solute}}$$

It is important to remember that in a molal solution, the amount of solvent is 1000 g. Thus in case of aqueous solution, 1 molal solution is obtained by dissolving

1 mole of the solute in 1000 ml (since Sp. Gravity = 1) of water. For example, for preparing 1 m Na_2CO_3 solution, dissolve 106 g of Na_2CO_3 (Mol. wt. of Na_2CO_3 = 106) in one kg of water.

1.2.1.3 Normality (N)

The normality of a solution is the number of gram equivalents of the solute per L of the solution.

A solution having one g equivalent of the solute per L of solution is called 1N solution. Therefore,

$$\text{Normality} = \frac{\text{Amount of a substance in g/L of solution}}{\text{Eq. wt. of substance}}$$

For preparing 0.1 N Na_2CO_3 (Eq. wt. of Na_2CO_3=53) solution, dissolve 5.3 g Na_2CO_3 in a final volume of 1 L of solution.

1.2.1.4 Mass concentration

Substances like proteins, nucleic acids etc. which do not have a uniformly defined composition, their concentration is expressed in terms of weight per unit volume rather than moles per unit volume. The unit of volume is L, so all concentrations should be expressed with L (g/L, mg/L, μg/L etc.). The term per cent (%) is also quite commonly used. However, to avoid any ambiguity it is necessary to properly define the basis of % solution as illustrated by the following example. A 5% solution of acetic acid could mean:

5 g of acetic acid per 100 g of solution (*w/w*)
5 g of acetic acid per 100 ml of solution (*w/v*)
5 ml of acetic acid per 100 ml of solution (*v/v*)

Thus, 1% (*w/v*) solution of casein would imply 1 g of casein dissolved in solvent to give a final volume of 100 ml of the solution.

1.2.1.5 Mass fraction

The mass fraction of a component in solution is the mass of that particular component per unit mass of the solution. If W_A and W_B represent the masses of components A and B in a solution, then

$$\text{Mass fraction of A} = \frac{W_A}{W_A + W_B}$$

$$\text{Mass fraction of B} = \frac{W_B}{W_A + W_B}$$

1.2.1.6 Mass percentage or % (w/w)

It is the weight of the component present in 100 parts by weight of the solution. In a solution containing 10 g sugar in 40 g of water, then

$$\text{mass \% of sugar} = \frac{10 \times 100}{(10 + 40)} = 20\%$$

1.2.1.7 Percentage by volume or % (v/v)

It is the volume of the component in 100 parts by volume of the solution. In a solution containing 20 ml alcohol in 80 ml of water, the % volume of alcohol will be $\frac{20 \times 100}{(20 + 80)} = 20\%$

1.2.1.8 Parts per million (ppm)

This is generally employed for those solutions in which a substance is present in a very small quantity. It represents gram of a solute per million grams of solution or the gram of a solute per million ml of the solution.

$$\text{ppm} = \frac{\text{mass of the component}}{\text{total mass of the solution}} \times 10^6$$

or

$$\text{ppm} = \frac{\text{g or ml of solute or substance}}{\text{g or ml of solution}} \times 10^6$$

Thus, 1 ppm of solution of NaCl in water represents

$$1 \text{ ppm} = 1 \text{ mg NaCl/L of solution}$$

or

$$= 1 \text{ mg NaCl/1000 ml of solution}$$

or

$$= 1 \mu\text{g NaCl/ml of solution}$$

1.2.2 Types of Solutions

1.2.2.1 Stock solution

Stock solution of a substance is the one having a concentration many folds higher than that actually required in the experiment. Stock solutions are prepared of the substances that are to be used frequently and are stable at higher condentration for several days and can be used after appropriate dilution just before use. It is sometimes convenient to weigh out a relatively large amount of the compound and prepare a stock solution of that compound from which small amounts can be withdrawn at convenience and added to solution of other components. The use of a stock solution thus cuts down on the amount of pipetting and at the same time reduces variability between a number of similar incubation mixtures, assay mixtures etc.

Frequently, a particular volume of a solution containing a desired concentration of a substance has to be prepared by using its stock solution. A simple mathematical equation can be used to calculate the volume of the stock solution needed to prepare a solution of required volume containing desired concentration of the compound:

$$N_1V_1 = N_2V_2$$

where N_1 = concentration of the solution to be prepared
V_1 = volume of the solution to be prepared
N_2 = concentration of the stock solution
V_2 = volume of the stock solution

For example, for preparing 150 ml of a solution containing 10 mM Tris from a stock of 75 mM Tris, the volume of the stock solution which should be used can be calculated as follows:

$$V_2 = \frac{N_1V_1}{N_2} = \frac{10 \times 150}{75} = 20 \text{ ml}$$

Thus by taking 20 ml of 75 mM stock solution of Tris and making its final volume to 150 ml (by adding water or other solutions as per the requirement of an experiment) a solution containing 10 mM Tris is obtained.

1.2.2.2 Standard solution

A solution of known concentration is referred to as a standard solution. In many experiments, for quantitative determination of a particular substance in a sample preparation, a standard or reference solution is required. For example, for quantitative determination of total proteins in a sample preparation, a standard solution of a protein like bovine serum albumin is used for comparison. The standard solution is prepared accurately by weighing a fixed or known amount of bovine serum albumin and adding precisely measured volume of solvent. The standard solution serves the purpose of a reference. Very often the standard solution containing different concentrations of a compound is used for preparing a standard curve. This can be done by making a stock solution from which solutions of graded concentrations are made e.g. 1, 5, 10, 50 and 100 μg protein/ml and their experimental values such as O.D. are determined. Then a graph of O.D. *versus* amount of protein is plotted to obtain a standard or reference curve.

1.2.2.3 Saturated solution

Solubility of any substance in a particular medium, say water, varies with temperature. When a solution contains the solute in an amount in excess of that which can completely be dissolved at a given temperature and the solute in solution is in equilibrium with the excess of undissolved solute, the solution is said to be saturated.

1.2.2.4 Solutions of acids

Since most of the commonly used acids like hydrochloric acid and sulphuric acid are not pure, their effective strength is low which ought to be taken into account while making their solutions of desired concentrations. The following formula can be applied to compute the volume of the commercially available concentrated acid required for preparing a solution of required normality:

$$V_1 = \frac{\text{Eq. wt. of acid} \times V_2 \times \text{normality} \times 100}{1000 \times \text{specific gravity} \times \text{purity (\%)}}$$

where V_1 = required volume of the concentrated acid
V_2 = total volume of the acid solution of desired normality to be prepared.

The use of above formula is illustrated by the following example.

Exercise: Prepare 750 ml of 1N sulphuric acid from concentrated acid (purity, 96%; specific gravity, 1.84; Mol. wt., 98).

Eq. wt. of sulphuric acid = 49 (since sulphuric acid is diprotic)

Putting various values in the above equation, we get

$$V_1 = \frac{49 \times 750 \times 1 \times 100}{1000 \times 1.84 \times 96} = 20.8 \text{ ml}$$

thus, take 20.8 ml of concentrated sulphuric acid and dilute it to 750 ml with water to get 1 N solution of sulphuric acid.

1.2.3 Precautions

(i) Aqueous solutions should always be made using pure water (deionized or double distilled).

(ii) Solution should remain clear after preparation.

(iii) Solution should always be properly stored, e.g. some solutions can be stored at room temperature while others require storage at low temperatures. Solutions which are sensitive to light should always be stored in brown bottles.

(iv) Always note down the Mol. wt. printed on the bottle and take into account hydration state of the substance, purity, specific gravity etc.

1.2.4 Practical exercises

Use of solutions of specified compositions is one of the fundamental requisites for obtaining accurate, reliable and reproducible results in almost all biochemical investigations. Incomplete or improper knowledge of solution preparation is bound to give erroneous results. A few 'Exercises' are given below with a view to provide some practice to the students for preparing various types of solutions that are commonly used in biochemical experiments:

1. How many g of NaOH would be required to make 50 ml of 0.5 mole/L solution of NaOH? (Mol. wt. of NaOH = 40)
2. How many mmole/ml of the solute are contained in the following solutions:
   (a) 0.20 M NaCl
   (b) 10 mM Glucose
   (c) 100 μM Glutamate
   (d) 30 g of urea in 100 ml of solution (Mol. wt. of urea = 60)
3. How many g of alanine are present in 20 ml of a 50 mM solution of this amino acid (Mol. wt. of alanine = 89).
4. Conc HCl is 37.5% HCl by wt. and has a density of 1.19. Describe how would you prepare 500 ml of 0.2 N HCl solution (Mol. wt. of HCl = 36.5).
5. Describe preparation of 300 ml of 0.5 N CH_3COOH (Acetic acid is 100% pure, has a density of 1.05 and its Mol. wt. is 60).
6. Calculate the volume of 200 mM glucose solution required to make 25 ml of 50 mM glucose solution.
7. You are provided with 95% (*v*/*v*) ethanol. Explain how would you obtain 150 ml of 70% (*v*/*v*) ethanol?

8. How would you prepare 85 ml of 0.02% (*w/v*) solution of ninhydrin in acetone?
9. You need 10 ml of 0.5 M maltose for an experiment. Explain how would you make this solution (Mol. wt. of maltose = 342).
10. According to an experimental procedure you are required to prepare 5 ml of a reaction mixture containing 20 mM $MgCl_2$. This amount of $MgCl_2$ has to be supplied by adding 0.1 ml of $MgCl_2$ solution to 4.9 ml of the reaction mixture. Calculate how much $MgCl_2$ would you take for preparing 1 ml of its stock solution to fulfill the above requirement (Mol. wt. of $MgCl_2$ = 95.3).

1.3 Buffers

Buffers are extensively used in biochemical studies since these aid in maintaining a near constant pH of the media while performing various laboratory operations such as during extraction, isolation and purification of various biomolecules. Selection of an appropriate buffer with optimal pH is important as it may have a profound influence on extractability, stability and even biological functioning of cell constituents. For example, all the biochemical reactions in a cell are catalysed by enzymes whose stability as well as activity is highly dependent on the pH of the system. In several reactions protons may be consumed or released during an enzyme reaction. In such cases it, therefore, becomes increasingly important to provide a system i.e. buffer which could stabilize the hydrogen ion concentration and obviate adverse effects due to change in pH on the enzyme under investigation. Before considering how buffers function, it is necessary to know some of the important terms associated with the concept of pH and buffers.

1.3.1 Acids and bases

The modern concept of acids and bases defines acid as proton donor and a base as proton acceptor. On ionization, an acid donates a proton and at the same time a corresponding base (which is capable of accepting a proton) is formed. Such a base is known as conjugate base:

$$\text{Acid} \rightleftharpoons H^+ + \text{Base}$$

Acids and bases are classified into two groups: strong and weak acids and bases. Strong acids or bases are those which get completely ionized in solution so that the concentration of free H^+ or OH^- is the same as the concentration of the acid or base. Example:

Strong acid: $HCl \rightarrow H^+ + Cl^-$

Strong base: $NaOH \rightarrow Na^+ + OH^-$

On the other hand, weak acids or bases dissociate only to a limited extent and the concentration of free H^+ and OH^- depends on the dissociation constant.

Weak acid: $CH_3COOH \rightleftharpoons CH_3COO^- + H^+$

1.3.2 Dissociation of water, its ionic product K_w and concept of pH

Water is weak electrolyte which dissociates only slightly to form H^+ and OH^- ions.

$$H_2O \rightleftharpoons H^+ + OH^-$$

The equilibrium constant of this dissociation reaction is 1.8×10^{-16} mole/L at 25°C

Thus $$K_{eq} = \frac{[H^+][OH^-]}{[H_2O]} = 1.8 \times 10^{-16}$$

1000 ml of pure water contains 1000 g of water. Since the molecular weight of water is 18 its molar concentration can be calculated and is equal to 1000/18 or 55.5 moles/L. Because the concentration of H_2O in dilute aqueous solution is essentially unchanged from that in pure H_2O, this figure may be taken as constant. Thus,

$$\frac{[H^+][OH^-]}{55.5} = 1.8 \times 10^{-16}$$

$$[H^+]\,[OH^-] = 1.8 \times 10^{-16} \times 55.5$$

$$= 1.01 \times 10^{-14}$$

or $$K_w = 1.01 \times 10^{-14}$$

where K_w is ion product of water which expresses the relationship between the concentration of H^+ and OH^- ions in aqueous solutions. This equation may be used to calculate the concentration of H^+ in pure water. Let x be the concentration of H^+ on dissociation of water. Since production of one H^+ is associated with generation of one OH^-, the concentration of OH^- will also be x.

Hence, $$[H^+]\,[OH^-] = 1.01 \times 10^{-14}$$

or $$x^2 = 1.01 \times 10^{-14}$$

$$x = 1.01 \times 10^{-7} \text{ moles/L}$$

A more convenient method of expressing concentration of H^+ ions is by means of pH which is defined as negative logarithm of H^+ ion activity.

Thus, $$pH = \log \frac{1}{aH^+} = -\log aH^+$$

where aH^+ is defined as the activity of H^+. In this context no distinction is made between activities and concentration. Hence,

$$pH = \log \frac{1}{H^+} = -\log [H^+]$$

It is important to note that the pH is a logarithmic function, thus when the pH of a solution decreases by one unit from 5 to 4, the H^+ ion concentration increases by ten fold, i.e. from 10^{-5} to 10^{-4} M. If we now apply the term pH to the ion product of water

$$[H^+]\,[OH^-] = 1.0 \times 10^{-14}$$

$$\log [H^+] + \log [OH^-] = \log (1.0 \times 10^{-14})$$

$$-\log [H^+] - \log [OH^-] = 14$$

$$pH + pOH = 14$$

or

$$pH = 7$$

Hence pH of pure water is 7.0

1.3.3 Henderson-Hasselbalch equation

The Henderson-Hasselbalch equation is simply another way of expressing dissociation constant of an acid

$$HA \rightleftharpoons H^+ + A^-$$

$$K_{ion} = k' = \frac{[H^+]\,[A^-]}{[HA]}$$

on rearranging,

$$[H^+] = k' \frac{[HA]}{[A^-]}$$

Taking negative logarithm on both sides

$$-\log [H^+] = -\log k' - \log \frac{[HA]}{[A^-]}$$

Substituting pH for – log [H^+] and pK′ for – log k′, where pK′ is – log of dissociation constant

$$pH = pK' - \log \frac{[HA]}{[A^-]}$$

or

$$pH = pK' + \log \frac{[A^-]}{[HA]}$$

This is known as Henderson-Hasselbalch equation which can be written in a more general form as

$$pH = pK' + \log \frac{[\text{conjugate base}]}{[\text{conjugate acid}]}$$

It is an extremely useful equation from which either pH of the solutions of various concentration ratios of a conjugate acid-base pair of known pK′ can be calculated or the ratio of conjugate acid-base of known pK′ to obtain a buffer of desired pH can be found out.

1.3.4 Buffer system

A buffer system is one that resists a change in pH on the addition of acid or alkali and constitutes of conjugate acid-base pair. Most commonly, the buffer solution consists of a mixture of a weak acid and its conjugate base, e.g. a mixture of acetic acid and sodium acetate is a buffer solution. A well known physiological buffer system is the carbonate-bicarbonate system of blood.

The basis for functioning of a buffer with regards to its ability to resist change in pH can be illustrated by the following example. Suppose we have a buffer containing 5 ml of 0.1 M sodium acetate and 4 ml of 0.1 M acetic acid. Let us examine the effect on pH on addition of 1 ml of 0.1 N HCl to this buffer:

(a) Before adding acid

Total volume of buffer is (5 ml + 4 ml) = 9 ml

$$\text{Conc of acetate} = 5/9 \times 0.1\ \text{M} = 0.5/9\ \text{M}$$

$$\text{Conc of acetic acid} = 4/9 \times 0.1\ \text{M} = 0.4/9\ \text{M}$$

$$\text{pK}' \text{ of acetic acid} = 4.76$$

$$\text{Thus, pH} = 4.76 + \log \frac{0.5/9}{0.4/9}$$

$$= 4.76 + (0.097) = 4.86$$

(b) After adding acid

HCl provides H^+ which combine with the acetate ion to give acetic acid. Thus, it reduces the amount of acetate ions present and increases the amount of undissociated acetic acid leading to an alteration in the salt/acid ratio and hence to a change in pH.

The final volume of buffer after acid addition = 10 ml

$$\text{Conc of acetate after acid addition} = 5/10 \times 0.1\ \text{M} - 1/10 \times 0.1\ \text{M}$$

$$= 0.04\ \text{M}$$

$$\text{Conc of acetic acid after acid addition} = 4/10 \times 0.1\ \text{M} + 1/10 \times 0.1\ \text{M}$$

$$= 0.05\ \text{M}$$

$$\text{Thus, pH} = 4.76 + \log 0.04/0.05$$

$$= 4.76 + (-0.097)$$

$$= 4.66$$

Thus, the pH changes marginally from 4.86 to 4.66 on addition of 1 ml of 0.1 N HCl. However, if the same amount of the acid is added to pure water its pH will change from 7.0 to 2.0.

1.3.5 Preparation of buffers

Information regarding preparation of most of the commonly used buffers is available in tabulated form in several books (see Methods in Enzymology Vol. 1 and Vol. 182). These tables have been computed using Henderson-Hasselbalch equation. It is, however, quite simple to do calculations for preparing a buffer of required pH as shown in an example as follows:

Suppose one wishes to prepare 1 L of 0.1 M acetate buffer of pH 5.22 (pK'_a of acetic acid is 4.74). First determine the required ratio of conjugate base (acetate ion) to weak acid (acetic acid) in this buffer solution using Henderson-Hasselbalch equation.

$$pH = pK' + \log\frac{[CH_3COO^-]}{[CH_3COOH]}$$

Putting the values $$5.22 = 4.76 + \log\frac{[CH_3COO^-]}{[CH_3COOH]}$$

or $$\log\frac{[CH_3COO^-]}{[CH_3COOH]} = 5.22 - 4.76 = 0.48$$

or $$\frac{[CH_3COO^-]}{[CH_3COOH]} = \text{antilog } 0.48 = 3$$

Thus, in this buffer solution the molar ratio of CH_3COO^- to CH_3COOH should be 3. In other words, 75% of the buffer components should be present as the conjugate base CH_3COO^- and 25% of the components as acetic acid. Since 1 L of 0.1 M acetate buffer should contain 0.1 mole of acetate plus acetic acid, the solution should have 0.075 moles of acetate ions and 0.025 moles of acetic acid. Finally calculate the amount in g of sodium acetate required to obtain 0.075 moles of acetate ions. Similarly, to obtain 0.025 moles of acetic acid 1.5 g of acetic acid would be required. Prepare these solutions, mix them and make the final volume to 1 L with water. This will give 1 L of buffer of the desired pH and concentration.

1.3.6 Criteria for selection of buffers

While choosing a buffer the following factors need to be taken into account. It should:

(i) possess adequate buffering capacity in the required pH range. Generally buffers are most effective over a range of one pH unit on either side of their pK′ value, e.g. Tris which has pK′ value of 8.3 has an effective pH range of 7–9. The pK′ values of some important buffers are given in Appendix (Table 6)

(ii) be chemically inert and not react or bind with biomolecules or other components, particularly, for assaying activities of enzymes which require a metal ion for their functioning

(iii) be available in high degree of purity and should not contain impurities which may interfere with estimations

(iv) be enzymically and hydrolytically stable

(v) maintain pH that is minimally influenced by temperature, ionic composition and concentration or salt effect of the medium

(vi) not be toxic

(vii) not absorb light in the visible or ultraviolet regions

1.3.7 Measurement of pH

By using pH indicators (natural or chemically synthesized organic compounds) an approximate value of pH of a solution can be obtained. The indicators dissociate like a weak acid on coming in contact with solution and give range of colours depending on pH. Dissociated and undissociated form of the indicator have different colours. pH papers coated with indicators of different pH range are commercially available. However, accurate pH can be measured using pH meter which measures e.m.f. of a concentration cell developed from a reference electrode, test solution and a glass electrode sensitive to H^+ ions. The combined electrode constituting of glass and reference electrode is dipped into the test solution for accurate measurement of pH.

1.3.8 Composition of some commonly used buffers

Buffers can be made in stock solutions and these are diluted before use. Preparation of some of buffers frequently used in biochemical studies is given below. Unless otherwise stated, by following these procedures 0.1 M buffer will be obtained.

(a) Acetate buffer

Stock solutions

A. 0.2 M acetic acid (11.55 ml/L)
B. 0.2 M sodium acetate (16.4 g of sodium acetate or 27.2 g of sodium acetate $3H_2O$ per L)

x ml of A + y ml of B, diluted to a total volume of 100 ml

x	*y*	*pH*	*x*	*y*	*pH*
46.3	3.7	3.6	25.5	24.5	4.6
44.0	6.0	3.8	14.8	35.2	5.0
41.0	9.0	4.0	10.5	39.5	5.2
36.8	13.2	4.2	8.8	41.2	5.4
30.5	19.5	4.4	4.8	45.2	5.6

(b) Phosphate buffer

Stock solutions

A. 0.2 M monobasic sodium phosphate (27.8 g in 1 L)
B. 0.2 M dibasic sodium phosphate (53.65 g of $Na_2HPO_4 \cdot 7H_2O$ or 71.7 g $Na_2HPO_4 \cdot 12\ H_2O$ in 1 L)

x ml of *A* + *y* ml of *B*, diluted to a total volume of 200 ml

x	*y*	*pH*	*x*	*y*	*pH*
93.5	6.5	5.7	56.5	43.5	6.7
90.0	10.0	5.9	39.0	61.0	7.0
85.0	15.0	6.1	16.0	84.0	7.5
77.5	22.5	6.3	5.3	94.7	8.0
68.5	31.5	6.5			

(c) Tris (hydroxymethyl) aminomethane buffer or Tris buffer:

Stock solutions

A. 0.2 M Tris (hydroxymethyl) aminomethane (24.2 g per L)
B. 0.2 M HCl

50 ml of A + y ml of B diluted to a total volume of 200 ml. 0.05 M Tris-HCl buffer will be obtained.

y	*pH*	*y*	*pH*
5.0	9.0	26.8	8.0
8.1	8.8	32.8	7.8
12.2	8.6	38.4	7.6
16.5	8.4	41.4	7.4
21.9	8.2	44.2	7.2

(d) Preparation of imidazole, MOPS and HEPES buffers

For preparing these buffers, the molar ratio of the protonated to non-protonated species for obtaining buffer of required pH can be accomplished by mixing the computed amount of free buffer and one of its salts (from Henderson-Hasselbalch equation). In the Table given below, buffer as free acid and its salt can be mixed in the ratio ranging from 90:10 to 10:90 to obtain the desired pH. For example, for preparing 100 ml 0.1 M HEPES buffer of pH 7.22, mix 70 ml of 0.1 M HEPES and 30 ml of 0.1 M sodium salt of HEPES.

% buffer (as free acid)	*% buffer (as free base)*	*pH*		
		Imidazole	*MOPS*	*HEPES*
90	10	6.09	6.19	6.60
80	20	6.44	6.54	6.95
70	30	6.68	6.78	7.18
60	40	6.87	6.97	7.37
50	50	7.05	7.15	7.55
40	60	7.18	7.32	7.73
30	70	7.22	7.52	7.92
20	80	7.35	7.75	8.15
10	90	8.00	8.05	8.50

In laboratory, these buffers are normally prepared by titrating a solution of the free buffer with base (NaOH or KOH) and the pH is continuously monitored on a pH meter. For example, if 1 L of 0.1 M MOPS buffer of pH 6.6 is needed, first calculate the amount of MOPS (free acid) required to make 0.1 M solution. Dissolve this amount in approximately 800 ml water. Titrate this solution with concentrated solution of NaOH (e.g. 1 M) till the desired pH is obtained and then make the volume to 1 L with water.

1.3.9 Laboratory Exercises

Using the information in the above Tables, students may gain practice and experience in preparation of the buffers. Some such experiments are suggested below:

EXPERIMENT 1: Prepare a 0.1 M phosphate buffer of pH 7.5

EXPERIMENT 2: Examine the effect on pH of using double strength solutions of the components A and B of Tris buffer and mixing them in a selected fixed ratio.

EXPERIMENT 3: Determination of effect of addition of a fixed volume of an acid to a buffer of different molarities. For this prepare 0.025, 0.05, and 0.1 M buffer having identical pH. Take 50 ml of each of these buffers and add 5 ml of 0.1 M HCl. Note the change in pH. Also observe whether there exists any relationship between molarity of the buffer and its capacity to resist the change in its pH.

Suggested further reading

Blanchard J.S. (1984). Buffers for enzymes. *In* Methods in Enzymology, Vol. 104 Part C, (Jackoby WB, ed.), pp 404–417, Academic Press Inc., New York.

Conn E.E. and Stumpf P.K. (1987). Outlines of Biochemistry, 4th ed., pp 3–24, Wiley Eastern Ltd., New Delhi.

Gabb M.H. and Latchem W.E. (1967). Handbook of Laboratory Solutions, Andre Deutsch Ltd., London.

Gomori G. (1955). Preparation of buffers for use in enzyme studies. *In* Methods in Enzymology, Vol. 1 (Colowick PS and Kaplan NO, eds.), pp 138–146, Academic Press Inc., New York.

Segel I.H. (1976). Biochemical Calculations, 2nd ed., Wiley, New York.

Stoll V.S. and Blanchard J.S. (1990). Buffers: Principles and Practice. *In* Methods in Enzymology, Vol. 182, pp 24–38 (Deutscher MP, ed.), Academic Press Inc., New York.

2. Carbohydrates

2.1 Introduction

Carbohydrates are important constituents of all living cells. These are defined as aldehydes or ketones of polyhydric alcohols and also include those biopolymers which yield these compounds on hydrolysis. They occur in animals, plants as well as in microorganisms and perform diverse structural and metabolic roles. Sugars such as glucose are amongst the major sources of energy whereas starch and glycogen function as storage polysaccharides in plants and animals, respectively. In addition, carbohydrates are structural components of cell walls, connective tissues in animals and exoskeletons of invertebrates. These are also constituents of vital biomolecules like nucleic acids, coenzymes such as NAD(P), FAD etc.

2.2 Classification of Carbohydrates

Carbohydrates referred to as saccharides (Greek: *Sakcharon*, meaning sugar) are classified as follows:

(i) Monosaccharides

These carbohydrates are a single unit of polyhydroxy aldehydes or ketones and cannot be hydrolysed into simpler compounds. Sugars like arabinose, ribose, glucose, fructose, galactose etc. are examples of monosaccharides.

(ii) Oligosaccharides

They contain 2–10 units of monosaccharides (monomers) linked to each other via glycosidic linkages, e.g. sucrose, lactose, maltose, raffinose, stachyose etc.

(iii) Polysaccharides

Carbohydrates which are composed of more than ten monosaccharide units linked to each other through glycosidic bonds are referred to as polysaccharides. These include carbohydrates like cellulose, starch, glycogen, fructans, arabinans etc.

2.3 Qualitative Tests for Carbohydrates

Difficulties are encountered in the qualitative and quantitative analysis of samples containing mixtures of carbohydrates, particularly the sugars, because of their structural and chemical similarity and also with respect to their stereoisomers. During biochemical investigations it may become necessary to establish whether a given sample, particularly of a purified preparation, contains carbohydrates or not. Several rapid tests are available to establish the presence or absence of a sugar or a carbohydrate in a sample. These tests are based on specific colour reactions typical for their group and are described below. For laboratory practicals, it may be advisable to perform these tests with the individual rather than mixture of sugars. Use of sugar solutions of different concentrations (0.1–1%) during these experiments would also provide a valuable information about the sensitivity of these tests. The type of carbohydrates detected by these tests are:

	Name of test	*Application*
1.	Molisch's Test	General test for carbohydrates
2.	Anthrone Test	General test for carbohydrates
3.	Iodine Test	For glycans (starch, glycogen)
4.	Barfoed's Test	To distinguish between monosaccharides from reducing disaccharides
5.	Seliwanoff's Test	For ketoses
6.	Fehling's Test	For reducing sugars
7.	Benedict's Test	For reducing sugars
8.	Picric acid Test	For reducing sugars
9.	Mucic acid Test	For galactose
10.	Bial's Test	For pentoses

2.3.1 Molisch's test

Principle

This is a general test for all carbohydrates. Conc H_2SO_4 hydrolyses glycosidic bonds to yield monosaccharides which in the presence of an acid get dehydrated to form furfural and its derivatives. These products react with sulphonated α-naphthol to give a purple complex. Polysaccharides and glycoproteins also give a positive reaction.

Reaction

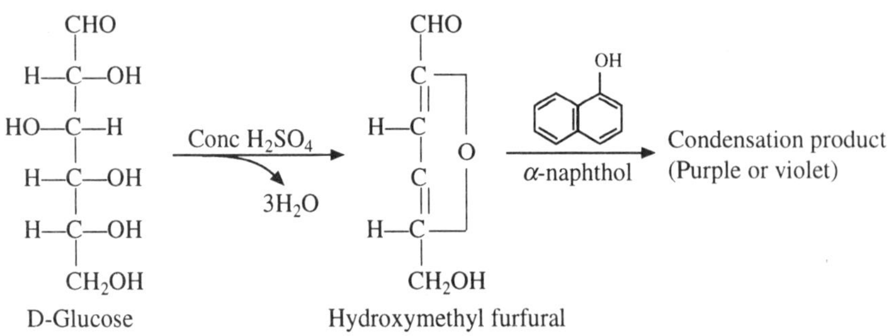

Reagents

1. Conc H_2SO_4
2. α-naphthol: 5% (*w/v*) in ethanol (prepare fresh)

Procedure and observations

Add 2–3 drops of α-naphthol solution to 2 ml of the test solution. Very gently pipette 1 ml conc H_2SO_4 along the side of the test tube so that the two distinct layers are formed. Carefully observe any colour change at the junction of two layers. Appearance of purple colour indicates the presence of carbohydrates in the sample preparation or the test solution.

Precautions

1. α-naphthol solution is unstable and should be prepared fresh.
2. Conc H_2SO_4 should be added along the sides of the test tubes causing minimal disturbance to the contents in the tube.

2.3.2 Anthrone test

Principle

Anthrone reaction is another general test for carbohydrates. In this the furfural produced reacts with anthrone to give bluish green coloured complex.

Reaction

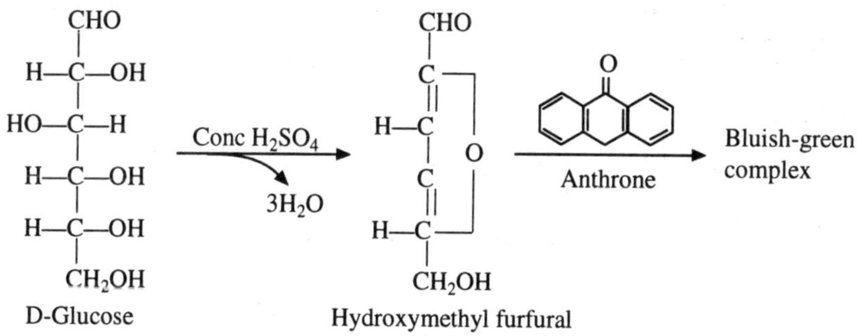

Materials and Reagents

1. Boiling water bath.
2. Conc H_2SO_4.
3. 0.2% (*w*/*v*) anthrone solution in conc H_2SO_4.

Procedure and observations

Add 0.5–1 ml of the test solution to about 2 ml of anthrone reagent and mix thoroughly. Observe whether the colour changes to bluish green. If not, examine the tubes again after keeping them in boiling water bath for 10 min.

2.3.3 Iodine test

Principle

Iodine forms coloured adsorption complexes with polysaccharides. Starch gives blue colour with iodine, while glycogen reacts to form reddish brown complex.

Hence it is a useful, convenient and rapid test for detection of amylose, amylopectin and glycogen.

Reagents

1. Iodine solution: Prepare 0.005 N iodine solution in 3% (*w*/*v*) potassium iodide solution.
2. 1% Test solutions of glucose, sucrose, starch, glycogen, cellulose etc.

Procedure and observations

Take 1 ml of the sample extract or test solution in a test tube. Add 4–5 drops of iodine solution to it and mix the contents gently. Observe if any coloured product is formed. Note the colour of the product.

2.3.4 Barfoed's test

Principle

This test is used for distinguishing monosaccharides from reducing disaccharides. Monosaccharides usually react in about 1–2 min while the reducing disaccharides take much longer time between 7–12 min to get hydrolysed and then react with the reagent. Brick red colour is obtained in this test which is due to formation of cuprous oxide.

Reaction

$$\underset{\text{Cupric acetate}}{(CH_3COO)_2Cu} + 2H_2O \rightarrow 2CH_3COOH + \underset{\text{Cupric hydroxide}}{Cu(OH)_2}$$

$$Cu(OH)_2 \rightarrow CuO + H_2O$$

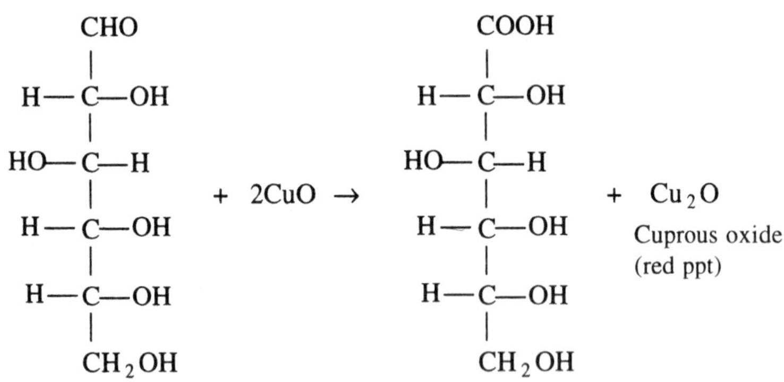

D-Glucose D-Gluconic acid

Materials and Reagents

1. Boiling water bath.
2. Barfoed's reagent: Dissolve 13.3 g of copper acetate in 200 ml water and add 1.8 ml of glacial acetic acid to it.

Procedure and observations

Take 2 ml of Barfoed's solution in a test tube and add 1 ml of sample solution to it. Keep the test tubes in a boiling water bath. A briskly boiling water bath should be used for obtaining reliable results. Look for the formation of brick red colour and also note the time taken for its appearance.

2.3.5 Seliwanoff's test

Principle

This test is used to distinguish aldoses from ketoses. Ketoses undergo dehydration to give furfural derivatives which then condense with resorcinol to form a red complex. Prolonged heating will hydrolyse disaccharides and other monosaccharides will also eventually give colour.

Reaction

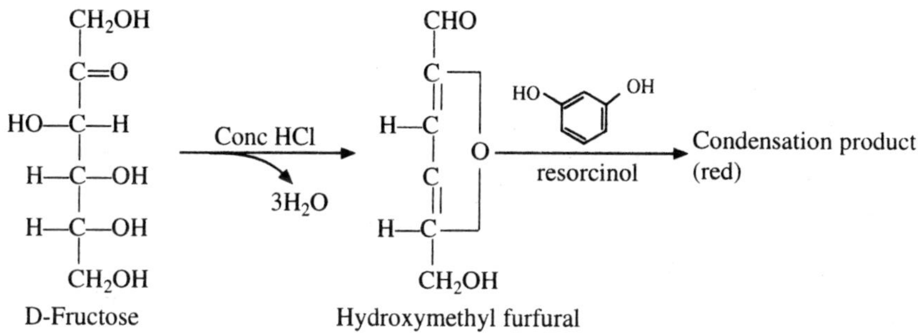

Materials and Reagents

1. Boiling water bath.
2. Seliwanoff's reagent: 0.05% (*w/v*) resorcinol in 3 N HCl.

Procedure and observations

Add 1 ml of the test solution to 2 ml of Seliwanoff's reagent and warm in a boiling water bath for 1 min. Note for the appearance of a deep red colour. This would indicate that the sample solution contains a keto sugar.

2.3.6 Fehling's test

Principle

Fehling's test is a specific and highly sensitive for detection of reducing sugars. Formation of yellow or red ppt of cuprous oxide denotes the presence of reducing sugars. Rochelle salt acts as the chelating agent in this reaction.

Reaction

$$CuSO_4 + 2KOH \rightarrow Cu(OH)_2 + K_2SO_4$$

$$Cu(OH)_2 \rightarrow CuO + H_2O$$

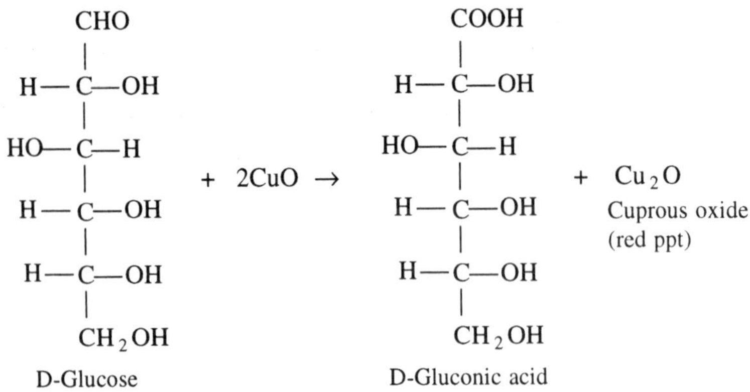

Materials and Reagents

1. Boiling water bath.
2. Fehling's solution A: Dissolve 35 g of $CuSO_4 \cdot 5H_2O$ in water and make the volume to 500 ml.
3. Fehling's solution B: Dissolve 120 g of KOH and 173 g Na-K tartrate (Rochelle salt) in water and make the volume to 500 ml.
4. Fehling's reagent: Mix equal volumes of Fehling's solution A and B. These solutions must be mixed immediately prior to use.

Procedure and observations

Add 1 ml of Fehling's reagent (Reagent No. 4) to 1 ml of aliquot of the test solution. Mix thoroughly and place the test tubes in vigorously boiling water bath. Look out for the formation of red ppt of cuprous oxide which would indicate the presence of reducing sugars in the test solution.

2.3.7 Benedict's test

Principle

Benedict's test is more convenient and this reagent is more stable. In this method sodium citrate functions as a chelating agent. Presence of reducing sugars results in the formation of red ppt of cuprous oxide.

Reaction

$$Na_2CO_3 + 2H_2O \rightarrow 2NaOH + H_2CO_3$$

$$2NaOH + CuSO_4 \rightarrow Cu(OH)_2 + Na_2SO_4$$

$$Cu(OH)_2 \rightarrow CuO + H_2O$$

$$\text{D-Glucose} + 2\ CuO \rightarrow \text{D-gluconic acid} + \underset{\text{(Red ppt)}}{Cu_2O}$$

Materials and Reagents

1. Boiling water bath.

2. Benedict's reagent: Dissolve 173 g of sodium citrate and 100 g of anhydrous Na_2CO_3 in 600 ml of hot H_2O. Dilute to 800 ml with water.
3. Dissolve 17.3 g of $CuSO_4 \cdot 5H_2O$ in 100 ml hot water. Cool and dilute to 100 ml.
4. Add Reagent No. 2 to Reagent No. 3 slowly with constant stirring. Make the final volume to 1 L.

Procedure and observations

Add 0.5–1 ml of the test solution or sample extract to 2 ml of Benedict's reagent (Reagent No. 4). Keep the test tubes in a vigorously boiling water bath. Observe for the formation of red precipitates whose appearance would suggest the presence of reducing sugars in the given solution or sample extract.

2.3.8 Picric acid test

Principle

It is another test for detection of reducing sugars. The reducing sugars react with picric acid to form a red coloured picramic acid.

Reaction

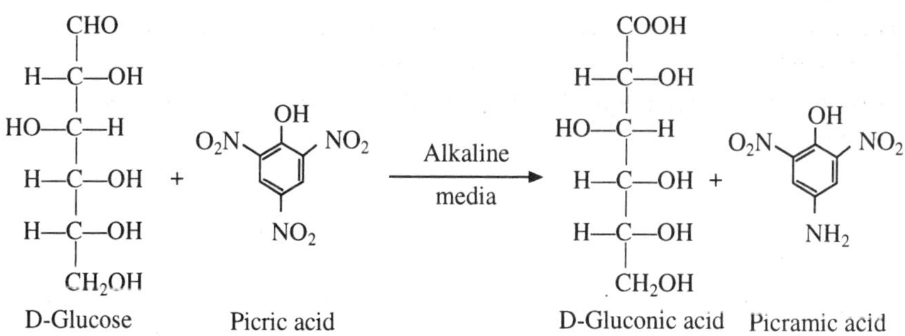

Materials and Reagents

1. Boiling water bath.
2. Saturated picric acid: Dissolve 13 g picric acid in distilled water, boil and cool.
3. 10% Na_2CO_3.

Procedure and observations

Add 1 ml saturated picric acid to 1 ml of sample solution followed by 0.5 ml 10% Na_2CO_3. Heat the test tubes in a boiling water bath. Appearance of red colour would indicate the presence of reducing sugars in the sample solution.

2.3.9 Mucic acid test

Principle

This test is highly specific for galactose. Hence this sugar can be distinguished from other monosaccharides by its reaction with conc HNO_3. Oxidation of other monosaccharides yields soluble dicarboxylic acids whereas galactose produces insoluble mucic acid.

Materials and Reagents

1. Boiling water bath
2. Solid galactose
3. Solid glucose
4. Conc HNO_3

Procedure and observations

Take about 50 mg galactose and 50 mg glucose separately in different test tubes. Add 1 ml of water and 1 ml conc HNO_3 to each tube. Heat the tubes in boiling water bath for 1.5 h. Add 5 ml water and keep them over night. Insoluble mucic acid will be formed in the case of galactose but not glucose or other sugars.

2.3.10 Bial's test

Principle

This test is useful in the determination of pentose sugars. Reaction is due to formation of furfural in the acid medium which condenses with orcinol in presence of ferric ions to give a blue-green coloured complex which is soluble in butyl alcohol.

Reaction

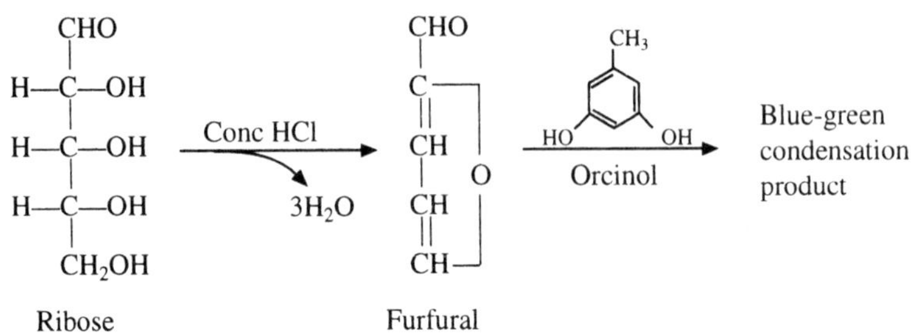

Materials and Reagents

1. Boiling water bath.
2. Dissolve 1.5 g of orcinol in 100 ml of conc HCl and add 20-30 drops of 10% ferric chloride solution to it.

Procedure and observations

To 2 ml of Bial's reagent add 4–5 drops of test solution. Heat in a boiling water bath. Observe for the formation of blue-green coloured complex.

2.3.11 Identification of the nature of carbohydrates in a given sample

Principle

Some of the qualitative tests for carbohydrates (described in Sections 2.3.1 to 2.3.10) are quite specific. Hence, by a judicious selection of the appropriate tests, a systematic experimental sequence can be employed to establish whether a given test solution or a sample preparation contains carbohydrates and, if so, its nature as well.

Procedure

The following tests may be carried out in succession for this purpose and the results be interpreted on the basis of the observations of response to each of the tests.

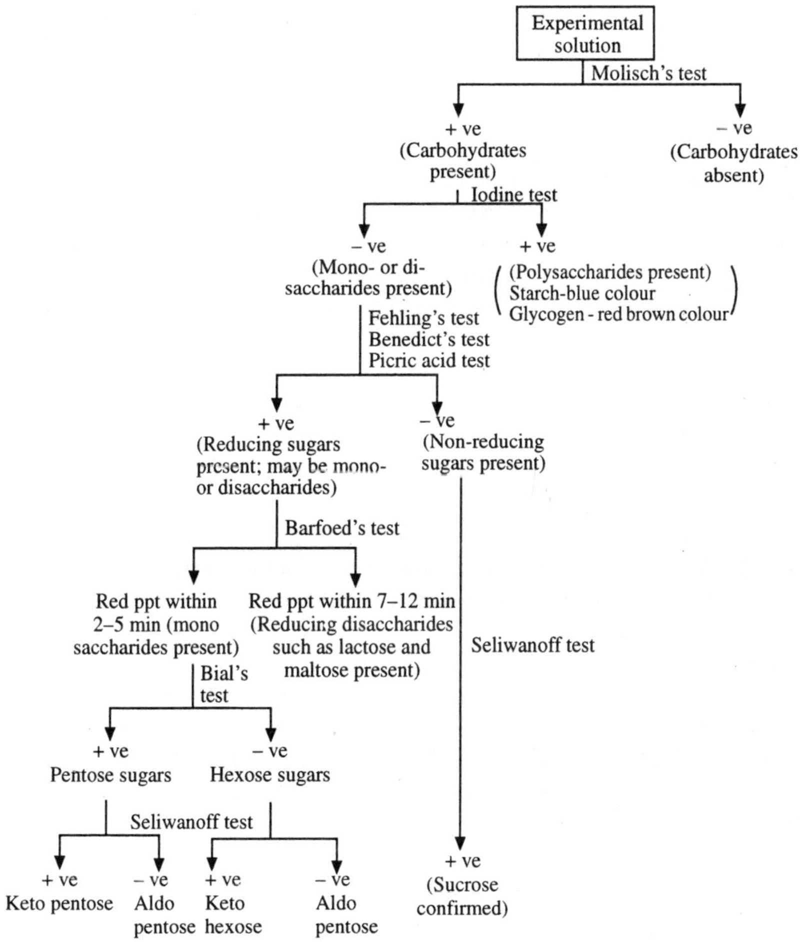

2.4 Quantitative Determination of Carbohydrates

There are several methods for the quantitative determination of various types of carbohydrates. For accurate determination of a particular sugar, it may be necessary to separate out the various carbohydrates in a mixture or a tissue extract by chromatographic techniques and then estimate them individually. Some of the methods which are employed for determination of different groups or types of sugars are:

Method	*Application*
Anthrone method	For total soluble sugars
Ferricyanide method	For total soluble sugars
Dubois method	For total soluble sugars
Somogyi-Nelson method	For reducing sugars
Picric acid method	For reducing sugars
High Pressure Liquid chromatography (HPLC)	For separation and determination of monosaccharides, disaccharides, and oligosaccharides
Enzymatic methods	For estimation of specific sugars

In HPLC method, sugars in an appropriately processed sample are first separated from each other by HPLC and their presence as well as the amount in the column effluent is continuously monitored and recorded by a suitable detector such as RI (refractive index) detector. The identity of sugars in individual peaks is established by comparison with elution profiles of authentic samples of different sugars.

Amount of some of the sugars in a sample can also be determined enzymatically. One main advantage of this technique is that prior separation of individual sugars is not essential since some of the enzymes exhibit absolute substrate specificity and hence will only utilize that particular sugar as the substrate. For example, amount of glucose can conveniently and accurately be determined, even if the sample contains other closely related sugars like galactose, mannose etc., by using glucose oxidase.

2.4.1 EXPERIMENT: Determination of total soluble sugars by ferricyanide method (colorimetric method)

Principle

This method is based on the principle that sugars are oxidised by alkaline potassium ferricyanide and the amount of ferrocyanide produced is then measured either photometrically or volumetrically (see Experiment 2.4.2). The amount of the total sugars can also be determined colorimetrically by Anthrone method (see EXPERIMENT 2.4.5).

Materials and Reagents

1. Colorimeter.
2. Table centrifuge.
3. Saturated neutral lead acetate solution.
4. Sodium oxalate (solid).

5. Sodium carbonate (solid).
6. Conc HCl.
7. Potassium ferricyanide solution: Dissolve 0.5 g potassium ferricyanide in 1 L of distilled water. Store it in a brown bottle.
8. Carbonate-cyanide reagent: Dissolve 5.3 g sodium carbonate and 0.65 g potassium cyanide in 1 L of distilled water.
9. Ferric ammonium sulphate solution: Dissolve 1.5 g ferric ammonium sulphate in 1 L of 0.05 N H_2SO_4.
10. Tungstate solution: Prepare 10% (*w*/*v*) sodium tungstate in distilled water.
11. Standard glucose solution: Dissolve 100 mg glucose in distilled water and make up the volume to 100 ml.
12. N/12 H_2SO_4: Take 2.26 ml of conc H_2SO_4 and make its volume to 1 L with distilled water.

Procedure

1. For extracting the sugars, suspend 1 g of finely powdered oven dried sample in 40 ml of distilled water and heat in boiling water bath for 30 min. Centrifuge at 3000 rpm for 20 min. Collect the supernatant. Again suspend the pellet in 20 ml of water and repeat this extraction step 6–8 times till the supernatant is free of sugars. Combine all supernatants.
2. Add 1–2 ml of saturated neutral lead acetate to the collected supernatant and after properly mixing it for 15 min, filter it through Whatman No.1 filter paper and make the volume to 250 ml with distilled water. Excess of lead acetate is then precipitated out with solid sodium oxalate.
3. For determination of reducing sugars the above extract (Step 2) is used directly. However, for estimating the amount of total sugars, take 25 ml of the above extract (obtained in Step 2) and hydrolyse it with 5 ml conc HCl for 8 min at 68°C. After cooling, neutralise the hydrolysate with solid sodium carbonate and make up the volume to 100 ml with distilled water.
4. To 1 ml of the preparation (Step 2) for reducing sugars and that obtained in Step 3 for total sugars, add 0.2 ml of N/12 H_2SO_4 and 9 ml of distilled water. Mix thoroughly, add 0.02 ml sodium tungstate and after shaking the contents allow to stand for 10 min. Centrifuge it at 3000 rpm for 20 min. Add 1 ml of ferricyanide solution to 1 ml of aliquot of this supernatant followed by addition of 1 ml of cyanide-carbonate solution. Heat this mixture for 15 min in a boiling water bath and cool immediately under running tap water. Now add 5 ml of ferric ammonium sulphate. Mix the solution and after 10 min record its O.D. at 690 nm on Spectronic 20.
5. Prepare a reference curve of glucose over a range of 0-100 μg as standard following the procedure described above (Step 4).

Calculations

Vol. of extract taken after tungstate treatment for colour development = *a* ml

O.D. of the extract = *y*

Assuming that from the standard curve, y O.D. units correspond to m μg of glucose

$$\text{Wt. of sample taken} = 1 \text{ g}$$

$$\text{Total vol. of extract} = 250 \text{ ml}$$

$$\text{Conc. of sugars in 1 ml extract} = \frac{m}{a} \times v \ \mu g$$

where v is the total vol. of the supernatant after treating with tungstate

$$\mu g \text{ reducing sugar/100 g sample} = \frac{m}{a} \times v \times 250 \times 100$$

$$\% \text{ reducing sugar} = \frac{m \times v \times 250 \times 100}{a \times 1000 \times 1000} = \frac{mv}{40a}$$

Similarly, calculate the % total sugar content of the sample after taking into account the dilution factors while processing the sample (as involved in Step 3 of *Procedure*) for estimation of total sugars.

2.4.2 EXPERIMENT: Determination of total soluble sugars by ferricyanide method (volumetric procedure)

Principle

Alkaline potassium ferricyanide oxidizes sugars. This method is based on reduction of the residual potassium ferricyanide by KI and the unreacted KI is volumetrically measured by titration against $Na_2S_2O_3$. The chemical reactions involved are as follows:

$$2K_3[Fe(CN)_6] + 2KOH \rightleftharpoons 2K_4[Fe(CN)_6] + H_2O + 1/2O_2$$

$$\underset{\text{Glucose}}{CH_2OH(CHOH)_4CHO} + 1/2O_2 \rightleftharpoons \underset{\text{Gluconic acid}}{CH_2OH(CHOH)_4COOH}$$

Excess of ferricyanide reacts with KI

$$K_3[Fe(CN)_6] + KI \rightleftharpoons K_4[Fe(CN)_6] + I$$

$$3ZnSO_4 + 2K_4[Fe(CN)_6] \rightleftharpoons \underset{\text{Potassium zinc ferrocyanide}}{K_2Zn_3[Fe(CN)_6]_2} + 3K_2SO_4$$

$$3I_2 + 6(OH) \rightarrow 5I' + IO_3' + 3H_2O$$

$$5I' + IO_3' + 6H^+ \rightarrow 3I_2 + 3H_2O$$

$$2Na_2S_2O_3 + I_2 \rightarrow \underset{\text{Sodium tetrathionate}}{Na_2S_4O_6} + 2NaI$$

Materials and Reagents

1. Burette.

2. Boiling water bath.
3. Potassium ferricyanide: Dissolve 8.25 g potassium ferricyanide and 10.6 g anhydrous sodium carbonate in 1 L of distilled water.
4. Iodine solution: Prepare by dissolving 12.6 g KI, 25 g $ZnSO_4$ and 125 g NaCl in 500 ml distilled water. Filter and store in coloured bottle.
5. Sodium thiosulphate solution: Make 0.01 N sodium thiosulphate solution by dissolving 2.5069 g of sodium thiosulphate in 1 L of distilled water.
6. Starch indicator solution: Suspend 1 g soluble starch in 20 ml distilled water and then add 60 ml of boiling distilled water. Add 20 g NaCl to this solution and make the volume 100 ml.
7. 5% glacial acetic acid.

Procedure

1. Take 5 ml of potassium ferricyanide and 5 ml of aliquot of the sample extract (as obtained in Step 2 of EXPERIMENT 2.4.1 for reducing sugars and in Step 3 for total sugars) in a test tube, heat for 15 min in boiling water bath and then cool it.
2. Add 5 ml of iodine-solution followed by 3 ml of 5% glacial acetic acid. The excess iodide is titrated against 0.01 N $Na_2S_2O_3$ till the colour of the solution turns pale yellow. Now add starch indicator solution, upon which the colour will change to blue.
3. Complete the titration till disappearance of blue colour.
4. Run a blank taking water instead of sugar solution or sample aliquot and proceed in the same manner. Volume of $Na_2S_2O_3$ used for the sample is deducted from that consumed for the blank.

Calculations

The amount of reducing sugars is calculated from the following relationship:

$$\text{mg of reducing sugar in 5 ml of sample extract} = \mu\,(x + 0.05)$$

where $\mu = 0.338$

x = vol. of 0.01 N $Na_2S_2O_3$ used for sample, i.e. vol. of $Na_2S_2O_3$ used in blank – vol. used in sample.

2.4.3 EXPERIMENT: Determination of reducing sugars by Nelson-Somogyi's method

Principle

The sample extract or the test solution is heated with alkaline copper reagent and the reduced copper formed then reacts with arsenomolybdate reagent to produce molybdenum blue resulting in the formation of a violet coloured complex which is estimated colorimetrically at 500 nm. Sodium sulphate is included in the copper reagent to prevent reoxidation of cuprous oxide by atmospheric oxygen.

Materials and Reagents

1. Boiling water bath.
2. Colorimeter.
3. 5% $ZnSO_4$ solution.
4. 0.3 N $Ba(OH)_2$.
5. Alkaline copper reagent:
   Solution A: Dissolve 25 g anhydrous Na_2CO_3, 25 g Rochelle salt (Na-K tartarate), 20 g $NaHCO_3$ and 200 g anhydrous Na_2SO_4 in 800 ml water and make the final volume to 1 L.
   Solution B: Prepare 15% (*w*/*v*) $CuSO_4 \cdot 5H_2O$ containing one or two drops of conc H_2SO_4.
   Mix 25 parts of solution A and 1 part of B just before use.
6. Arsenomolybdate colour reagent: Make by dissolving 25 g ammonium molybdate in 450 ml distilled water and adding 21 ml conc H_2SO_4 to it. 3.0 g of disodium hydrogen arsenate ($Na_2HAsO_4 \cdot 7H_2O$) is dissolved in 25 ml water and then added with stirring to the acidified molybdate solution. It is kept at 37°C for 24–48 h before use.
7. Standard glucose solution (stock solution): Prepare a solution containing 1.0 g of anhydrous glucose in 10–15 ml of 0.2% benzoic acid and make its volume to 100 ml with distilled water.

Procedure

1. Add 0.2 ml of 0.3 N $Ba(OH)_2$ to 1 ml of sample extract (prepared as described in EXPERIMENT 2.4.1, Step 2) and mix well. Then add 0.2 ml $ZnSO_4$, shake thoroughly and after 10 min filter it through Whatman No.1 filter paper. During this step, proteins are precipitated by $Zn(OH)_2$ to give a protein-free sample extract.
2. Add 1 ml of alkaline copper reagent to 1 ml aliquot of the filtrate. Cover the mouth of tubes with marble and place the tubes in boiling water bath for 20 min.
3. Cool the tubes and add 1 ml of arsenomolybdate solution.
4. Record the absorbance at 500 nm using reagent blank containing 1 ml distilled water in place of sample extract or test solution to set Spectronic 20.
5. Prepare a standard curve following the same procedure with standard glucose solution (0–50 μg).

Calculations

$$\mu\text{g of glucose (a)} = \frac{A_{500} \text{ of the sample solution}}{A_{500} \text{ of known standard}} \times \mu\text{g glucose in standard}$$

$$\text{g of glucose (b)} = \frac{a}{10^6}$$

$$\%\text{ glucose} = \frac{\text{b} \times \text{dilution factor}}{\text{wt. of sample (g)}} \times 100$$

2.4.4 EXPERIMENT: Enzymatic method for quantitative estimation of glucose in blood plasma using glucose oxidase

Principle

Glucose oxidase is found in the growth media of *Penicillium notatum*. It catalyses oxidation of β-D-glucopyranose to D-glucono-1,5-lactone accompanied by formation of hydrogen peroxide. D-glucono-1,5-lactone is converted to D-gluconic acid. This enzyme is highly specific for β-D-glucopyranose. The reaction mixture for estimation of glucose contains peroxidase which utilizes H_2O_2 to oxidise *o*-toluidine to give a blue coloured product. These reactions are rapid but the colour is not very stable and the readings are recorded usually after 8 min. The sequence of events involved in this method is depicted below:

$$\text{D-Glucose} + O_2 + H_2O \xrightarrow{\text{Glucose oxidase}} H_2O_2 + \text{D-gluconic acid}$$

$$H_2O_2 + o\text{-toluidine} \xrightarrow{\text{Peroxidase}} \text{oxidised toluidine}$$

(Blue coloured product)

Materials and Reagents

1. Table centrifuge.
2. Colorimeter.
3. Water bath.
4. 10% (*w*/*v*) $ZnSO_4$ solution.
5. Isotonic sodium sulphate (93 mM): Dissolve 13.206 g of Na_2SO_4 and make the volume to 1 L with water.
6. Sodium sulphate-Zinc sulphate reagent: Take 55 ml of 10% $ZnSO_4$ solution (Reagent 4) and add isotonic Na_2SO_4 (Reagent 5) to make the volume 1L.
7. 0.5 N NaOH.
8. 0.15 M acetate buffer (pH 5.0).
9. Glucose oxidase: 30 International Units (one I.U.= I μ mole of substrate transformed per min).
10. Peroxidase: 30 International Units.
11. *o*-toluidine: Dissolve 1 g of *o*-toluidine in 100 ml of absolute ethanol and store in dark bottle.
12. Combined *o*-toluidine reagent: Prepare a mixture containing the ingredients—150 ml of acetate buffer (Reagent no. 8), 30 I.U. of glucose oxidase, 30 I.U. of peroxidase and 1 ml of *o*-toluidine solution. Store the mixture in brown bottle and it is stable for several weeks.
13. Glucose standard: Prepare a fresh solution containing 10 mg of D-glucose in 100 ml of water.

Procedure

1. Pipette 0.1 ml of blood sample into 1.8 ml of sodium sulphate-zinc sulphate reagent in a centrifuge tube and add 0.1 ml of 0.5 N NaOH to it.

2. Centrifuge at 5000 rpm for 15 min. Take 1 ml of the supernatant and for blank use 1 ml of sodium sulphate-zinc sulphate reagent. Add 5 ml of the combined *o*-toluidine reagent and mix thoroughly.
3. Incubate the tubes at 30°C and exactly after 8 min read the colour intensity at 625 nm.
4. In another set of test tubes take suitable aliquots of glucose standard and proceed as described above (Steps 1–3) and prepare a standard curve.
5. From the standard curve determine the quantity of glucose in the sample tube and calculate the amount of glucose per ml of the blood sample.

2.4.5 EXPERIMENT: Determination of starch in plant tissues

Principle

Sugars are first extracted by treating the finely powdered dried grains or leaves sample repeatedly with hot 80% (*v*/*v*) alcohol. The residue is then treated with cold perchloric acid to solubilise starch. After filtration, starch in the perchloric acid extract is hydrolysed to glucose in hot acidic medium which undergoes dehydration to hydroxymethyl furfural. This derivative then condenses with anthrone to give a blue coloured complex and is determined quantitatively by anthrone-sulphuric acid procedure given below.

Materials and Reagents

1. Boiling water bath.
2. Table centrifuge.
3. Refrigerated centrifuge.
4. Colorimeter.
5. Anthrone-sulphuric acid reagent: Dissolve 2 g anthrone in 1 L of cold 95% conc H_2SO_4. Store at 4°C. This solution should be prepared fresh after 2 days.
6. 80% (*v*/*v*) ethyl alcohol: Dilute 1.68 L of 95% ethyl alcohol to 2 L with water.
7. 52% Perchloric acid: Add 270 ml of 72% perchloric acid to 100 ml of water.
8. Glucose standard: Dissolve 0.1 g of anhydrous glucose in 100 ml of water containing 0.1% benzoic acid. This standard solution contains 1 mg glucose per ml.

Procedure

1. For extraction of sugars and starch take 0.2 g of finely ground sample in a 50 ml centrifuge tube and add 20 ml of hot 80% alcohol. Shake the tubes for 5–10 min and after centrifuging at 3000 rpm for 10 min decant the supernatant. Repeat this extraction with 80% hot ethanol 5–6 times till the supernatant is free of sugars as judged by a negative test with anthrone reagent (see Step 3).
2. Cool the residue in ice water and add 6.5 ml of 52% perchloric acid while

stirring the contents with a glass rod. Let it stand for 15 min with occassional stirring and then centrifuge at 4°C. Collect the supernatant and repeat this extraction step 4–5 times. Combine the supernatant fractions and make up the volume to 100 ml with water.

3. Dilute the above extract so that it contains 5 to 20 μg of glucose per ml. Take 5 ml aliquot of this diluted extract and place the tubes in a cold water bath. Add 10 ml freshly prepared anthrone reagent, mix properly and transfer the tubes to boiling water bath for 7.5 min. After cooling the tubes under running tap water note the absorbance of these solutions at 630 nm.
4. Prepare a standard curve using 0 to 100 μg glucose as per the procedure described above. Calculate the amount of glucose in the sample aliquot.

Calculations

1. From the standard curve determine the amount of glucose in 5 ml of the diluted aliquot of sample extract.
2. Calculate the quantity of glucose in the original extract by applying the dilution factor.
3. Multiply the obtained value by 0.9 for conversion of glucose value to starch.
4. Calculate the amount of starch in 1 g of the powdered sample.

2.4.6 EXPERIMENT: Determination of glycogen in liver

Principle

Glycogen is the principal storage polysaccharide in animals and it is stored in substantial quantities in liver and to a lesser extent in skeletal muscles as well. Glycogen content in liver is determined after digesting the tissue with hot conc KOH, precipitation of glycogen with ethanol, hydrolysing extracted glycogen with acid to glucose and then estimating the latter by Nelson-Somogyi method (see Experiment 2.4.3).

Materials and Reagents

1. All materials and reagents listed in EXPERIMENT 2.4.3
2. Refrigerated centrifuge
3. 30% KOH solution
4. Ethanol (95%)
5. Ethanol (65%, *v*/*v*)
6. 2 N Sulphuric acid
7. 6 N NaOH
8. Phenol red indicator

Procedure

1. Dissect liver from an animal and remove excess of blood by dabbing it with several folds of a filter paper.

2. Weigh the liver and put it in a stoppered tube containing 2 ml of 30% KOH solution per g of the liver sample.
3. Transfer the tube containing the sample to a boiling water bath and after 1.5 h remove and cool it in an ice cold water.
4. Add 2 volumes of 95% ethanol and heat it in a boiling water bath till the contents just start boiling. Leave the tube in cold overnight.
5. Centrifuge the contents at 8000 rpm for 20 min and dissolve the precipitate in 5–10 ml of warm water. Reprecipitate the extracted glycogen with 2 volumes of 95% ethanol.
6. Collect the precipitate by centrifugation at 5000 rpm for 15–20 min at 4°C and wash it several times with 65% (*v*/*v*) ethanol.
7. Add 2 ml of 2 N H_2SO_4 per g of liver to the precipitate and hydrolyse glycogen by keeping the tubes in a boiling water bath for 3 to 4 h.
8. After cooling, neutralize the contents with 6 N NaOH using phenol red as an indicator, make to a fixed volume and filter it.
9. Determine glucose in an appropriate aliquot by Nelson-Somogyi method (for details see EXPERIMENT 2.4.3).

Calculations

Use a factor of 0.93 to convert glucose to glycogen and calculate the amount of glycogen per g of liver.

Suggested Further Reading

Chyakin S. (1973). Biochemistry Laboratory Techniques, Wiley Eastern, New Delhi.

Diamond P.S. and Denman R.F. (1973). Laboratory Techniques in Chemistry and Biochemistry, 2nd ed., Butterworths, London.

Plummer D.T. (1971). An Introduction to Practical Biochemistry, Tata McGraw-Hill Publishing Company Ltd., Bombay.

Pomeranz Y. and Meloan C.E. (1996). Food Analysis, CBS Publishers and Distributors, New Delhi.

Sadasivam S. and Manickam A. (1996). Biochemical Methods, II ed., New Age International (P) Limited Publishers, New Delhi.

3. Lipids

3.1 Introduction

Lipids are non-polar organic biomolecules that are totally or nearly insoluble in water but are quite soluble in non-polar organic solvents like ether, chloroform or benzene. They are rather heterogenous group of compounds differing in their structure and perform diverse biological functions. They are the chief storage and transport form of metabolic fuel. They serve as major structural components of the membranes and also form a protective coating on many organisms. Some of the vitamins and hormones are lipids. There are several different classes of lipids but all of them derive their distinct properties due to the hydrocarbon nature of their structure. Lipids have been classified in several ways but the most acceptable classification is the one based on the structure of their backbone. Based on this, they are divided into two groups: *complex lipids* and *simple lipids*. The complex lipids are esters of fatty acids. In various forms of complex lipids like acylglycerides, phospholipids, sphingolipids and waxes, fatty acids are covalently joined via an ester linkage to a trihydroxy alcohol, glycerol, or its derivative. Since the complex lipids yield soap on alkaline hydrolysis, they are also called *saponifiable* lipids. The simple lipids, on the other hand, do not contain fatty acids and are therefore called *non-saponifiable* lipids. They include compounds like terpenes, sterols etc.

3.2 Qualitative Tests of Lipids

Complete characterization of a lipid involves detailed analysis of its various chemical constituents, length and degree of unsaturation of fatty acid chains etc. which can be time consuming. However, preliminary information about the overall nature of lipids may be obtained from relatively simple qualitative tests, some of which are described below:

3.2.1 EXPERIMENT: Solubility test for lipids

Triacylglycerols having small chain fatty acids are somewhat soluble in water but those containing non-polar long chain fatty acids are insoluble and they form emulsions in water. All triacylglycerols are soluble in diethyl ether, chloroform and benzene. They are slightly soluble in cold methanol, ethanol and acetone but their solubility increases on warming. Understanding of their solubility characteristics is helpful in developing efficient procedures for extraction of various lipids from the biological materials.

Materials and Reagents

1. Fatty acids (butyric, palmitic and oleic acids)
2. Fats and oils (desi ghee, butter, margarine, cod liver oil, phospholipids)
3. Solvents (water, acetone, ethyl alcohol, chloroform, diethyl ether)

Procedure

1. Take small amount of different lipids in various test tubes and add water. Shake well and check their solubility.
2. Note the change, if any, in solubility on warming the above tubes in water bath at 50°C for 5 min.
3. Repeat the solubility test using different solvents.
4. Record the observations regarding solubility of these lipids and conclude about their solubility characteristics.

3.2.2 EXPERIMENT: Acrolein test for glycerol

Principle

When glycerol, either in free form or as an ester of fatty acids, is heated with potassium hydrogen sulphate it gets dehydrated to an unsaturated aldehyde called *acrolein*. Acrolein can be identified by its characteristic pungent smell.

$$\begin{array}{c} CH_2OH \\ | \\ CHOH \\ | \\ CH_2OH \end{array} + KHSO_4 \xrightarrow{\text{Heat}} \begin{array}{c} CHO \\ | \\ CH \\ \| \\ CH_2 \end{array} + 2H_2O$$

Glycerol   Acrolein

Materials and Reagents

1. Lipid samples (butter, olive oil, stearic acid, glycerol)
2. Anhydrous potassium hydrogen sulphate

Procedure

1. Take approximately 1.5 g potassium hydrogen sulphate in a test tube and add 5 drops of the liquid test sample or an approximately equivalent weight of the test sample if it is solid. Cover the test sample completely by adding more of solid potassium hydrogen sulphate on top of it.
2. Heat the test tube slowly on burner and note the odour of the fumes evolved from the tube.

3.2.3 EXPERIMENT: Qualitative test for the presence of fatty acids by titrimetric method

Principle

The presence of non-esterified fatty acids in a given sample can be determined by titrating it with an alkali using phenolphthalein as an indicator.

Materials and Reagents

1. Lipid samples (butter, olive oil, stearic acid dissolved in 50% alcohol)
2. Phenolphathalein: Prepare 1% solution in alcohol
3. 0.1N NaOH
4. Erlenmeyer flasks

Procedure

1. Take 10 ml of 0.1 N NaOH in an Erlenmeyer flask and add a drop of phenolphthalein solution which will give permanent pink colour.
2. To this, with pipette add the test solution drop by drop with constant shaking of the flask.
3. Disappearance of pink colour indicates the presence of free fatty acids in the given test compound.
4. Repeat this with other lipid samples.

Observations
By comparing the results, is it possible to infer any thing about relative free fatty acid content of the various samples?

3.2.4 EXPERIMENT: Test for unsaturation of fatty acids in lipid sample

Principle
The fatty acids present in animal fats are usually fully saturated whereas those occurring in vegetable oils contain one or more double bonds in their hydrocarbon chain. A semiquantitative estimate about the degree of unsaturation of lipid samples can be deduced since halogens are readily added across the double bonds and this reaction results in decolorization of bromine water or iodine solution.

$$CH_3(CH_2)_7CH{=}CH(CH_2)_7COOH + Br_2$$

Oleic acid

$$\rightarrow CH_3(CH_2)_7\underset{\displaystyle Br}{\underset{|}{C}H}-\underset{\displaystyle Br}{\underset{|}{C}H}(CH_2)_7COOH$$

Dibromostearic acid

Materials and Reagents

1. Test solutions (olive oil, corn oil, coconut oil, oleic acid, stearic acid).
2. Bromine water: Add 5 ml of bromine to 100 ml of water. Shake the mixture and keep it in dark bottle.

Procedure

1. Take approximately 5 ml of test solution in a test tube and slowly add bromine water dropwise and shake the tube after each addition.
2. Keep on adding bromine water till it fails to get decolorized and retains its colour.
3. Note the amount of bromine water added.
4. Repeat the experiment with other test solutions.

Observations

From the volume of bromine water used, what can you deduce about the relative degree of unsaturation of the various lipids?

3.3 Preliminary Characterization of Fats and Oils

3.3.1 EXPERIMENT: Determination of acid value of fats and oils

Principle

Different fat samples may contain varying amount of fatty acids. In addition, the fats often become rancid during storage and this rancidity is caused by chemical or enzymatic hydrolysis of fats into free acids and glycerol. The amount of free fatty acids can be determined volumetrically by titrating the sample with potassium hydroxide. The acidity of fats and oils is expressed as its *acid value* or *number* which is defined as mg KOH required to neutralize the free fatty acids present in 1 g of fat or oil. The amount of free acids present or acid value of fat is a useful parameter which gives an indication about the age and extent of its deterioration.

Materials and Reagents

1. Burette.
2. Conical flasks.
3. Test compounds (olive oil, butter, margarine etc; fresh and the samples that have been stored at room temperature for several days may be used for comparison).
4. 1 % phenolphthalein solution in 95% alcohol.
5. 0.1 N Potassium hydroxide: Weigh 5.6 g of KOH and dissolve it in distilled water and make the final volume to 1 L. Standardize this solution by titrating it with a known volume of 0.1 N oxalic acid (Prepare by taking 630 mg oxalic acid in 100 ml water) using phenolphthalein as an indicator till a permanent pink colour appears. Calculate the actual normality (N_2) of KOH solution from equation $N_1V_1 = N_2V_2$, where N_1 and V_1 are normality and volume of oxalic acid taken for titration and V_2 is the volume of KOH solution used.
6. Fat solvent (95% ethanol: ether 1:1, *v*/*v*).

Procedure

1. Take 5 g of fat sample in a conical flask and add 25 ml of fat solvent (Reagent No. 6) to it. Shake well and add a few drops of phenolphthalein solution and again mix the contents thoroughly.
2. Titrate the above solution with 0.1 N KOH until a faint pink colour persists for 20–30 sec.
3. Note the volume of KOH used.
4. Repeat the steps 1 to 3 with a blank (Reagent No. 6) which does not contain any fat sample.

Calculations

0.1N KOH solution used for blank = x ml
0.1N KOH solution used for sample = y ml
Titre value for sample = $(y - x)$ ml.

$$\text{Acid value (mg KOH/g fat)} = \frac{\text{Titre value} \times \text{Normality of KOH} \times 56.1}{\text{weight of sample (g)}}$$

1 ml of 1 N KOH contains 56.1 mg of KOH. Hence a factor of 56.1 is incorporated in the numerator in the above equation to obtain weight of KOH from the volume of 0.1 N KOH solution used during this titration.

3.3.2 EXPERIMENT: Determination of saponification value of fats and oils

Principle

Hydrolysis of fat with an alkali results in the formation of salts of fatty acids (also called soap) and glycerol. This process is called saponification. From the amount of potassium hydroxide utilized during hydrolysis, the saponification value of a given fat sample can be calculated. The saponification value is defined as mg of KOH required to saponify 1 g of the given fat.

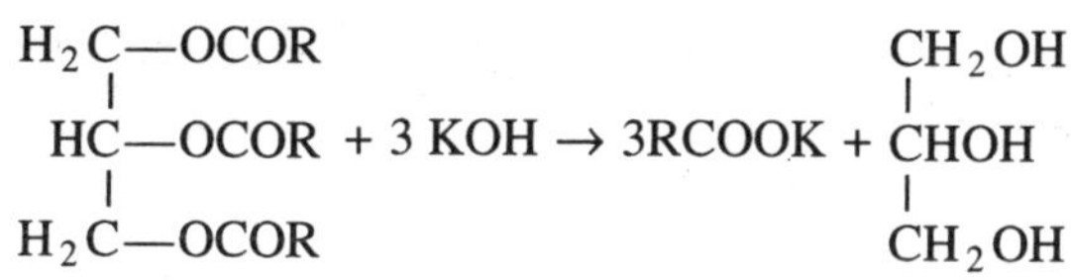

Triacylglycerol — Potassium salt of fatty acid — Glycerol

From the above reaction, it may be noted that 3 molecules of KOH are consumed for saponification of each molecule of triacylglycerol irrespective of chain length of fatty acid. Evidently, each g of a triacylglycerol with shorter chain fatty acids will contain larger number of molecules of the triacylglycerol and will thus require much more KOH. The saponification value is thus an indication of average molecular weight of the fatty acids in an acylglyceride.

The procedure involves refluxing of known amount of fat or oil with a fixed but an excess of alcoholic KOH. The amount of KOH remaining after hydrolysis is determined by back titrating with standardized 0.5 N HCl and the amount of KOH utilized for saponification can thus be calculated.

Materials and Reagents

1. Reflux condenser
2. Boiling water bath
3. Burette
4. Test compounds (tristearin, coconut oil, corn oil, butter)
5. Fat solvent: A mixture of 95% ethanol and ether (1:1, v/v)
6. 0.5 N alcoholic KOH: Prepare 0.5 N solution of KOH by dissolving 28.05

g of KOH pellets in 20 ml water and making the volume to 1 L with 95% ethanol
7. 1% Phenolphthalein solution in 95% alcohol
8. 0.5 N HCl

Procedure

1. Weigh accurately 1 g of the fat sample in a conical flask and dissolve it in about 3 ml of the fat solvent (Reagent No. 5).
2. Add 25 ml of 0.5 N alcoholic KOH, attach a reflux condenser to it and reflux the contents on boiling water bath for 30 min.
3. Cool it to room temperature and add a few drops of phenolphthalein into the flask.
4. Titrate the contents of the flask with 0.5 N HCl till the pink colour disappears.
5. Similarly, run a blank by refluxing 25 ml of 0.5 N alcoholic KOH without any fat sample.

Calculations

$$\text{0.5N KOH in blank} = x \text{ ml}$$
$$\text{0.5N KOH in test sample} = y \text{ ml}$$
$$\text{Titre value for sample} = (x - y) \text{ ml}$$

$$\text{Saponification value} = \frac{28.05 \times \text{titre value}}{\text{weight of sample (g)}}$$

The multiplication factor of 28.05 in the above equation is included since 1 ml of 0.5 N KOH contains 28.05 mg of KOH.

Precautions

1. Electric heater should be used during refluxing as alcohol is highly inflammable.
2. During refluxing, effective cooling of condenser is required so that alcohol does not get evaporated during saponification.

3.3.3 EXPERIMENT: Determination of iodine number of a fat sample

Principle

The most important analytical determination of an oil/fat is the measurement of its unsaturation. The generally accepted parameter for expressing the degree of carbon to carbon unsaturation of fat, oil or their derivatives is *iodine value*. Iodine value or iodine number is defined as g of iodine absorbed by 100 g of fat. It is a useful parameter in studying oxidative rancidity of triacylglycerols since, higher the unsaturation, greater is the possibility of rancidity.

Estimation of iodine number is based on treatment of a known weight of fat or oil with a known volume of standard solution of iodine monochloride and then determining the amount of unused iodine monochloride from iodine liberated on addition of excess of KI. The released iodine is titrated against 0.1 N sodium thiosulphate solution using starch as an indicator.

$$CH_3(CH_2)_7CH{=}CH(CH_2)_7COOH + ICl$$
Oleic acid

$$\rightarrow CH_3(CH_2)_7\underset{\displaystyle I}{\underset{|}{CH}}-\underset{\displaystyle Cl}{\underset{|}{CH}}(CH_2)_7COOH$$

$$ICl + KI \rightarrow KCl + I_2$$

$$I_2 + 2Na_2S_2O_3 \rightarrow 2NaI + Na_2S_4O_6$$

Materials and Reagents

1. Stoppered bottles.
2. Burette (25 ml).
3. Test compounds: 2% solution of corn oil, olive oil and butter in chloroform.
4. Wij's solution: Dissolve 8.5 g of iodine and 7.8 g of iodine trichloride separately in 450 ml acetic acid each. Mix both the solutions and make the volume upto 1 L.
5. 0.1 N sodium thiosulphate: Dissolve 24.82 g of $Na_2S_2O_3 \cdot 5H_2O$ in 1 L water. To check its normality, take 20 ml of 0.1 N potassium dichromate, add 10 ml of 15% KI and then 5 ml of HCl. Dilute to 100 ml with water and titrate with thiosulphate solution till the yellow colour appears. Now add a few drops of starch solution (Reagent No. 7) and continue the titration till the blue colour disappears. Note the volume of thiosulphate solution and calculate its exact normality ($N_1V_1 = N_2V_2$ where N_1 is the normality, V_1 is the volume of dichromate solution taken for titration and V_2 is the volume of thiosulphate solution used).
6. 10 % Potassium iodide solution: Dissolve 10 g of KI crystals in water and make up the volume to 100 ml.
7. 1% Starch indicator: Take 1 g of starch and dissolve it in 100 ml water, boil for a min, cool and centrifuge to get a clear solution.

Procedure

1. Take 10 ml of fat solution into stoppered bottles and add 25 ml of Wij's solution. Shake thoroughly and allow it to stand in dark for 1 h.
2. Similarly prepare a blank in which fat solution is replaced by chloroform.
3. After the reaction time of 1 h in dark, rinse the stopper and neck of the bottle with 50 ml of water and add 10 ml of potassium iodide solution.
4. Titrate the liberated iodine with standard sodium thiosulphate solution till the content of the flask become pale yellow in colour.
5. Add a few drops of starch solution and continue to titrate it further with sodium thiosulphate solution till the blue colour disappears.

Calculations

The difference between the blank and test readings gives the amount of 0.1 N sodium thiosulphate required to react with an equivalent volume of iodine. One L of 0.1 N iodine solution contains 12.7 g of iodine. The iodine number can thus be calculated as follows:

Volume of 0.1 N sodium thiosulfate used for blank = x ml
Volume of 0.1 N sodium thiosulfate used for sample = y ml

$$\text{Iodine number} = \frac{(x - y) \times 12.7}{1000} \times \frac{100}{\text{wt. of sample (g)}}$$

Precautions

1. The bottles must be shaken thoroughly throughout the titration to ensure that all the iodine is expelled from the chloroform layer.

3.4 Extraction, Separation and Estimation of Lipids

3.4.1 EXPERIMENT: Extraction and estimation of total lipid content in the given sample of oil seed

Principle
Lipids are soluble in some organic solvents. This property of specific solubility in non-polar solvents is utilized for extracting lipids from tissues. In biological materials, the lipids are generally bound to proteins and they are, therefore, extracted either with a mixture of ethanol and diethyl ether or a mixture of chloroform and methanol. Inclusion of methanol or ethanol in the extraction medium helps in breaking the bonds between the lipids and proteins.

Materials and Reagents

1. Iodometric flasks
2. Separatory funnel
3. Oil seeds (sunflower, peanuts or soybean)
4. Anhydrous sodium sulphate
5. Chloroform:methanol mixture (2:1, *v*/*v*)
6. 1% sodium chloride

Procedure

1. Take 1 g of the oil seed and grind it in presence of 5 g of anyhdrous sodium sulphate in a pestle and mortar. A small amount of acid washed sand may be used as an abrasive if the seed material is tough.
2. Add 20 ml of chloroform-methanol mixture to it and transfer to an air tight glass stoppered iodometric flask. Shake the content of the flask on a mechanical shaker for one hour and then filter it through a glass sintered funnel. Repeat the extraction of the residue twice and pool the filtrates.
3. Remove the solvent from the residue by distilling under vacuum. Since the residue left after drying contains crude lipids, extract it again with 10 ml of chloroform-methanol mixture containing 1 ml of 1% sodium chloride.
4. Take the pooled fractions in a separatory funnel, shake it thoroughly and allow it to stand for 5 min. The lipids will be recovered in the lower

chloroform layer while soap, glycerol and other water insoluble impurities move into the upper layer.

5. Drain out the lower layer and treat the upper layer again 3–4 times with 5–10 ml of chloroform-methanol mixture to extract any residual lipid from it.
6. Collect the lipid containing fractions in a pre-weighed beaker.
7. Evaporate the solvent by keeping the beaker in warm water bath (50°C) with a constant blowing of a slow stream of nitrogen gas over the surface.
8. Record the weight of beaker and determine the amount of crude lipids in the sample by subtracting the weight of empty beaker.
9. Express the results in terms of % crude lipids in the given sample of the oil seed.

Precautions

1. The sample should not be exposed to high temperature or light as some lipids get polymerized or decomposed on exposure to light, heat and oxygen.

3.4.2 EXPERIMENT: Separation and identification of various lipids by column chromatography

Principle

The extracted lipids are fractionated into neutral, non-polar and polar lipids by adsorbing them on solid adsorbent in non-polar solvents. They are then eluted stepwise with solvents of increasing polarity. The individual components in neutral and polar fractions obtained by this method may further be separated by thin layer chromatography (see EXPERIMENT 3.4.3).

Materials and Reagents

(a) For column chromatography

1. Lipid preparation (extracted as in EXPERIMENT 3.4.1)
2. Silicic acid
3. Chloroform
4. Acetone
5. Methanol
6. Glass column (2.5 cm dia × 15 cm length)
7. Glass wool

(b) For estimation of galactose (Reagents 8–9)

8. Anthrone reagent: Dissolve 0.2 g anthrone in 100 ml of conc H_2SO_4.
9. Standard galactose solution: Dissolve 25 mg galactose in water and make the volume to 250 ml. This solution contains 100 μg galactose/ml.

(c) For estimation of phosphorus (Reagents 10–14)

10. Digestion mixture: Sulphuric acid and 72% Perchloric acid (9:1, *v/v*)

11. Aminonaphthol sulphonic acid (ANSA) reagent: Dissolve 3.43 g of sodium metabisulphite, 0.063 g of 1-amino-2-naphthol-4-sulfonic acid and 0.125 g of sodium sulphite separately in small volume of water. Mix these solutions, filter through Whatman No. 1 filter paper and make the volume to 25 ml with water.
12. 0.26% ammonium heptamolybdate.
13. ANSA-molybdate reagent: Mix 22 ml of ANSA reagent (Reagent No 11) and 500 ml of ammonium heptamolybdate just before use. This reagent is stable for 1 h only.
14. Standard phosphorus: Dissolve 34 mg of KH_2PO_4 in water and make the volume to 250 ml. This solution contains 1 μmole of phosphorus/ml.

Procedure

1. Wash silicic acid thrice with double distilled water to remove fine particles and activate it at 125°C for 14 h.
2. Suspend the activated silicic acid in chloroform and carefully pour the slurry into the glass column having glass wool plug at its bottom. Allow the adsorbent to settle and add more slurry till a column with a bed height of 9 cm is obtained.
3. Wash the packed column with three column volumes of chloroform.
4. Apply lipid preparation (dissolved in chloroform) to the silicic acid column and elute it with 200 ml of chloroform at constant flow rate of 2 ml/min. The neutral lipids i.e. triacylglycerols, diacylglycerols, sterols and sterol esters get eluted during this step. The different lipid components in these fractions may further be separated by Thin Layer Chromatography (see EXPERIMENT 3.4.3).
5. The polar lipids, which remain adsorbed in the column during previous step are then eluted with 120 ml of acetone. These fractions contain glycolipids and sterol glycolipids.
6. The residual glycolipids and all the phospholipids are finally eluted from the column with 280 ml of methanol. Collect fractions of 5 ml each and examine fractions for the presence of galactose (Step 7) and phosphorus (Step 8).
7. Take 1 ml of the fraction in a test tube and place it in a water bath at 70°C to evaporate acetone. Add 3 ml of diluted sulphuric acid (1 ml of H_2SO_4 diluted to 13 ml with water) and keep the sample for hydrolysis in boiling water bath for 1.5–2.0 h to release galactose. Allow to cool and extract the liberated fatty acids in 5 ml chloroform. The aqueous layer is used for estimation of galactose. Take 1 ml of the aqueous extract in a test tube and keep it in ice bath. Add 5 ml of anthrone reagent (Reagent No. 8) and mix the contents. For reagent blank use 1 ml of water in place of the aqueous phase of the hydrolysed lipid sample. Place the tubes in boiling water bath for 10 min and record optical density at 625 nm against reagent blank after cooling them under running tap water. Prepare a standard curve as described above

over a range of 0–100 μg of galactose. Determine the amount of galactose in the sample tubes from the standard curve.

8. Pipette 1 ml of the phospholipid fractions (containing 0.5–3.0 μg phosphorus) in test tubes and evaporate methanol by heating them in water bath at 70°C. Now add 0.4 ml of digestion diacid mixture (Reagent No. 10) and heat at 180°C for about 45 min to release phosphorus from phospholipids. When the solution in test tubes becomes clear, remove the tubes. After allowing them to cool, add 9.6 ml of ANSA—molybdate reagent (Reagent No. 13) and keep them in boiling water bath for 10 min. Simultaneously prepare a reagent blank by replacing acid hydrolysate with water and also a standard curve using standard solution of KH_2PO_4 (Reagent No. 14) containing 0–1.0 μmole of phosphorus. Cool the tubes and note the optical density at 640 nm against the reagent blank. Determine the amount of phosphorous in each fraction from the standard curve.

3.4.3 EXPERIMENT: Separation of various components in the different lipid fractions by thin layer chromatography

The qualitative and quantitative analysis of non-polar and polar lipids can be effectively done after separating the components in these fractions by thin layer chromatography (TLC). It is a chromatographic technique in which different components in a sample get separated during their passage along a very thin layer, usually of 0.20–0.25 mm thickness, of a suitable chromatographic media which is spread as a uniform layer on a glass plate (see Section 9.4 for details of TLC). The chromatographic material used for this experiment is Silica Gel-G. Different lipids are adsorbed onto activated Silica Gel-G with varying degrees of strength. Those components which are not adsorbed or adsorbed with lesser strength tend to move faster alongwith the mobile phase (solvent system), whereas those which are held more firmly by the adsorbent (stationary phase) travel slowly. This differential distribution of various components between the mobile and stationary phases determines the rate of migration of different compounds in a mixture and results in their separation from each other.

Materials and Reagents

1. Activated TLC plates	Glass plates (20 × 20 cm) onto which a 0.2 mm thick layer of Silica Gel-G has been layered with the help of a spreader. The prepared thin layer plates are dried at room temperature and then activated at 110°C for 45 min before use.
2. Thin layer chromatographic tanks	
3. Developing mixtures	(a) Chloroform : methanol : water (65 : 25 : 4) for separation of phospholipids and galactolipids from polar lipids.

(b) Toluene : ethyl acetate : ethanol (2 : 1 : 1) for separation of galactolipids from phospholipids.
(c) Hexane : diethylether : acetic acid (80 : 20 : 1) for separation of various neutral lipids.

4. Spraying reagents for location of spots on TLC plates
   (a) Sulphuric acid : 50% H_2SO_4 in water (*v*/*v*).
   (b) Perchloric acid : 20% Perchloric acid in water (*v*/*v*).
   (c) Ferric chloride spray: Dissolve 50 mg $FeCl_3 \cdot 6H_2O$ in 90 ml water and add 5 ml each of glacial acetic acid and conc sulphuric acid.
   (d) Ninhydrin spray: 0.2 g ninhydrin dissolved in 100 ml ethanol.
   (e) Anthrone spray: Dissolve 0.2 g anthrone in 100 ml conc H_2SO_4.
   (f) Molybdate spray: Prepare a solution containing 16 g ammonium molybdate in 120 ml water. To 80 ml of this solution add 40 ml conc HCl and 10 ml mercury. Shake it for 30 min. To this solution add 200 ml conc H_2SO_4 and remaining 40 ml of ammonium molybdate. Cool the mixture and make the volume to 1 L with water.
   (g) Iodine spray: 1% iodine in chloroform.
5. Various lipid fractions prepared in EXPERIMENT 3.4.2.
6. Standard lipids: Make 0.2% solution of cholesterol, tristearin, phosphatidyl ethanolamine, palmitic acid, stearic acid, lecithin, galactosyldiacylglycerol.

Procedure

1. Apply 10–20 μl fraction of the lipid sample in form of a spot at a distance of 2 cm starting from the left bottom edge of the activated TLC plate. For details of TLC technique see Sections 9.6.4 and 9.6.5.
2. Develope the plates in the appropriate developing mixture (depending upon the types of lipids to be separated) in an air tight chromatographic glass tank till the solvent front moves upto 4 cm below the top edge of the glass plate.
3. Take out the plates, dry them for 5 min in air and spray the plates with the required detection reagent. The specificity and use of different spraying reagents is given below.

4. Locate the position of the lipid spots on the glass plate and measure the distance travelled by the individual lipid component. Calculate their R_f values and compare them with those of standards.

Specification of various spraying reagents

1. Sulphuric acid spray: This reagent forms charred black spots with all the lipids. Spray the plates with H_2SO_4 and keep the plates in oven at 120°C for 30 min. Mark the spots on the plate with pencil.
2. Iodine vapours: This is also a general detection reagent for all lipids. On keeping the plates in a jar or development tank containing a trough filled with iodine crystals, yellowish or brown spots appear and then fade away after some time on exposure of plates to air.
3. Perchloric acid spray: Heat the developed chromatogram at 100°C for 15 min and spray with perchloric acid. Note the brown spots which are formed by the lipids.
4. Ferric chloride spray: Spray the developed plates with ferric chloride and heat them at 100°C for 2–3 min in an oven. Lipid samples of cholesterol and cholesteryl esters will produce red or violet spots.
5. Ninhydrin spray: For detection of lipids containing amino groups, spray the developed plates with ninhydrin reagent and heat them in oven at 100°C for 5 min. Violet spots will indicate their presence and location on the plates.
6. Anthrone spray: Spray the chromatogram with anthrone reagent and heat the plate in oven at 100°C for 10 min. Formation of green spots will be due to its reaction with glycolipids and violet spots due to sulpholipids.
7. Molybdate spray: Phospholipids are identified as violet spots on spraying the developed plates with molybdate reagent or iodine solution. Heat the plates in oven at 100°C for 5 min after spraying them with molybdate reagent.

3.4.4 EXPERIMENT: Quantitative estimation of different lipid fractions separated by thin layer chromatography

Principle

Various fractions of lipids are separated by thin layer chromatography by using an appropriate solvent system as described in EXPERIMENT No. 3.4.3. These plates are then exposed to iodine vapours for location of the lipids. An outline of all the visible spots is marked with a needle. The spots are carefully scraped from the plates and triacylglycerols, glycolipids and phospholipids in the collected gel particles are quantitatively estimated using specific colorometric procedures.

Materials and Reagents

1. Table centrifuge.
2. TLC plates: Prepare and develope them as described in EXPERIMENT 3.4.3.
3. Hexane-diethylether (80:20, *v*/*v*).

4. Alcoholic KOH: Dissolve 2 g of KOH in 50 ml of 95% redistilled ethanol.
5. 0.4 N sulphuric acid: Take 5.5 ml of conc H_2SO_4 and dilùte it to 500 ml with distilled water.
6. 0.5 M Sodium arsenite: Dissolve 6.5 g of $NaAsO_2$ in 100 ml of distilled water.
7. 0.01 M sodium metaperiodate: Weigh 1.07 g of sodium metaperiodate and dissolve in 500 ml of distilled water.
8. Chromotropic acid (0.2%): Prepare by dissolving 2 g of chromotropic acid in 1 L of 66% H_2SO_4.
9. All reagents required for the estimation of galactose (see EXPERIMENT 3.4.2).
10. All reagents needed for the estimation of phosphorus (see EXPERIMENT 3.4.2).

Procedure

1. Take out the plates from the chromatographic tanks after performing thin layer chromatography. Dry the plates at room temperature till the solvent has completely evaporated.
2. Place the plates in a chamber containing iodine vapours for the detection of lipids (see EXPERIMENT 3.4.3). Remove the plates and mark an outline of all the visible spots by using a needle. Keep the plates in open to facilitate the removal of iodine by exposure of the plates to air.
3. Quatitative estimation of triacylglycerols: Thoroughly scratch the gel containing triacylglycerols spots and quantitatively transfer the collected powder to 15 ml centrifuge tubes. For blank use gel scraped from the zone of plate which after treatment with iodine vapours did not have any visible spots. To each tube add 10 ml of a mixture of hexane:diethylether (80:20) and shake vigorously for 5 min. Centrifuge the tubes àt 5,000 × g for 15 min and collect the supernatant. Repeat the extraction step twice using 5 ml of hexane:diethyl ether mixture each time. Pool the supernatants and evaporate to dryness at 70°C on a water bath. Now add 0.5 ml of 4% alcoholic KOH to each tube and keep at 70°C for 20 min. Neutralize this hydrolysate with 0.4 N H_2SO_4. For determining the amount of glycerol liberated, add 0.5 ml of 0.01 M sodium metaperiodate to each tube and keep it for 10 min for oxidation. Neutralize the excess of sodium metaperiodate with 0.1 ml of 0.5 M sodium arsenite. Then add 8 ml of 0.2% chromotropic acid and mix the contents thoroughly. Place the tubes in a boiling water bath for 30 min. Record absorbance of the developed colour at 570 nm.
4. Quantitative determination of phospholipids: Scrape the spots containing phospholipids and collect the silica gel scraped from individual spots in different 15 ml centrifuge tubes. Digest the material and develop the colour for phospholipids as described in EXPERIMENT 3.4.2, Step 8. Centrifuge the tubes at 3,000 × g for 10 min to remove silica gel and record colour intensity of the supernatant at 640 nm against reagent blank.

5. Estimation of glycolipids: Scrape the identified spots of glycolipids and transfer the collected silica gel into 15 ml centrifuge tubes. Take two blanks by taking silica gel from lipid free areas of the plate. Add 1.5 ml of methanol to each tube and shake the tubes vigorously. Centrifuge the tubes at 3000 × g for 10 min and collect the supernatant by decantation. Repeat this extraction step thrice. Keep the pooled supernatants of each spot in a water bath at 70°C till a dry residue is obtained. Proceed as described in Step 7 of EXPERIMENT 3.4.2 for estimating galactose content in each tube.

3.4.5 EXPERIMENT: Identification and quantification of fatty acids by gas liquid chromatography (GLC)

Principle

The quantitative composition of fatty acids in a fat can be evaluated by identification of fatty acids present in it. Though the fatty acid esters can be identified by TLC, their quantification is best done by gas liquid chromatography (see Section 9.5.5 for details of GLC). For this fatty acids are first converted into their respective volatile esters by transesterification i.e. by converting them from glycerol esters to methyl esters. Esterification can also be done after saponification (see EXPERIMENT 3.3.2). The esters are identified and quantified by injecting the processed sample into GLC and identified by comparing with chromatographic patterns of a set of standard esters.

Materials and Reagents

1. Table centrifuge.
2. Water bath at 65°C.
3. GLC: GLC fitted with a stainless steel column (10′ × 1/8″) packed with 20% diethylene glycol succinate on Chromasorb W (60–80 mesh). Detector used is flame ionization detector (FID).
4. Carrier gas: N_2 gas at flow rate of 35 ml/min.
5. Carbon disulphide (CS_2).
6. 0.4 N sodium methylate: Cut 1 g metallic sodium into small pieces under petroleum ether and add to 100 ml methanol. Add only small pieces of metal at a time. The amount of the metal added should be in slight excess of that required for desired normality.
7. Mixture of silica gel and calcium chloride (1:1).
8. Standards: Authentic samples of methyl esters of various fatty acids.

Procedure

1. Take 1–30 mg of lipid sample in a centrifuge tube and add 0.25 ml of 0.4 N sodium methylate. Flush the tubes with N_2 gas.
2. Cover mouth of the tubes with aluminium foil, heat in water bath at 65°C for 2 min and shake it on vortex for 30 sec. Appearance of homogenous solution will indicate the completion of esterification.

3. Heat the contents of the tubes for another 1.5 min without shaking.
4. Cool the tubes and add 0.6 g mixture of silica gel and calcium chloride quickly and stirr the contents of the tube vigorously with a glass rod.
5. Add 3 ml of carbon disulphide, cover the tubes with aluminium foil and shake them for 1–2 min.
6. Centrifuge the tubes at 1200 rpm for 2 min.
7. Decant the clear layer of carbon disulphide into small test tubes. This contains the methyl esters of fatty acids.
8. Operate GLC as per operator's manual.
9. Inject an aliquot (1–2 μl) into preconditioned GLC in which the temperature of injection port is 280°C and column temperature 165°C.
10. After all the components of the sample have come out of the column, inject the standard methyl esters separately and note the retention time of esters of different fatty acids.

Calculation

1. Identify the individual peaks of the samples by comparing their relative position or retention time with the standards. Esters appear in order of increasing number of carbon atoms and of increasing degree of unsaturation for the same number of carbon atoms present in the fatty acids.
2. Calculate the peak area of each acid by using the formula base/2 × peak height in the sample and compare with that of standard. From the quantity of sample or standard injected and attenuation, the amount of fatty acid can be calculated (consult operator's manual of the GLC equipment).

3.4.6 EXPERIMENT: Estimation of cholesterol content by Liebermann-Burchard method

Principle

Cholesterol is the most abundant sterol in animal tissues. It is a derivative of saturated tetracyclic hydrocarbon-perhydrocyclopentanophenanthrene ring.

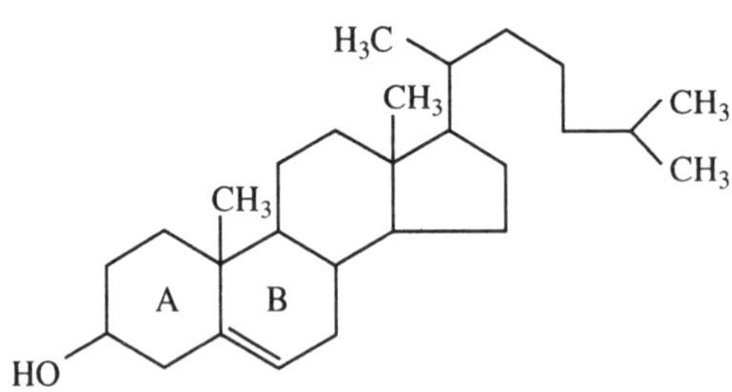

Cholesterol

Blood cholesterol value is an important diagnostic parameter from clinical point of view. The Liebermann-Burchard test is based on the fact that the sterols with unsaturation in the ring A or B react with sulphuric acid in the presence of acetic anhydride to give blue colour which immediately changes to green. During the reaction, dehydration, condensation and isomerization take place with the formation of halochromic salt which is green in colour. The exact nature of the chromophore is not known but reaction probably includes esterification of the hydroxyl group at 3-position and sterols become activated and fleeting colours are produced due to shift in double bond.

Materials and Reagents

1. Colorimeter
2. Cholesterol in chloroform (100 μg/ml)
3. Acetic anhydride-sulphuric acid mixture: Prepare just before use by mixing acetic anhydride and H_2SO_4 (30:1, *v/v*).
4. Chloroform

Procedure

1. For preparation of standard curve, pipette out solution containing 10–100 μg cholesterol in different test tubes and make the volume to 1 ml with chloroform.
2. Take a suitable volume of the extract in which cholesterol has to be estimated and make its volume upto 1 ml.
3. Add 5 ml of acetic anhydride-sulphuric acid mixture carefully to each test tube. Mix well and note the change in colour of the solution. Cover the test tubes and leave them in dark for 15 min.
4. Measure the absorbance of the solution at 640 nm and draw the standard curve for cholesterol.
5. From the standard curve, calculate the amount of cholesterol present in 1 ml of the sample preparation.

Precautions

1. All the apparatus used for this experiment should be absolutely dry.
2. Acetic anhydride and sulphuric acid reagent should be colourless.

Suggested further reading

Gurr M.J. and James A.T. (1977). Lipid Biochemistry, Chapman Hall, London.

Jayaraman J. (1992). Laboratory Manual in Biochemistry, Wiley Eastern Limited, New Delhi.

Kates M. (1986). Techniques in Lipodology–Isolation, Analysis and Identification of Lipids, 2nd ed, Elsevier, New York.

Pearson D. (1973). Laboratory Techniques in Food Analysis, Butterworths, London.

Pelick N., Bolliger H.R. and Mangold H.K. (1966). *In* Advances in Chromatography, (Giddings J.C. and Keller R.A., eds), Vol. 3, p. 85, Marcel-Dekker, New York.

Renkonen O. and Varo P. (1967). *In* Lipid-Chromatographic Analysis (Marinetti G.V., ed), Vol. 1, Marcel-Dekker, New York.

Skipski, V.P. and Barclay M. (1969). Thin Layer Chromatography of Lipids. *In* Methods in Enzymology, (Lowensterin J.M., ed), Vol. III-LIPIDS, pp 531–598.

4. Amino Acids and Proteins

4.1 Introduction

Amino acids are organic compounds that contain amino and carboxyl groups and therefore possess both basic and acidic properties. Proteins are polymers of amino acids and contain about 22 different amino acids, with the exception of proline and hydroxyproline, which strictly speaking are imino acids. General formula of an amino acid is as shown below:

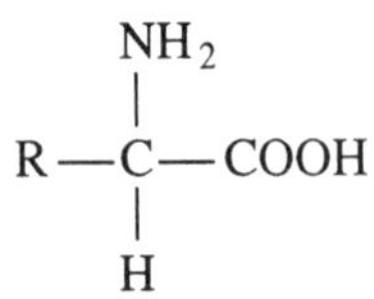

Amino acids exist as *zwitter ions* in a neutral solution.

Uncharged *zwitter ion*

All amino acids with the exception of glycine show optical activity due to presence of asymmetric carbon atom.

Amino acids are the basic building blocks of proteins. In proteins, amino acids are linked by peptide (—CO—NH—) bond, formed by the condensation of the α-COOH of one amino acid with the α-NH_2 group of the next amino acid as illustrated below:

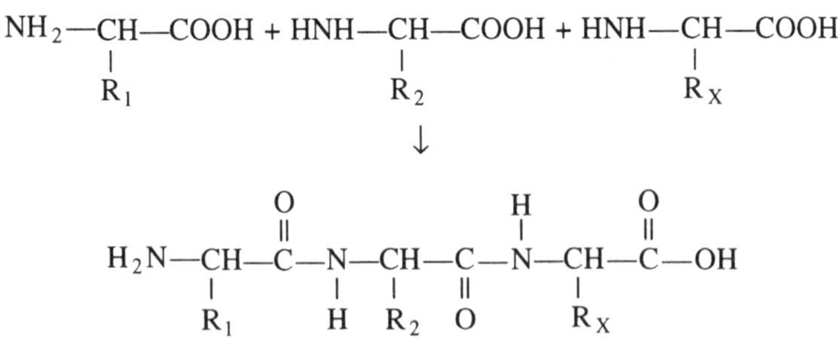

Proteins are linear polymers of about 20 different types of amino acids. Arbitrarily the lower limit of protein is fixed at 10,000 daltons and those having a molecular mass lesser than this are referred to as peptides.

Proteins (Greek: *Proteios*, meaning holding first place) are one of the most important group of biomolecules. The quantity and type of proteins shows a wide variation between not only among different organisms but also in different organs or parts of the same organism. Proteins perform very vital and diverse roles in organisms. A simple bacterial cell such as *Escherichia coli* contains more than 5000 different types of proteins. A few of the functions of proteins are enumerated as follows:

1. Enzymes: With the exception of ribozymes all the enzymes are proteins and these act as biocatalysts in a cell.
2. They are constituents of biomembranes where some of them function as carriers or translocators, thus facilitating selective transport of metabolites across the cell membranes.
3. Antibodies and some of the toxins and antibiotics are either proteins or peptides and form an integral part of defence systems of living organisms.
4. Some of the proteins, such as insulin, act as hormones.
5. Transport proteins like heamoglobin, ferretin.
6. Structural proteins, e.g. hair, nail, proteins of cell wall and cell coat.
7. Storage proteins, e.g. zein in maize, casein in milk.

In this chapter some of the methods employed for qualitative as well as quantitative determination of amino acids and proteins in biological materials are described.

4.2 Qualitative Tests for Amino Acids and Proteins

The various qualitative tests for detection of amino acids in a sample preparation are described in this section. Some of these tests are general and are given by all the amino acids while others are specific for a particular amino acid. The latter type of tests can thus be employed for identification of a particular amino acid. Many of these tests give positive response irrespective of the fact whether the concerned amino acid exists in free form in a solution or happens to be constituent of a protein molecule. Hence, these tests can also be employed for the purpose of detection of proteins in a sample. A negative test for a particular amino acid, however, need not necessarily denote absence of protein in a purified sample preparation. Such a situation can be encountered if the isolated protein totally lacks or contains very few residues of that particular amino acid. For example, failure to get a positive result for Hopkins-Cole test (which is specific for tryptophan) for purified preparation of gelatin is not due to absence of protein but because this protein does not contain any tryptophan residue. However, all proteins do give positive reaction with ninhydrin and biuret and thus serve as reliable qualitative test for preliminary establishment of presence of protein in a sample.

The colour tests for various amino acids are listed in Table 4.1 and some of these have been described in detail in later part of this section.

4.2.1 Ninhydrin test

Principle

This test is due to a reaction between α-amino group of free amino acid and

Table 4.1 Colour tests for proteins and various amino acids

Test/reaction	*Reactants*	*Specificity of the test*	*Colour obtained*
Ninhydrin	Ninhydrin	All amino acids, peptides, proteins	Purple except in case of proline and hydroxyproline which give yellow colour and also asparagine where brown colour is produced.
Biuret	Alkaline $CuSO_4$	Peptides, Proteins	Purple
Millon's Test	$Hg(NO_3)_2$ in HNO_3 with traces of nitrous acid	Tyrosine	Brick red colour
Xanthoproteic reaction	Boiling conc HNO_3	Tyrosine Tryptophan Phenylalanine (gives weak test)	Yellow in acid which turns orange yellow
Pauly's Test	Diazotized sulphanilic acid in alkaline solution	Tyrosine Tryptophan Histidine	Red
Hopkins-Cole Test	Glyoxylic acid in conc H_2SO_4	Tryptophan	Violet ring at interface
Ehlrich's Test	*p*-dimethyl-amino-benzaldehyde in conc HNO_3	Tryptophan	Blue
Sakaguchi reaction	α-naphthol and sodium hypochlorite	Arginine	Red
Sodium nitroprusside Test	Sodium nitroprusside in dil HNO_3	Cysteine	Red
Sullivan reaction	Sodium-1,2-naphthaquinone-4-sulfonate and sodium hydrosulphite	Cysteine	Red

ninhdyrin (triketohydrindene hydrate). Ninhydrin is a powerful oxidising agent and in its presence, amino acids undergo oxidative deamination liberating ammonia, carbon dioxide, a corresponding aldehyde and reduced form of ninhydrin. The ammonia formed from α-amino group reacts with ninhydrin and its reduced product (hydrindantin) to give a blue substance diketohydrin (Ruhemann's purple). However, in case of imino acids like proline, a different product having a bright yellow colour is formed. Asparagine which has a free amide group reacts to give a brown coloured product. This test is also given by proteins and peptides.

Reaction

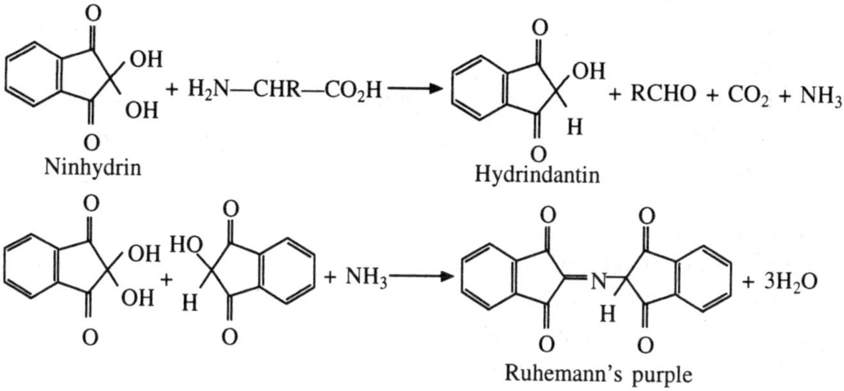

Materials and Reagents

1. Boiling water bath.
2. Ninhydrin: 0.2% solution prepared in acetone.
3. Test solution: Prepare solution containing 50 μg/ml of individual amino acids.

Procedure and observations
Add 2–5 drops of ninhydrin solution to 1 ml of test solution or sample preparation. Mix and keep for 5 min in boiling water bath and observe the development of a pink, purple or violet-blue colour. Imino acids like proline and hydroxyproline give a yellow coloured complex.

4.2.2 Xanthoproteic test

Principle
Amino acids containing an aromatic nucleus (tyrosine, tryptophan, phenylalanine) form yellow nitro derivatives on heating with conc HNO_3. The salts of these derivatives are orange in colour. Proteins containing these amino acids also give a positive response to this test.

Reactions

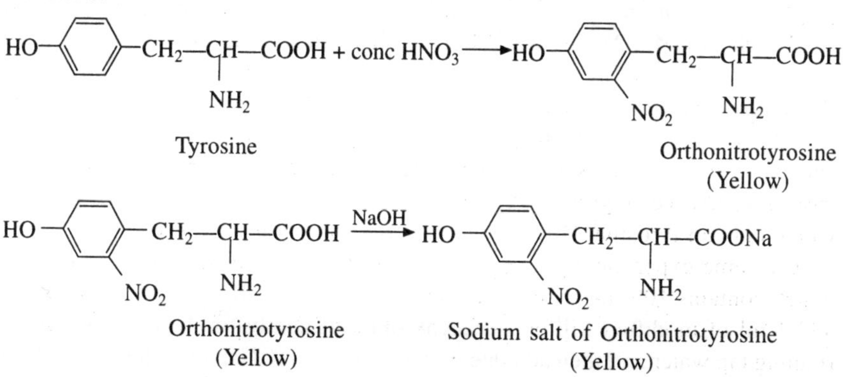

Materials and Reagents

1. Conc. HNO_3.
2. NaOH solution (40%, *w*/*v*): Dissolve 40 g of NaOH in water and make the final volume to 100 ml.
3. Test solutions: Prepare separate solutions containing 50 μg/ml of amino acids like tyrosine, glycine, tryptophan, phenylalanine, lysine, cysteine, leucine etc.

Procedure and observations

Add 1 ml conc. HNO_3 into 1 ml of test solution. Mix the contents and keep in boiling water bath for 1 min. Cool it and then slowly pipette NaOH till the solution becomes alkaline. Note whether the mixture turns orange red in colour. Appearance of orange red colour denotes presence of aromatic amino acids.

4.2.3 Millon's test

Principle

Amino acids or compounds containing hydroxybenzene radical react with Millon's reagent which constitutes of mercurous and mercuric nitrates containing HNO_3, to form a red complex. Thus this test is specific for tyrosine.

Reaction

$$Hg + 4HNO_3 \rightarrow Hg(NO_3)_2 + 2NO_2 + 2H_2O$$

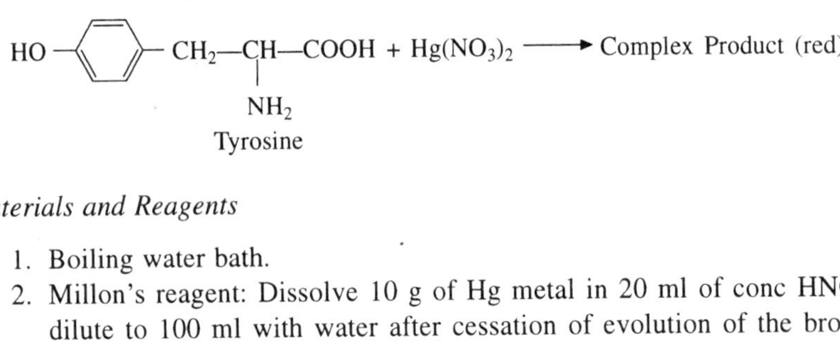

Tyrosine

Materials and Reagents

1. Boiling water bath.
2. Millon's reagent: Dissolve 10 g of Hg metal in 20 ml of conc HNO_3, dilute to 100 ml with water after cessation of evolution of the brown fumes.
3. Test solutions of different amino acids made as in the preceding test (Section 4.2.2). Use a solution of phenol also as a test sample.

Procedure and observations

Add 2–5 drops of Millon's reagent into 1 ml of test solution. Shake the contents and place the test tubes in boiling water bath. Development of a red colour indicates presence of tyrosine. A positive reaction will also be obtained for proteins which contain tyrosine. Does phenol react positively to this test?

The same experiment may be carried out with a modified Millon's reagent which contains 10% mercuric sulphate in 10% H_2SO_4. To 1 ml of test solution, add 1 ml of modified Millon's reagent and boil gently for 1 min. Cool under running tap water and then add one or two drops of 1% $NaNO_2$ solution and heat slightly. Appearance of red colour denotes presence of tyrosine. The advantage of using modified Millon's test lies in its being less susceptible to interference from inorganic salts.

4.2.4 Hopkins-Cole test

Principle

The indole group of tryptophan reacts with glyoxylic acid in the presence of conc H_2SO_4 to give a purple colour. Glyoxylic acid is prepared by reducing oxalic acid with magnesium powder or sodium amalgam. Glacial acetic acid which has been exposed to the sunlight also contains glyoxylic acid and can thus be used for this test.

Reaction

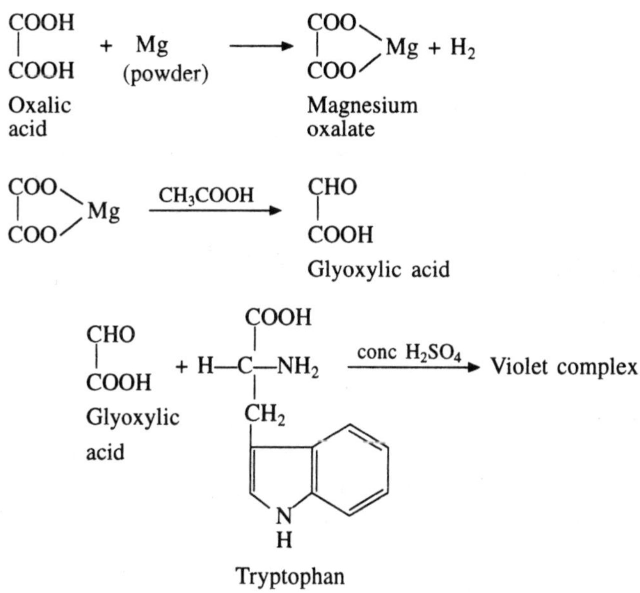

Materials and Reagents

1. Test reagent: To 10 g of magnesium powder add enough water to submerge it. Then slowly add 250 ml of cold saturated oxalic acid solution. Keep cool. Filter magnesium oxalate formed. Acidify the filtrate with 25 ml of glacial acetic acid and dilute it to 1 L with water. Alternatively, glacial acetic acid which has been kept in sunlight for a few minutes may be used as a test solution.
2. Test solutions: Prepare solution containing 50–100 μg/ml of individual amino acids such as tryptophan, tyrosine, phenylalanine, glutamic acid etc.
3. Conc H_2SO_4.

Procedure and observations

Add 1 ml of Hopkins-Cole test reagent to 1 ml of test solution. Mix and pour with pipette conc H_2SO_4 along the sides of test tube. Observe for appearance of blue or violet ring at the interface of the liquids which would indicate presence of tryptophan.

4.2.5 Biuret test

Principle

This test is a general test for compounds having a peptide bond. Alkaline copper sulphate reacts with compounds containing two or more peptide bonds to give a violet or pinkish coloured product which is due to formation of coordination complex of cupric ions with unshared electron pairs of peptide nitrogen and oxygen of water.

Reaction

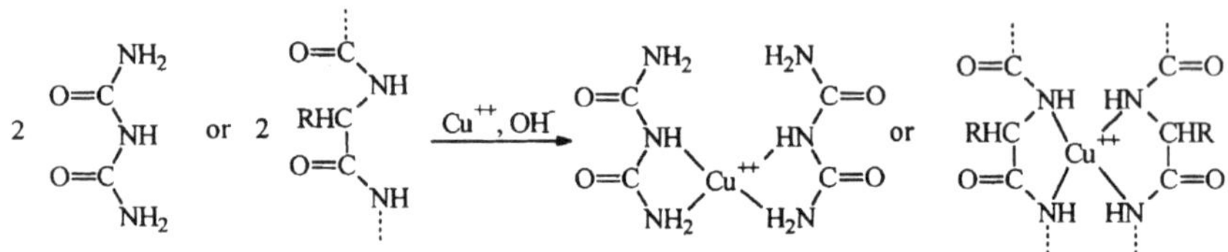

Biuret Peptide Cu^{++}-Coordination complexes

Materials and Reagents

1. Copper sulphate: 1% (*w/v*) of $CuSO_4.5H_2O$ in water.
2. 40% (*w/v*) solution of NaOH.
3. Test solutions:
   (a) Prepare 0.5% (*w/v*) solution of a protein like casein or bovine serum albumin in NaOH and also a tripeptide glutathione.
   (b) Prepare 0.5% (*w/v*) solutions of glutamate, cysteine and glycine.

Procedure and observations

Add 0.5 ml NaOH to 1 ml of test solution and mix well. Then add 2–5 drops of copper sulphate and observe for the formation of pink or violet colour. Appearance of pink or violet colour shows presence of peptides or proteins in the sample. Try this test with solutions of glutathione and its constituent amino acids and record the observations.

4.2.6 Pauly's test

Principle

Diazotized sulphanilic acid couples with amines, phenols and imidazole to form highly coloured compounds. The diazonium compound is only formed in the cold, so all solutions must be cooled in ice before diazotization.

Reaction

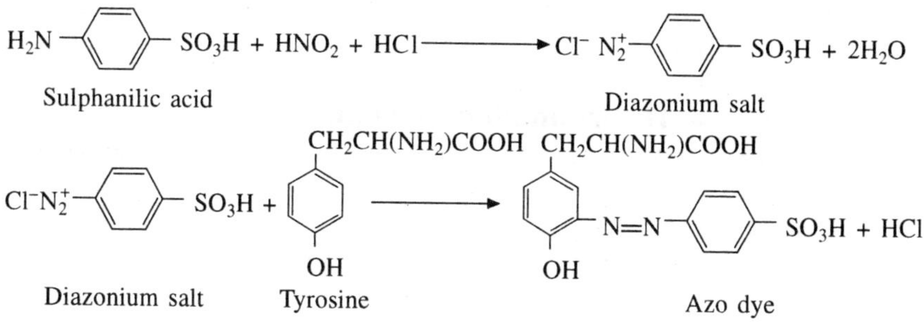

Materials and Reagents

1. Amino acids: 0.1% solutions of glycine, tyrosine, histidine, tryptophan.
2. Sulphanilic acid: Make 1% solution of sulphanilic acid in 10% HCl.
3. 5% (*w/v*) sodium nitrite solution in water.
4. 1% (*w/v*) solution of sodium carbonate in water.

Procedure and observations
Mix 1 ml of sulphanilic acid with 2 ml of the test solution cooled in ice, add 1 ml of sodium nitrite solution and leave in the cold for 3 min. Make the solution alkaline by the addition of 2 ml of sodium carbonate solution. Note whether any coloured products are formed or not.

4.2.7 Sakaguchi Reaction

Principle
This reaction is given by guanidinium compounds. The only amino acid containing the guanidinc group is arginine, and this reacts with α-naphthol and an oxidizing agent such as bromine water or sodium hypochlorite to give a red coloured product. The other guanidine containing non-amino acid compounds also give this reaction.

Materials and Reagents

1. Amino acids: 0.1% solution of amino acids like glycine, arginine, lysine, tyrosine etc.
2. Guanidines: 0.1% solutions of glycocyamine, methylguanidine and creatine.
3. 0.1% urea solution.
4. NaOH: 40% (*w/v*).
5. α-Naphthol: 1% in alcohol.
6. Bromine water: Add a few drops of bromine to 100 ml water and shake. This should be done in a fume chamber. (*Caution*: Bromine can cause severe burns).

Procedure and observations
Mix 1 ml of NaOH with 3 ml of the amino acid or test solution and add 2 drops

of α-naphthol. Mix thoroughly and add 4 to 5 drops of bromine water. Watch for the formation of red colour which would indicate presence of arginine or a guanidinium compound.

4.3 Quantitative Determination of Amino Acids

As mentioned earlier, an organism contains several thousands different types of proteins. Elucidation of properties, structure and role of these proteins is an integral part of biochemistry. These proteins differ from each other in their size i.e. amino acid content, amino acid composition and more importantly their amino acid sequence, which eventually governs their three dimensional structure and biological function. Determination of amino acid composition constitutes one of the initial steps in the complete chemical characterization of proteins.

Investigations on protein content and amino acid composition are also of immense value from point of view of animal and human nutrition as they provide valuable information about the total protein content as well as the amount of essential amino acids in various food commodities. A few exercises on quantitative estimation of total amino acid content and determination of amino acid composition of protein samples are described as follows.

4.3.1 EXPERIMENT: Quantitative determination of free amino acid content in germinating seeds

Principle

Though bulk of amino acids in an organism occur as constituents of cellular proteins, a small proportion exists in free form which represents soluble amino acid pool. The free pool supplies the various amino acids for biosynthesis of proteins as well as for other metabolic requirements. In germinating seeds, the storage or reserve proteins are rapidly hydrolysed to maintain a sustained supply of amino acids for *de novo* synthesis of proteins and hence they contain fairly high level of free amino acids. The quantity of free amino acids in biological materials can be estimated by their extraction with 70% (*v*/*v*) ethanol and then determining their content in the ethanolic extract by ninhydrin method.

Ninhydrin (triketohydrindene hydrate), a powerful oxidising agent, reacts with all amino acids between pH 4 and 8 to give a purple coloured complex with absorbance maxima at 570 nm. The imino acids (proline and hydroxyproline), however, give yellow colour which is measured at 440 nm. The reactions involved between the amino acids and ninhydrin are given in Section 4.2.1.

Materials and Reagents

1. Table centrifuge.
2. Boiling water bath.
3. Colorimeter.
4. Germinating pea seeds: Surface sterilize by dipping the pea seeds in 0.1% mercuric chloride for 1–2 minutes. Wash the seeds several times with distilled water and keep them at 25°C in Petri plates lined with moistened

Whatman No. 1 filter paper and if necessary add water periodically till they start germinating.

5. 0.2 M citrate buffer (pH 5.0): Dissolve 21 g citric acid in 200 ml water and mix with 200 ml of 1 N NaOH. If necesssary adjust the pH to 5.0 and make the volume to 500 ml with distilled water.
6. Ethanol: 70% (*v*/*v*).
7. KCN—Acetone ninhydrin solution: Take 50 ml of 5% ninhydrin solution in acetone and mix it with 250 ml of KCN-acetone solution (prepared by diluting 5 ml of 0.01 M KCN in 60% ethanol to 250 ml with acetone). *Alternatively*, ninhydrin reagent may be prepared by dissolving 2 g of ninhydrin in 25 ml of methyl cellosolve and then adding 25 ml of 0.2 M acetate buffer (pH 5.5).
8. Standard amino acid solution: Prepare a solution containing 100 μg of glycine/ml of 0.2 M citrate buffer (Reagent No. 5).

Procedure

1. Free amino acids are extracted by grinding 0.5 to 1.0 g of germinating seeds in 10 ml of 70% ethanol in a pestle and mortar. Centrifuge the homogenate at 5000 rpm for 10 min. Decant the supernatant. Repeat the extraction 4–5 times and combine the supernatants and make the volume to 50 ml with 70% (*v*/*v*) ethanol.
2. Take an appropriate volume (5–10 ml) of this ethanolic extract of germinating seeds, evaporate it to dryness on a boiling water bath and dissolve the residue in 5 ml of 0.2 M citrate buffer (pH 5.0).
3. Pipette 2 ml of the above sample preparation in a test tube. For a reagent blank, take 2.0 ml of 0.2 M citrate buffer (pH 5.0) in place of the sample preparation. In another set of tubes, take graded concentrations of glycine (0–100 μg) and make the total volume to 2 ml with 0.2 M citrate buffer (pH 5.0). Add 1 ml of KCN-acetone ninhydrin reagent or that prepared in methyl cellosolve to each tube and mix thoroughly.
4. Keep the test tubes in boiling water bath for 20 min, cool under running tap water and make the volume to 10 ml with distilled water.
5. Note the absorbance at 570 nm after setting the instrument with reagent blank.
6. Prepare a standard curve by plotting A_{570} against amount of glycine.
7. Find the amount of free amino acids content in the sample test tubes from the above reference curve.
8. Express the free amino acid content in seeds in terms of mg of glycine equivalents per g of seeds.

4.3.2 EXPERIMENT: Determination of amino acid composition of proteins of wheat grain by an automatic amino acid analyzer

Principle

Analysis of amino acid composition of a protein sample essentially involves two distinct steps: hydrolysis of protein to liberate the constituent amino acids followed by separation and quantitative estimation of the individual amino acid in the hydrolysed sample. The hydrolysis of a protein preparation can be carried out with an acid or an alkali. However, both the methods suffer from drawback of decomposition of the more labile amino acids. For example, acid hydrolysis of proteins results in complete destruction of tryptophan and partial loss of serine and threonine and glutamine and asparagine are converted to their respective amino acids due to the removal of their amide groups. On the other hand, the alkaline hydrolysis causes decomposition of cysteine, cystine, serine, threonine and arginine. Another disadvantage of alkali hydrolysis is that it yields a racemic mixture of the released amino acids. The most commonly used method is hydrolysis with 6 N HCl in evacuated sealed tubes at 110°C for 12–96 h. Alkali hydrolysis is generally carried out with 2–4 N NaOH at 100°C for 4 to 8 h and is particularly useful for determination of tryptophan in the protein sample.

The amino acids in the hydrolysed sample are then separated by techniques such as paper chromatography, paper electrophoresis or ion exchange chromatography before the individual amino acids can be quantitatively determined by their reaction with ninhydrin. However, Automatic Amino Acid Analysers are commercially available which are programmed to separate and quantitatively estimate the individual amino acids in the sample hydrolysate. HPLC is also used for quantitative estimation of amino acids.

The separation of different amino acids in the processed sample in an automatic amino acid analyser is accomplished by chromatography on columns of ion exchange resins. The theory of ion exchange chromatography is discussed in detail in Section 9.5.3. The chromatographic media used for separation of amino acids by ion exchange is a polymer of sulphonated polystyrene or a sulphonated copolymer of polystyrene and divinyl benzene which function as a strong cation exchangers. The sample of protein hydrolysate is loaded on the top of the column at pH 3.0, a condition under which all the amino acids exist in their positively charged state and thus get bound to acidic group of the resin via electrostatic forces. At this pH, the basic amino acids are held most strongly followed by neutral and then acidic amino acids. Due to this variation in the binding strength, the different amino acids are then selectively released from the resin and these travel down the column at different rates resulting in their separation from each other. To accomplish this the column is developed with elution buffer of gradually increasing pH and ionic strength. During chromatography, the acidic amino acids emerge out of the column first followed by neutral and then lastly the basic amino acids. The amount of each amino acid is determined by their colour reaction with ninhydrin (see Section 4.3.1).

A schematic diagram showing essential features of an automatic amino acid analyser is given in Fig. 4.1.

The basic components of an amino acid analyser include a set of buffer reservoirs connected through an automatic buffer valve and buffer pump to an ion exchange column, a ninhydrin reagent reservoir which is connected to the column effluent tubing through a pump, a high temperature water bath, two photometers for monitoring absorbance at 570 nm and 440 nm respectively and a recorder. For ensuring precision and reproducibility of this automatic procedure, both the elution buffers and ninhydrin reagents are delivered at constant and controllable rates by reliable pumps. This mixture of effluent and the colour developing reagent then flows through a coil of flexible teflon tubing immersed in a hot water bath, where the amino acid reacts with ninhydrin to yield a coloured product. This stream of liquid then passes through photometer units which monitor the absorbance at 570 and 440 nm respectively and the signals from the photometers are printed in form of peaks by a recorder on a chart paper. The identity of amino acid in each peak is established by referring to caliberated elution profiles of pure amino acid standards. The amount of amino acid in each peak is directly proportional to the peak area.

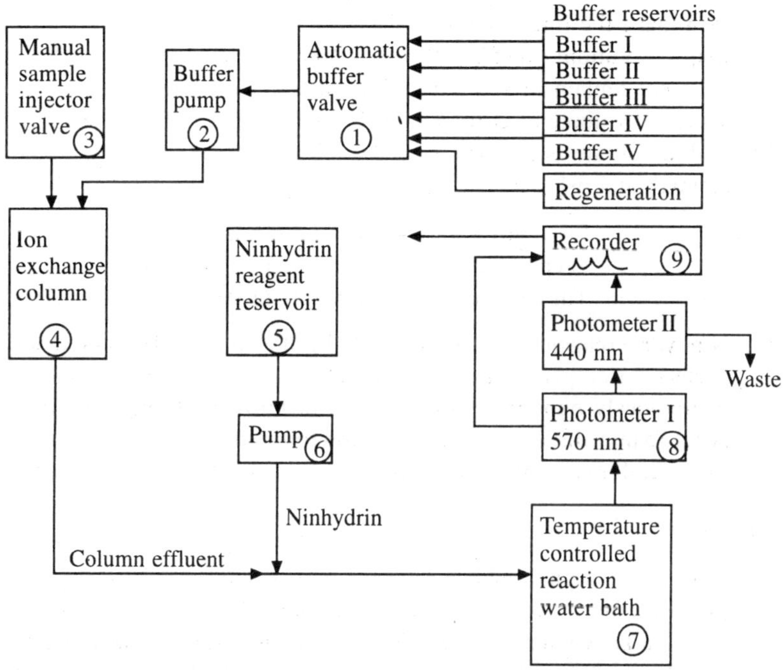

Fig. 4.1 Diagramatic representation of basic components of a single column automatic amino acid analyser.

Although, single column amino acid analysers are adequate in accomplishing complete and satisfactory separation of all individual amino acids, most of the analysers are equipped with two columns. The basic amino acids which are held most strongly are analysed on a small column and the neutral and acidic amino acids on a longer column. The two column analyser offers the advantage of considerably reducing the time required for analysis of a sample. With such two column analysers complete analysis of hydrolysate of a protein sample can be accomplished in about 2 h.

Materials and Reagents

1. Amino acid analyser, fitted with a column (54 × 0.9 cm) and packed with cation exchanger polystyrene sulphonic acid (Dowex-50).
2. Flash evaporator.

3. Boiling water bath.
4. Nitrogen cylinder.
5. Vacuum pump.
6. Soft glass ampules or test tubes.
7. Petroleum ether.
8. 6 N HCl.
9. NaOH pellets.
10. Sample dilution buffer (pH 2.2; Na^+, 0.2 N): Dissolve 19.6 g of sodium citrate (dihydrate) in small volume of water and add 16.5 ml conc HCl, 5.0 ml of thiodiglycol, 2 ml of Brij-35 solution, 0.1 ml *n*-caprylic acid and then make the final volume to 1 L with water.
11. Brij-35 solution: Add 50 g Brij-35 to 100 ml of hot water. If the detergent does not remain in solution, dilute to 150 ml and use 1.35 times the specified volume of Brij solution for preparation of. Reagent No. 10.
12. Sodium citrate buffer: Prepare the following three buffers of indicated strength and pH

First buffer	—	pH 3.49 (0.2 N)
Second buffer	—	pH 4.12 (0.4 N)
Third buffer	—	pH 6.4 (1.0 N)

13. Ninhydrin reagent: Take 3 L of methyl cellosolve and 1 L of 4 M sodium acetate buffer (pH 5.5) and stirr magnetically in a 4 L aspirator bottle while continuously bubbling a slow stream of nitrogen through the mixture. After 15 min, dissolve 80 g of ninhydrin followed by 1.6 g of $SnCl_2 \cdot 2H_2O$. Transfer this reagent to the ninhydrin reservoir of the amino acid analyser.

Procedure

1. Take wheat grains which have been dried in an oven at 70°C for 24 h and prepare a fine powder by grinding in a Willey Mill or a pestle and mortar. Weigh a known amount (1 g) of the powdered sample and add 20 ml of petroleum ether and keep it for 2 to 3 h. Carefully decant the solvent into a pre-weighed beaker and repeat this extraction step 4–5 times. Evaporate the pooled solvent, which was collected after decantation, on a boiling water bath. Record the weight of the beaker and determine the fat content, to facilitate calculation of amino acid content in the original non-defatted sample.
2. Weigh a known amount (containing 100–200 μg protein) of the above defatted sample of wheat grain and transfer it into a soft glass test tube or an ampule. Add 10 ml of 6 N HCl to the test tube containing the sample and then with the help of a burner carefully draw out a narrow neck of the test tube of approximately 2–3 mm diameter ensuring that the glass wall does not become too thin at any place. During this operation avoid any contact of the sample in the tube with the hot areas of the glass.
3. Blow a moderate stream of purified nitrogen into the liquid in the test tube using a thinly drawn out capillary tube connected via rubber tubing to the outlet control valve of the nitrogen cylinder. Foaming and rising of HCl

should be strictly avoided during this step. After flushing with nitrogen, completely evacuate the tubes with a vacuum pump and then properly seal off the tube at the neck with a suitable flame. Adequate flushing with nitrogen and complete evacuation of the tube is important to minimize the losses due to air oxidation of amino acids during hydrolysis of the sample.

4. Keep the evacuated sealed tubes upright in an oven at 110°C ± 1°C for 18–24 h. Then remove the tubes from the oven, holding them upright so that the liquid does not wet the top of the tubes.
5. Break the top of the tube by making a deep scratch with a triangular file and then touching the edge of the scratch with tip of a heated glass rod.
6. In case of any visible precipitate (humin) in the hydrolysate, remove it by centrifugation or filtration through a nitrogen free filter paper (Whatman No. 1) which has previously been washed firstly with 1% HCl, then twice with water and finally dried. This washing removes ammonia or other ninhydrin positive compounds from the filter paper.
7. Remove HCl from the hydrolysate by evaporating the samples to dryness on a boiling water bath. However, when the total volume of remaining HCl is 2 ml or less, its removal can more conveniently be effected by keeping the tubes at a slight inclination at the bottom of a vacuum dessicator plate and placing a small shallow dish containing NaOH pellets. Close the dessicator and evacuate overnight with an efficient water aspirator. Remove traces of HCl by adding about 2 ml distilled water to the residue and drying it again in a flash evaporator.
8. Dissolve the residue in 0.5 ml of distilled water and bring pH of the solution to about 6.5 with 0.5 ml of 0.2 M sodium phosphate buffer (pH 6.5) as no loss of ammonia occurs at this pH. The solution is allowed to stand for 4 h to permit oxidation of any cysteine to cystine. Now readjust pH of the above preparation to 2.2 by adding 0.06 ml of freshly prepared 1 N HCl. Transfer the solution to a 5 ml volumetric flask and make up the volume with the dilution buffer.
9. Inject 0.25 ml of this hydrolysate onto the top of the column bed of the automatic amino acid analyser through the sample injector valve and, maintaining column temperature at 40°C, elute the sample with buffer at a flow rate of 70 ml/h till the sample is completely eluted. Consult the Manual supplied by the manufacturer for details of operating the instrument.
10. Calibrate the instrument by loading solution containing a mixture of known amounts of various amino acids.
11. Using the method given under *Calculations* determine content of individual amino acid per 100 g of the original grain sample.

Calculations

Most of the automatic amino acid analysers are provided with a microprocessor based integrator which records amount of amino acid in an individual peak. However, the amount of amino acid can be calculated as follows:

$$\frac{\mu g \text{ of amino acid in injected}}{\text{volume of sample hydrolysate}} = \frac{H \times W\,(\text{sample}) \times \mu g \text{ amino acid in standard}}{H \times W\,(\text{standard})}$$

$$= a$$

where H = peak height
W = width of peak at half height of the peak

$$\mu g \text{ of amino acid in 100 g of defatted sample} = a \times \frac{V_2}{V_1} \times \frac{100}{W_1}$$

where V_1 = Volume of hydrolysate injected for analysis
V_2 = Total volume of the sample hydrolysate
W_1 = g of defatted sample taken for analysis

μg of amino acid in 100 g of original non-defatted sample

$$= \frac{a \times V_2}{V_1} \times \frac{100}{W_1} \times \frac{W_2}{W_3}$$

where W_2 = sample weight after defatting
W_3 = weight of the sample prior to defatting

4.4. Quantitative Determination of Proteins

Various methods have been employed for quantitatively estimating the amount of proteins in biological samples and these are listed below. Each of these methods offers some advantages and also suffers from certain limitations.

1. Micro-Kjeldahl's method
2. Lowry's method
3. Bradford's method
4. Biuret method
5. Dye binding method
6. Spectrophotometric method
7. Turbiditimetric method

4.4.1 EXPERIMENT: Determination of crude protein by Micro-Kjeldahl's method

Principle

The amino nitrogen in various nitrogenous compounds in biological samples such as plant material (leaves, seeds etc.) is converted to ammonium sulphate on digestion with concentrated sulphuric acid in presence of K_2SO_4 and $CuSO_4$. K_2SO_4 is included in digestion mixture to raise the boiling temperature whereas $CuSO_4$ acts as a catalyst. Alternatively, selenium dioxide can be used as a catalyst. On distillation of the digested sample with NaOH, NH_3 is liberated and trapped in boric acid containing a mixture of bromocresol green and methyl red as an indicator. Ammonia reacts with boric acid to form ammonium borate which is then determined volumetrically by titrating against standardized HCl and the amount of nitrogen determined. Since proteins contain about 16% nitrogen, the protein content of the sample is calculated by multiplying its nitrogen content by 6.25. The value obtained is referred to as crude protein content since this procedure estimates total organic nitrogen rather than true proteins. This method is commonly used for determination of crude protein content of grains, forage and animal feeds.

Materials and Reagents

1. Long necked digestion flasks (30–50 ml capacity)
2. Digestion rack
3. Micro Kjeldahl apparatus (also known as Markham distillation apparatus)
4. Microburette (5 or 10 ml capacity)
5. Conc H_2SO_4
6. Catalyst: Either selenium dioxide powder or a mixture of $CuSO_4 \cdot 5H_2O$ and K_2SO_4 in the ratio of 2:1
7. 40% NaOH: Take 400 g of NaOH, dissolve it in water and make the volume to 1 L
8. 2% boric acid: Dissolve 10 g of reagent grade boric acid in about 470 ml of hot distilled water, cool and add 2 ml of 0.1% alcoholic solution of bromocresol green and 4 ml of 0.1% methyl red solution. Make the final volume to 500 ml
9. 0.01 N HCl

Procedure

1. Take 1 g of finely powdered grain sample in a long necked digestion flask. Add 10 ml of conc H_2SO_4 and 200 mg of catalyst mixture (Reagent No. 6) and digest the sample on heater or Kjeldahl's heating unit till the solution becomes clear. Cool it and transfer to 50 ml volumetric flask. Rinse the digestion flask several times with small amount of water and pour the washings in the volumetric flask. Finally make up the volume to 50 ml with distilled water.
2. Take 10 ml of boric acid solution in 100 ml conical flask. Place this receiving flask in such a way that outlet of the condenser of micro-Kjeldahl's distillation apparatus dips into the boric acid solution.
3. Transfer 5 ml of acid digested sample to the steam chamber of micro-Kjeldahl's apparatus. Add about 5–8 ml of 40% NaOH to the aliquot of the digested sample. Immediately close the stopcock and pass the steam through the steam chamber to distill ammonia till about 30–40 ml of distillate is collected in the receiving boric acid containing flask.
4. Remove the receiving flask, rinse the condenser outlet tip into the receiving flask with water.
5. Titrate the contents of receiving flask against 0.01 N HCl till the bluish green colour changes to pink.
6. Run a blank preparation which has been identically prepared except that it does not contain the sample.

Calculations

1 ml of 0.01 N HCl = 0.00014 g nitrogen

Since average nitrogen content of most proteins is 16%

$\therefore$ 1 g of nitrogen = 100/16 g of protein.

Volume of 0.01 N HCl used for blank = v ml.

Volume of 0.01 N HCl used for sample = y ml.

Titre volume of sample = $y - v$ ml

Vol. of sample taken for distillation = 5 ml

Total volume made of the digested sample = 50 ml.

Nitrogen present in 5 ml of digested sample = $(y - v) \times 0.00014$ g

$$\text{Nitrogen present in 50 ml digested sample} = \frac{(y - v) \times 0.00014 \times 50 \text{ g}}{5}$$

$$\text{Nitrogen present in 1g sample} = \frac{(y - v) \times 0.00014 \times 50 \text{ g}}{5}$$

$$\text{Amount of nitrogen present in 100 g sample} = \frac{(y - v) \times 0.00014 \times 50 \times 100 \text{ g}}{5}$$

$$\text{\% Protein content in sample} = \frac{6.25 \times (y - v) \times 0.00014 \times 50 \times 100}{5}$$

4.4.2 EXPERIMENT: Estimation of Protein by Lowry's method

Principle

It is the most commonly used method for determination of proteins in cell free extracts because of its high sensitivity and quantities as low as 20 μg of proteins can be measured. The —CO—NH— (peptide bonds) in polypeptide chain react with copper sulphate in an alkaline medium to give a blue coloured complex. In addition, tyrosine and tryptophan residues of proteins cause reduction of the phosphomolybdate and phosphotungstate components of the Folin-Ciocalteau reagent to give bluish products which contribute towards enhancing the sensitivity of this method. It is, however, important to remember that several compounds like EDTA, Tris, carbohydrates, NH_4^+, K^+, Mg^{++} ions, thiol reagents, phenols etc. interfere with the colour development and it should be ensured that such substances are not present in sample preparations.

Materials and Reagents

1. Phosphate buffer (0.1M, pH 7.6): See Section 1.3.8 for preparation of the buffer.
2. Alkaline Na_2CO_3 reagent: Dissolve 2.0 g Na_2CO_3 in 0.1 N NaOH and make up the volume to 100 ml with 0.1 N NaOH.

3. Copper sulphate reagent: Prepare 0.5% $CuSO_4 \cdot 5H_2O$ in 1% sodium potassium tartarate solution.
4. Alkaline copper sulphate reagent: Add 1 ml of Reagent 3 to 50 ml of Reagent 2. This mixture is unstable and should be prepared fresh.
5. Folin's reagent: Dilute the reagent appropriately so that it is 1 N in respect of its acid content.
6. 20% (*w/v*) TCA: Dissolve 20 g of trichloroacetic acid in water and make the volume to 100 ml.
7. Acetone.
8. 0.1N NaOH.
9. Bovine serum albumin (BSA): 100 μg/ml solution in distilled water.

Procedure

1. Sample extract: Weigh 1 g sample (germinating seeds, leaves, bacterial cells), macerate the sample in pestle mortar in 5 ml of phosphate buffer and transfer the material to centrifuge tubes. Centrifuge the homogenate at 8000 rpm for 20 min. Collect the supernatant and repeat the extraction 4–5 times. Combine the supernatants and make the volume to 50 ml with phosphate buffer.
2. Take 1 ml of the above extract and add 1 ml of 20% TCA. Keep it for half an hour and centrifuge at 8000 rpm for 20 min. Wash the pellet with acetone twice and again centrifuge it. Discard the supernatant.
3. Dissolve the pellet in 5 ml of 0.1 N NaOH and mix well till it gets dissolved.
4. Take suitable aliquot (1 ml) of above solution and add to it 5 ml of freshly prepared alkaline copper sulphate reagent. Mix properly and after 10 min add 0.5 ml of Folin's reagent. Mix the contents instantaneously. Allow the colour to develop for 30 min.
5. Record the absorbance at 660 nm after setting the instrument with reagent blank which contains 1 ml of 0.1 N NaOH instead of the sample aliquot.
6. In another set of tubes take suitable aliquots of BSA solution (in a range of 0–100 μg) make the total volume to 1 ml with 0.1 N NaOH and develop the colour as described in Steps 4 and 5. Draw a standard curve of absorbance at 660 nm *versus* μg of BSA. From this standard curve determine the amount of protein in the sample tube.
7. Calculate the amount of protein per g of the sample.

Precautions

1. It is important that contents of the tube be mixed immediately after the addition of Folin-Ciocalteau reagent. Proper colour development will be affected if this step is unduly delayed.
2. Compounds like Tris, EDTA, thiols like 2-mercaptoethanol, phenols and inorganic ions including K^+, NH_4^+ and Mg^{++} interfere in this procedure of protein estimation.

4.4.3 EXPERIMENT: Determination of proteins by Bradford method

Principle

This is a rapid, simple and sensitive method for estimation of proteins in a sample extract. The colour development is virtually complete in 2 min and the colour is stable for about 1 h. Unlike Lowry's method, metal ions such as NH_4^+, Na^+, K^+, phenols and carbohydrates such as sucrose do not interfere in this assay. The procedure is based on interaction of a dye, Coomassie Brilliant Blue, with proteins. The unbound dye has an absorbance maxima at 465 nm. However, on interaction with proteins the dye turns blue and its absorbance maxima is displaced to 595 nm. Thus from the absorbance at 595 nm the amount of protein in a sample solution can be quantitatively estimated. However, as in Lowry's procedure, detergents such as SDS, Triton X-100 etc., interfere in estimation of proteins by this method.

Materials and Reagents

1. Colorimeter.
2. Sample extract: Prepare as given in Step 1 of EXPERIMENT 4.4.2.
3. Bradford reagent: Dissolve 100 mg of Coomassie Brilliant Blue G 250 in 50 ml of ethanol, add 100 ml of 85% of phosphoric acid and make the volume to 1 L with water.
4. 0.1 M Phosphate buffer (pH 7.5): See Section 1.3.8 for preparation of this buffer.
5. Standard protein solution: Dissolve 5 mg of bovine serum albumin in 50 ml of 0.1 M phosphate buffer. This solution contains 100 μg protein/ml.

Procedure

1. Take 0.1 ml of sample solution (see Step 1, EXPERIMENT 4.4.2) and make the volume to 1 ml with 0.1 M phosphate buffer (pH 7.5).
2. Pipette appropriate aliquots of bovine serum albumin solutions containing 0–100 μg protein. Make the volume to 1 ml with 0.1 M phosphate buffer (pH 7.5) in all the tubes.
3. Add 5 ml of Bradford reagent to all the tubes and mix thoroughly.
4. Record the absorbance at 595 nm against the reagent blank.
5. Plot a standard curve of A_{595} *versus* μg of proteins in the standards (Step 2).
6. Determine the protein content in the sample extract from the standard curve.
7. Calculate the amount of protein per ml of the sample preparation.

Suggested further reading

Blackburn S. (1968). Amino acid determination: Methods and Techniques, Marcel-Dekker Inc., New York.

Gaspar L. (1984). General Laboratory Methods. *In* Methods of Protein Analysis (Kerese I. and Chalmers R.A., eds) pp 30–86, Ellis Horwood Ltd., Chichester.

Marshak D.R. (ed), (1997). Techniques in Protein Chemistry VIII, Academic Press, New York.

Moore S. and Stein W.H. (1963). Chromatographic determination of amino acids by the use of automatic recording equipment. *In* Methods in Enzymology, Vol. VI, pp 819–831. (Colowick, S.P. and Kaplan N.O., eds), Academic Press, New York.

Lehninger A.L., Nelson D.L. and Cox M.M. (1993). Principles of Biochemistry, CBS Publishers and Distributors, New Delhi.

Spackman D.H., Stein W.H. and Moore S. (1958). Automatic recording apparatus for use in chromatography of amino acids. *Anal. Chem.* 30: 1190–1206.

5. Nucleic Acids

5.1 Introduction

Nucleic acids are nitrogen containing high molecular weight compounds that are present in all living organisms. Two types of nucleic acids are known to occur in cells, deoxyribonucleic acid (DNA) and ribonucleic acid (RNA). Hydrolysis of DNA and RNA under controlled conditions yields nucleotides, which are the basic monomeric units of nucleic acids. Each nucleotide constitutes of a heterocyclic nitrogenous base, a sugar and phosphoric acid. There are two types of nitrogenous bases in nucleic acids, namely, purines and pyrimidines. The two purine bases are adenine (A) and guanine (G) whereas pyrimidine bases are thymine (T), uracil (U) and cytosine (C). In a nucleotide the sugar is linked through its C-1 to nitrogen at position 9 of purines and at position 1 of pyrimidines via β-N-glycosidic linkage. The phosphate group is attached to C-5 of the pentose sugar by an ester linkage. RNA differs from DNA in that, firstly, it contains uracil as one of the pyrimidine bases in place of thymine and, secondly, the pentose sugar in RNA is D-ribose instead of 2-deoxyribose. Both the sugars are in their furanose forms in the nucleotides. Structures of these pentose sugars are as follows:

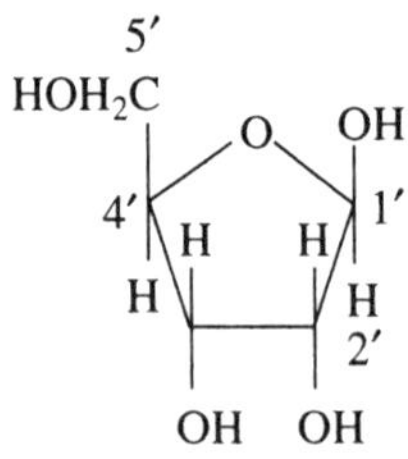

β-D-ribose

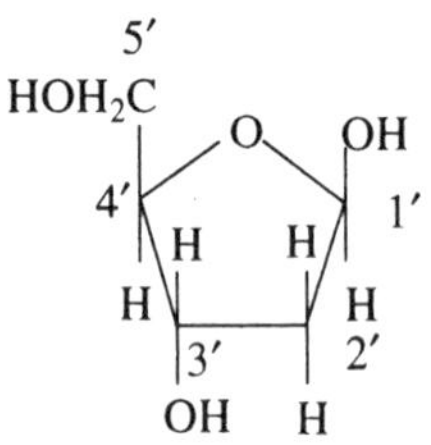

β-2-deoxy-D-ribose

DNA and RNA are polynucleotides in which the successive nucleotides are covalently joined together via 3′, 5′-phosphodiester bonds as shown in Fig. 5.1.

Most DNAs occur in a double helical form though some viral DNAs exist as single strands. Double stranded DNA consists of two polynucleotide chains wound together into a right handed helix and are held together by H-bonding between complementary bases (A pairs with T and G with C). The nitrogenous bases are stacked inwards perpendicular to the central axis of the molecule, with sugar-phosphate forming the backbone of the polymer on the outside (Fig. 5.2). The A-T base pair has two H-bonds whereas G-C base pair has three H-bonds and hence the later base combination is more stable. DNA under physiological conditions occurs mainly in the above form which is referred to as B-form of

DNA but under certain conditions DNA occurs in alternate forms viz. A, C and Z (left handed) DNA.

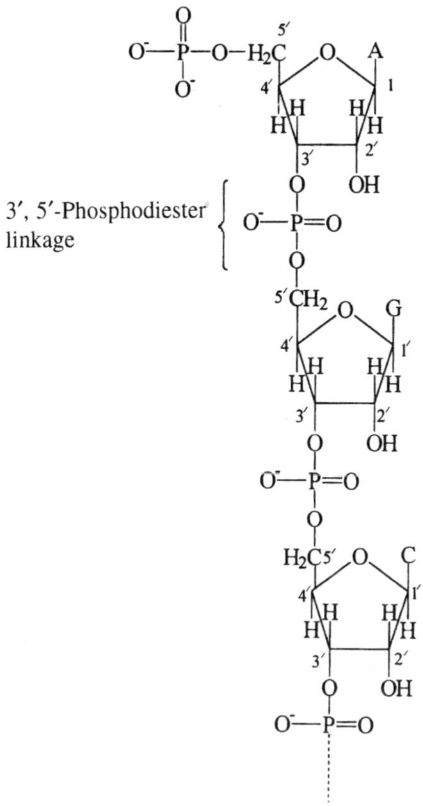

Fig. 5.1 A portion of polyribonucleotide chain showing 3′, 5′-phosphodiester internucleotide linkage between adjoining nucleotides.

DNA is the hereditary material in all living organisms except in a few viruses which contain only RNA as their genetic material. The function of DNA is to store and express the genetic information in an organism. It serves as a template for synthesis of various types of RNAs which participate in synthesis of proteins. The primary structure of protein, which ultimately determines its biological activity, is dictated by mRNA. Prokaryotic chromosome is a single large circular DNA. In addition to this, many bacteria also contain small, circular extrachromosomal elements which are capable of autonomous replication and these are called plasmids. Eukaryotic cells, however, contain several chromosomes and hence several DNA molecules and these exist as nucleoprotein complexes due to interaction of basic proteins, histones, with DNA. Bulk of DNA in eukaryotes is located in the nucleus but small amounts are also present in other organelles such as mitochondria and chloroplasts. RNA molecules present in the cells are of three different types i.e. messenger RNA (mRNA), ribosomal RNA (rRNA) and transfer RNA (tRNA). All the three types of RNAs exist as single polyribonucleotide strands and play an important role in protein synthesis. Besides these three forms, certain RNAs are associated with specific proteins to form ribonucleoproteins that participate in post-transcriptional processing of other RNAs.

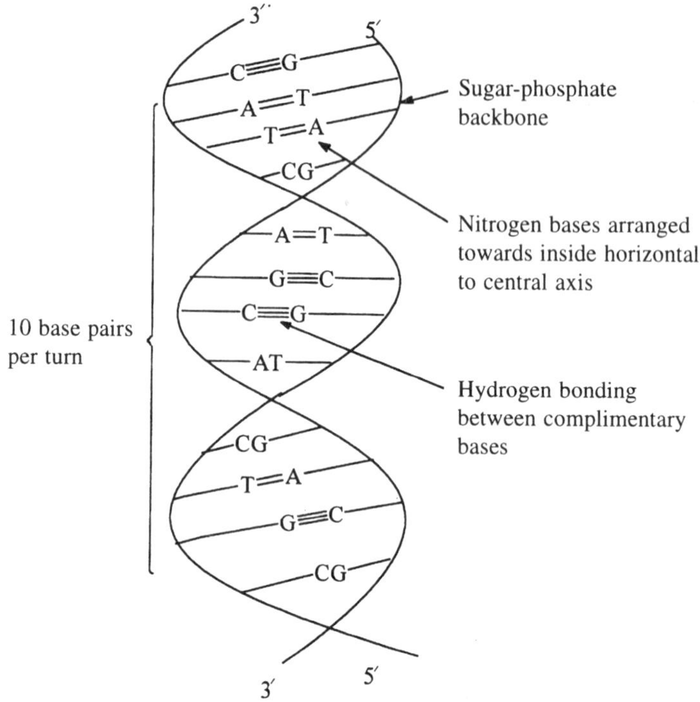

Fig. 5.2 Double helical structure of DNA molecule showing hydrogen bonding between complementary base pairs of the two polydeoxyribonucleotide chains.

In this chapter laboratory exercises on isolation of nucleic acids from tissues, their quantitative estimation and also their preliminary characterization are described. More advanced practicals on nucleic acids such as improved techniques for their isolation, their fractionation, restriction enzyme digestion, blotting and hybridization and polymerase chain reaction (PCR), cloning and sequence determination have been presented in Chapter 14.

5.2 Quantitative Estimation of Nucleic Acids and their Preliminary Characterization

5.2.1 EXPERIMENT: Extraction of total nucleic acids from plant tissues

Principle

Nucleic acids are extracted by grinding the tissues with ice cold trichloroacetic acid followed by centrifugation to separate them from the acid soluble compounds (nucleosides and nucleotides etc.). The precipitated nucleic acids are then extracted with ethanol-ether to remove the interfering lipid material. RNA is selectively solubilized from the residue with alkali and DNA subsequently recovered. Extracted DNA and RNA are then quantitatively estimated by their colour reactions with diphenylamine and orcinol, respectively, as outlined in EXPERIMENTS 5.2.2 and 5.2.3.

Materials and Reagents

1. Table centrifuge
2. Boiling water bath
3. Water bath at 37°C
4. Plant tissue: any tissue such as leaves, roots or embryos from germinating seeds can be used.
5. Trichloroacetic acid (10%, *w/v*)
6. Ethanol : ether (2 : 1)
7. 0.5N NaOH
8. 5% $HClO_4$

Procedure

1. Macerate 2.5 g plant tissue in 10 ml of ice cold 10% trichloroacetic acid (TCA).
2. Centrifuge the homogenate at 3000 rpm for 10 min.
3. Discard the supernatant and suspend the pellet in 5 ml of cold 10% TCA.
4. Centrifuge the suspension at 3000 rpm for 10 min, discard the supernatant and suspend the precipitate in 5 ml of ethanol : ether mixture.
5. Again centrifuge as in Step 4 and suspend the precipitate in 5 ml of 0.5 N NaOH. Mix well and keep it for 18 h at 37°C.
6. Centrifuge this alkali treated preparation at 3000 rpm for 10 min. Decant the supernatant, which contains RNA in the hydrolyzed form as well as proteins, in a test tube.
7. To the precipitate, which contains DNA and some proteins, add 1 ml of 5% $HClO_4$ and after mixing the contents well, heat it in a boiling water bath for 1 h. Cool the preparation and centrifuge it at 3000 rpm for 10 min. Collect the supernatant, which contains deoxyribonucleotides released from hydrolysis of DNA, make up to a known volume with 5% $HClO_4$ and using diphenylamine reaction estimate the amount of DNA in a suitable aliquot of this preparation as described in EXPERIMENT 5.2.2.
8. To the supernatant obtained in step 6 (which contains hydrolyzed RNA), add an equal volume of 10% TCA. Centrifuge at 3000 rpm for 10 min. Collect the supernatant. Make upto known volume with 5% $HClO_4$ and estimate RNA using orcinol method as described in EXPERIMENT 5.2.3.

5.2.2 EXPERIMENT: Estimation of DNA by diphenylamine reaction

Principle

This is a general reaction given by deoxypentoses. The 2-deoxyribose of DNA, in the presence of acid, is converted to ω-hydroxylevulinic aldehyde which reacts with diphenylamine to form a blue colored complex with absorbance maxima at 600 nm. Compounds such as furfuryl alcohol and arabinal, which can be converted into ω-hydroxylevulinic aldehyde will also give this reaction. In DNA, since only deoxyribose of purine nucleotides is released, the value obtained represents half of the total deoxyribose in the sample. The reactions leading to the formation of the coloured complex are as follows:

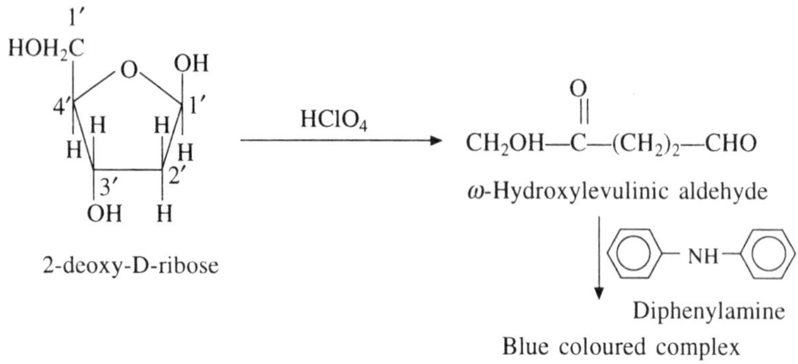

Materials and Reagents

1. Colorimeter or spectrophotometer.
2. Standard DNA solution: Dissolve calf thymus DNA (100 μg/ml) in 1 N $HClO_4$ by heating at 70°C for 15 min. Make different dilutions of this stock solution ranging from 20–100 μg DNA/ml using 0.5 N $HClO_4$.
3. 1.6% (*w/v*) acetaldehyde: Prepare by dissolving 1 ml of ice cold acetaldehyde in 50 ml of distilled water.
4. Diphenylamine solution: Dissolve 1.5 g of diphenylamine in 100 ml of glacial acetic acid and 1.5 ml of conc H_2SO_4.
5. Diphenylamine reagent: Prepare by mixing 0.5 ml of 1.6% of acetaldehyde and 100 ml of diphenylamine solutions. This solution must be prepared fresh.

Procedure

1. Take 2.0 ml aliquot of the sample in which DNA has to be estimated in a test tube.
2. In another set of test tubes, pipette 2.0 ml of standard DNA solution of different dilutions. In one of the test tubes take 2.0 ml of 0.5 N $HClO_4$ as a reagent blank.
3. Add 4.0 ml diphenylamine reagent (Reagent No. 5) to all the tubes, mix the contents properly and keep at room temperature in dark for 16–18 h or overnight. Alternatively, keep the tubes in boiling water for 10 min and cool them under running tuning tap water.
4. Record the absorbance at 600 nm.
5. Draw a standard curve of A_{600} *vs* DNA concentration. From absorbance of the sample, determine the amount of DNA in it. Express the results as mg of DNA/g fresh weight of tissue.

5.2.3 EXPERIMENT: Determination of RNA by orcinol method

Principle

This is a general method for estimation of pentoses. Acid hydrolysis of RNA releases ribose which in the presence of strong acid undergoes dehydration to yield furfural. Orcinol, in the presence of ferric chloride as a catalyst, reacts with furfural producing a green coloured compound with absorbance maxima at 665 nm. DNA gives a limited positive reaction with orcinol test. The reactions leading to the formation of a green coloured complex are as follows:

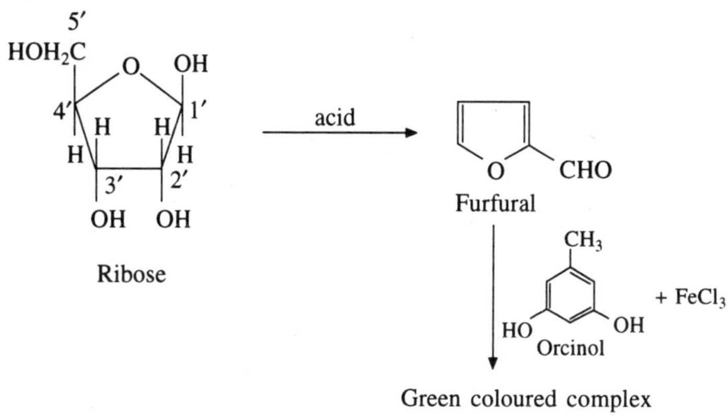

Materials and Reagents

1. Colorimeter or spectrophotometer.
2. Boiling water bath.
3. 5% $HClO_4$.
4. Standard RNA solution: Dissolve yeast RNA (500 μg/ml) in 5% $HClO_4$. Make different dilutions to obtain solutions containing 100–500 μg RNA/ml with 5% $HClO_4$.
5. Orcinol reagent: Dissolve 100 mg of ferric chloride ($FeCl_3 \cdot 6H_2O$) in 100 ml of conc HCl and then add 3.5 ml of 6% solution of orcinol prepared in alcohol.

Procedure

1. Take 2.0 ml solution of each of the dilutions of RNA standard solutions, test sample and 2.0 ml of 5% $HClO_4$, as a blank, in different test tubes.
2. Add 3.0 ml orcinol reagent to all the tubes and mix properly.
3. Keep the test tubes in a boiling water bath for 20 min.
4. After cooling them, add 7.0 ml of *n*-butanol to each tube and measure the A_{665} against blank.
5. Plot a graph between A_{665} vs amount of RNA and from this standard curve determine the amount of RNA in the provided sample.

5.2.4 EXPERIMENT: Quantitative determination of DNA and RNA by spectrophotometric method

Principle

Both DNA and RNA exhibit strong absorption of ultra-violet light due to presence of conjugated double bonds of the constituent purine and pyrimidine bases. These have a characteristic absorbance maxima at 260 nm which is linearly related with concentration of DNA or RNA in the solution upto an O.D. value of 2.0. This spectrophotometric method can also be employed for judging purity of DNA and RNA preparations. Proteins are usually the major contaminants in nucleic acid extracts and these have absorption maxima at 280 nm. The ratio of absorbance at 260 and 280 nm, hence, provides a rough idea about the extent of contaminants in the preparations. A ratio between 1.8–2.0 is indicative of fairly pure DNA and RNA preparations but values less than 1.8 signify presence of proteins as impurities.

Materials and Reagents

1. UV- spectrophotometer.
2. Saline sodium citrate (SSC) solution: Prepare 0.015 M solution of sodium citrate (pH 7.0) and dissolve NaCl so that its final concentration in solution is 0.15 M.

Procedure

1. Switch 'ON' the spectrophotometer and allow it to warm up for about 10 min. Adjust the wavelength at 260 nm and put 'ON' the UV-lamp.
2. Set the instrument at zero absorbance with SSC solution.
3. Read absorbance of solution of the provided sample. If O.D. is too high, appropriately dilute the sample solution with SSC and again take the reading.
4. Calculate concentration of DNA and RNA in the sample from following formulae:

   (i) For double stranded DNA:
   Concentration of DNA in sample solution (μg/ml) $= 50 \times A_{260} \times$ Dilution factor

   (ii) For RNA:
   Concentration of RNA in sample solution (μg/ml) $= 40 \times A_{260} \times$ Dilution factor

5.2.5 EXPERIMENT: Quantitative determination of individual bases in DNA

Principle

DNA is hydrolysed to constituent bases by heating with 72% perchloric acid for 1 h. The resulting bases are then separated by paper chromatography and located by viewing the chromatogram under UV light. After eluting them from the filter paper, the amount of different bases is determined from their absorbance values obtained at appropriate wavelengths (see Table 5.1).

Materials and Reagents

1. Paper chromatography chamber
2. Whatman No. 1 filter paper

3. Micropipette (20 μl)
4. Ultraviolet lamp
5. UV-spectrophotometer
6. Deoxyribonucleic acid: Commercially available DNA sample may be used
7. Perchloric acid
8. *Iso*-propanol
9. Conc HCl
10. Standard bases: Prepare separately the aqueous solutions of adenine, guanine, cytosine and thymine to contain 10 μg of the base per 50 μl of the solution
11. 0.1 N HCl
12. Solvent system: Prepare the solvent system for separation of bases by mixing *iso*-propanol, water and conc HCl in the ratio of 130 : 37 : 33

Procedure

1. Dissolve 10 mg of DNA in 1 ml of 72% perchloric acid and heat on a boiling water bath for 1 h. Cover the tube with a marble to prevent evaporation.
2. Cool the tube, add 1 ml of water and centrifuge the contents at 3000 rpm for 10 min. Use the clear supernatant for separation of bases by paper chromatography (see Section 9.3 for details of this chromatographic technique).
3. Using a micropipette, accurately apply 50 μl spot of the supernatant on the starting line of Whatman No. 1 filter paper sheet. Also spot 50 μl of each of the bases as the reference standards about 2.5 cm apart from each other on the starting line.
4. Allow the spots to dry completely and using *iso*-propanol : water : conc HCl solvent system, develop the chromatogram in the chromatographic chamber. When the solvent has reached 8–10 cm from the other end of the paper, remove it from the chromatographic chamber and dry in a current of cold air for 4 h. Detect the spots by examining the dried paper under UV light. Draw an outline of the dark black spots due to the bases with a lead pencil.
5. Identify each spot by comparing its R_f value (see Section 9.3.3) with the standard bases.
6. For quantitation, cut the area of the chromatogram containing individual spots and transfer each piece into a separate tube containing 0.1 N HCl. Allow the tubes to stand overnight for elution of the spots. Next morning, drain out the liquid from the tubes and note its absorbance on UV spectrophotometer at a wavelength corresponding to the absorbance maxima of each base (Table 5.1).
7. Further identify the bases by measuring and comparing their absorption spectra with the standard bases. Determine amount of each type of base in the DNA sample from the value of absorption coefficients given in Table 5.1.

Table 5.1 Absorption of purine and pyrimidine bases in 0.1 N HCl

Nucleotide bases	*Absorption maxima (nm)*	*Millimolar extinction coefficient**
Adenine	260	13.3
Guanine	250	11.2
Thymine	265	7.9
Cytosine	275	10.5
Uracil	260	7.9

*Absorbance value of 1 mM solution of the compound when measured in a cuvette of 1 cm length path.

Calculations

Suppose A_{260} for adenine is y O.D. units

$$\therefore \ \mu \text{ moles of adenine per mg of DNA sample} = \frac{y}{13.3} \times \frac{V_2}{V_1} \times \frac{V_3}{W}$$

where V_1 = volume in ml of hydrolysate spotted on chromatogram
V_2 = Total volume of supernatant in Step 2
V_3 = ml of 0.1 N HCl used for elution of spot from chromatogram
W = mg of DNA taken for hydrolysis

Similarly, calculate the values for other bases.

Precautions

UV radiation is extremely harmful for eyesight. Care should be taken to ensure that UV light does not fall directly on eyes. The chromatogram, which is placed at the base of a UV chamber should be viewed from the top. The upper surface of the UV light should be well covered so that the rays are directed only towards the base of the chamber. Wearing of UV protectant glasses is also advised.

5.2.6 EXPERIMENT: Determination of melting temperature (T_m) and base composition of DNA from thermal denaturation characteristics

Principle

When a dilute aqueous solution of double stranded DNA is heated, the two strands get separated due to disruption of H-bonds between the complementary bases. Such DNA is referred to as denatured DNA and the process is called denaturation.

Denaturation is a reversible process. If the above solution is allowed to cool slowly to room temperature, the complementary strands of DNA reassemble to give a duplex DNA. DNA is now said to be reannealed or renatured and process is called reannealing or renaturation of DNA.

Upon denaturation, the absorbance of DNA at 260 nm increases by 30–40% due to exposure of bases. This is called hyperchromic effect. Heat treatment at temperatures of upto 80°C causes only a very slight increase in absorbance at 260 nm and this is followed by a sharp increase within a narrow temperature range (80–90°C) and then A_{260} becomes constant when DNA is fully denatured (Fig. 5.3). A graph between absorbance at 260 nm and temperature gives a thermal denaturation curve or a melting curve of DNA. The temperature corresponding to the midpoint of the curve is called melting temperature (T_m) of DNA and it denotes the temperature at which 50% of the DNA has undergone denaturation. T_m is largely dependent on the (G+C) content of DNA, higher is the (G+C) content, higher will be the T_m. Under standard conditions of ionic strength and pH, (G+C) content is related with T_m in accordance with the following equation:

$$\% (G + C) = (T_m - 69.3) \times 2.44$$

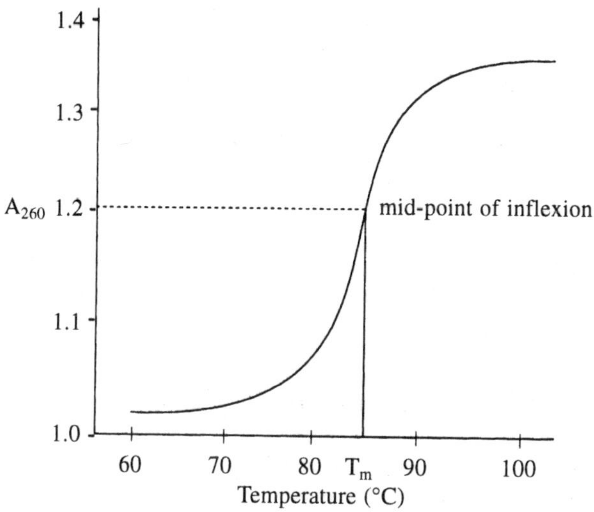

Fig. 5.3 Typical melting curve of DNA sample.

Materials and Reagents

1. UV-spectrophotometer with thermoprogrammer.
2. Saline sodium citrate (SSC): Prepare 0.015M sodium citrate solution (pH 7.0) and dissolve NaCl in it so that its final concentration in solution is 0.15M.
3. DNA: Dissolve 50 μg DNA/ml of SSC.

Procedure A

1. Switch 'ON' the spectrophotometer and after allowing sufficient period for warming up, set it to zero absorbance at 260 nm with SSC.
2. Measure A_{260} of the DNA sample at 25°C.
3. Heat the DNA solution at a rate of 1°C rise/min upto 100°C with the help of a thermoprogrammer. Record the absorbance values manually or with the printer.
4. Calculate $A_{260}(T°C)/A_{260}$ at 25°C for each of the following temperatures: 25, 35, 50, 70, 75, 80, 85, 90, 95 and 100°C and plot the absorbance ratio against the temperature.

5. Determine midpoint of increase in absorbance and by extrapolation find the corresponding temperature which represents T_m for the DNA sample.
6. Calculate % (G+C) content of the DNA using the following equation:

$$\% (G + C) = (T_m - 69.3) \times 2.44$$

Procedure B

If UV-spectrophotometer with thermoprogrammer is not available, then the following procedure can be adopted:

1. Arrange a series of constant temperature water baths maintained at 25, 50, 75, 80, 85, 90, 95 and 100°C respectively.
2. Record absorbance at 260 nm of DNA sample solution (50 μg/ml) kept at 25°C.
3. Distribute the above solution into eight test tubes and put one test tube each in the water baths maintained at different temperatures. Allow the tubes to stand for 15 min. After incubation, quickly cool all the tubes (except one at 25°C) by placing them in ice bath for 10 min. (*Note*: Reannealing occurs only if the sample is cooled gradually but reannealing is negligible if the cooling is instantaneous.)
4. Record the A_{260} of these samples.
5. Then proceed as in Steps 4 to 6 of the procedure A described earlier.

Suggested further reading

Adams R.L.P., Knowler J.T. and Leader D.P. (1992). The Biochemistry of Nucleic Acids. 11th edn., Chapman and Hall Ltd., London.

Ashwell G. (1957). Colorimetric analysis of sugars. *In* Methods in Enzymology, (Colowick S.P. and Kaplan N.O., eds), Vol. III, pp 73–104, Academic Press, New York.

Bendich A. (1957). Methods for characterization of nucleic acids by base composition. *In* Methods in Enzymology, (Colowick S.P. and Kaplan N.O., eds), Vol. III, pp 715–723, Academic Press, New York.

Burton K. (1968). Determination of DNA concentration with diphenylamine. *In* Methods in Enzymology, (Grossman L. and Moldave K., eds), Vol. 12 Part B, pp 163–167, Academic Press, New York.

Mandel M. and Marmur J. (1968). Use of ultraviolet absorbance-temperature profile for determining the guanine plus cytosine content of DNA. *In* Methods in Enzymology, (Grossman L. and Moldave K., eds), Vol 12. Part B, pp 195–206, Academic Press, New York.

Parish J.H. (1972). Principles and Practice of Experiments with Nucleic Acids, Longman Group Ltd., London.

Wilson K. and Walker J. (1996). Practical Biochemistry–Principles and Techniques, 4th edn., Cambridge University Press, Cambridge, London.

6. Minerals and Vitamins

6.1 Introduction

Minerals and vitamins play an important role in cell metabolism and are essential for normal growth and development of an organism. A large number of minerals and trace metals are present in plants, animals and microorganisms. Some of these form part of body structural components while others act as a part of catalytic agents for various biochemical reactions. Bones and skeleton are made up of mainly calcium, magnesium and phosphorus and iron is a component of blood. Minerals like zinc, molybdenum, copper, manganese and magnesium are either an integral part of the enzyme molecules or are required as activating ions for functioning of a large number of enzymes. Iodine is a constituent of hormone thyroxine. Sodium and potassium are important elements present in fluids within and outside the cells and along with ions like chloride, bicarbonate and carbonate maintain an appropriate water and acid-base balance. On an average man excretes daily 20–30 g of mineral salts consisting of sodium, potassium, magnesium, calcium, chloride and phosphates and these must be replenished by an adequate intake of these mineral salts through the diet.

A. MINERALS

6.2 Analytical Techniques

The techniques used for estimation of minerals include volumetric, gravimetric, colorimetric, flame photometric and atomic absorption spectrophotometric methods. The principles underlying the latter two techniques are briefly described as follows.

6.2.1 Flame photometry

Flame photometry is a technique by which concentration of sodium and potassium ions in the given solution can be determined by measuring the light emission on introducing processed sample into a non-luminous flame. Emission of a radiation of specific wavelength by ions of a particular element forms the basis of flame photometric estimation of sodium and potassium. The test solutions containing these elements are introduced in form of a fine mist into the flame. The various parts of a flame photometer are shown in Fig. 6.1. The instrument essentially consists of a burner, an atomizer which disperses the solution as a fine spray into the flame, specific filter system or diffraction grating prism that isolates from the spectrum that portion of emitted light which is characteristic of the substance under examination, a photo cell which converts the light energy into electric current, an amplifier to amplify the feeble current and a galvanometer to measure the electric signal.

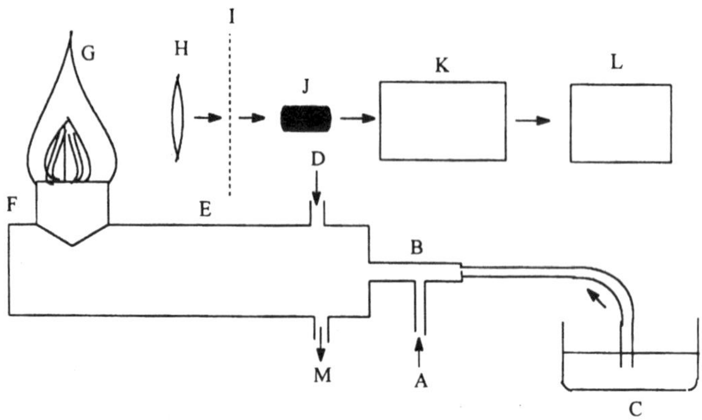

Fig. 6.1 Schematic diagram of flame photometer.
A, air inlet; B, atomizer; C, solution or sample container; D, gas inlet; E, mixing chamber; F, burner; G, flame; H, lens system; I, filters; J, detector; K, amplifier and signal processor; L, read out system; M, drainage outlet for sample waste.

6.2.2 Atomic absorption spectroscopy

All modern atomic absorption instruments are capable of measuring both atomic absorption and atomic emission. The events that occur in each of these photoprocesses are described as follows.

6.2.2a. Atomic absorption

Every element has a specific number of electrons associated with its nucleus. The normal and most stable orbital configuration of an atom is known as its ground state. When light energy of a specific wavelength is applied to an atom, it is absorbed and the outer electron is elevated to higher energy orbital leading to a short lived less stable configuration known as the excited state. Since this state is unstable, the electron tends to immediately return to its ground state releasing light energy. The events associated with atomic absorption are schematically represented by

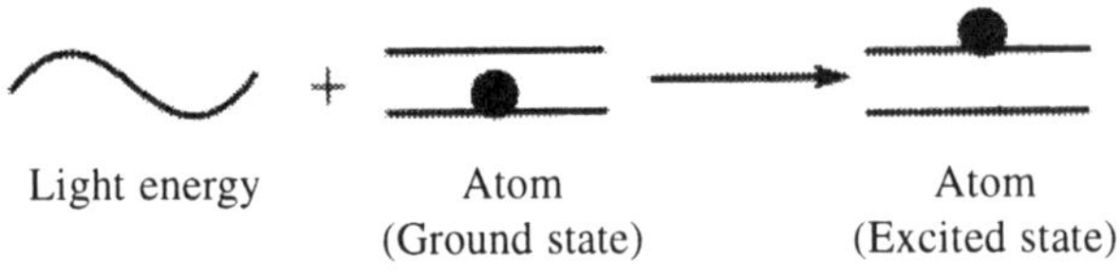

As the number of atoms in the light path increases, greater amount of light energy is absorbed. Thus by measuring the amount of light absorbed, amount of the analyte can be quantitatively determined. Use of special light sources alongwith a careful selection of wavelengths allows specific determination of individual elements in the sample.

6.2.2b. Atomic emission

The process of excitation and decay are both involved in atomic emission as shown in Fig. 6.2.

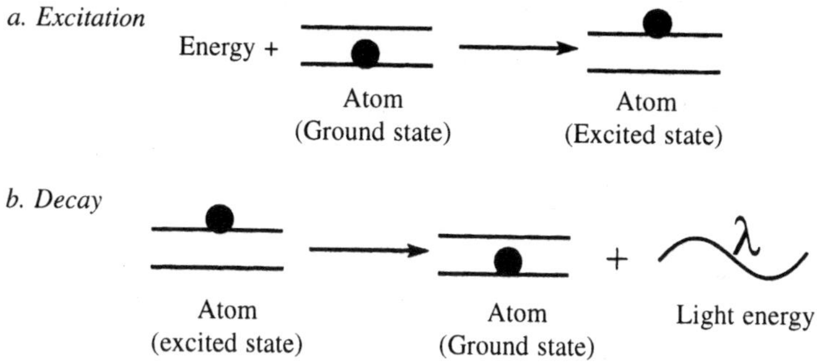

Fig 6.2 Excitation and Decay phenomenon involved during atomic emission.

The sample is subjected to a higher energy thermal environment in order to convert the atoms to their excited state. The environment can be provided by a flame or according to a relatively recent method, by a plasma. However, since the excited state is unstable, the atoms spontaneously return to their ground state and emit light. The emission spectra consists of a collection of emission wavelengths called emission lines because of the discrete nature of the emitted wavelengths. The intensity of an emission line increases as the number of excited atoms of the element returning to their ground state increases.

In both atomic absorption and emission, a flame is used. In atomic emission, the flame serves dual purpose: it converts the sample aerosol into an atomic vapour and thermally elevates the atoms to their excited states. While returning to ground state, they release excessive energy in the form of light which can be detected by appropriate instrumentation. The intensity of the light emitted is related to concentration of the element of interest in the solution. In atomic absorption, the only function of the flame is to convert the sample aerosol into atomic vapour which then absorb light from the primary light source (hollow cathode lamp or electrodeless discharge lamp).

6.2.2c. Atomic absorption spectrophotometer

The five basic components of an atomic absorption spectrophotometer are: (i) a light source that emits the radiation spectrum of the element of interest; (ii) an absorption cell in which atoms of the sample are produced (flame, graphite furnace); (iii) a monochromator for light dispersion; (iv) a detector, which measures the light intensity; and (v) a display that shows the reading after it has been processed by the instrument. The essential components of an atomic absorption spectrophotometer are shown in Figs. 6.3 and 6.4.

In a single beam atomic absorption spectrophotometer, the light source emits a spectrum of light specific for the element of which it is made and it is focussed through the sample cell into the monochromator. The light source is electronically modulated or mechanically chopped to differentiate between the light from the

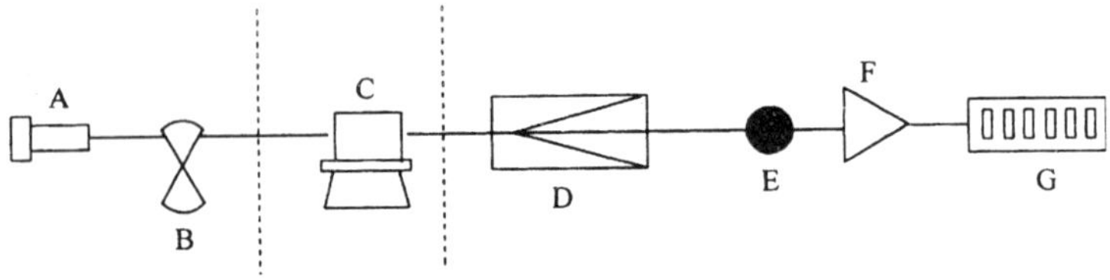

Fig 6.3 A single beam atomic absorption spectrophotometer.
A, light source; B, chopper; C, flame and sample cell; D, monochromator; E, detector; F, electronics; G, readout.

source and that emitted from the sample cell. The monochromator disperses the light and isolated light of the specific wavelength travels to the detector which is usually a photomultiplier tube. An electric current produced, which is directly related with the light intensity, is processed by the instrument electronics. The electronics measure amount of light attenuation in the sample cell and converts the readings to actual concentration of that particular element in the sample. With single beam system, a short warm up period is required for allowing the source lamp to stabilize.

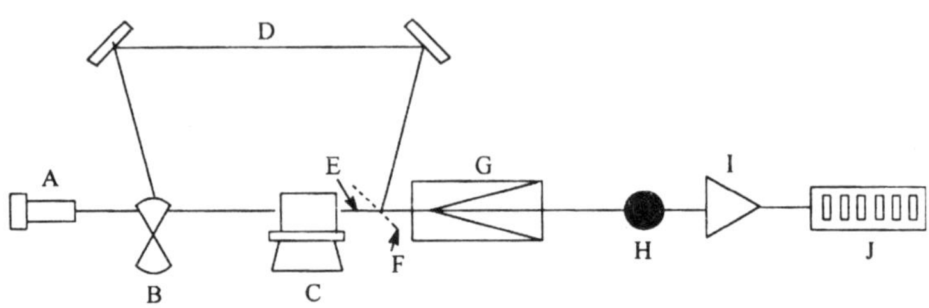

Fig 6.4 A double beam atomic absorption spectrophotometer.
A, light source; B, chopper; C, sample cell; D, reference beam; E, sample beam; F, beam recombiner; G, monochromator; H, detector; I, electronics; J, readout.

In a double beam instrument, the light from source lamp is equally split into two paths: a sample beam which is focussed onto the sample cell and a reference beam which passes through reference cell totally circumventing the sample cell. In a double beam system, the read out represents ratio of the sample and reference beams. Therefore, fluctuations in instrument readout and stability are automatically compensated. Further no warm up period is required and hence analysis can be performed almost immediately.

Light sources used for atomic absorption are either hollow cathode lamps or electrodeless discharge lamps. The hollow cathode lamp is an excellent, bright and a stable line source for most elements. However, for some volatile elements where low intensity and short lamp lifetime are a problem, electrodeless discharge lamps are available. These lamps are more intense than hollow cathode lamps and, therefore, offer better precision and lower detection limits for some elements.

A hollow cathode lamp (Fig. 6.5) is made up of a hollowed out cylinder constructed entirely or in part of the metal whose spectrum is to be produced. The anode and cathode are sealed in a glass cylinder filled with neon or argon. The glass cylinder has a quartz or UV glass window for optimum transmittance of the emitted radiation. The filling gas *i.e.* neon or argon, is chosen taking into account that it gives the best lamp intensity with minimum spectral interferences. Lamps filled with neon give a reddish glow, while argon filled lamps have a blue glow. Hollow cathode lamps are available for more than 60 different elements. When an electric potential is applied across the anode and cathode, some atoms of the filled gas get ionized. The positively charged ions collide with the negatively charged cathode and dislodge metal ions due to sputtering. Sputtered metal ions are further excited to emission through impact with the filled gas. Hollow cathode lamp has a finite life time. The sputtering process removes some of the metal atoms from the cathode, which get deposited elsewhere. Filled gas is absorbed in the sputtered metal, on the glass walls and also absorbed into the glass from bombardment. Lamps for volatile elements age faster due to more rapid sputtering of cathode.

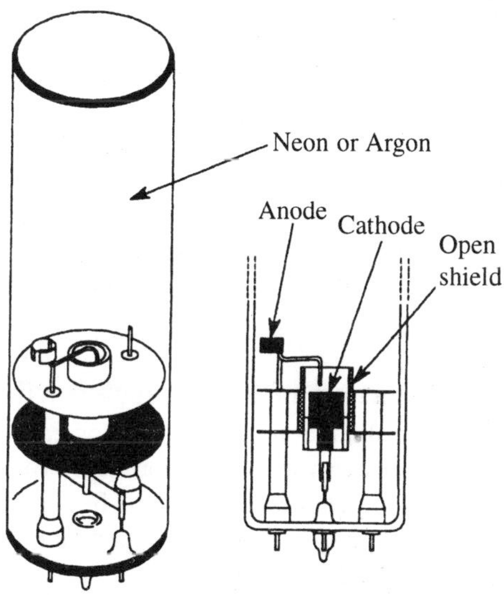

Fig. 6.5 Hollow cathode lamp

Two types of hollow cathode lamps are available. These are single element and multielement lamps. The cathode of a single element cathode lamp is generally constructed from a highly pure metal resulting in a pure emission spectrum. These lamps can obviously be used for estimation of only a particular element in the sample. It is, however, possible to construct a cathode from a mixture or alloy of several metals. The resulting multielement lamp can be used as a source for all the metals contained in the cathode. There is a wide variety of multielement lamp combinations available. Not all metals can be used in combination due to their metallurgical properties or spectral limitations. The emission intensity of a particular element in a multielement lamp is rather low. This consequently results in a poorer signal/noise ratio which may influence the precision of analysis and the detection limits. Hence, when working close to the detection limit or when high precision is required, a single element lamp is preferred. Because of cost considerations, for routine analysis one multielement lamp can serve the purpose of several single element lamps.

6.2.2d. Characteristic concentration

In atomic absorption characteristic concentration, which is quite oftenly referred to as sensitivity, is defined as the concentration of an element (expressed in mg/L) which produces a signal of 1% absorption (which is equivalent to 0.0044 absorbance units). In the linear working range characteristics concentration can experimentally be determined from the following relationship after measuring the absorbance by a known (standard) concentration of the element:

$$\text{Characteristic concentration} = \frac{\text{Conc of standard} \times 0.0044}{\text{Absorbance}}$$

Information about characteristic concentration of elements is helpful in predicting the expected absorbance range for known concentrations of various elements. Another commonly used term, check value, is the concentration of element (in mg/L) that will produce a signal of approximately 0.2 absorbance units under optimum conditions at the suitable wavelength. This value enables one to ascertain whether instrumental parameters are properly optimized and whether instrument performance is upto the desired specifications.

The characteristic concentrations and check values are determined under standard conditions. These standard conditions include atomic absorption wavelengths, optimum slit width at each wavelength, relative noise, characteristic concentration, characteristic concentration check value and approximate linear range. The standard conditions for the quantitative determination of various minerals are given in Table 6.1.

6.3 Quantitative Estimation of Few Selected Minerals

In this Section determination of mineral content in biological materials by atomic absorption spectrometer as well as a few simple colorimetric/volumetric procedures have been described. The biological sample must first be appropriately processed before it can be subjected to mineral analysis. The commonly used methods for sample preparation are described as follows.

6.3.1 Sample processing

The three methods which have generally been employed for processing the sample prior to mineral analysis are:

Direct solutions

Liquid samples can be analysed directly without any pretreatment after dilution with water or some other suitable reagent. For example, serum is diluted 5 times with 80% (*v*/*v*) butanol in deionized water for mineral determination. Butanol should be checked for its mineral impurities by running a parallel control which does not contain any sample. Fruit juices and beverages such as tea, fresh fat-free milk and wine can similarly be analysed after their proper dilution.

Ashing

According to this procedure the entire organic matter of the tissue is destroyed and the non-combustible material is recovered as ash. The minerals are then

Table 6.1 Standard conditions for estimation of some important minerals by atomic absorption spectrophotometer

Element	*Wavelength (nm)*	*Slit band width (nm)*	*Flame gas mixture check (a)**	*Characteristic concentration*	*Remarks*
Al	309.3	0.7	N—Ac	50.0	(b)**
Ca	422.7	0.7	A—Ac	4.0	—
Cd	228.9	0.7	A—Ac	1.5	(e)***
Co	240.7	0.2	A—Ac	7.0	(e)
Cr	357.9	0.7	A—Ac	4.0	(e)
Cu	324.8	0.7	A—Ac	4.0	(e)
Fe	248.3	0.2	A—Ac	5.0	(e)
K	766.5	0.7	A—Ac	2.0	(b, e)
Li	336.0	0.2	N—Ac	250.0	(b)
Mg	285.2	0.7	A—Ac	0.3	(e)
Mn	279.5	0.2	A—Ac	2.5	—
Mo	313.3	0.7	N—Ac	30.0	(c)$^+$
Na	589.0	0.2	N—Ac	700.0	(c)
Ni	232.0	0.2	A—Ac	7.0	(e)
Se	196.0	0.7	A—Ac	30.0	(e)
Zn	360.1	0.2	N—Ac	300.0	(c)

A—Ac = Air-acetylene
N—Ac = Nitrous oxide-acetylene
(a)* Metal concentration (mg/L) in aqueous solution which will give a reading of approximately 0.2 absorbance units
(b)** Addition of alkali salt (e.g. K, La or Cs as chlorides) recommended to control ionization
(c)$^+$ 0.2% Al and 1% HF added to improve linearity and sensitivity
(e)*** Use of the impact bead improves sensitivity by about 2 fold.

extracted from the ash with an acid (usually dil HCl), filtered and diluted to a known volume with deionized water and finally estimated quantitatively. In this method the organic compounds in a sample are decomposed by incineration at high temperatures (500–600°C) for 4–12 h using a muffle furnace.

Wet digestion

Organic matter in plant or animal samples can be oxidized and destroyed by prolonged boiling with mixtures of H_2SO_4, $HClO_4$ and HNO_3. A quite frequently used mixture is that containing HCl and 70% $HClO_4$. The main advantage of the wet digestion procedure is that it almost totally eliminates elemental losses by volatilization as it is carried out at relatively low temperatures. However, it requires continued attention of the analyst, is tedious and also susceptible to reagent contamination.

6.3.2 Laboratory exercises

6.3.2.1 EXPERIMENT: Preparation of sample for mineral analysis by ashing method

Principle
See Section 6.3.1.

Materials and Reagents

1. Silica crucible
2. Bunsen burner
3. Muffle furnace
4. Desiccator containing fused $CaCl_2$ at its bottom
5. Boiling water bath
6. Volumetric flasks (100 ml capacity)
7. Whatman No 40 filter paper
8. Dilute HCl: Prepare by adding 1 ml of conc HCl to 4 ml of water

Procedure

1. Take a known weight (5–10 g) of an oven dried sample in a tared silica crucible and heat it first on a Bunsen burner on a low flame till the contents get charred.
2. Now transfer the crucible to a muffle furnace and heat at dull red heat (500–550°C) till it is completely converted to white ash.
3. Remove the crucible and keep it in a desiccator till it cools down.
4. Weigh the crucible and calculate weight of the ash obtained. Express the results in terms of g of ash per 100 g of dry weight of the sample.
5. Moisten the ash with a small amount of glass distilled water (0.5–1.0 ml) and add 5 ml of dilute HCl (1:4) to it.
6. Evaporate the mixture to dryness on a boiling water bath and then add another 5 ml of HCl and evaporate the solution to dryness again.
7. Now add 4 ml of HCl and a few ml of water and warm the solution on a boiling water bath.
8. Filter the extract into a 100 ml volumetric flask using Whatman No.40 filter paper, allow it to cool and make the volume upto 100 ml. Suitable aliquots of this extract are then used for the estimation of minerals like calcium, phosphorus, sulphur, iron, copper, zinc, sodium, potassium etc.

6.3.2.2 EXPERIMENT: Wet digestion procedure of sample preparation for mineral analysis

Principle
See Section 6.3.1

Materials and Reagents

1. Digestion flasks
2. Acid-washed glass beads
3. Digestion rack or electric heater
4. Whatman No 40 filter paper disks
5. Volumetric flasks (100 ml capacity)
6. Triacid digestion mixture: Prepare by mixing conc HCl, $HClO_4$ and conc H_2SO_4 in ratio of 3:2:1
7. Dilute HCl: Add 1 ml of conc HCl to 19 ml of water

Procedure

1. Weigh about 3 g of dried and finely ground sample and transfer it to a 100 ml Kjeldahl flask. Add 25 ml of a mixture of conc HNO_3, conc $HClO_4$ and conc H_2SO_4 (3:2:1) and shake well ensuring that no dry lumps are left behind. Drop a clean acid-washed glass bead into the flask to prevent bumping during digestion.
2. Allow it to stand for 3–4 h or preferably overnight in a fume cupboard. Heat it on a digestion heater and watch out for foaming during the first hour especially in samples with high fat content. In case of excessive foaming which tends to overflow, immerse the bulb of the flask in cold water till the froth subsides. Continue heating till the digestion is complete.
3. Allow to cool, filter the digested sample through Whatman filter paper No. 40 into a 100 ml volumetric flask. Make sure that the contents of digestion flask are quantitatively transferred by rinsing the flask 3–4 times with deionised water. Wash the residue of silica on the filter paper with very dilute HCl (1:19) diluted with water in order to wash down the salts completely. Make up the volume of the collected filtrate to 100 ml.
4. If necessary this mineral solution may be transferred to pre acid-washed polythene bottles and stored in a cool place till use.

6.3.2.3 EXPERIMENT: Quantitative estimation of calcium (Volumetric procedure)

Principle

Calcium is precipitated as calcium oxalate and is titrated with standard potassium permanganate which also acts as self indicator. The reactions involved are:

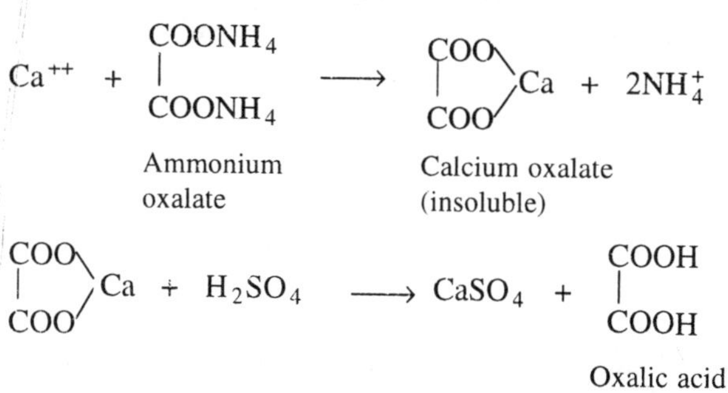

$$2KMnO_4 + 3H_2SO_4 \rightarrow K_2SO_4 + 2MnSO_4 + 3H_2O + 5O$$

$$\begin{matrix} COOH \\ | \\ COOH \end{matrix} + O \longrightarrow 2CO_2 + H_2O$$

Oxalic acid

$$NH_4OH + HCl \rightarrow NH_4Cl + H_2O$$

Materials and Reagents

1. Hot plate or bunsen burner.
2. Burette (50 ml capacity).
3. Whatman No. 40 filter paper disks.
4. 4% ammonium oxalate solution: Weigh 4 g of ammonium oxalate and dissolve in water and make the volume to 100 ml.
5. Dilute ammonia solution: Take 2 ml of liquor ammonia and add 98 ml water to it.
6. 1 N H_2SO_4: Take 1 ml of conc. H_2SO_4 and add 35 of water to it.
7. 0.01 N potassium permanganate solution weigh 0.316 g of $KMnO_4$ (A. R. grade) and dissolve it in water and nake the volume to 1 L.
8. 0.01 N oxalic acid: Dry sodium oxalate by keeping in an oven at 100–105°C for 12 h. Weigh exactly 0.67 g and dissolve it in redistilled water, add 5 ml conc. H_2SO_4 and after the solution has cooled down make its volume upto 1 L with water.
9. Methyl red indicator: Weight 0.1 g methyl red indicator and dissolve it in 95% ethanol and make volume to 100 ml with 95% ethanol.

Procedure

1. Standardization of potassium permanganate solution: Pipette 25 ml of 0.01 N oxalic acid into an Erlenmeyer flask. Add one ml of conc. H_2SO_4, warm to about 70°C and titrate against $KMnO_4$ solution till faint pink colour remains. Record the volume of $KMnO_4$ used.
2. Take 50 ml of solution of the processed sample (see EXPERIMENT 6.3.2.1) in a 250 ml beaker. Heat it to boiling and add 10 ml of 4% ammonium oxalate solution and a drop of methyl red indicator.
3. Add ammonia to the contents till the colour becomes faint pink, boil until the precipitate is granular and allow to stand for at least 4 h.
4. Filter through Whatman No. 40 filter paper and wash the residue with hot water till the precipitate is free from soluble oxalates and chlorides.
5. Pierce tip of the filter paper cone with a glass rod and wash down the precipitate into the same beaker in which calcium precipitation was carried out. Dissolve the precipitate in 1 N H_2SO_4 and add hot water and heat to about 60°C.
6. Now titrate against 0.01 N $KMnO_4$. Finally add the filter paper also and complete the titration till a definite pink colour, which persists for at least 1 min, is obtained.

7. Simultaneously run a blank containing distilled water in place of sample, as described in Steps 2 to 6.

Calculations

ml of 0.01 N $KMnO_4$ used for sample = x

ml of 0.01 N $KMnO_4$ used for blank = b

1 ml of 0.01 N $KMnO_4$ is equivalent to 0.2004 mg of Ca^{2+}

∴ mg of calcium/100 ml mineral solution of sample = $(x - b) \times 0.2004 \times 2$

Precautions

Potassium permanganate solution needs to be frequently standardised.

6.3.2.4 EXPERIMENT: Determination of phosphorus content in plant material (Colorimetric method)

Principle

For the determination of acid soluble, lipid or total phosphorus, the sample in which organic matter has been destroyed by tri-acid mixture, is used. The phosphate containing solution is treated with sulphomolybdic acid to produce phosphomolybdic acid. This is then reduced by stannous chloride giving a blue coloured complex whose colour intensity is proportional to the amount of phosphate in the preparation.

Materials and Reagents

1. Hot plate.
2. Whatman No. 1 filter paper.
3. Volumetric flasks.
4. Colorimeter or spectrophotometer.
5. Sulphomolybdic acid (2.5%): Dissolve 25 g of ammonium molybdate in 200 ml of distilled water and after warming it to 60°C, filter through Whatman No.1 filter paper. In a separate beaker take 275 ml of conc H_2SO_4 and dilute to 750 ml with distilled water. Allow both the solutions to cool down. Pour ammonium molybdate solution slowly with constant stirring into H_2SO_4 solution. After this mixture has cooled down to room temperature, dilute the resultant solution to 1 L with distilled water and store in dark coloured reagent bottle.
6. 2,4-dinitrophenol indicator: Weigh 0.1 g of 2, 4-dinitrophenol and dissolve it in 95% ethanol and make volume to 100 ml with 95% ethanol.
7. 4 N ammonia solution: Take 270 ml of liquor ammonia and add 730 ml of water to it.
8. 2 N H_2SO_4: Take 2 ml of conc H_2SO_4 and add 34 ml of water to it.
9. Stannous chloride: Dissolve 0.5 g $SnCl_2$ in 2.5 ml of conc HCl and boil till the solution is clear. Cool and make the volume to 100 ml with distilled water.
10. Standard phosphorus solution (50 ppm): Dissolve 0.2195 g of KH_2PO_4 in

water and make the volume to 1 L. Dilute this solution ten times to get working solution containing 5.0 ppm of phosphorus.

Procedure

1. Transfer 2 ml of digested sample extract into 25 ml volumetric flask. Add a few drops of 2,4-dinitrophenol indicator and neutralise the contents with 4 N ammonia. Neutralise any excess of ammonia with 2 N H_2SO_4 and make the volume to about two third of the flask with water.
2. Dispense 1 ml of sulphomolybdate solution into it, wash the neck of the volumetric flask with distilled water and add 0.5 ml of freshly prepared stannous chloride solution.
3. Thoroughly mix the contents and make the volume to 25 ml. Then within 4 to 20 min record the absorbance at 660 nm using a spectrophotometer.
4. Following the above procedure (Steps 1–3) prepare a standard curve containing 0.2 ppm to 1.0 ppm phosphorus.
5. Find out the amount of phosphorus in the sample from the standard curve and express the results as mg P/100 g dry wt. of the sample after taking into account the dilution factors.

Precautions

The blue colour is stable upto 20 min. If the number of samples for phosphorus assay is large, develope colour in manageable lots so that their absorbance can be recorded within this time limit.

6.3.2.5 EXPERIMENT: Colorimetric estimation of iron in foodstuffs by α', α'-dipyridyl method

Principle

Iron is determined in food stuffs after either dry (See EXPERIMENT 6.3.2.1) or wet ashing (see EXPERIMENT 6.3.2.2) of the food sample. A suitable aliquot of the mineral extract containing 0.1 to 0.5 μg of iron is made to react with α', α'-dipyridyl in the presence of hydroxylamine hydrochloride to give a pink colour with absorbance maxima at 520 nm.

Materials and Reagents

1. Volumetric flasks.
2. Colorimeter.
3. α',α'-dipyridyl (0.1%) in glass distilled water.
4. Hydroxylamine hydrochloride (10%): Dissolve 10 g of hydroxylamine hydrochloride in glass distilled water and make final volume to 100 ml.
5. Acetate buffer (0.2 M, pH 4.7): Weigh 8.3 g sodium acetate and dissolve it in about 50 ml of water. Add 12 ml of acetic acid to it and make the final volume to 100 ml with glass distilled water.
6. Standard iron solution: Dissolve 3.512 g of ferrous ammonium sulphate in glass distilled water, add a few drops of conc HCl and dilute to 500 ml

with glass distilled water in a volumetric flask. Prepare a working standard by taking 10 ml of this stock solution and make its volume to 100 ml with double distilled water. This solution contains 0.1 μg iron per ml.

Procedure

1. Take graded aliquots of 0–5 ml of the working standard iron solution in different test tubes and add double distilled water to make the final volume to 5 ml in each tube.
2. Add sequentially 3 ml of acetate buffer, 2 ml of hydroxylamine hydrochloride and 2 ml of α',α'-dipyridyl reagent and mix thoroughly.
3. Similarly develop the colour using a suitable volume of the mineral extract of food sample.
4. Measure the absorbance at 520 nm against a reagent blank which has been processed in the same way as the sample except that it does not contain any sample.
5. Prepare a standard curve and from it find out the amount of iron in aliquot of the food sample.

Calculations

Calculate iron content in the food sample in terms of mg iron/100 g dry wt. by taking into consideration the amount of sample taken and the extent of dilution made while preparation of sample extract. The following formula may be used:

$$\text{mg iron/100 g sample} = \frac{a \times y \times 100}{v \times w}$$

where a = amount of iron in mg in the sample aliquot
v = ml of sample aliquot taken for determination of iron
y = final volume of the mineral extract of the sample
w = g of the food sample used for preparation of mineral extract

Precautions

1. Prepare all the reagents in glass distilled water.
2. Use of iron grinders should be avoided while preparing powder of sample. Porcelain pestle and mortar should be used for this purpose.

6.3.2.6 EXPERIMENT: Determination of sodium and potassium content in blood serum samples by flame photometer

Principle

Flame photometric method is one of the most extensively used techniques employed for quantitative estimation of Na^+ and K^+ in diverse type of samples such as water, soil, plant material, body fluids etc. It has the advantage of being rapid, convenient, reliable and sensitive for determination of these two minerals. Detailed principle of flame photometry has been described in Section 6.2.1. It is essentially based on light emission of characteristic wavelengths for Na^+ and K^+ on introducing the sample as a fine mist into a non–luminous flame. Since over a certain range, the intensity of emitted light is proportional to the amount of these metal ions, their actual amount can be determined by measuring the intensity of light emitted at their characteristic wavelengths.

Materials and Reagents

1. Table centrifuge.
2. Flame photometer.
3. Cooking gas (LPG) cylinder as fuel for the flame.
4. Sodium stock solution (100 mEq/L): Dissolve 5.85 g of NaCl in 1 L of deionized double distilled water.
5. Potassium stock solution (5 mEq/L): Dissolve 3.73 g of KCl in 1 L of deionized double distilled water.
6. Combined working standard solution of Na^+ and K^+: This solution has a good similarity for Na^+ and K^+ with regards to their amounts generally found in serum and compensates the interference of one element with the other. It is prepared as given below by pipetting of stock solutions of Na^+ and K^+ into 500 ml volumetric flasks and the final volume is made with deionized double distilled water. These working standards are quite stable and may be stored in polythene bottles.

Solution No.	*Sodium stock solution (ml)*	*Potasium stock solution (ml)*	*Na^+ conc (mEq/L)*	*K^+ conc (mEq/L)*
1	1.4	2	0.28	0.02
2	1.5	3	0.30	0.03
3	1.6	4	0.32	0.04
4	1.7	5	0.34	0.05
5	1.8	6	0.36	0.06
6	1.9	7	0.38	0.07
7	2.0	8	0.40	0.08
8	2.1	9	0.42	0.09

7. Blood serum: Collect blood sample into a 15 ml test tube and allow it to clot keeping the tube in a slanting position. When a firm clot has been formed, place the tube to stand upright and then gently drain off the serum into a centrifuge tube and centrifuge at low to medium speed.

Procedure

1. Arrange two rows of 15 ml test tubes and appropriately number the tubes in the first and second row according to the sample number of serum sample to be analysed. Include one tube in each row for a reagent blank.
2. Add 9.9 ml of deionised double distilled water in the first row of the test tubes and 0.1 ml of the resultant serum sample into the tube.
3. Mix thoroughly and take 2.5 ml of the solution from the first row and add

it to the corresponding numbered tubes in the second row and make the volume to 10 ml with deionised double distilled water and mix well.

4. Take 10 ml of deionised double distilled water in the blank tube.
5. Out of series of the standard solutions, select two solutions so that one of them has a slightly higher and other slightly lower reading than that of the sample.
6. Operation of the instrument: Use the flame photometer as per the instructions given in Operator's Manual. Connect the instrument to an air compressor and a cooking gas cylinder. Switch on the air compressor and adjust air pressure passing through the instrument to 10 psi. Open outlet of the gas cylinder and ignite the burner. Adjust the flame in such a way that cones seen in the flame do not fuse with one another and the flame remains non-luminous. Dip the suction tube in the blank containing only deionised double distilled water. Ensure that water is sprayed as a mist by the atomizer into the flame. Adjust the galvanometer so that background radiation is zero when the sodium filter is at the proper place. Remove the beaker containing water and keep the container having standard solution of the highest concentration of Na^+ and K^+ and adjust the digital display to 100. Again adjust with double distilled glass water and recheck with standard solution. The instrument is now ready for the analysis of the samples.
7. Feed the diluted serum sample (Step 3) into the instrument through the suction tube and note down the reading. Then feed two standard solutions separately, one which gives a lower and other slightly higher reading than the serum sample. Repeat the same procedure for different serum samples. Proceed as above for estimation of potassium in different serum samples, but now using a potassium filter.

Calculations

The method for calculating Na^+ and K^+ content in the sample is illustrated below with the help of an example.

Reading of the first serum sample for Na^+ = 70 (RT)

Reading of the standard solution with lower value = 62.5 (RS_1)

Reading of the standard solution with higher value = 75.0 (RS_2)

Suppose, sodium equivalent in the standard solution giving reading lower than serum sample = 144 (CS_1)

Sodium equivalent in standard solution, with reading slightly higher than the serum sample = 152 (CS_2)

$$\text{Sodium conc (mEq/L) in serum sample} = \frac{CS_1 + (RT - RS_1)\,(CS_2 - CS_1)}{RS_2 - RS_1}$$

$$= 144 + \frac{7.5 \times 8}{12.5} = 148.8$$

Potassium conc in the test serum sample is also calculated in a similar manner as described for Sodium.

Precautions

1. Always use double distilled water or deionised water.
2. Mix the solutions using glass stoppers and avoid use of thumb and fingers since these minerals are also secreted through skin.
3. Serum preparation of the blood samples should be made as early as possible to minimize possible errors due to leakage of intracellular potassium. Care should also be taken to prevent haemolysis.

6.3.2.7 EXPERIMENT: Mineral analysis of plant tissue using atomic absorption spectrophotometer

Principle

Details of the atomic absorption spectrophotometry have been described in Section 6.2.2. The plant tissue, however, must first be properly processed before its introduction into the atomic absorption spectrophotometer. The commonly used methods for the preparation of sample have been dealt with in Sections 6.3.1, 6.3.2.1 and 6.3.2.2. Some additional information pertaining to sample preparation for atomic absorption spectrophotometry is given below.

Dry ashing (see Section 6.3.1 and 6.3.2.1) can effectively be used for determination of Na, K, Ca, Mg, Cu, Mn and Fe in plant tissues. For Ca and Mg, after dissolution of the ash in acid, the presence of 1% lanthanum in the finally diluted extract is helpful in overcoming potential anionic interferences. An alcohol-water mixture may be used for dilution of the dissolved ash for increasing the sensitivity for Cu, Fe and Cd. If the expected concentrations of some minerals are too low, as in case of Co, it may be complexed with appropriate reagent and then extracted in smaller volumes of organic solvents thus leading to its increased concentration in the solution. In sample processed according to the wet digestion method (see Sections 6.3.1 and 6.3.2.2) the mineral extract, diluted to a specific volume can directly be used or it may also be further extracted into an organic solvent if the minerals to be analysed are present in too low concentrations. For example, such a solvent extraction after wet digestion may be followed for estimation of elements such as Fe, Mn, Zn, Cu, Cr, Co, Mo, Cd which occur in biological materials in minute quantities.

Materials and Reagents

1. Atomic absorption spectrophotometer.
2. Gases: Cylinders of acetylene and air fitted with regulators.
3. Tissue extract: Take 0.5 to 2.0 g of the oven dried sample and prepare the extract for analysis by either dry ashing (see Sections 6.3.1 and 6.3.2.1) or wet ashing procedure (Section 6.3.2.2). Make the final volume to 50 ml with double distilled or deionised water.
4. Standard solutions: Prepare separate suitable working standard from the stock solutions (1000 mg/L) individually for each mineral. Prepare reagent blanks or deionised water containing any additives (e.g. ethanol for Cu, Fe, Cd or lanthanum etc for Ca and Mg) at the concentration equivalent

to that contained in the sample extract. Concentration of a particular mineral should be such that it falls within linear response range of the instrument.

Procedure

1. Switch 'ON' the atomic absorption spectrophotometer and allow the instrument to warm up as per the instructions in the User's Manual.
2. Use the appropriate combination of flame gases for a specific mineral to be assayed (see Table 6.1) using appropriate gas pressure so as to get an optimum height of a non-luminous flame.
3. Check that the recommended hollow cathode tube pertaining to the mineral to be analysed is fitted into the instrument.
4. Introduce one by one different volumes of the standard solutions for preparation of a reference curve using the standard conditions for the mineral to be analysed.
5. Aspirate the sample mineral extract to be analysed into the instrument and record the observations after compensating for the blanks.
6. Take readings of the standard solutions periodically in between the samples to ensure proper functioning and reproducibility of the instrument response.

Calculations

The method for calculating concentration of the mineral in the sample is illustrated as follows with an example.

Suppose,

Reading of the test sample for a mineral = s
Reading of the standard solution of the mineral = y
Suppose the mineral content in the standard solution = Z μg per ml
Dilution involved = A
Wt. of the sample taken for ashing in g = w (g)
Volume of mineral solution prepared after dissolving the ash = 50 ml

$$\text{Mineral content (ppm)} = \frac{Z \times s \times A}{y \times 50 \times w}$$

Precautions

It is essential that the impurities from the reagents be maintained at the minimal level. Deionised or distilled water, which is further subjected to deionisation, should be used. Even these do tend to contain traces of minerals such as Na, Cd, Zn etc. Ultrapure water prepared by system such as Millipore system is, in fact, recommended. The acids used for wet digestion of the sample or any other additives used should be of very high purity grade.

6.3.2.8 EXPERIMENT: To estimate calcium, magnesium, sodium and potassium in milk sample by atomic absorption spectrophotometric method

Principle

Estimation of calcium, magnesium, sodium and potassium in milk samples is carried out by precipitating the milk proteins, including casein, by trichloroacetic acid. The samples are then filtered and the filtrate is analysed by atomic absorption spectrophotometer.

Materials and Reagents

1. Atomic absorption spectrophotometer.
2. Gases: Cylinders of acetylene and air, fitted with regulators.
3. Whatman No. 1 filter paper.
4. Trichloroacetic acid (24%): Weigh 24 g of tricholroacetic acid, dissolve in water and make the final volume to 100 ml.
5. Lanthanum oxide (5%): Dissolve 5 g lanthanum oxide in deionised water and make the final volume to 100 ml (stock solution).
6. Standard mineral solution: Prepare a mixed standard solution containing 5.0 mg/L Ca, 0.6 mg/L Mg, 1.6 mg/L Na, 5.0 mg/L K, 500 mg/L La and 1.2% (*w*/*v*) trichloroacetic acid (TCA).
7. Reagent blank: Prepare a reagent blank containing 500 mg/L La and 1.2% TCA.

Procedure

1. Sample preparation: Pipette 5 ml aliquot of milk into a 100 ml volumetric flask and add 50 ml of 24% (*w*/*v*) TCA. Make the final volume with deionised water. Shake the sample at 5 min intervals for 30 min and then filter through Whatman No. 1 filter paper. Transfer 5 ml aliquot of the filtrate into a 50 ml volumetric flask, add 1 ml of 5% (*w*/*v*) lanthanum solution and make up the volume with deionised water.
2. Follow the directions given in the Operating Manual for setting the wavelength, slit width, flame gases (air-acetylene) as specified in Table 6.1.
3. Using the recommended hollow cathode tube separately each specific for Na, K, Ca and Mg, respectively, calibrate the instrument using the standard solution.
4. Record the reading of the solution of the processed milk sample separately for each of the above minerals. Also note the reading for the sample blank.

Calculations

The method for calculating amounts of sodium in milk is illustrated below. The same can be employed for calculating the contents of Ca, Mg and K.

Let,

Atomic absorption spectrophotometer reading for standard Na = s
Atomic absorption spectrophotometer reading for 1 ml milk sample = y
Concentration of Na in standard solution = 5 μg/ml

$$\text{Sodium concentration in diluted milk } (\mu\text{g/ml}) = \frac{5 \times y}{s}$$

$$\text{Sodium concentration in undiluted milk } (\mu\text{g/ml}) = \frac{5 \times y \times \text{Dilution}}{s}$$

$$\text{Sodium concentration in given milk sample } (\mu\text{g/100 ml}) = \frac{5 \times y \times 200}{s \times 1000}$$

B. VITAMINS

6.4 Introduction

Vitamins are a diverse group of organic substances occurring in small amounts in all living organisms where they perform many vital functions. Many of them are involved in utilization of the major nutrients like proteins, fats and carbohydrates. Animals and human beings are unable to synthesize vitamins and these have to be provided through diet. Although these are needed in small quantities, they are essential for proper growth and health of animals and their inadequacy results in appearance of deficiency symptoms. Vitamins can be broadly classified into two groups: water soluble and fat soluble vitamins. Members of the vitamin B complex and ascorbic acid belong to the former group while vitamins A, D, E and K are the fat soluble vitamins. Water soluble vitamins are not stored in the body and are readily excreted while fat soluble vitamins are accumulated in the body and the excessive intake of vitamin A and D can prove to be harmful. In view of their role in human and animal nutrition, information about their content in different foods and feeds is important. Vitamins in biological material are generally determined by colorimetric, spectrophotometric and spectrofluorometric procedures. In addition, microbiological techniques are also available for estimation of vitamins. More recently, methods based on GLC and HPLC have also been introduced. The salient features of the microbial techniques are described below (Section 6.4.1).

6.4.1 Microbiological assays of vitamins

Some species or strains of microorganisms are unable to synthesize a specific essential biomolecule, which may be a vitamin, nitrogenous base or an amino acid, either due to their inherent metabolic incapability or due to certain biochemical lesions. As a consequence of this metabolic deficiency, such microbial strains are almost absolutely dependent on the external supply of that specific biomolecule. This property constitutes the basis for quantitative microbiological estimation of that particular metabolite in the extracts of biological materials. Microbial assays of vitamins, particularly the members of vitamin B group, as well as some other metabolites have been used quite frequently for analysis of foods, feeds and animal tissues. This technique offers an advantage of greater specificity, sensitivity and precision over the chemical methods. The microbiological procedures essentially involve selection of a microbial strain which is absolutely dependent on an exogenous supply of vitamin to be estimated and culturing it in a basal medium supplemented with extract of the sample in which vitamin content has to be determined as well as in the media to which known amount of authentic sample of that vitamin has been added. Since the growth of the test organism would be proportional to the amount of that vitamin in the media, the vitamin content in the sample extract can be determined by comparing the growth rates with that in the media containing reference standards. Some of the microbial strains which have been used for estimation of vitamins and their coenzymic derivatives in biological materials are listed in Table 6.2.

Table 6.2 Microorganisms used for assay of vitamins and coenzyme derivatives of vitamins

Organism	*Strain*	*Vitamin*	*Assay range (ng)*
Lactobacillus plantarium	ATCC 8014	Nicotinic acid (Niacin)	0-125
Lactobacillus	ATCC 12706	Thiamine	0-25
Saccharomyes carlbergensis	ATCC 9080	B_6	0-10
Lactobacillus liechmannii	ATCC 7830	B_{12}	0-0.2
Lactobacillus plantarium	ATCC 8014	Pantothenic acid	0-100
Lactobacillus plantarium	ATCC 10100	Biotin	0-0.25
Lactobacillus casei	ATCC 7469	Riboflavin	0-0.10
Streptococcus faecalis	ATCC 8043	Riboflavin	0-200
Lactobacillus casei	ATCC 7419	Folic acid	0-1

With a few exceptions, vitamins generally exist as conjugates associated with macromolecules. This requires their prior extraction from the tissues in water soluble form so that these can be utilized by the test organism. This is normally accomplished by either acid or alkali hydrolysis of the sample. For vitamins which are labile in an acid and alkali, enzymatic procedures for the extraction are employed. Vitamins such as thiamine, riboflavin, nicotinic acid, biotin and inositol are extracted by acid hydrolysis whereas for pantothenic acid and folic acid, enzymatic procedures are adopted. A brief description of the type of procedures generally used for extraction of various vitamins from biological samples is given in Table 6.3.

While estimating vitamins or other metabolites by microbial assay, it is important not to overlook the fact that the sample extract may contain some other compound(s) which might stimulate or inhibit growth of the test organism. Some of the compounds reported to have such effects are listed in Table 6.4. Generally, care is taken to remove these stimulatory or inhibitory compounds by precipitating them by adjusting pH of the sample extract to 4.5. One of the ways of evaluating the presence or absence of such interfering compounds is the use of an internal

Table 6.3 Procedures for extraction of different vitamins from samples

Vitamin	*Procedure*
Thiamine	The sample is steamed for 30 min in 0.1 N H_2SO_4, followed by enzymatic hydrolysis with papain and takadiastase at pH 4.5 for 18–24 h at 37°C.
Riboflavin	Hydrolyse the sample by autoclaving at 15 psi for 15 min in presence of 0.1 N HCl and treat the autoclaved sample with mixture of enzyme (papain and takadiastase) at pH 4.5 to 4.7 at 37–45°C for 24 h. This procedure can be employed for extraction of almost all acid stable vitamins.
Nicotinic acid	Suspend the powdered sample in 0.1 N H_2SO_4 and autoclave at 15 psi for 30 min.
Pantothenic acid	Hydrolyse the finely ground experimental material with calf intestinal phosphatase and pigeon/chick liver enzyme preparation in 0.02 M $NaHCO_3$ buffer (pH 7.2) and incubate at 37°C for 4 h.
Vitamin B_6	Autoclave the sample in presence of 0.5 M HCl at 15 psi for 4 h.
Vitamin B_{12}	Suspend the sample in extraction solution (pH 4.5) consisting of 13 g anhydrous $NaHPO_4$, 1.2 g citric acid and 1.0 g anhydrous Na metabisulphite per 100 ml and autoclave at 15 psi for 10 min.
Biotin	Hydrolyse by autoclaving with 6 N H_2SO_4 at 15 psi for 2 h.
Folic acid	Liberation of folic acid from conjugated form is achieved enzymatically using one of the following conjugases: Conjugases from hog kidneys (acetone dried powder) at pH 4.5 for 48 h at 37°C or conjugases from rat liver/chicken pancreas at pH 7.0 for 48 h at 37°C.
Inositol	Hydrolyse with 18% HCl by refluxing for 6 h.
Choline	Autoclave in 3% H_2SO_4 at 15 psi for 2 h.

Table 6.4 Substances in sample extracts shown to exert stimulatory or inhibitory effect on growth of test organism

Vitamin assayed	*Test organism*	*Stimulators*	*Inhibitors*
Riboflavin	*Lactobacillus casei*	Free starch, certain free fatty acid e.g. stearic and oleic acid	Linoleic acid
Niacin	*Lactobacillus arabinosus*	Fatty acids such as oleic acid and linoleic acid	—
Pantothenic acid	-do-	-do-	
Folic acid	*Streptococcus faecalis*	Thiamine	—
Vitamin B_6	*Neurospora sitophila*	Aneurin	—
Thiamine	*Lactobacillus fermentii*	TPP*	—
Vitamin B_{12}	*Lactobacillus leichmannii*	Deoxyribosides, Factors A, B, C and pseudo vitamin B_{12}	—

*TPP—Thiamine pyrophosphate.

standard. In this approach, a known amount of the vitamin is added to the media alongwith the limiting amounts of the extract and then ascertaining whether the differences in the growth between the sample extract and that containing internal standard is fully accounted by the internal standard.

6.5 Laboratory Exercises

6.5.1 EXPERIMENT: Estimation of β-carotene in carrots

Principle

β-Carotenes are precursors of vitamin A and hence their estimation in foodstuffs is important from nutritional point of view. Carotenoids are separated on a column of calcium hydroxide after their extraction from the experimental material by organic solvents. The individual carotenoids are then determined in the column eluent spectrophotometrically. The values for their respective vitamin A potency are used to derive the total vitamin A content of the foodstuffs.

Materials and Reagents

1. Glass column (1 × 30 cm)
2. Colorimeter
3. Petroleum ether-acetone (1:1) mixture
4. Calcium hydroxide
5. Sodium sulphate (anhydrous)
6. Petroleum ether
7. β-Carotene standard: Prepare a solution containing 20 mg β-carotene in 100 ml of petroleum ether

Procedure

1. Preparation of the sample: Take 25 g of freshly ground carrot sample and allow it to stand overnight in 100 ml mixture of petroleum ether-acetone (1:1). Filter the extract through sintered crucible and wash the residue on the sintered crucible twice with successive 500 ml of the extraction mixture till all the yellow colour is extracted. Pool the filtrate, shake it with 50 ml of water and discard the water washings. Repeat this twice. Take the organic solvent layer and dehydrate it by adding a small amount of anhydrous sodium sulphate. Concentrate this sample under vacuum to a final volume of 4 ml.
2. Take a glass column (1 × 30 cm) and pack it with 6 g of calcium hydroxide under gentle suction. Place about 1–2 g of anhydrous sodium sulphate on top of the packed column.
3. Pass sufficient volume of petroleum ether to wet the column bed. Then load 2 ml of the concentrated extract containing about 40–100 μg of total carotenes onto the column and, if necessary, a mild suction may be employed to facilitate the flow of solvent through the column.
4. Now develop the column by passing 1% acetone in petroleum ether and collect fractions of 25 ml each.

5. Transfer suitable aliquot (about 10 ml) of each fraction into a 50 ml volumetric flask and dilute it with petroleum ether-acetone mixture (1:1) to the mark.
6. To facilitate quantitative estimation, take graded volumes 0.2 to 2.0 ml of standard β-carotene solution in 50 ml volumetric flask, dilute them to 50 ml with petroleum ether-acetone (1:1) mixture.
7. Record the absorbance of standards and sample fractions at 450 nm and prepare the standard curve. Determine the quantity of β-carotene in each of the collected fraction of the sample from the standard curve.

Calculations

Add values of β-carotenes of all the fractions of the sample taking into consideration the portion of the column effluent of each fraction used for monitoring β-carotenes. The true vitamin A value in sample is expressed in terms of International Units. One International Unit of vitamin A is equivalent to 0.6 μg of β-carotene.

Precautions

The preliminary treatment of the sample, such as extraction of carotenes, should be carried out in a diffused light since vitamin A is quite sensitive to light.

6.5.2 EXPERIMENT: Quantitative determination of thiamine in cereals and foods

Principle

Thiamine is vitamin B_1 and its deficiency causes a disorder known as beri-beri. Thiamine undergoes oxidation on treatment with mild oxidizing agents such as ferricyanide to form thiochrome which is a fluorescent compound. This property is used for the determination of vitamin B_1 in biological samples.

Materials and Reagents

1. Table centrifuge.
2. Spectrofluorometer.
3. Separating funnels.
4. Sulphuric acid (0.1 N): Take 0.1 ml of conc H_2SO_4 and add 35.9 ml of water to it.
5. Potassium ferricyanide (0.03%): Weigh 30 mg of potassium ferricyanide and dissolve in 100 ml of distilled water.
6. Sodium hydroxide (15%): Dissolve 15 g of NaOH in water and make the final volume to 100 ml.
7. *Iso*-butanol.
8. Sodium sulphate (anhydrous).
9. Stock thiamine solution: Dissolve 5 mg of thiamine hydrochloride in 100 ml of water.
10. Working thiamine solution: Take 2 ml of the above solution and dilute it to 100 ml with water. This solution contains 1 μg thiamine hydrochloride per ml.

Procedure

1. Grind 2 g of cereal or food material in a pestle and mortar in presence of 20 ml of 0.1 N H_2SO_4. Centrifuge the suspension at 4000 rpm for 20 min. Decant the supernatant and extract the residue again with 10 ml of the acid. Pour the collected supernatant fractions into 50 ml volumetric flask. Make the final volume to the mark with 0.1 N H_2SO_4.
2. Add 3 ml of the potassium ferricyanide reagent to a suitable portion of thiamine extract followed by 3 ml of 15% NaOH. Ensure that the solution is alkaline. Shake the the contents for 10 min.
3. Add 10 ml of *iso*-butanol and shake the mixture thoroughly. Allow it to stand till the two phases separate out completely.
4. Discard the lower aqueous layer and remove traces of water in the upper organic layer by adding a small amount of solid anhydrous sodium sulphate.
5. Pipette graded volumes of standard solution containing 0 to 5 μg thiamine hydrochloride and proceed as in Steps 2 to 4.
6. Operate the fluorometer as per manufacturer's instruction manual and measure the fluorescence intensity of the standards and samples using a primary filter of 365 nm and a secondary filter of 435 nm.
7. From the reference curve, determine the amount of thiamine in sample and express the results in terms of mg of thiamine/100 g of sample.

6.5.3 EXPERIMENT: Estimation of ascorbic acid in lemon juice

Principle

Vitamin C (L-ascorbic acid) gets oxidized to its dehydro form by air especially at alkaline pH. However, it is stable in an acidic solution. Therefore, vitamin C is extracted in metaphosphoric acid or in a mixture of metaphosphoric and dilute acetic acid. Its estimation in the extract is carried out by titrating it against 2,6-dichlorophenol indophenol solution. Oxidized form of this dye is blue in colour in an alkaline medium and red in an acidic medium. Reduced form of the dye, on the other hand, is colourless and is termed as its leuco form. The redox reaction occurring during the titration is

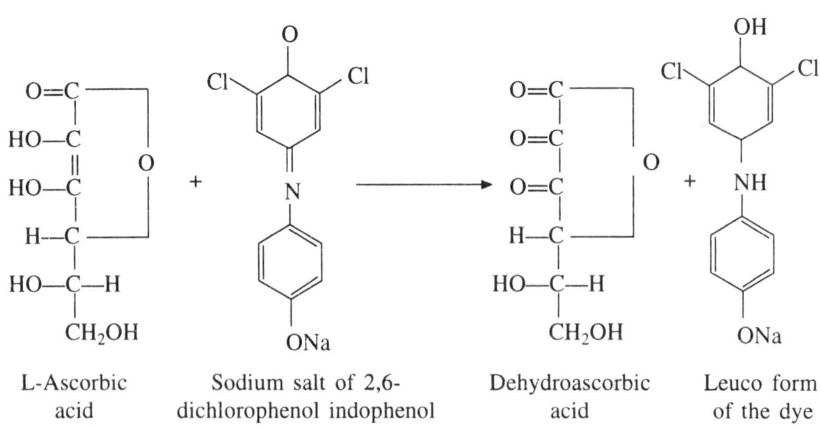

Materials and Reagents

1. Burette
2. Lemons
3. 2,6-Dichlorophenol indophenol solution: Dissolve 52 mg of sodium salt of the dye and 42 mg of sodium bicarbonate in water. Make up the final volume to 500 ml.
4. Standard vitamin C solution: Dissolve 10 mg of vitamin C in 6% (*w/v*) metaphosphoric acid and make up the final volume with metaphosphoric acid to 1 L.

Procedure

1. Squeeze fresh lemons to obtain lemon juice. Strain the juice through muslin cloth and immediately dilute (10 folds) with 6% metaphosphoric acid.
2. Transfer 20 ml standard vitamin C solution in Erlenmeyer flask. Titrate against the dye solution till the appearance of a light pink colour. Note down the volume of the dye used.
3. Similarly, titrate 20 ml of test solution (diluted lemon juice) against the dye solution and record the volume of dye used.

Calculations

Conc of Vit. C in standard solution = 10 μg/ml
Amount of Vit. C in 20 ml of standard solution = 200 μg

Suppose v ml of dye is required for the oxidation of Vit. C in 20 ml of the standard solution, then 1 ml of the dye causes oxidation of $200/v$ μg of Vit C. If 20 ml of diluted lemon juice is oxidized by y ml of dye,

Then,

$$\text{Amount of Vit. C in 20 ml of diluted juice} = \frac{200}{v} \times y \ \mu\text{g}$$

$$\text{Vit. C in 100 ml of diluted juice} = \frac{200}{v} \times y \times \frac{100}{20} \times \frac{1}{1000} = \frac{y}{v} \text{ mg}$$

$$\text{Vit. C in 100 ml of undiluted juice} = \frac{y}{v} \times 10 \text{ mg}$$

6.5.4 EXPERIMENT: Measurement of riboflavin in human urine

Principle

Riboflavin is used in veterinary and medical practices for supplementation of animal feeds and as natural colouring agent in food products. It is eliminated in urine and with an average diet, the daily losses amount to 12 per cent of the intake. A fall in the level of riboflavin (Vit. B_2) excretion occurs before deficiency symptoms of Vit. B_2 are noticed. Under protein deficiency the urinary output of Vit. B_2 increases. Riboflavin deficiency can be assessed by measurement of urinary excretion of riboflavin. Urine is collected in a dark container using glacial acetic acid as preservative. The interfering impurities are removed by oxidation with potassium permanganate. This is followed by removing excess of potassium permanganate by reacting it with hydrogen peroxide. Riboflavin is extracted in to a butanol-pyridine mixture and fluorescence intensity is measured in a fluorometer using excitation light of 450 nm and recording emission at 545 nm.

Materials and Reagents

1. Fluorometer
2. Acetic acid (glacial)
3. Potassium permanganate (5% *w/v*)
4. Hydrogen peroxide (3%)
5. Sodium sulphate (anhydrous)
6. Butanol-pyridine mixture (92:8)
7. Riboflavin standard: Dissolve 25 mg of riboflavin in 500 ml distilled water, add 1.2 ml of glacial acetic acid and make volume to 1 L. Dilute 5 ml of this solution to 50 ml with distilled water to get working solution containing 2.5 μg riboflavin/ml

Procedure

1. Transfer 2 ml urine sample in a test tube and add 0.5 ml glacial acetic acid. Then pipette 0.5 ml of 5% potassium permanganate solution, shake well and leave for one min.
2. Add 0.5 ml of 3% H_2O_2 solution until the purple colour of excess potassium permanganate disappears completely.
3. Now add 10 ml of butanol-pyridine mixture and shake for 30 sec. Keep the tubes undisturbed for 10 min and then discard the aqueous layer.
4. Add 1.5 g sodium sulphate anhydrous with small dropping funnel to remove the traces of moisture. Shake and keep in dark for 10–15 min.
5. Read fluorescence intensity at 450 nm excitation light and 545 nm emission light.
6. Prepare a blank and standards as described above and record their fluorescence.

Calculations

Let the fluorescence intensity of sample be f and fluorescence intensity of standard containing z μg riboflavin be y and volume of urine taken for estimation be v ml

$$\text{Riboflavin content (mg/ml)} = \frac{z}{y} \times \frac{f}{v}$$

Precautions

1. As far as possible, carry out all the steps in a diffused light to prevent destruction of riboflavin.
2. Sometimes a fine precipitate of $KMnO_4$ may be produced. This should be removed by centrifugation.

6.5.5 EXPERIMENT: Microbial assay for estimation of niacin in food

Principle

Niacin can be determined microbiologically using *Lactobacillus plantarium* (ATCC 8014). The organism provides a highly specific analysis and responds equally to nicotinic acid and nicotinamide but not to intermediate precursors of nicotinic acid such as picolinic and quinolinic acids. Even though *Lactobacillus plantarium* also requires an exogenous supply of pantothenic acid (see Table 6.2), this complication is overcome by using a culture media which contains sufficiently high amounts of pantothenic acid so that growth of the organism is dependent exclusively on niacin content of the test material. Niacin from the food material is extracted by suspending the sample in H_2SO_4 and then autoclaving. After filtration and precipitation of protein at pH 6.8, the filtrate is diluted to get niacin concentration of about 0.1 μg niacin/ml. The assay organism is then grown in basal medium under specified conditions in presence of standard concentrations of niacin and sample extract. Growth of the microorganism is measured by monitoring the turbidity at 660 nm. By comparing growth rates in media containing the sample extract with those in the presence of graded amounts of this vitamin the niacin content in the sample can be determined.

Materials and Reagents

1. Flash evaporator.
2. Autoclave.
3. Colorimeter.
4. Non-adsorbent cotton.
5. Niacin stock solution: 50 mg of niacin is dissolved in 500 ml of 50% alcohol to give a solution containing 100 μg niacin/ml.
6. Working standard: Dilute the stock solution to obtain a solution containing 0.1 μg niacin/ml.
7. Acid hydrolysed casein: To 100 mg of casein, add enough 95% ethanol and reflux the mixture for 2–3 h and filter it. Collect the residue and repeat this step twice. Remove the alcohol from the refluxed residue at room temperature under vacuum, add 250 ml conc. HCl and 250 ml distilled water and reflux for 8–10 h. Distill off HCl under vacuum below 80°C. Remove HCl completely by repeating the process 2–3 times. Dissolve the hydrolysed casein in 600 ml distilled water and adjust the pH to 3.5 with 40% NaOH and filter the preparation.
8. L-cystine: Add 4.0 g L-cystine in a small volume of water and dissolve it using minimum volume of HCl and make the volume to 500 ml with distilled water.
9. DL-Tryptophan: Dissolve 2.0 g DL-tryptophan in water using a few drops of liquor ammonia. Adjust the pH to 3.5 and make the volume to 500 ml with distilled water.
10. Biotin: Dissolve 25 mg biotin in 500 ml of 50% ethanol.
11. Calcium pantothenate: Dissolve 54.4 mg calcium pantothenate in distilled water.

12. Vitamin solution: Dissolve 20 mg riboflavin, 10 mg thiamine HCl, 10 mg *p*-aminobenzoic acid, 40 mg pyridoxine HCl in 1 L of 0.02 N HCl.
13. Adenine, Guanine, Uracil (AGU) solution: Prepare a solution containing 0.2 g each of adenine sulphate, guanine hydrochloride and uracil in distilled water using a few drops of conc HCl to aid dissolution of these components and make the volume to 100 ml.
14. Salt solution A: Make a salt solution containing 25 g each of K_2HPO_4 and KH_2PO_4 in 250 ml water.
15. Salt solution B: Dissolve the following salts in distilled water and make the volume to 250 ml. A few drops of conc HCl may have to be added to obtain a clear solution.

$MgSO_4 \cdot 7H_2O$	10.0 g
NaCl	0.5 g
$FeSO_4 \cdot 7H_2O$	0.5 g
$MnSO_4 \cdot 4H_2O$	0.5 g

16. Basal medium: Prepare the basal medium by mixing the specified volume of the various solutions mentioned below.

Casein hydrolysate	10 ml
DL-tryptophan	5 ml
L-cystine	5 ml
AGU solution	1 ml
Vitamin solution	2 ml
Salt solution A	1 ml
Salt solution B	1 ml
Calcium pantothenate	0.2 ml
Biotin	0.8 ml

Then add 2 g of glucose (anhydrous) and 3.32 g of $CH_3COONa \cdot 3H_2O$ to the above mixture and adjust the pH to 6.8 and make the final volume to 100 ml. Filter the mixture through Whatman No. 1.

Procedure

1. Preparation of sample: Weigh sufficient amount of the powdered food material so as to contain approximately 0.1 mg niacin and transfer it to a 100 ml conical flask. Add 50 ml 1 N H_2SO_4. Autoclave the mixture for 30 min at 15 psi and adjust its pH to 4.5. After cooling, filter it through Whatman No 1 and make the volume to 100 ml. Take suitable volume of the aliquot and dilute it so that it contains about 0.1 μg/ml niacin.
2. To 15 ml test tubes, take volumes varying from 0.5 to 3.0 ml of the sample extract. For each volume of the aliquot, take atleast two test tubes.
3. Make the final volume of the contents in the test tubes to 5 ml with water. Then add 5 ml of the basal media to each tube. Plug the tubes with non-absorbent cotton.
4. Similarly, prepare in duplicate, test tubes containing 0, 0.5, 1.0, 1.5, 2.0, 2.5, 3.0 and 4.0 ml of standard solution which contains 0.1 μg of niacin/

ml. Add appropriate volume of water to each of the tubes to make the final volume to 5.0 ml followed by addition of 5 ml of the basal media and close the tubes with a cotton plug.

5. Autoclave the tubes containing the sample aliquot as well as the standards at 15 psi for 15 min.
6. Allow the tubes to cool and pipette aseptically 0.1 ml of the inoculum of the broth.
7. Incubate the tubes at 37°C for 18 h and record absorbance of each tube at 660 nm.

Calculations

1. Prepare a standard curve by plotting a graph of A_{660} *vs* ng niacin for tubes containing the reference standards.
2. From the standard curve, determine the niacin content in each of the sample tubes.
3. Calculate the amount of niacin per 100 g of food sample by taking into consideration the dilution factors involved during preparation of the sample extract and the amount of aliquot taken for estimation of the vitamin.

Suggested further reading

Augustin J., Klein B.P., Becker D. and Venugopal P.B. (eds), (1985). Methods of Vitamin Assay, 4th edn, John Wiley & Sons, New York.

Brubacher G. and Muller-Mulot W. (eds), (1985). Methods for the Determination of Vitamins in Foods Recommended by Costal, Elsevier Applied Science Publishers, London.

Jeffery G.H., Basset J., Mendham J., Denney R.C. and Vogel S. (1989). Text Book of Quantitative Chemical Analysis, 5th edn, Longman Group UK Ltd., Harlow.

Kirk R.S., Swager R. and Pearson S. (1991). Composition and Analysis of Foods, 9th edn, Longman Group UK Ltd., Harlow.

Linder M.C. (1985). Nutritional Biochemistry and Metabolism with Clinical Applications, Elsevier Science Publishing Company Inc., New York.

Lindon G. (1996). Analytical Techniques for Foods and Agricultural Products, 2nd edn, VCH Publishers (UK) Ltd., United Kingdom.

Mutton J.L. (1997). Analysis of Food Constituents, 2nd edn, Wiley-VCH, New York.

Pomeranz Y. and Meloan C.E. (1996). Food Analysis—Theory and Practice, 3rd edn, CBS Publishers and Distributors, New Delhi.

Talwar G.P., Srivastava L.M. and Moudgil K.D. (1989). Text Book of Biochemistry and Human Biology, 2nd edn, Prentice-Hall of India Pvt. Ltd., New Delhi.

7. Enzymes

7.1 Introduction

Enzymes are the catalytic proteins operative in biological systems. Their catalytic power is extremely high, often far greater than that of synthetic catalysts. They are highly specific for their substrates and function in aqueous media under very mild conditions of temperature and pH. They accelerate the rate of chemical reaction without altering the equilibrium of the chemical reaction. Therefore, the enzyme accelerates the forward and reverse reaction by precisely the same factor. In biological systems, they generally act in organized sequences catalyzing hundreds of stepwise reactions in metabolic pathways. A few of these enzymes are regulatory in nature, whose activity responds to various metabolic signals.

Study of enzymes has an immense practical importance. In some of inherited genetic disorders there may be deficiency or even a total absence of one or more enzymes in the tissues. Excessive activity of a specific enzyme can also cause abnormal conditions. Measurement of the activity of certain enzymes in the blood plasma, erythrocytes or tissue samples is a valuable aid in diagnosis of certain diseases. Besides their usefulness in medicine, the enzymes are becoming important practical tools in the chemical industry, in food processing and in agriculture. Even in every day activities the enzymes play a part in homes in cleansing agents such as laundry detergents etc.

With the exception of a small group of catalytic RNA molecules, all enzymes are proteins and their catalytic activity is intimately linked with the integrity of their native protein conformation. On denaturation or on dissociation into subunits, the enzyme loses its catalytic activity indicating that the primary, secondary, tertiary and quarternary structures of protein enzyme are essential for its catalytic activity. Their molecular weight, in general, ranges from 12,000 to over one million. Some of these enzymes require no chemical group other than the amino acid residues for their functioning. Others depend on additional chemical component, called cofactor, for their activity. A cofactor may be either one or more inorganic ions such as Fe^{2+}, Mg^{2+}, Mn^{2+} or Zn^{2+} or a complex organic non-proteinaceous molecule called a *coenzyme*. In some cases both a coenzyme and one or more metal ions are needed for activity. An organic compound or a metal ion that is covalently bound to the enzyme protein is called a *prosthetic group*. A complete catalytically active enzyme together with its prosthetic group is referred to as *holoenzyme*. The protein part of such an enzyme is known as an *apoenzyme*.

Based on the nature of the chemical reaction catalyzed, the enzymes have been grouped into six major classes. These classes are oxidoreductases, transferases, hydrolases, lyases, isomerases and ligases. Oxidoreductases transfer hydrogen atom, oxygen atom or electrons from one substrate to another. Transferases transfer chemical groups between substrates. Hydrolases catalyze hydrolytic reactions. Lyases cleave substrates by reactions other than hydrolysis. Isomerases interconvert isomers by intramolecular rearrangement and ligases catalyze covalent bond formation with concomitant breakdown of a nucleoside triphosphate.

A comprehensive study of an enzyme involves investigations on its molecular structure (primary, secondary, tertiary, quaternary, cofactors), physical properties (molecular weight, sub-unit composition, isoelectric point, electrophoretic mobility, pH and temperature stability, spectroscopic properties), enzyme properties (specificity, reversibility, kinetics), thermodynamics (activation energy, free energies and entropies), active site (number, molecular nature and mechanism), and biological properties (cellular location, isoenzymes, metabolic relevance of the reaction catalyzed). To carry out the above detailed studies, the enzyme has to be isolated in pure form and studied *in vitro*. The study of purified enzyme is fundamental to biochemistry as the data generated contribute towards understanding and then manipulating the cellular situation *in vivo*. These studies may help in designing selective inhibitors that can be used as drugs or biocides. Such information may also be employed in development of biosensors etc.

7.2 Measurement of Reaction Velocity

The progress curve of most enzyme reactions is as shown in Fig. 7.1, in which the velocity declines with time. This may be due to various reasons such as: (i) the products of the reaction may inhibit the enzyme, (ii) the degree of saturation of the enzyme with substrate or coenzyme may fall because of their conversion to product,(iii) the reverse reaction may become more important as the concentration of the product increases, (iv) the enzyme (or a coenzyme) may undergo some inactivation at the temperature or pH of the reaction owing to instability etc. In view of these, initial velocity is always measured when the various factors mentioned above have not yet had time to operate. Accordingly, in enzymic work, the effect of one factor is studied at a time keeping all other conditions constant. The enzyme activity is influenced by a large number of factors such as substrate and

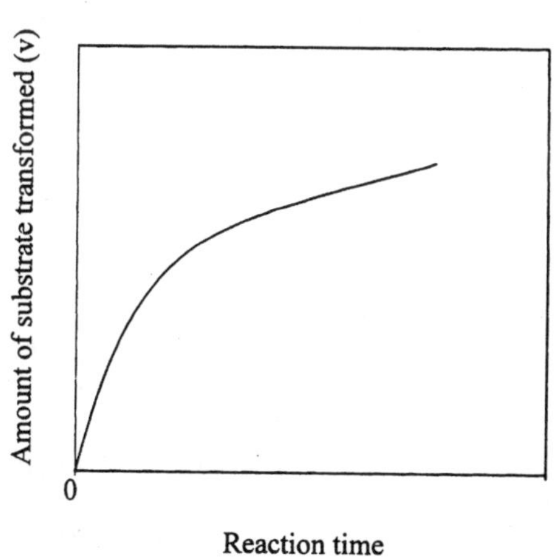

Fig. 7.1 Effect of time of incubation on velocity of an enzyme catalyzed reaction.

enzyme concentration, pH, temperature, ionic strength of media and the presence of activators or inhibitors. In view of this, it is recommended that the conditions in the reaction mixture must be maintained as constant as possible during the observations. The reaction must be carried out in a thermostated system and all the reagents must be brought to the required temperature before the reaction is started by their addition. To prevent changes in pH, adequate buffering must be provided, and conditions chosen should be such that the enzyme is reasonably stable; the temperature should not be too high (most assays are carried at about 25 to 30°C), and the pH and ionic strength should be within the range of the maximal stability of enzyme.

7.3 General Handling of Enzymes

Enzymes are relatively fragile and undergo denaturation under unfavourable conditions. It is, therefore, essential that to avoid inactivation, mild conditions of temperature and pH be used. Hence, except for cold labile enzymes, the extraction and purification of enzymes are carried out at 0–4°C. While adjusting pH of an enzyme solution as during pH precipitation with acid or alkali or ammonium sulfate fractionation, these reagents should be added in such a way that they run slowly down the side of the vessel and the solution is continuously stirred to prevent formation of localized regions of unfavourable pH or high salt concentration. Many enzymes are denatured at surfaces. Hence while stirring or transferring an enzyme solution from one vessel to another, formation of froth should be avoided. Organic solvents inactivate enzymes at room temperature. Fractionation with these solvents must be carried out at lowest possible temperature. Since in many cases, an enzyme undergoes slow inactivation on standing even under most favourable conditions, the enzymic studies must be carried out without undue delay. Storage of enzymes without loss of activity is a commonly encountered problem. If the enzyme solution can be frozen and thawed without loss of activity, it is advisable to keep the preparation in a deep freeze. Some enzymes can, however, be stabilized by high concentration of salts. Drying an enzyme solution under high vacuum at low temperatures from the frozen state is a valuable method often yielding an active and soluble powder which can be stored at 0–5°C or sometimes even at room temperature.

7.3.1 Enzyme extraction

Bringing an enzyme into the solution from its source is called extraction. For this the cell wall or cell membrane has to be ruptured. The extraction medium generally used is a suitable buffer which may contain additives such as EDTA, Mg^{2+}, PVP (polyvinyl pyrrolidone), DTT (dithiothreitol), glycerol and protease inhibitors etc. This may help in obtaining stable enzyme extract. Buffer helps to maintain the pH as on rupturing of tissue, certain acids etc may be released. EDTA is included for chelating heavy metals which may be present in traces in water and other reagents etc. Many enzymes require Mg^{2+} for their activity so Mg^{2+} is quite often added in the extraction medium. In case of plant tissues rich in phenols, PVP helps in binding of these compounds and protects the enzyme against inactivation. DTT stabilizes the enzyme by keeping the thiol groups in their reduced state. β-Mercaptoethanol, ascorbate, reduced glutathione and cysteine etc. can also be used for the same purpose. Upon disintegration of cellular structure, the proteolytic enzymes present in the tissue come in contact with the enzyme and may degrade it. To avoid this proteolytic cleavage, use of protease inhibitors is recommended. An ideal extraction medium must, therefore, contain all the components which contribute towards stabilization or protection of the enzyme.

It is easier to extract an enzyme from animal tissues than from plant or microbial sources. The methods employed for extraction include mechanical breaking of the cells by grinding with sand or shaking at high speed with fine glass beads, ultrasonic or sonic oscillations, freezing and thawing, treatment with solvents such as acetone, autolysis (either alone or with toluene, ethyl acetate or sodium sulfite), or lysis with added enzymes. The extraction of enzyme from subcellular organelles is rather difficult. Some enzymes are more firmly bound to the structural framework of these organelles probably as the lipoprotein complexes and require special methods for their extraction. For instance, in case of mitochondria, these methods include drying with acetone, treatment with butanol-water mixtures, extraction of freeze-dried mitochondria with various organic solvents, extraction with aqueous solutions of detergents, use of hydrolytic enzymes such as lipases, proteases and nucleases etc.

The extract obtained after rupturing of the cells is generally referred to as crude extract which is cleared off of the insoluble material by centrifugation at 4°C for about half an hour at $10{,}000 \times g$. The clear supernatant which contains soluble or solubilized enzymes, is used for enzyme assays. Sometimes, this extract is too dilute and may be concentrated either by addition of powder of gel filtration material of low fractionation range such as Sephadex G-25, by lyophilization or by ultrafiltration. Water could also be soaked by covering the dialysis sac containing the enzyme preparation with sucrose or dry flakes of polyethylene glycol.

7.3.2 Enzyme purification

Purification is the first step in the study of physico-chemical, kinetic and biological properties of an enzyme. Purification generally involves a series of fractionations by which the enzyme protein is separated from the other proteins. The type and sequence in which these fractionation techniques are employed are generally determined by hit and trial. However, rapid and more effective purification is achieved by alternating the different fractionation methods than by repetition of the same. At each fractionation step the enzyme activity and protein content of the preparation are determined. The two factors which affect purification considerably i.e. pH and electrolyte concentration must be given proper attention. In each fractionation step, the proteins in the preparation are separated into a series of fractions in such a way that a greater part of enzyme of interest is recovered, if possible, in one particular fraction. A successful fractionation is recognized by a fraction with high specific activity and hence fold purification (specific activity in fraction/original specific activity) and high yield (units of enzyme in fraction/units of enzyme in original preparation). The major techniques employed for enzyme purification include: (i) fractional precipitation by change of pH, (ii) fractional denaturation by heating, (iii) fractional precipitation by organic solvents, (iv) fractionation by inorganic salts such as $(NH_4)_2SO_4$ and (v) column chromatography involving gel filtration, ion exchange and affinity chromatography etc.

The results of purification are presented in a standard tabular form, an example of which is given in Table 7.1.

Table 7.1 Standard form of the table for presenting data on purification of enzymes

Fraction or Purification step	*Volume (ml)*	*Total protein (mg)*	*Total activity (nkats)*	*Specific activity (nkats/mg protein)*	*Fold purification*	*Recovery %*
Crude extract	500	2250	83	0.04	1	100
$(NH_4)_2SO_4$ fraction (40–70%)	110	572	64	0.11	3	77
Acetone fraction	30	258	59	0.23	6	71
Sephadex G-200	60	16.3	37	2.27	57	45
DEAE-cellulose	30	6.6	29	4.39	110	34

The purity of the enzyme preparation is generally tested by native gel electrophoresis. However, the most reliable test for determining homogeneity is use of immunological techniques.

7.3.3 Expression of enzyme activity

The enzyme activity is generally expressed in terms of units. The most widely accepted unit of enzyme activity (International Unit) is defined as that amount which causes transformation of 1.0 μmole (10^{-6} moles) of substrate per minute at 25°C under specified conditions of assay. The specific activity is units of enzyme activity per mg protein and is a measure of enzyme purity. The specific activity increases during purification and becomes maximum and constant when the enzyme is in a pure state. The Enzyme Commission has recommended use of *katal* (abbreviated as kat) as a new international unit of enzyme activity. One kat is defined as the amount of enzyme that transforms one mole of substrate per second. The activity can also be expressed in smaller units such as microkatals (μkat), nanokatals (nkat) or picokatals (pkat). In this case the specific activity is denoted in terms of katals per kg of protein or μkatals per mg protein.

7.3.4 Techniques for measurement of enzyme activity

An enzyme catalyzes the reaction via the formation of ES complex as follows:

$$E + S \rightleftharpoons ES \rightleftharpoons E + P$$

Either the rate of utilization of the substrate or the rate of product formation is monitored for quantitative determination of enzyme activity. Two types of methods used to determine enzyme activity are: the sampling and continuous methods. In sampling method, samples are withdrawn at regular intervals from the reaction mixture and are analyzed for either the disappearance of substrate or for the appearance of product while in the continuous method such changes in amount of substrate, coenzyme or product are followed continuously in the reaction mixture while the reaction is in progress. In both the methods, a control assay (which generally is same as the test assay in all the respects but lacks either the enzyme or the substrate) should also be carried out simultaneously. Any changes in the control would give an idea of the background reaction. The following analytical methods are used for determining either the amount of substrates, coenzymes or the products in the assay mixtures.

7.3.4.1 Visible and ultraviolet spectrophotometric methods

The fact that substrate and the product don't absorb at the same wavelength, forms the basis of many enzyme assays. For example, the enzyme fumarate hydratase catalyzes the conversion of fumarate to L-malate (fumarate absorbs strongly at 300 nm but malate does not) and xanthine oxidase catalyzes the conversion of hypoxanthine to urate (urate absorbs strongly at 290 nm but hypoxanthine does not), so decrease in absorbance at 300 nm in the former case and/or increase in absorbance at 290 nm in the later case could be used for determining the rate of reaction catalyzed by these enzymes. Similarly, there are other substances like NAD and NADP which show strong absorption at 340 nm when in the reduced state but not in their oxidized state. This fact has been utilized to measure the activity of a large number of dehydrogenases, e.g.

$$\text{Malate} + \text{NAD}^+ \xrightleftharpoons[\text{dehydrogenase}]{\text{Malate}} \text{Oxaloacetate} + \text{NADH} + \text{H}^+$$

$$\text{Pyruvate} + \text{NADH} + \text{H}^+ \xrightleftharpoons[\text{dehydrogenase}]{\text{Lactate}} \text{Lactate} + \text{NAD}^+$$

Increase and decrease in absorbance at 340 nm, respectively, can be used for measuring the activity of malate dehydrogenase and lactate dehydrogenase.

In some of the cases where the reaction catalyzed by an enzyme does not involve such spectrophotometrically measurable interconversions directly, they can be linked to other reactions which utilize NAD^+-NADH system, e.g. assay of 3-phosphoglycerate kinase (3 PGA kinase) is linked to glyceraldehyde-3-phosphate dehydrogenase.

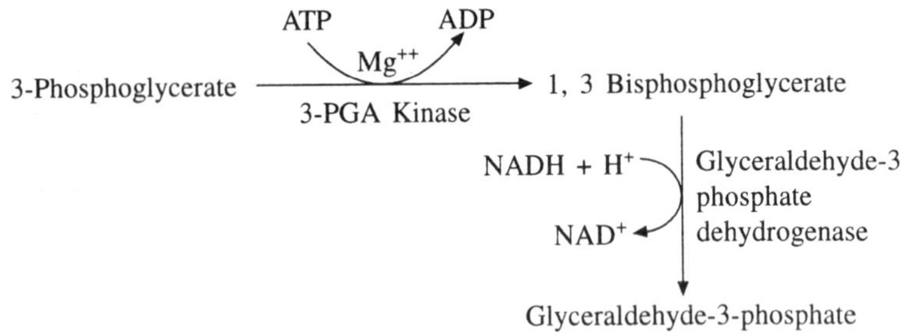

The assay mixture would, therefore, contain 3-phosphoglycerate, ATP, Mg^{2+} and in addition NADH and glyceraldehyde-3-phosphate dehydrogenase and measurement of the rate of NAD^+ production, i.e. decrease in absorbance at 340 nm thus enables the determination of activity of 3-phosphoglycerate kinase in the tissue extract added to the assay mixture. In such assay systems, the linking enzyme and cofactors required by it are always used in non-limiting amounts.

In other cases where neither the substrate nor the product absorbs light in UV region, artificial substrates can be used so that they form coloured products whose amount can quantitatively be measured in the visible region. For example the enzyme phosphatase can be assayed by using *p*-nitrophenyl phosphate as a substrate and the product *p*-nitrophenol is yellowish in colour.

$$p\text{-nitrophenyl phosphate} \xrightarrow[\text{phosphatase}]{\text{Acid or alkaline}} \underset{\text{(yellow)}}{p\text{-nitrophenol}} + \text{Pi}$$

Similarly, certain dyes, e.g. tetrazolium dyes, methylene blue, 2,6-dichlorophenol indophenol which have different colours in their oxidized and reduced form can also be used for oxidation-reduction reactions. Alternatively, a small amount of the reaction mixture can be withdrawn and treated with a specific reagent to give coloured derivative of the substrate or the product, which can be determined colorimetrically.

7.3.4.2 Spectrofluorimetric methods

This method can be used for assaying those enzymes in whose cases either the substrate or the product is a fluorescent compound, e.g. NAD(P)H is fluorescent and can be used to assay many enzymes directly or by coupling with another reaction in which NAD(P)H is either utilized or produced.

7.3.4.3 Luminescence method

The enzyme catalyzed reactions which emit light can be assayed by determining the intensity of the emitted light, e.g. the enzyme luciferase emits light when it oxidizes luciferin in the presence of ATP.

$$\text{Luciferin} + \text{ATP} + O_2 \xrightarrow{\text{Luciferase}} \text{Oxyluciferin} + \text{AMP} + PP_i + CO_2 + \text{Light}$$

7.3.4.4 Radioisotopic method

This method is highly sensitive but has the limitation of separating the radiolabelled substrate from the product. This problem is not encountered if one of the substrates or products is a gas because the rate of production/utilization of radiolabelled gas can be easily measured. In other cases, substrate or the product may be separated by an appropriate method such as chromatography, electrophoresis etc. and radioactivity measured in either of these in a scintillation counter.

7.3.4.5 Manometric methods

The reaction rate of the enzyme catalyzed reactions associated with net utilization or release of a gas such as O_2 or CO_2 can be monitored by the Warburg manometer or the Gilson respirometer.

7.3.4.6 Microcalorimetric methods

Several enzyme-catalyzed reactions are accompanied by changes in heat. Hence the instruments which can quantitatively detect small changes in temperature of the reaction mixture can be utilized to determine the rate of such enzymic reactions.

Of all the methods enumerated above, spectrophotometric procedures either in UV or visible region are the most popular for enzyme assays.

7.3.5 Factors affecting enzyme activity

As stated earlier, the main factors determining the initial velocity of an enzymatic reaction are enzyme concentration, substrate concentration, pH, temperature and the presence of activators or inhibitors. A brief description of each is given as follows:

7.3.5.1 Enzyme concentration

In general, the velocity is directly proportional to enzyme concentration, i.e. two molecules of an enzyme acting independently in solution will transform twice as much substrate in a given time as compared to one molecule. However, deviation may sometimes occur from the linear relationship. Accordingly, it is recommended that relationship between velocity and enzyme concentration be established before undertaking further studies on kinetics. Departure from linearity could be due to one or the other following reasons: (a) the presence of small amount of some highly toxic impurities in any one of the components of the incubation mixture other than the enzyme solution itself; (b) the presence of a dissociable activator or coenzyme in enzyme preparation (in such a case, the percentage of the enzyme which is in the active form increases as the concentration of the activator in the incubation mixture increases); (c) a limitation in the capacity of the method of estimation rather than a true decrease in the activity of the enzyme; (d) when an enzyme has an appreciable affinity for one of the components of the system (e.g. coenzyme) and (e) when an enzyme preparation contains a reversible inhibitor which combines with the enzyme to give an inactive enzyme-inhibitor complex (in these cases, percentage of the enzyme which is in the inactive form increases as the concentration of inhibitor in the incubation mixture increases). Despite these exceptions, in normal cases, one generally gets proportionality between enzyme concentration and initial velocity (Fig. 7.2). This is the basic assumption in most of the kinetic studies.

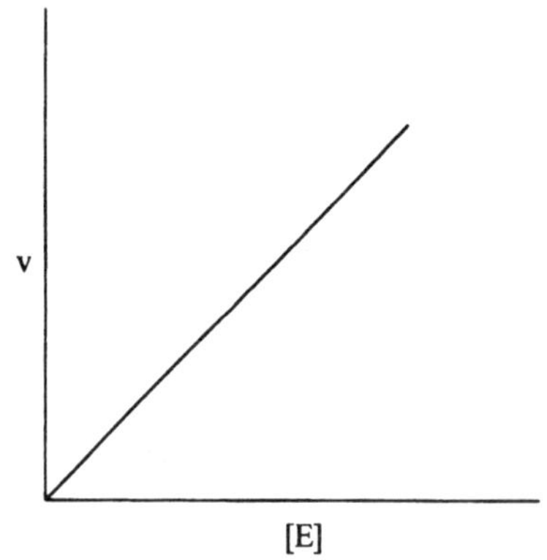

Fig. 7.2 Typical effect of enzyme concentration on reaction velocity.

7.3.5.2 Substrate concentration

The most important factor affecting the rate of an enzyme catalyzed reaction *in vitro* is the amount of substrate present. The effect of varying substrate on initial velocity of an enzyme catalyzed reaction is shown in Fig. 7.3. At relatively low substrate concentration, initial velocity increases almost linearly with an increase in the substrate (in this region of the curve the reaction follows first order kinetics). At higher substrate concentration, initial velocity increases by smaller and smaller extent in response to increases in substrate (in this region of the curve, the reaction is of mixed order type). Finally a point is reached beyond which there is small and/or no increase in the velocity with increase in the concentration of the substrate (the reaction follows zero order kinetics in this region). This plateau is called the V_{max}. At this stage, the enzyme is fully saturated with the substrate.

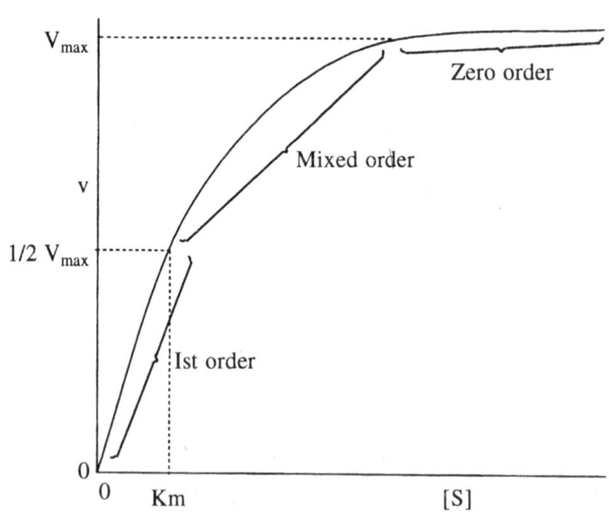

Fig. 7.3 Effect of substrate concentration on the initial velocity of an enzyme catalyzed reaction.

7.3.5.3 Effect of pH

Enzymes are active only over a limited range of pH and in most cases they exhibit a definite optimum pH. This optimum may be due to (a) a true reversible effect on velocity itself, (b) an effect of pH on the affinity of enzyme for the substrate. The fall on either side of the optimum could be due to decreased saturation of the enzyme with the substrate because of decreased affinity. This effect can be eliminated by using a sufficiently high concentration of substrate to saturate the enzyme at all pHs, or (c) the effect of pH on the stability of the enzyme. The enzyme may get denatured on one or both sides of the optimum. These effects may occur in combination, i.e. the fall on one side may be because of a fall in affinity and on the other to instability of the enzyme. However, these effects may easily be distinguished experimentally, e.g. the irreversible destruction of the enzyme could be tested by exposing the enzyme to a range of pH values and then testing the activity after readjusting the pH to optimum (Fig. 7.4, curve B). Figure 7.4 shows standard pH curve for an enzyme. The fall on the alkaline side is due to destruction of enzyme.

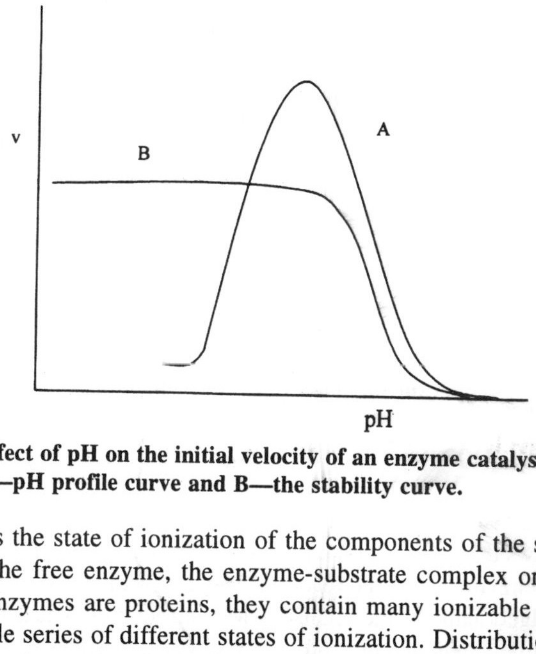

Fig. 7.4 Effect of pH on the initial velocity of an enzyme catalysed reaction. A—pH profile curve and B—the stability curve.

pH changes the state of ionization of the components of the system, which may involve the free enzyme, the enzyme-substrate complex or the substrate itself. Since enzymes are proteins, they contain many ionizable groups which exist in a whole series of different states of ionization. Distribution of the total enzyme among the various ionic forms depends on the pH and the ionization constants of the various groups. But, since the catalytic activity is usually confined to a narrow range of pH, only one of the ionic forms of the enzyme (or rather of the active center) is catalytically active.

7.3.5.4 Temperature

The rate of an enzyme catalyzed reaction increases with rise in temperature till it reaches the maximum. The velocity decreases on further increasing the temperature as shown in Fig. 7.5. This effect of temperature may be due to several reasons including an effect on the stability of the enzyme; an effect on the actual velocity of the breakdown of the complex determined by the heat of activation of the reaction; an effect on the enzyme substrate affinity; an effect on the pH functions of any or all of the components due to an alteration of their pK_as as influenced by the heats of ionization; or an effect on the affinity of an enzyme for the activators or inhibitors, if any. In a system involving two or more enzymes with different temperature coefficients, temperature may affect transfer of rate limiting functions from one enzyme to another. Like pH, the effects due to instability of the enzyme could be studied by exposing the enzyme to various temperatures for a definite period, and then measuring its activity at optimum temperature. The effects on the affinity could be eliminated by using sufficiently high concentration of substrate or activator to saturate the enzyme. In multienzyme systems, each enzyme may be studied separately. In this way the limiting effects of other enzymes may be identified.

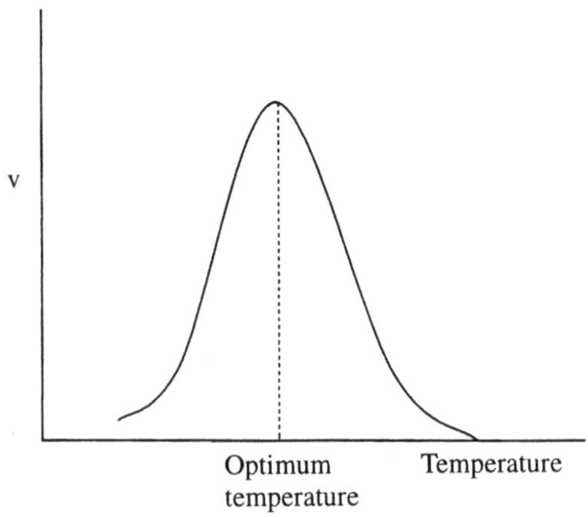

Fig. 7.5 Effect of temperature on the initial velocity of an enzyme catalyzed reaction.

7.4 Determination of V_{max} and K_m

7.4.1 Michaelis-Menten equation

Figure 7.3 shows the relationship between substrate concentration and the initial velocity of an enzyme catalyzed reaction. The hyperbolic shape of this curve can be expressed algebrically by the Michaelis-Menten equation. The equation assumes the interaction between enzyme and substrate in a reversible manner to give ES complex, which then dissociates to enzyme and substrate or is transformed further (irreversibly) to give enzyme and product. This sequence of events is

$$E + S \underset{K_2}{\overset{K_1}{\rightleftharpoons}} ES \xrightarrow{K_2} E + P$$

Initial velocity is determined by the breakdown of ES to give the product (Eq. 1)

$$v = K_2 [ES] \qquad (1)$$

ES cannot be measured easily. Accordingly, a term $[E_t]$ is introduced which represents the total enzyme concentration (sum of the free and the substrate bound enzyme). Free or unbound enzyme can then be represented by $[[E_t]-[ES]]$. At the same time, since the substrate is in far greater excess than $[E_t]$, the amount of the substrate bound by the enzyme at any given time is negligible compared with the total substrate, [S]. With these assumptions in mind, the rate of formation and breakdown of ES can be expressed as:

$$\text{Rate of ES formation} = K_1\ ([E_t] - [ES])\ [S] \tag{2}$$

$$\text{Rate of ES breakdown} = K_{-1}\ [ES] + K_2[ES] \tag{3}$$

Under steady state conditions [ES] remains constant, i.e. the rate of formation of ES is equal to the rate of its breakdown. Hence under these conditions, Eqs. 2 and 3 can be equated giving

$$K_1([E_t] - [ES])\ [S] = K_{-1}[ES] + K_2[ES] \tag{4}$$

On further simplification, we get

$$K_1[E_t][S] - K_1[ES][S] = (K_{-1} + K_2)\ [ES] \tag{5}$$

Adding the term $K_1[ES][S]$ to both sides of the equation and simplifying, we get

$$K_1[Et][S] = K_1[ES][S] + (K_{-1} + K_2)\ [ES]$$

Solving this equation for [ES], we get

$$[ES] = \frac{K_1 [E_t]\ [S]}{K_1 [S] + (K_{-1} + K_2)} \tag{6}$$

on dividing the numerator and denominator by K_1

$$[ES] = \frac{[E_t]\ [S]}{[S] + \left(\frac{K_{-1} + K_2}{K_1}\right)} \tag{7}$$

The term $\left(\frac{K_{-1} + K_2}{K_1}\right)$ is defined as the Michaelis-Menten constant K_m.

Substituting this in the above equation, we get

$$[ES] = \frac{[E_t]\ [S]}{[S] + K_m} \tag{8}$$

Initial velocity (v) and maximum velocity (V_{max}) can now be expressed in terms of [ES] and $[E_t]$, respectively. Substituting these terms in above equation, we get:

$$v = \frac{V_{max}\ [S]}{K_m + [S]} \tag{9}$$

This is the Michaelis-Menten equation which gives quantitative relationship between the initial velocity (v), the maximum velocity (V_{max}) and the substrate concentration [S], all related through the Michaelis-Menten constant K_m. If the initial velocity is half of the maximum velocity (V_{max}) then Eq. 9 can be rewritten as:

$$\frac{V_{max}}{2} = \frac{V_{max}\,[S]}{K_m + [S]}$$

On dividing by V_{max} we obtain

$$\frac{1}{2} = \frac{[S]}{K_m + [S]}$$

Solving for K_m, we get

$$K_m + [S] = 2[S]$$

$$K_m = 2[S] - [S] = [S]$$

Hence K_m is the substrate concentration at which the initial velocity is half of the maximum velocity and it is always expressed in the same units as that of the substrate.

The Michaelis-Menten equation can be transformed into different forms that are useful in the determination of V_{max} and K_m. The important transformations are discussed as follows:

7.4.2 The Lineweaver-Burk plot

The graph of the Michaelis-Menten equation (v vs [S], Fig 7.3) is not entirely satisfactory for the determination of V_{max} and K_m. The hyperbolic curve does not give an accurate value of V_{max} and hence of K_m until and unless there are three consistent points on the plateau of the curve. Furthermore, the graph being a curve cannot be accurately extrapolated upwards. These problems are overcome if we plot reciprocals of v and [S] as is the case in Lineweaver-Burk Plot. The Michaelis-Menten equation can be converted to its reciprocal form as follows:

$$v = \frac{V_{max}\,[S]}{[S] + K_m}$$

$$\frac{1}{v} = \frac{[S] + K_m}{V_{max}\,[S]} = \frac{[S]}{V_{max}\,[S]} + \frac{K_m}{V_{max}\,[S]}$$

$$\therefore \quad \frac{1}{v} = \frac{K_m}{V_{max}} \cdot \frac{1}{[S]} + \frac{1}{V_{max}}$$

A plot of 1/v vs 1/[S] which gives a straight line is shown in Fig. 7.6. It is also known as a double reciprocal plot since reciprocals of v and [S] are used for preparing the graph.

The graph being linear can be extrapolated even if no determinations have been made at saturating substrate concentrations. K_m and V_{max} can be easily calculated from the extrapolated graph from the intercepts on X- and Y-axis, respectively.

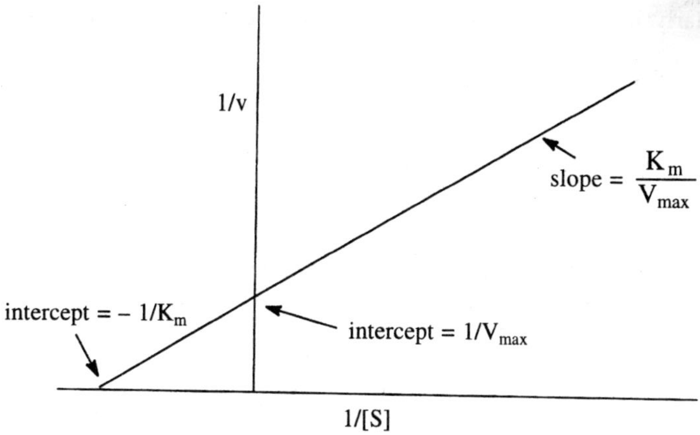

Fig. 7.6 Lineweaver-Burk plot.

7.4.3 Eadie-Hofstee and Hanes plots

The Lineweaver-Burk plot also suffers from certain drawbacks. Sometimes the extrapolation across the 1/v axis to determine the value of $-1/K_m$ reaches the edge of the graph before reaching the 1/[S] axis. It also gives undue weight to measurements made at low substrate concentrations. Thirdly, departures from linearity are less obvious than in some other plots. Some of these complications are obviated by using Eadie-Hofstee plot.

By multiplying the Lineweaver-Burk equation by a factor $v \cdot V_{max}$ on both the sides, we get:

$$\frac{1}{v} \cdot v \cdot V_{max} = \frac{K_m}{V_{max}} \cdot \frac{1}{[S]} \cdot v \cdot V_{max} + \frac{1}{V_{max}} \cdot v \cdot V_{max}$$

On simplifying,

$$V_{max} = K_m \cdot \frac{v}{[S]} + v$$

$$v = -K_m \frac{v}{[S]} + V_{max}$$

A plot of v vs v/[S] is shown in Fig. 7.7 and from this graph, V_{max} and K_m can be determined.

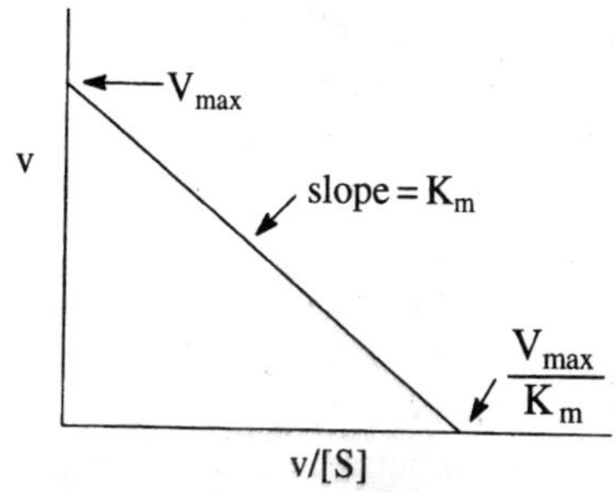

Fig. 7.7 Eadie-Hofstee plot.

Similarly, by multiplying Lineweaver-Burk equation by [S] on both sides, we get

$$\frac{1}{v} \cdot [S] = \frac{K_m}{V_{max}} \cdot \frac{1}{[S]} \cdot [S] + \frac{1}{V_{max}} \cdot [S]$$

Therefore,

$$\frac{[S]}{v} = \frac{1}{V_{max}} [S] + \frac{K_m}{V_{max}}$$

So the plot between [S]/v vs [S] gives a straight line from which V_{max} and K_m can be obtained (Fig. 7.8).

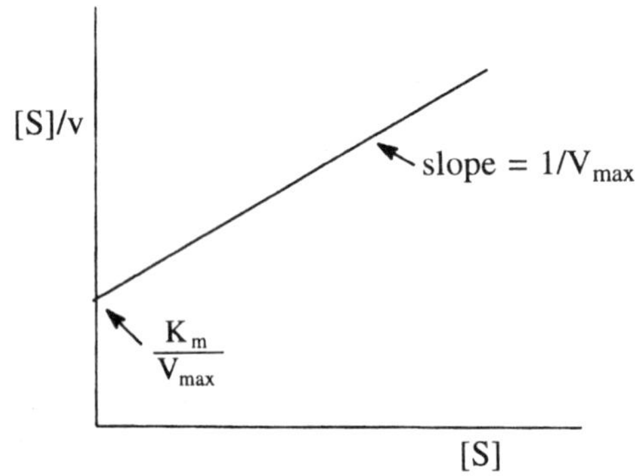

Fig. 7.8 Hanes plot.

7.4.4 Eisenthal and Cornish-Bowden plot

Eisenthal and Cornish-Bowden in 1974 gave another graph based on Michaelis-Menten equation, the form of which is as follows:

$$\frac{1}{v} = \frac{K_m + [S]}{V_{max}\,[S]}$$

$$\frac{V_{max}}{v} = \frac{K_m + [S]}{[S]} = \frac{K_m}{[S]} + 1$$

Therefore, at constant v and [S], a plot of V_{max} against K_m is linear. When K_m = 0, V_{max} = v and when V_{max} = 0, K_m = [–S]. Therefore, each v, [S] pair can be used to generate a line by marking v on the V_{max} axis and [–S] on the K_m axis. These two points are then joined and the line extrapolated. The line for all v and [S] pairs at constant [E], must pass through true values of K_m and V_{max}, so these values are given by the point of interception of all the lines. However, because of the experimental error, the lines are likely to intersect over a range of values (Fig. 7.9).

7.5 Multisubstrate Reactions

Most biochemical reactions involve at least two substrates. The kinetics of such a reaction is more complex. However, by plotting simple Lineweaver-Burk plot, preliminary information can sometimes be obtained regarding the possible mechanisms for these reactions. Bisubstrate reactions catalyzed by transferases,

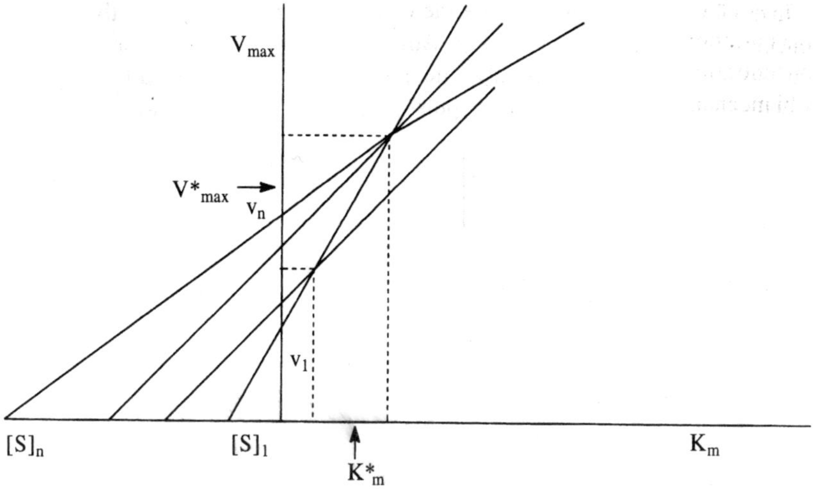

Fig. 7.9 Typical Eisenthal-Cornish-Bowden plot at constant [E]. V^*_{max} is the best estimate of V_{max} and K^*_m is the best estimate of K_m.

kinases and dehydrogenases in which two substrates are converted into two products may follow different mechanisms such as sequential, in which both substrates bind to give ternary complex before the products are formed. Sequential reactions which are referred to as single displacement reactions may be compulsory ordered (Fig. 7.10a), in which case the two substrates bind in a definite sequence, or random ordered (Fig. 7.10b), in which case either substrate can bind first. In both cases, there are separate binding sites for each substrate. Alternatively, the reaction may be non-sequential. In such cases, one product is released before the second substrate gets bound to the enzyme. Such reactions are also known as double displacement reactions. One example of this type of mechanism is a Ping-Pong reaction, which proceeds via a modified form of the enzyme (Fig. 7.10c) as in the case of transaminases.

$$E + S_1 \rightleftharpoons ES_1 \overset{S_2}{\rightleftharpoons} ES_1S_2 \rightleftharpoons EP_1P_2 \rightleftharpoons P_2 + EP_1 \rightleftharpoons P_1 + E$$

(a) Sequential compulsory-ordered mechanism.

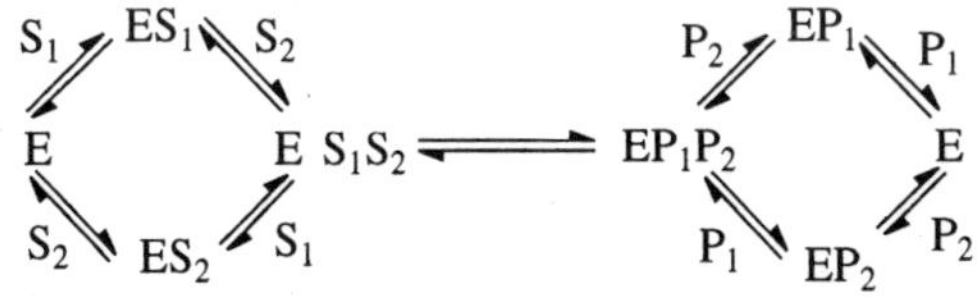

(b) Sequential random-ordered mechanism.

$$E + S_1 \rightleftharpoons ES_1 \rightleftharpoons EP_1 \rightleftharpoons E^* + P_1$$

(modified enzyme)

$$E^* + S_2 \rightleftharpoons E^*S_2 \rightleftharpoons EP_2 \rightleftharpoons E + P_2$$

(c) Non-sequential Ping-Pong mechanism.

Fig. 7.10

In double reciprocal plots, when the variation of the initial rate with increasing concentration of one substrate is investigated in the presence of a series of fixed concentrations of the second substrate, parallel lines are obtained for Ping-Pong bi-bi mechanism (Fig. 7.11). A compulsory ordered and a random-ordered ternary

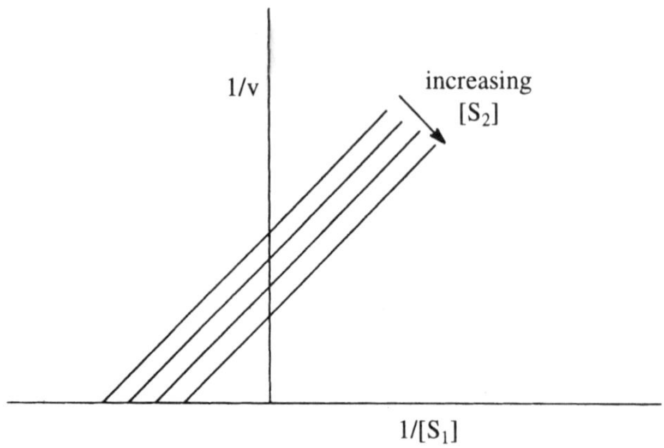

Fig. 7.11 Steady-state kinetic analysis of bisubstrate reaction following a Ping-Pong or double displacement pathway.

complex mechanisms give non-parallel double reciprocal plots with a progressively smaller interecept on the 1/v axis as the concentration of the fixed second substrate is increased (Fig. 7.12). V_{max} and K_m for the bisubstrate reaction can be determined from Lineweaver-Burk plots which could be obtained by holding constant the concentration of one of the two substrates and studying the influence on the initial rate of varying the concentration of the second substrate. In such cases, V_{max} is the maximum initial rate when both substrates are saturating, while the K_m for a particular substrate is the concentration of that substrate which gives

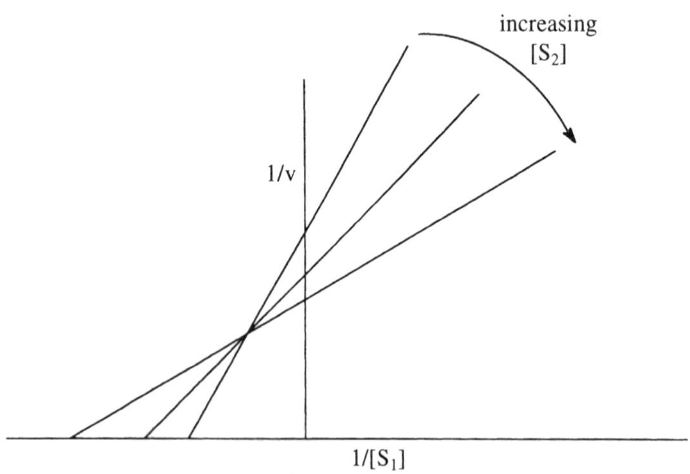

Fig. 7.12 Steady-state kinetic analysis of bisubstrate single displacement reaction in which a ternary complex is formed.

half V_{max} when the other substrate is saturating. A double reciprocal plot is drawn for each concentration of the second substrate, thus giving a series of straight lines. Such simple double reciprocal plot represents primary plot. A secondary plot is then drawn of the 1/v intercepts of the primary Lineweaver-Burk plots against the reciprocal of the second (fixed) substrate. This gives a straight line slope K_m (for the second substrate)/V_{max} and intercept $1/V_{max}$. Experiment is then repeated reversing the role of two substrates.

7.6 Enzyme Inhibition

Inhibitors are the substances which decrease the rate of enzyme catalyzed reaction. Therefore, irreversible destruction of an enzyme activity caused due to denaturation by heat, strong acid, or other means, is not usually regarded as inhibition. Inhibitors acting via a definite chemical reaction may be either reversible or irreversible. In case of reversible inhibition, the inhibitor binds to an enzyme in a reversible fashion and the activity is regained on merely removing free inhibitor by dialysis or other means. In these cases, equilibrium is established between the free inhibitor and the enzyme-inhibitor complex. In case of irreversible inhibitors, no such equilibrium exists and the activity is not recovered on dialysis. In reversible inhibition, the equilibrium constant is the measure of the affinity of the inhibitor and the effectivness of the inhibitor is expressed by the term K_i, which is the reciprocal of the enzyme-inhibitor affinity. In this system, a definite degree of inhibition is obtained, depending on the inhibitor concentration. Usually this is reached fairly rapidly and thereafter it is independent of the time provided that the inhibitor is stable. On the contrary, irreversible inhibition is characterized by a progressive increase in inhibition with time ultimately causing a complete inhibition even with very small amounts of the inhibitor provided that the inhibitor is in excess of the enzyme present. In these cases, the effectiveness of the inhibitor is expressed by a velocity constant, which is determined by the fraction of the enzyme inhibited in a given period of time by a certain concentration of the inhibitor. Action of cyanide on xanthine oxidase and of the nerve gases on choline esterase are the few examples of irreversible inhibition, whereas the action of cyanide on cytochrome *c* oxidase and of malonate and oxaloacetate on succinate dehydrogenase are the examples of reversible inhibition. Reversible inhibitor may affect the apparent K_m, V_{max} or both these constants. Those inhibitors which act only by increasing the effective K_m are termed as competitive inhibitors, while those having no effect on K_m but act simply by reducing V_{max} are known as non-competitive inhibitors. Those inhibitory compounds which alter both K_m and V_{max} are either the mixed type inhibitors or the uncompetitive inhibitors.

Competitive inhibitors generally are the structural analogues of the substrate and because of this structural similarity, they compete for the same substrate binding site of the enzyme, e.g. malonate is the competitive inhibitor of succinate dehydragenase. The amount of inhibition will depend on the amount and relative affinities of the inhibitor and the substrate for the enzyme. Accordingly, with fixed amount of the inhibitor, the extent of inhibition decreases with increasing concentration of the substrate and at very high substrate concentration, inhibition becomes negligible. Hence V_{max} for the reaction remains unchanged and K_m for the substrate is increased by a competitive inhibitor. The type of Lineweaver-Burk plot for a competitive inhibitor is shown in Fig. 7.13.

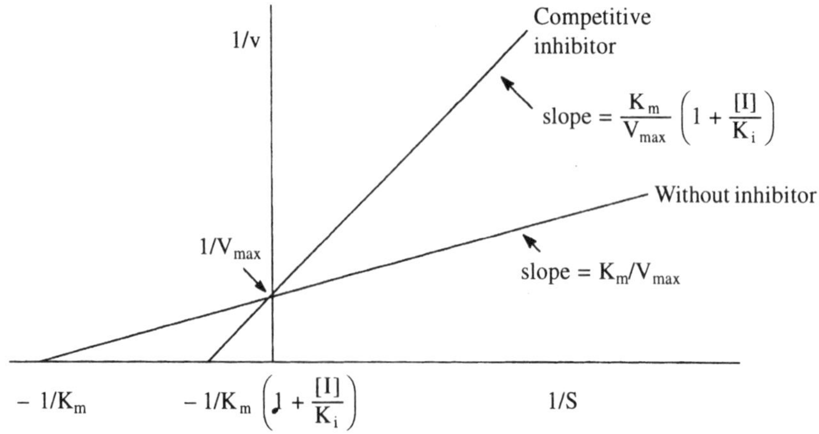

Fig. 7.13 Lineweaver-Burk plot showing the effect of competitive inhibitor.

In the non-competitive type of inhibition, the inhibitor does not affect the formation of complex between the substrate and the enzyme but affects only the V_{max} of the reaction. The Lineweaver-Burk plot for the non-competitive type of inhibition is given in Fig. 7.14.

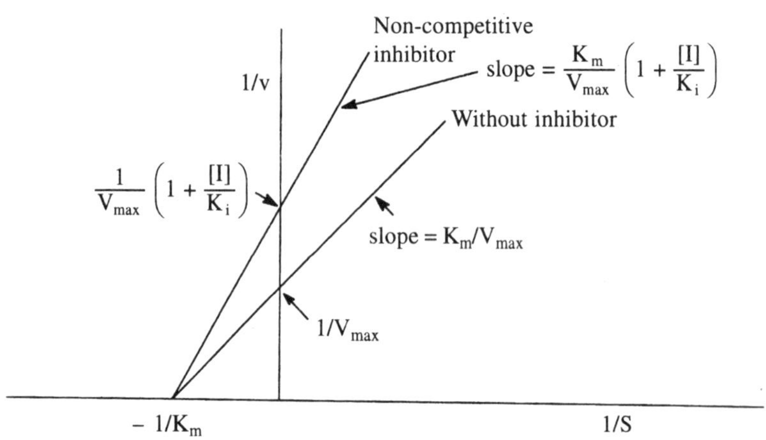

Fig. 7.14 Lineweaver-Burk plot showing the effect of a non-competitive inhibitor.

Mixed type of inhibitor will affect both the apparent K_m and V_{max} of the reaction. In this type of inhibition, the apparent K_m is increased and V_{max} is decreased (Fig. 7.15).

Uncompetitive inhibitors bind to the enzyme-substrate complex and not to the free enzyme. Hence, in this type of inhibition, there is decrease in both K_m and V_{max} as may be observed from Fig. 7.16.

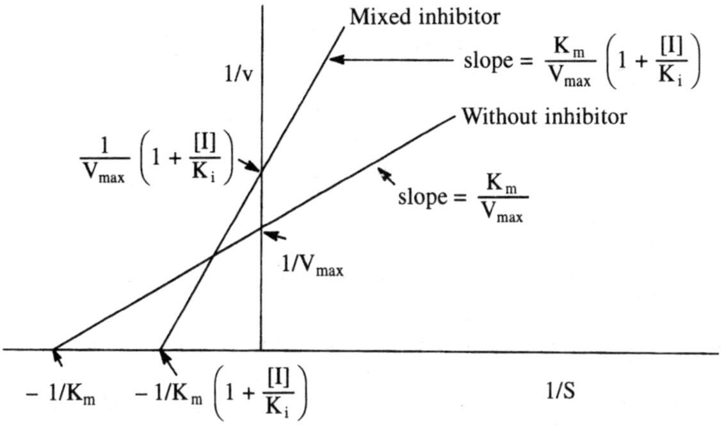

Fig. 7.15 Lineweaver-Burk plot showing the effect of mixed type of inhibitor.

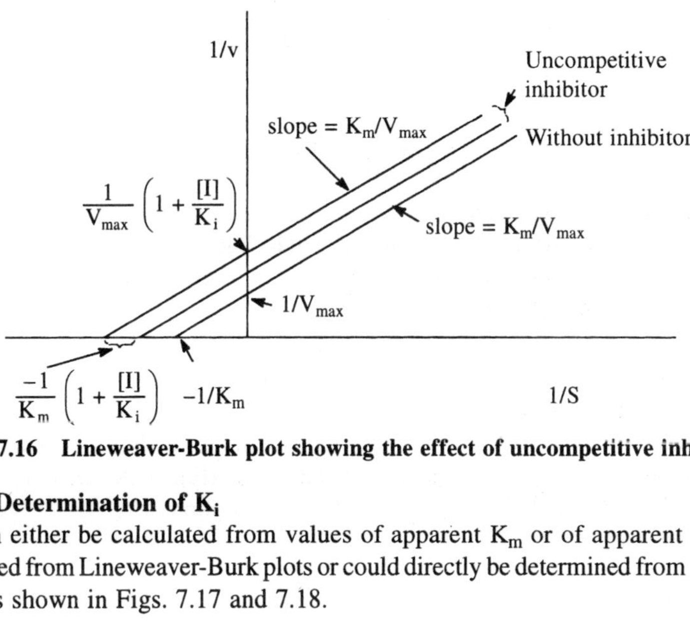

Fig. 7.16 Lineweaver-Burk plot showing the effect of uncompetitive inhibitor.

7.6.1 Determination of K_i

K_i can either be calculated from values of apparent K_m or of apparent V_{max} as obtained from Lineweaver-Burk plots or could directly be determined from Dixon's plot as shown in Figs. 7.17 and 7.18.

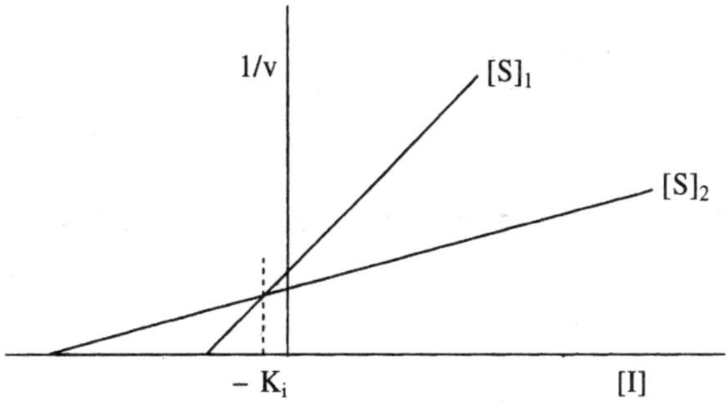

Fig. 7.17 Dixon plot for competitive inhibition.

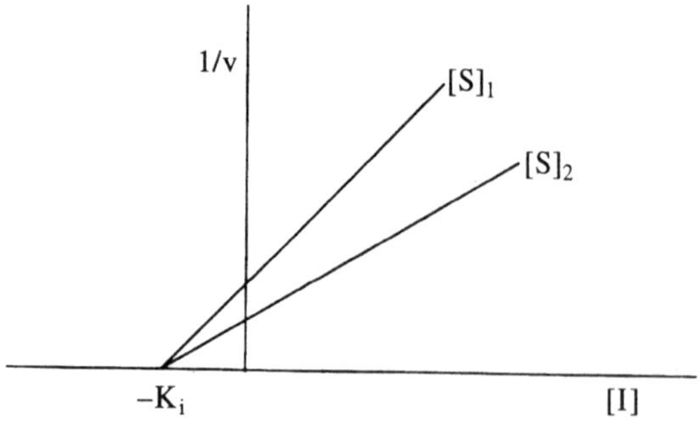

Fig. 7.18 Dixon plot for non-competitive inhibition.

K_i could also be determined from secondary plots. In case of competitive inhibition, K_i could be obtained from the following expression:

$$K'_m = K_m\left(1 + \frac{[I]}{K_i}\right)$$

On simplification, it gives

$$K'_m = \frac{K_m}{K_i}[I] + K_m$$

Therefore, a plot of K_m (determined from the intercept on the 1/[S] axis of the primary Lineweaver-Burk plot) against [I] will be linear, with the intercept on the [I] axis giving $-K_i$. Similarly, since the slope of the Lineweaver-Burk plot in the presence of a competitive inhibitor is:

$$\text{Slope} = \frac{K_m}{V_{max}}\left(1 + \frac{[I]}{K_i}\right)$$

a graph of slope of the primary plot against [I] will also be linear. The intercept on the [I] axis will give $-K_i$. These secondary plots (Fig. 7.19) are constructed from the data obtained at a fixed substrate concentration.

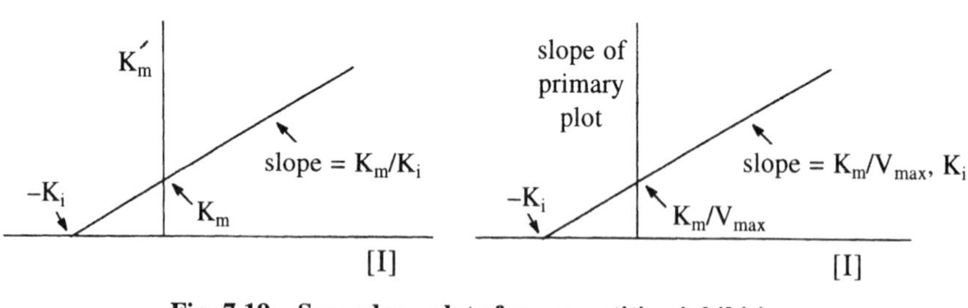

Fig. 7.19 Secondary plots for competitive inhibition.

For non-competitive type of inhibition,

$$\frac{1}{V'_{max}} = \frac{1}{V_{max}}\left(1 + \frac{[I]}{K_i}\right)$$

and the slope of the primary Lineweaver-Burk plot is

$$\frac{K_m}{V_{max}}\left(1+\frac{[I]}{K_i}\right)$$

The plots of $1/V_{max}$, against [I] and of slope of the primary plot against [I] will be linear, and the intercept on the [I] axis will give $-K_i$ (Fig. 7.20).

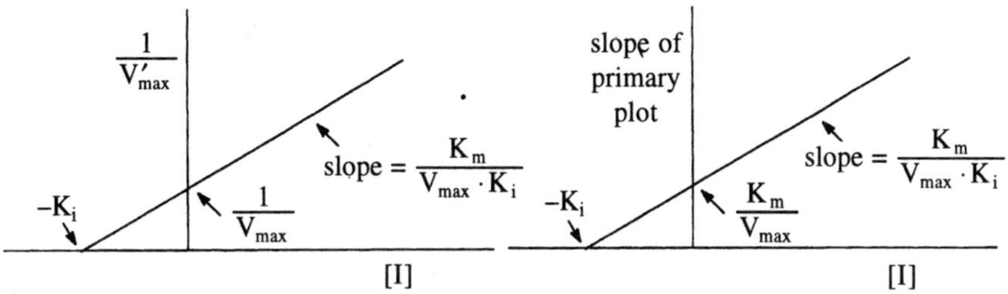

Fig. 7.20 Secondary plots for non-competitive inhibition.

For uncompetitive inhibition,

$$\frac{1}{V'_{max}}=\frac{1}{V_{max}}\left(1+\frac{[I]}{K_i}\right)$$

and
$$\frac{1}{K'_m}=\frac{1}{K_m}\left(1+\frac{[I]}{K_i}\right)$$

Hence plots of $1/V'_{max}$ or $1/K'_m$ (obtained from intercepts on 1/v and 1/[S] axis, respectively of primary plot) against [I] will be linear, and the intercept on the [I] axis will give $-K_i$ (Fig. 7.21).

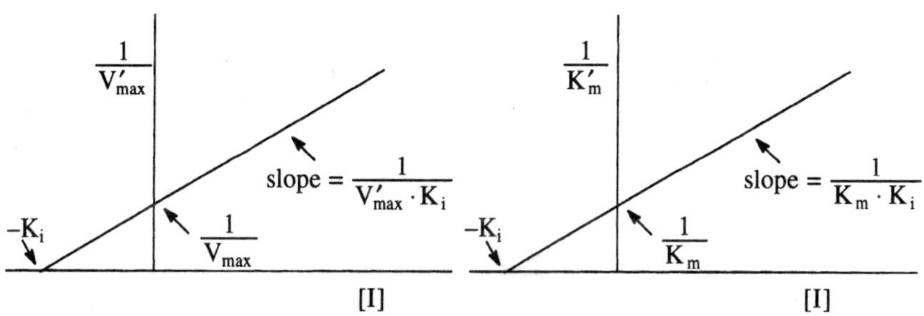

Fig. 7.21 Secondary plots for uncompetitive inhibition.

For mixed inhibition,

$$\frac{1}{V'_{max}}=\frac{1}{V_{max}}\left(1+\frac{[I]}{K_i}\right)$$

and slope for inhibited reaction is equal to the product of slope for uninhibited reaction and $(1+[I]/K_i)$. Hence, a secondary plot of $1/V'_{max}$ against [I] and slope of primary plot against [I] will be linear, and the intercept on the [I] axis will give $-K_i$ (Fig. 7.22).

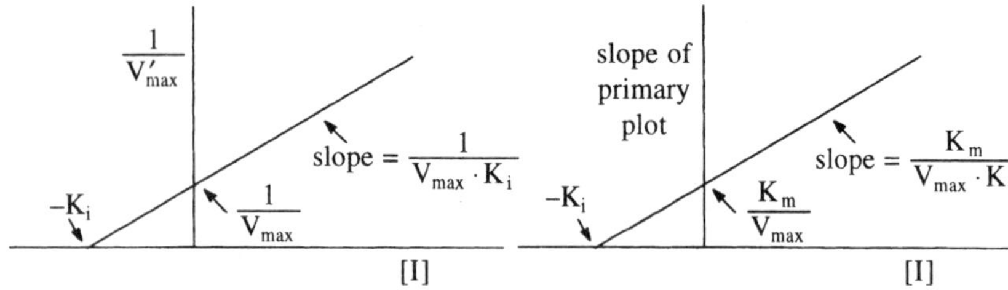

Fig. 7.22 Secondary plots for mixed inhibition.

7.7 Enzyme Immobilization

Immobilization is the technique by which the movement of enzymes, cells, organelles etc. is completely or severely restricted, usually resulting in a water-insoluble form of the enzyme. This has wide applications in the industry, as it enables re-use or continuous use of the biocatalyst. This also helps in making the enzyme more stable and maintaining a constant environment for the immobilized enzyme. In addition, the use of immobilized system prevents the enzyme or cell from contaminating the product or polluting a water stream. Immobilized enzymatic processes are easier to automate and they enable exploiting the advantages of various reactor configurations including rapid pH and temperature control, good gas transfer in stirred reactors, and the minimization of product inhibition in packed bed reactors. Immobilized enzymes are used in synthetic chemistry, in enzyme electrodes, and other analytical devices, and are potentially useful in therapy in the form of extra corporeal shunts. Immobilization has also been used to study the tertiary and subunit structures of proteins and as models of membrane-bound proteins and enzymes. A brief outline of various methods employed for enzyme immobilization is as follows.

7.7.1 Methods of enzyme immobilization

This section describes the features of the main techniques employed for immobilization of enzymes.

7.7.1.1 Entrapment

In this method the enzyme is entrapped within the lattice of a polymer matrix or a membrane which while allowing the penetration of substrates, the release or dispersion of protein in the reaction media is prevented. The method has wide applications and can be employed for entrapment of any kind of enzymes, microbial cells, and organelles of different sizes and properties. The wide applicability is mainly because of the fact that in this method, the enzyme does not bind in any way to the gel matrix or membrane. Based on the material used, the method sometimes is subclassified into gel entrapment, fibre entrapment and micro-encapsulation etc. In gel entrapment method, the enzyme is entrapped within the interstitial spaces of cross-linked water insoluble polymer gels, while in fibre entrapment method, the enzyme is simply entrapped within the microcavities of synthetic fibres. In the later case, by using very fine fibres, one can get high surface area for enzyme binding. Furthermore, the fibres are resistant to weak acids and alkalis, high ionic strengths, and some organic solvents, and depending on the polymer used, can show good resistance to microbial attack. Cellulose acetate is the most commonly used polymer in this category.

In microencapsulation, the enzyme is enclosed within spherical semipermeable polymer membranes having diameters in the range of 1–100 μm. This allows enzymes to be physically retained within the membranes, while substrate and product molecules are free to diffuse across the membrane provided that their molecular sizes are sufficiently small to allow this. The method suffers from certain disadvantages such as occasional inactivation of the enzyme during microencapsulation, the high enzyme concentration required; and the possibility of enzyme incorporation into the membrane wall and subsequent leakage in case of some of the methods used for microencapsulation. However, in this system one gets extremely large surface area for the contact of substrate and enzyme within a relatively small volume. Moreover, many enzymes could be immobilized simultaneously in a single step.

7.7.1.2 Carrier binding

Enzymes could also be immobilized by attachment to an insoluble carrier. Based on the binding mode of the enzymes, the methods could further be subclassified into physical adsorption, ionic binding, chelation or metal binding and covalent binding etc. However, the carrier of choice is the one which would increase substrate binding, decrease product inhibition, shift the apparent pH optimum to the desired value, discourage microbial growth, and enable ready recovery of the enzyme for re-use. Furthermore, the carrier should be mechanically rigid, stable in solution and should not deteriorate under the reaction conditions. For immobilized enzymes to be used in medicine, the carrier should not evoke immune responses or clotting reactions. Some of the carriers generally used for immobilization of enzymes include: *organic carriers* (natural)-activated carbon, agar, agarose, albumin, cellulose, chitin, chitosan, collagen, dextran, gelatin, glucomannans, silk, starch; *organic carriers* (synthetic)-acrylamide-based polymers, acrylic/methacrylic acid copolymers, maleic anhydride copolymers, nylon, polystyrene; and *inorganic carriers*-alumina, bentonite, calcium phosphate, titanium phosphate, celite, ceramics, cordiorite, expanded clays, hydrous metal oxides, hydroxyapatite, kaolinite, magnetite, pumice, sand, silica gel, titania, zirconia etc. Methacrylate or silicone rubber appears to be the best for medical use, whereas inorganic carriers are better for industrial applications.

Physical adsorption is the easiest way of immobilizing enzymes, as it simply involves physical adsorption of the enzyme molecules onto the surface of solid matrices. This is achieved by allowing an aqueous solution of the enzyme to come into contact with the carrier by any of the methods such as static procedure, electrodeposition, reactor-loading process and mixing or shaking bath loading. On laboratory scale, the last procedure is preferred in which the carrier is placed in the enzyme solution and mixed with a stirrer. Alternatively, continuous agitation in a shaking water bath may also help in uniform loading of the enzyme. The adsorption in these cases depends on the experimental conditions such as pH, nature of the solvent, ionic strength, quantity of enzyme and adsorbent, time and temperature. The method suffers from the major disadvantage of disorption of the enzyme from the carrier during use owing to the weakness of the involved binding forces, with subsequent loss of catalytic activity and contamination of products.

In ionic binding, immobilization is achieved by ionic binding of the enzyme molecules to solid supports containing ion-exchange residues. The supports mainly used in this method are ion-exchangers prepared most frequently from organic supports. However, in some cases inorganic forms (especially silica) with similar ion-exchange residues have also been used. The most frequently used anion exchangers include DEAE, TEAE-and ECTEOLA-derivatives, while CM-derivatives are the most commonly used cation-exchangers.

The technique of chelation or metal-binding involves the use of transition metal compounds, which activate the surface of the support and allow direct coupling of the enzyme without prior derivatization of the activated support, through formation of chelates. The major supports used include glass, chitin, celite, alginic acid, gelatin and cellulose etc.

In covalent binding, the enzyme is covalently attached to water-insoluble matrices. In this method, covalent bonds are formed, which result into stable immobilized-enzyme preparations from which the enzyme does not leach into solution in the presence of substrate. High ionic concentration preparations are formed in almost all cases. However, use of any reagent which could affect the binding and active sites of the enzyme has to be avoided. Accordingly, covalent attachment must involve only functional groups of the enzyme that are not essential for catalytic action. This is achieved by any of the following means such as (i) covalent attachment of the enzyme in the presence of a competitive inhibitor or substrate, (ii) a reversible covalently linked enzyme-inhibitor complex, (iii) a chemically modified soluble enzyme whose covalent linkage to the matrix is achieved by newly incorporated residues, (iv) a zymogen precursor.

Another method termed as cross linking is based on the formation of covalent bonds between enzyme molecules by means of bi-or multifunctional reagents. As a result, three dimensional crosslinked aggregates are formed which are completely insoluble in water. This is carried out by adding appropriate amount of crosslinking agent to an enzyme solution under conditions that give rise to the formation of multiple covalent bonds. The reagents used for crosslinking generally contain carboxyl groups, diazo groups, isocyanate groups, alkyl iodide groups or iodoacetamide groups.

The methods described above for immobilization involve modification of the enzyme or its microenvironment, which results in alterations in pH and temperature profiles and kinetic characteristics, reducing enzyme activity relative to the corresponding free enzyme. For using an enzyme in its native (soluble) form continuously over a long period, methods of physically confining the enzyme are available which utilize semipermeable membranes, hollow bore fibres or ultrafiltration membranes. Sometimes immobilization of enzymes with greater efficiency can be achieved by a combination of more than one specific method. For example, enzyme immobilized by adsorption could further be cross-linked. Similarly, the efficiency of the entrapment method of immobilization can be improved by use of intramolecular cross-linking reactions.

Though a large number of methods for enzyme immobilization are described here, no particular method can be regarded as the universally applicable for all enzymes. This is mainly due to different chemical characteristics and composition of enzymes, different properties of substrates and products, and the different uses to which the product can be applied. Therefore, for each application it is necessary to find a procedure that is simple and inexpensive to perform and gives a product in large amounts and provides with good retention of activity and high operational stability of the enzyme.

7.8 Laboratory Exercises

This section describes a few experiments to illustrate the general approach employed for optimizing conditions for assaying activity of enzymes, determination of V_{max} and K_m values, procedures for measuring activities of some selected enzymes, use of enzymes for estimation of metabolites and finally immobilization of enzymes.

7.8.1 EXPERIMENT: To study time course of the reaction catalysed by alkaline phosphatase (EC 3.1.3.1)

Principle

Phosphatases is a broad term used for non-specific phosphomonoesterases which hydrolyse organic phosphate esters liberating an alcohol derivative of the substrate and inorganic phosphate (P_i). These enzymes catalyse the following reaction:

$$\text{Orthophosphoric monoester} + H_2O \rightarrow \text{Alcohol} + P_i$$

Depending on their pH optima, phosphatases have been classified into two groups viz. acid phosphatases and alkaline phosphatases. Acid phosphatases function optimally at acidic pH (4.0–5.5) whereas alkaline phosphatases give maximum activity at alkaline pH (8–10). For assaying phosphatases, *p*-nitrophenyl phosphate can be used as substrate which is hydrolysed to *p*-nitrophenol and P_i. *p*-nitrophenol is colourless at acidic or neutral pH but at alkaline pH of 11 it gives yellowish colour with absorbance maxima at 410 nm. Hence the activity of alkaline or acid phosphatases can be conveniently determined colorimetrically by determining the amount of *p*-nitrophenol produced. The reaction catalysed by alkaline phosphatase is:

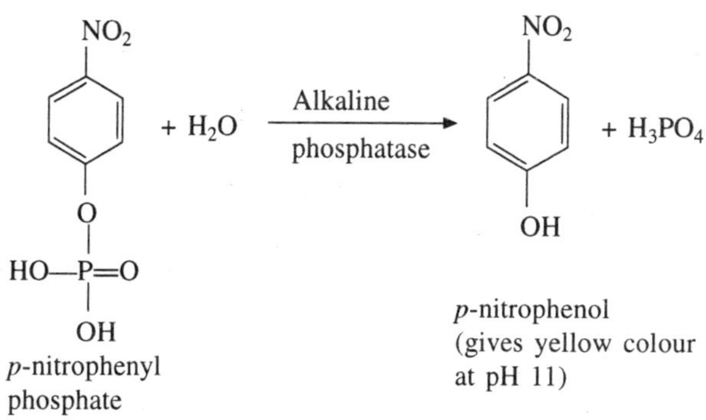

For determination of activity in tissue extract it is important to ensure that the incubation period is such that it is within the range where the activity of the enzyme is proportional to the reaction time.

Materials and Reagents

1. Water bath at 37°C.
2. Colorimeter.
3. Refrigerated centrifuge.
4. Glycine-NaOH buffer (0.05 M, pH 10.5): Dissolve 375 mg glycine in small volume of water, add 42 ml of 0.1N NaOH solution and adjust the pH to 10.5.
5. NaOH solution (0.085N): Prepare by dissolving 340 mg of NaOH in 100 ml of distilled water.
6. $MgCl_2$ solution (10.5 mM): Weigh 10 mg of $MgCl_2$ and dissolve it in 10 ml of distilled water. 0.1 ml of this solution is used in 3.5 ml of reaction mixture so that the final conc. of Mg^{++} during the assay is 0.3 mM.
7. *p*-nitrophenyl phosphate (35 mM): Take 38.8 mg of *p*-nitrophenyl phosphate and dissolve it in 5 ml of 0.05 M glycine-NaOH buffer pH 10.5.
8. Standard solution of *p*-nitrophenol (100 mM): Weigh 69.75 mg of *p*-nitrophenol and dissolve it in 5 ml of distilled water. This solution contains 100 μmoles *p*-nitrophenol/ml.
9. Plant material: Use 5 days old germinating pea seeds which have been germinated at 25°C in Petri plates lined with two layers of Whatman No. 1 filter paper. Ensure that the Petri plates contain sufficient amount of water throughout this period.

Procedure

1. All the operations for preparation of the tissue extract have to be carried out in cold at 0–4°C. Weigh 1 g of the germinating seeds and grind them in chilled pestle and mortar in presence of 10 ml of glycine-NaOH buffer (0.05 M, pH 10.5). A small amount of acid washed river sand may be used as an abrasive to facilitate complete breakage of the cells.
2. Centrifuge the homogenate in a refrigerated centrifuge at 10,000 × *g* for 20 min. Decant the supernatant and use it as the enzyme preparation.
3. Take nine numbered test tubes and add 3.0 ml of glycine-NaOH buffer, 0.1 ml of $MgCl_2$ and 0.3 ml of the enzyme preparation into each of them.
4. Transfer these tubes to a water bath maintained at 37°C. After 3 min start the reaction in seven of the above tubes by adding 0.1 ml *p*-nitrophenyl phosphate. Note down the time of starting the reaction for each tube.
5. Exactly after 5, 10, 15, 20, 25, 30 and 45 min stop the reaction by adding 9.5 ml of 0.085 N NaOH to each tube. In the eighth tube add NaOH followed by 0.1 ml of *p*-nitrophenyl phosphate. This represents 0 min control. In the ninth tube, instead of *p*-nitrophenyl phosphate add 0.1 ml of 0.05 M glycine-NaOH buffer (pH 10.5) and this serves as the reagent blank.

6. Adjust the colorimeter at 410 nm to 100% transmission with the reagent blank (Tube 9) and record absorbance of the other eight tubes.
7. To prepare a standard curve, take 0–1 ml (0–100 μmoles) of *p*-nitrophenol. Add 3 ml of 0.05 M glycine-NaOH buffer (pH 10.5) to all the tubes and make the final volume to 3.5 ml with distilled water. Pipette 9.5 ml of 0.085 N NaOH into each tube, mix the contents and record the absorbance at 410 nm using the tube without *p*-nitrophenol for setting the instrument to 100% transmission or zero absorbance.
8. Plot a graph of A_{410} vs μmoles of *p*-nitrophenol to obtain a standard curve.
9. Determine the amount of μmoles of *p*-nitrophenol in the tubes 1–8 of Step 5.
10. Draw a graph of μmoles of *p*-nitrophenol produced *vs* reaction time.
11. From this graph note down the maximum assay period upto which the production of *p*-nitrophenol is linear. In all the subsequent experiments the duration of reaction should be well within this time limit.
12. Using the experimental data plot another graph of velocity of reaction (i.e. μmoles of *p*-nitrophenol produced/min) *vs* reaction time.

Observations

Do both the graphs (prepared according to Steps 11 and 12 of procedure) look alike? If not, explain the reason for the difference in terms of reaction velocity and reaction time.

7.8.2 EXPERIMENT: To examine the effect of enzyme concentration on the rate of an enzyme catalysed reaction

Principle

Concentration of enzyme in the assay mixture is one of the important factors which determines the velocity of the reaction. Generally, the velocity increases proportionally with increasing concentration of the enzyme, but starts to decline beyond its certain concentration. For obtaining quantitatively reliable values for the enzyme activity in a sample, it is imperative that the amount of the enzyme in the assay mixture be within the range where proportional increase in activity is obtained. Assays performed with enzyme concentration in excessive amounts than this would underestimate the enzyme activity in the experimental material. The optimum concentration of the enzyme which should be used can be determined by measuring the activity at varying concentrations of the enzyme in the reaction mixture. This experiment is designed to study the effect of different amounts of extract on activity of alkaline phosphatase on velocity of the reaction. The principle for assaying the activity of alkaline phosphatase has already been discussed in the preceding experiment (EXPERIMENT 7.8.1).

Materials and Reagents

1. All the equipment and reagents listed under EXPERIMENT 7.8.1.

Procedure

1. Prepare the cell free enzyme extract of germinating pea seeds as described in EXPERIMENT 7.8.1 (Steps 1 and 2).
2. Take two sets of numbered test tubes and pipette 0, 0.05, 0.10, 0.15, 0.20, 0.25, 0.30, 0.35 and 0.40 ml of the enzyme preparation to successive tubes of both the sets. Add appropriate amount of 0.05 M glycine-NaOH buffer (pH 10.5) to each tube so that the combined volume of the extract and the buffer is 3.4 ml. Then pipette 0.1 ml of $MgCl_2$ into each tube.
3. After 3 min incubation of the tubes at 37°C in a water bath, start the reaction by adding 0.1 ml of *p*-nitrophenyl phosphate to each tube of the first set.To the second set of tubes, in place of *p*-nitrophenyl phosphate, pipette 0.1 ml of 0.05 M glycine-NaOH buffer (pH 10.5) and these tubes serve as minus substrate blanks for the corresponding tubes of the first set. Perform the assay for 30 min and then stop the reaction with 9.5 ml of 0.085 N NaOH.
4. Note down the absorbance at 410 nm of all the tubes after setting the colorimeter to 100% transmission using the tube which did not contain the enzyme preparation.
5. Deduct the absorbance values of the minus substrate blanks from their corresponding tubes of the first set of tubes.
6. Using the standard curve determine the amount of *p*-nitrophenol formed in the assay tubes.

Observations

1. Draw a graph of μmoles *p*-nitrophenol produced vs volume of the enzyme extract in the reaction mixture. Determine the highest volume of extract upto which the formation of *p*-nitrophenol is linear.
2. Plot another graph for velocity of reaction (μmoles of *p*-nitrophenol formed/min) vs volume of the enzyme preparation. Note down the maximum volume of extract beyond which the velocity starts to decline.
3. Are the curves obtained in these two graphs identical? If not, can you explain why?

7.8.3 EXPERIMENT: To determine temperature optima for alkaline phosphatase

Principle
All the enzymes have a narrow temperature range for their efficient functioning. The reasons for decline in velocity at temperature beyond optimum temperature have been discussed in Section 7.3.5.4. For determining enzyme activity in the sample, assay should be carried out at optimum temperature. So before undertaking further investigations on an enzyme, temperature optima for its activity ought to be determined in a preliminary experiment. While carrying out any enzyme assay it is essential that appropriate controls must be run simultaneously. For example, for this experiment it would be important to determine whether any

p-nitrophenol is produced due to non-enzymatic hydrolysis of *p*-nitrophenyl phosphate particularly at the higher temperatures. For this zero minute control can be used. The control tubes contain all the components of the reaction mixture except the enzyme preparation and are incubated under identical conditions as the assay tubes. After the stipulated period, the reaction is terminated in all the tubes with the reagent which stops the reaction (0.085 N NaOH in the present case) followed by addition of the enzyme preparation to the control tubes. These controls will provide the requisite information about the extent of non-enzymatic formation of the product under the experimental conditions being employed for assaying the enzyme activity.

Materials and Reagents

1. All the requirements listed in EXPERIMENT 7.8.1.
2. In addition several water baths maintained at the chosen temperatures (say 20, 25, 30, 35, 40, 45 and 50°C) will be needed.

Procedure

1. Prepare the enzyme extract as outlined in EXPERIMENT 7.8.1 (Steps 1 and 2).
2. Take two sets of tubes (one set for zero minute control and the other set for actual enzyme assays) and pipette 3.0 ml of 0.05M glycine-NaOH (pH 10.5) and 0.1 ml of $MgCl_2$ into each tube.
3. Add 0.3 ml of the enzyme extract to the assay tubes. Keep one assay tube and one control tube at the temperature at which the enzyme activity has to be determined (i.e. 20, 25, 30, 35, 40, 45 and 50°C) and after 3 min add 0.1 ml of *p*-nitrophenyl phosphate to all the tubes.
4. Allow the reaction to proceed for 30 min or any other suitable time period which falls within linear region of curve of the graph constructed in EXPERIMENT 7.8.1 and stop the reaction with 9.5 ml of 0.085N NaOH. After this add 0.3 ml of the enzyme preparation to control tubes.
5. Record the absorbance of all the tubes at 410 nm.
6. Deduct the obtained A_{410} for the controls from A_{410} of the corresponding assay tubes and from the standard curve determine the amount of *p*-nitrophenol formed in the assay tubes.
7. Draw a graph of amount of *p*-nitrophenol produced at different temperatures.
8. From the graph determine the optimum temperature for the activity of alkaline phosphatase.

7.8.4 EXPERIMENT: To examine the effect of pH on activity of alkaline phosphatase

Principle

The influence of pH on activity of enzymes has been discussed in detail in Section 7.3.5.3. For measurement of enzyme activity in biological sample, the assays ought to be conducted at its optimum pH. Suitable controls must be included alongwith the enzyme assays to make corrections for any non-enzymatic formation of the product under the particular experimental conditions employed. As in case of the effect of temperature zero minute controls may be used here also. The results obtained in EXPERIMENTS 7.8.1–7.8.3 should be used in selecting the optimal conditions (i.e. reaction time, amount of enzyme and temperature) for assay of alkaline phosphatase from germinating pea seeds.

For determining the pH optima the enzyme activity is measured at varying pH of the reaction mixture. Each buffer has a limited range over which it can effectively fulfill the buffering role. In many instances two different buffers may have to be tried to ascertain effect over a broader range of pH. In such cases, it is advisable to select at least one common pH for both these buffers so that from this overlapping pH the variation in enzyme activity due to nature of the buffer can be distinguished from the actual effect of pH.

Materials and Reagents

1. All the requirements as in EXPERIMENT 7.8.1 except that 0.05 M glycine-NaOH of different pH (8.6, 9.0, 9.6, 10.0, 10.6, 11.0, 11.5) by adjusting the pH to the desired level with increasingly higher volumes of 0.1 M NaOH to 50 ml 0.1 M solution of glycine and then making the final volume to 100 ml with water.
2. 0.05 M Tris-HCl buffer: Prepare 0.05 M Tris-HCl of pH 7.2, 7.6, 8.0, 8.6 and 9.0 as described in Section 1.3.8.

Procedure

1. Prepare the enzyme extract according to the method given in Steps 1 and 2 of EXPERIMENT 7.8.1.
2. Arrange two sets of tubes (one for assay and the other for zero min control). Pipette 3.0 ml of different buffers in succession to control and assay tubes followed by 0.1 ml $MgCl_2$ to all the tubes.
3. Add 0.3 ml of enzyme preparation in the assay tubes and incubate them at 37°C (or more appropriately at the optimum temperature as determined in EXPERIMENT 7.8.3). Start the reaction by pipetting 0.1 ml of *p*-nitrophenyl phosphate to all the tubes.
4. After 30 min, stop the reaction with 9.5 ml of 0.085N NaOH in all the tubes. Now add 0.3 ml of the enzyme extract in the control tubes.
5. Note the absorbance at 410 nm against distilled water.
6. Deduct absorbance of the controls from the corresponding assay tubes and using the standard curve (see EXPERIMENT 7.8.1, Step 7) determine the amount of *p*-nitrophenol formed in the assay tubes.
7. Since the enzyme extract was prepared in a buffer of pH 10.5, the actual pH of the reaction mixture might deviate from that of the assay buffer. Hence prepare a mixture containing extraction buffer and assay buffer in the same ratio as used for measuring enzyme activity (1:10 or 1.5 ml of extraction buffer and 15 ml of assay buffer) and using pH meter determine its pH which will represent the actual pH of the reaction mixture.

8. Plot a graph of µmoles of *p*-nitrophenol produced vs pH of the reaction mixture and find out the optimum pH for activity of the enzyme. Also observe whether at a given pH of Tris-HCl and glycine-NaOH buffers the enzyme activity is identical or different.

7.8.5 EXPERIMENT: To study the effect of substrate concentration on activity of alkaline phosphatase and determine the K_m and V_{max} of the reaction

Principle

The enzyme activity is measured at varying concentrations of the substrate under optimal conditions on the basis of the information obtained in EXPERIMENT 7.8.1–7.8.4. The K_m and V_{max} of the reaction are then determined by processing the data and drawing Lineweaver-Burk (1/v *vs* 1/[S]), Eadie-Hofstee (v vs v/[S]), Hanes ([S]/v vs v) or Eisenthal and Cornish-Bowden (v *vs* [S]) plots (See Section 7.4 for details).

Materials and Reagents

All the requirements as mentioned in EXPERIMENT 7.8.1 except for Reagent 7. For this experiment prepare 5 mM solution by dissolving 38.8 mg *p*-nitrophenyl phosphate in 35 ml of 0.05 M glycine-NaOH buffer (pH 10.5).

Procedure

1. Prepare the enzyme extract from the germinating pea seeds as outlined in EXPERIMENT 7.8.1 (Steps 1 and 2).
2. Take eight numbered test tubes and add 0.0, 0.05, 0.10, 0.15, 0.20, 0.25, 0.30, 0.40 and 0.50 ml of *p*-nitrophenyl phosphate (which corresponds to 0–2.5 µmoles) in successive tubes.
3. Add calculated amount of 0.05 M glycine-NaOH (pH 10.5) so that the total volume of *p*-nitrophenyl phosphate and the buffer in each tube is 3.0 ml. Pipette 0.1 ml of 10 mM $MgCl_2$ to the tubes and keep them in water bath at 37°C.
4. After letting them to warm up for 3 min, start the reaction with 0.3 ml of the enzyme extract in each tube.
5. Stop the reaction after 30 min with 9.5 ml of 0.085 N NaOH.
6. Record the absorbance of the colour formed at 410 nm against distilled water.

Calculations

1. From the standard curve determine the amount of *p*-nitrophenol formed in each tube.
2. Plot a graph of µmoles of *p*-nitrophenol produced/30 min against molar conc of the substrate (*p*-nitrophenyl phosphate) in the assay mixture.
3. Calculate µmoles of *p*-nitrophenol formed/min for each conc of the substrate and replot the results of reaction velocity against molar conc of

the substrate. Are the two plots identical? If not, explain reason for the difference between them.

4. Compute the data in the following forms:
   (i) Reciprocals of velocity (1/v) and molar (in term of mM) conc of the substrate (1/[S]).
   (ii) Ratio of v and [S], i.e. v/[S]
   (iii) Ratio of [S] and v, i.e. [S]/v
5. From the above processed data plot the following graphs (see Section 7.4):
   (i) Lineweaver-Burk plot (1/v vs 1/[S])
   (ii) Hanes plot ([S] vs [S]/v)
   (iii) Eadie-Hofstec plot (v vs v/[S])
   (iv) Eisenthal and Cornish-Bowden plot (v vs [S])
6. From the above plots, calculate the K_m value and V_{max} of this reaction.

7.8.6 EXPERIMENT: Determination of malate dehydrogenase activity (EC 1.1.1.37) in plant tissues

Principle

Malate dehydrogenase (MDH) is an important enzyme of TCA cycle and catalyses the following reaction which is readily reversible.

$$\text{Oxaloacetate} + \text{NADH} + H^+ \underset{}{\overset{Mg^{++}}{\rightleftharpoons}} \text{Malate} + \text{NAD}^+$$

In addition to mitochondria, the enzyme also occurs in cytosol. Another process in which the enzyme is involved is CO_2 assimilation in C_4 plants where it utilizes NADPH rather than NADH as the coenzyme.

MDH is, thus, an oxidoreductase which utilizes the reduced form of nicotinamide adenine dinucleotide (NADH) for reduction of oxaloacetate to L-malate. During this reaction NADH is oxidized to NAD. NADH exhibits absorbance maxima at 340 nm whereas at this wavelength NAD has a negligible absorbance. Hence the decrease in absorbance at 340 nm due to utilization of NADH provides a convenient parameter for following the progress of the reaction catalysed by MDH.

Materials and Reagents

1. Plant material: Freshly harvested leaves of any plant like wheat, pea, rice etc.
2. Acid washed river sand: Suspend the sand in 1 N HCl and keep it for several hours with occasional stirring. Drain out the acid and wash it thoroughly with distilled water till it is completely free of the acid.
3. Refrigerated centrifuge.
4. Spectrophotometer.
5. Extraction buffer: Prepare 50 mM imidazole-HCl buffer containing

10 mM dithiothreitol, 20 mM $MgCl_2$ and 2 mM EDTA (refer to Section 1.3.8 for preparation of imidazole buffer).

6. PVP (polyvinyl pyrrolidone).
7. Assay buffer: Make 50 mM Tris-HCl buffer (pH 8.0) containing 1 mM EDTA.
8. NADH solution: Make 6 mM NADH solution in 50 mM Tris-HCl (pH 8.0). Store this solution in ice.
9. Oxaloacetate solution: Prepare 0.3M solution of oxaloacetate in water and neutralize it by adding solid $NaHCO_3$ till gas bubbles cease to evolve.

Procedure

1. Prepare the enzyme extract by homogenizing 0.5 g of leaves, at 0–4°C, in chilled pestle and mortar using a small amount of acid washed sand as an abrasive in presence of 1% (*w/v*) PVP and 10 ml of the ice-cold extraction buffer which should be added in small amounts at a time.
2. Centrifuge the leaf homogenate at 10,000 × *g* for 20 min in a refrigerated centrifuge at 0–4°C.
3. Decant the supernatant so obtained and use it as the enzyme preparation. Keep this preparation in ice till the enzyme assays are performed.
4. Switch 'ON' the spectrophotometer and after a few min switch 'ON' the UV-lamp and set the wavelength at 340 nm. Allow it to warm up for about 20 min, set to 'zero'. Take 1.5 ml of the assay buffer and 0.1 ml of the enzyme extract in a quartz cuvette (glass cuvette should not be used since it absorbs UV light) and adjust 100% transmission with it.
5. Take out the cuvette and keep it in water bath at 37°C for about 2–3 min. Add 0.05 ml of NADH and mix the contents by inverting the cuvette and place it immediately in spectrophotometer and note the initial reading.
6. Record the decrease in absorbance at intervals of 15 sec upto 3 min. This serves as a blank.
7. Start the reaction by adding 0.05 ml of oxaloacetate solution and note the decrease in absorbance upto 3 min at regular intervals of 15 sec.
8. Prepare a standard curve of NADH over a range of 0–0.3 μmoles. Make sure that total volume of the solution in the cuvette for preparation of the standard curve is the same as during the assay (i.e. 1.7 ml) and this is done by adding calculated volume of assay buffer.

Calculations

The enzyme activity is generally expressed as μmoles of malate formed/g tissue/min and is calculated as follows.

Suppose

Weight of the tissue taken	=	0.5 g
Final volume of enzyme extract	=	10 ml
Enzyme extract taken for assay	=	0.1 ml
Decrease in A_{340} in blank	=	0.01 O.D. units/min
Decrease in A_{340} in sample	=	0.06 O.D. units/min

$\therefore$ Decrease in A_{340} due to MDH activity = 0.06 – 0.01
= 0.05 O.D. units/min

Let it be assumed that from the standard curve, O.D. of 0.12 corresponds to 40 nmoles of NADH.

Then

0.12 O.D. units = 40 nmoles of NADH

0.05 O.D. units = 40/.12 × .05 = 16.66 nmoles of NADH

In this reaction one molecule of NADH is oxidized for conversion of one molecule of oxaloacetate to give one molecule of malate. Therefore, the number of molecules of malate produced will be equal to the number of NADH molecules oxidized.

Hence,

Total enzyme activity as nmoles/min in 10 ml of enzyme preparation = 16.66/0.1 × 10

Total enzyme activity in 1 g of tissue = 16.66/0.1 × 10 × (1.0/0.5)
= 3332 nmoles/min

If one unit of enzyme activity is defined as the amount of enzyme which forms 1 μmole of malate/min, then the tissue contains 3.332 units of MDH activity/g fresh weight.

7.8.7 EXPERIMENT: Measurement of phosphoenol pyruvate (PEP) carboxylase (EC 4.1.1.32) activity in leaves

Principle

PEP carboxylase catalyses the following reaction:

$$\text{PEP} + \text{CO}_2 \xrightarrow[\text{Mg}^{++}]{\text{PEP carboxylase}} \text{Oxaloacetate} + \text{P}_i$$

In C_4 plants, photosynthetic CO_2 is fixed by PEP carboxylase (PEPCase) which has a high affinity for CO_2. Activity of this enzyme can be measured using a coupled enzyme assay with malate dehydrogenase as the coupling enzyme. The later enzyme reduces oxaloacetate to malate using NADH as the reductant which consequently is converted to its oxidized form in this reaction. The rate of formation of oxaloacetate and, hence the activity of PEPCase, can be monitored by recording the change in absorbance at 340 nm. It must be ensured that the coupling enzyme is present in large excess and does not limit the rate of the reaction. The reactions involved in this coupled enzyme system are as follows:

$$\text{PEP} + \text{CO}_2 \xrightarrow{\text{PEP carboxylase}} \text{Oxaloacetate} + \text{P}_i$$

$$\text{Oxaloacetate} + \text{NADH} \xrightarrow{\text{Malate dehydrogenase}} \text{Malate} + \text{NAD}^+$$

Materials and Reagents

1. All the materials and reagents required for activity of malate dehydrogenase as listed in EXPERIMENT 7.8.6 except for the assay buffer (Reagent No. 7).
2. Plant material: Freshly harvested leaves of any species of C_4 plant (maize, sorghum, *Pennisetum glaucum*, sugar cane etc.).
3. Assay mixture: Prepare an assay mixture containing 100 mM HEPES buffer (pH 7.0), 10 mM $MgCl_2$, and 10 mM $MnCl_2$, 20 mM $NaHCO_3$ and 2 units of malate dehydrogenase (1 unit = 1 μmole of NADH oxidized/min).
4. NADH solution: Make 6 mM NADH solution in 50 mM Tris-HCl (pH 8.0). Store this solution in ice.
5. 0.3 M phosphoenol pyruvate (PEP) solution in 100 mM HEPES buffer (pH 7.0).

Procedure

1. Prepare the enzyme extract from the leaves as described in Steps 1-3 of EXPERIMENT 7.8.6.
2. Proceed as given in Step 4 (EXPERIMENT 7.8.6) for setting of the spectrophotometer with a solution containing 1.5 ml of assay mixture and suitable aliquot of the leaf extract (0.1–0.2 ml).
3. Take out the cuvette and warm it up at 37°C for about 2–3 min. Add 0.05 ml NADH and after immediately mixing the contents properly, place the cuvette in spectrophotometer and note down initial reading.
4. Continue to record the decrease in absorbance at intervals of 15 sec upto 3 min. This gives the rate of endogenous NADH oxidation and serves as the blank.
5. Start the reaction with addition of 0.1 ml of PEP and note the decrease in absorbance at 15 sec intervals upto 3 min.
6. Prepare a standard curve for NADH as outlined in Step 8 of EXPERIMENT 7.8.6.

Calculations

In the reaction catalysed by PEPCase one molecule of oxaloacetate is produced for carboxylation of every molecule of PEP. Subsequently, in the coupling reaction the reduction of one molecule of oxaloacetate involves concomitant oxidation of one molecule of NADH. Hence the rate of NADH oxidation in this coupled system corresponds to the rate at which PEP molecules are carboxylated by PEPCase. The method, for calculating the rate of oxidation is given under *Calculations* in EXPERIMENT 7.8.6. Express the activity of PEP carboxylase in terms of μmoles of product formed/g fresh wt/min.

7.8.8 EXPERIMENT: Determination of pyruvate phosphate dikinase (EC 2.7.4.1) activity in plant tissue

Principle

Pyruvate phosphate dikinase catalyzes the following reaction converting pyruvate to phosphoenol pyruvate (PEP) and P_i to PP_i. Since the enzyme brings about phosphorylation of two substrates (i.e. pyruvate to PEP and of P_i to PP_i), it is referred to as a dikinase. Activity of pyruvate phosphate dikinase can be assayed by using an appropriate coupled enzyme system. PEP formed in this reaction can be quantitatively estimated by coupling the reaction to PEP carboxylase and malate dehydrogenase during which NADH undergoes oxidation to NAD^+. The sequence of reactions involved in this coupling system are as under:

$$\text{Pyruvate} + P_i + \text{ATP} \xrightarrow{\text{Pyruvate-Pi dikinase}} \text{PEP} + \text{AMP} + PP_i$$

$$\text{PEP} + HCO_3^- \xrightarrow{\text{PEP carboxylase}} \text{Oxaloacetate}$$

$$\text{Oxaloacetate} + \text{NADH} \xrightarrow[\text{dehydrogenase}]{\text{Malate}} \text{Malate} + NAD^+$$

Since all the three reactions bear stoichiometric relationship of unity with the preceding reaction, the rate of formation of PEP can be determined from the rate of oxidation of NADH.

Materials and Reagents

1. All the equipment and reagents listed in EXPERIMENT 7.8.6.
2. 5 mM ATP solution in the extraction buffer. Store the solution in ice.
3. NADH solution : Prepare 6 mM solution of NADH in 50 mM Tris-HCl (pH 8.0). Store the solution in ice.
4. Assay mixture : Prepare an assay mixture which comprises of 100 mM Tris-HCl buffer (pH 8.0), 10 mM $MgCl_2$, 0.5 mM EDTA, 5 mM dithiothreitol, 20 mM $NaHCO_3$, 1.25 mM pyruvate, 2.5 mM K_2HPO_4 and 2 units each of malate dehydrogenase and PEP carboxylase.

Procedure

1. Extract the enzyme from maize leaves as described in Steps 1-3 of EXPERIMENT 7.8.6.
2. Follow the instructions given in Step 2 (EXPERIMENT 7.8.7) for setting of the spectrophotometer with a solution containing 1.5 ml of the assay mixture and 0.1–0.2 ml suitably diluted aliquot of the leaf extract.
3. Take out the cuvette and incubate it for 2–3 min at 37°C. Pipette 0.05 ml of NADH, mix the contents properly and immediately place it in the spectrophotometer and note the initial reading.
4. Continue to record the change in absorbance after every 15 sec upto 3 min. This represents the substrate blank (minus ATP).
5. Start the reaction by adding 0.1 ml of ATP and again note the change in absorbance upto 3 min at 15 sec intervals.

Calculations

Calculate the rate of NADH oxidation by complete assay mixture after deducting the value for blank (Step 4) as outlined in EXPERIMENT 7.8.6. Express the enzyme activity in terms of μmoles of product formed/g fresh wt/min.

7.8.9 EXPERIMENT: Quantitative estimation of catalase (EC 1.11.1.6) activity in mung beans

Principle

H_2O_2 produced in several reactions in the cell is highly toxic and must be scavenged promptly to avoid injury to metabolic machinery of the tissue. The enzyme catalase decomposes H_2O_2 as:

$$2H_2O_2 \xrightarrow{\text{Catalase}} 2H_2O + O_2$$

The estimation of catalase activity is based on the fact that its substrate, H_2O_2, shows strong absorbance at 240 nm. Hence decomposition of H_2O_2 by catalase is accompanied by decrease in absorbance at this wavelength. This method is, however, suitable for only those enzyme preparations which contain negligible or low amounts of other interfering substances with significant absorbance at 240 nm.

Materials and Reagents

1. Refrigerated centrifuge.
2. Spectrophotometer.
3. Acid washed sand (see EXPERIMENT 7.8.6).
4. Experimental material: About $1–1^1/_2$months old nodulated field grown plants of mung bean.
5. Phosphate buffer (0.067 M, pH 7.0).
6. H_2O_2-phosphate buffer: Dilute 0.16 ml of H_2O_2 to 100 ml with phosphate buffer (Reagent No. 5). After its preparation it should be confirmed that its absorbance at 240 nm is not more than 0.5 O.D. units. This reagent should be prepared just before use.

Procedure

1. Detach the nodules from roots, thoroughly wash them first with tap water and then with distilled water. Dab them dry with several folds of filter paper.
2. Weigh 1 g of the nodules and grind them in chilled pestle and mortar in presence of 10 ml of cold .0067 M phosphate buffer (dilute 10 times the Reagent No. 5 for this step). A small amount of sand may be used during grinding to facilitate complete breakage of the cells.
3. Transfer the homogenate to centrifuge tube and centrifuge at 10,000 × g for 20 min at 4°C. Repeat extraction of the pellet twice with cold buffer and combine the supernatants obtained after each centrifugation.
4. Take 3 ml of phosphate buffer (Reagent No. 5) and 0.01 to 0.04 ml of the

nodule extract into a quartz cuvette. Mix the contents gently by inversion. Use this mixture as blank for adjusting the spectrophotometer to zero absorbance or 100% transmission at 240 nm.

5. For assaying the enzyme activity, pipette 3 ml of H_2O_2-Phosphate buffer solution (Reagent No. 6) and add the same volume of nodule extract as used in the blank (Step 4). Mix the contents and immediately record the absorbance against blank and note the time taken for decrease in O.D. by 0.05 units. This time should not exceed 60 sec otherwise repeat the assay with less diluted enzyme preparation.

Calculations

Let us suppose 1 g tissue was homogenized in total volume of 30 ml. One ml of the extract was diluted to 10 ml with water and 0.01 ml was taken for assay. The absorbance at 240 nm decreased from 0.5 to 0.45 in 14 sec.

$$\frac{17}{14} = 1.21 \text{ units in the assay mixture}$$

$$\frac{1.21}{0.01} \times 10 = 1210 \text{ units/ml of original extract}$$

or

$$1210 \times 30 = 36300 = 3.63 \times 10^4 \text{ units/g tissue}$$

7.8.10 EXPERIMENT: Determination of β-amylase (EC 3.2.1.2) activity in germinating barley seeds

Principle

Amylases hydrolyse starch and glycogen into sugars. There are two types of amylases: α-amylase and β-amylase. α-amylase is an endohydrolase which hydrolyses the α-1 $\rightarrow$ 4 linkage within the chain and breaks these polymers into small fragments. In contrast β-amylase is an exohydrolase and it starts acting from non-reducing end removing two glucose molecules at a time. The products thus formed are a disaccharide, β-maltose, and limit dextrin. The hydrolytic action of β-amylase results in increasing reducing power of the starch solution due to release of a large number of molecules of the reducing sugar viz. maltose. The reducing sugars can be determined quantitatively by their reaction with dinitrosalicylic acid which undergoes reduction to a brown coloured product, nitroaminosalicylic acid with absorption maxima at 570 nm.

Materials and Reagents

1. Water bath at 37°C.
2. Boiling water bath.
3. Refrigerated centrifuge.
4. Experimental material: Use barley seeds which have been kept for germination for 2–3 days at 25°C.

5. Dinitrosalicylic acid: Dissolve 1 g dinitrosalicylic acid, 200 mg crystalline phenol and 50 mg sodium sulphite in 100 ml of 1% NaOH.
6. Sodium potassium tartarate (40%): Take 40 g of sodium-potassium tartarate and dissolve it in small amount of water and make the final volume to 100 ml with water.
7. Starch solution (1%, *w/v*): Weigh 1 g of starch and add 100 ml of boiling water to it. Continue boiling for 2–3 min. Let the solution cool down and then filter it. Use the filtrate as the substrate solution.
8. Sodium phosphate buffer (0.05 M, pH 7.0) containing 0.5 M NaCl.
9. Standard maltose solution (0.01%) : Dissolve 10 mg of maltose in 100 ml of distilled water. This solution contains 100 μg maltose/ml.

Procedure

1. Extraction of the enzyme should be done in cold at 0–4°C. Macerate 500 mg of washed germinating seeds in a chilled pestle and mortar with 5 ml of 0.05 M phosphate buffer (pH 7.0) containing 0.5 M NaCl.
2. Centrifuge the homogenate at 10,000 × g for 15 min in a refrigerated centrifuge at 4°C. Decant the supernatant and use it as the enzyme preparation.
3. Pipette 1 ml of starch solution and 1 ml of an appropriately diluted enzyme preparation into two tubes (control and experimental).
4. In one of the tubes, which serves as a zero min control, terminate the reaction immediately by adding 2 ml of dinitrosalicylic acid reagent.
5. Incubate the other test tube at 37°C for 15 min and then stop the reaction by adding 2 ml dinitrosalicylic acid reagent.
6. Keep both the tubes in a boiling water bath for 5 min and then add 1 ml of sodium potassium tartarate to each of the tubes.
7. Cool the tubes under running tap water. Make the volume to 10 ml with water.
8. Set the colorimeter to zero absorbance at 570 nm with the zero min control and then record the absorbance of the second tube.
9. Prepare a standard curve of maltose over a range of 0–100 μg as described in Steps 5–7.

Calculations

1. From the standard curve find out the amount of maltose in the experimental tube.
2. Calculate the activity of enzyme in terms of µmoles of maltose produced/ g fresh wt/min.

7.8.11 EXPERIMENT: Quantitative estimation of glucose-6-phosphate, glucose-1-phosphate and fructose-1,6-bisphosphate by enzymatic method

Principle

Enzymes have been employed quite extensively for estimation of metabolites in biological samples. Majority of the methods for enzymatic determination of biomolecules are based on use of enzymes or coupled enzyme systems which ultimately result in oxidation of NAD(P)H or reduction of NAD(P). The oxidation or reduction of these nicotinamide adenine dinucleotides can readily be monitored from the change in absorbance at 340 nm.

Determination of glucose-1-phosphate, fructose-6-phosphate and fructose-1,6-bisphosphate involves their conversion to glucose-6-phosphate by auxiliary enzymes and then estimation of glucose-6-phosphate from reduction of NADPH by the enzyme glucose-6-phosphate dehydrogenase. The enzymes involved in estimation of glucose-6-phosphate and conversion of the other above listed metabolites to glucose-6-phosphate are:

(i) $$\text{Glucose-6-phosphate} + \text{NADP}^+ \xrightarrow[\text{dehydrogenase}]{\text{Glucose-6-phosphate}} \text{6-Phosphogluconic acid} + \text{NADPH} + \text{H}^+$$

(ii) $$\text{Glucose-1-phosphate} \xrightarrow{\text{Phosphoglucomutase}} \text{Glucose-6-phosphate}$$

(iii) $$\text{Fructose-6-phosphate} \xrightarrow[\text{isomerase}]{\text{Hexose phosphate}} \text{Glucose-6-phosphate}$$

(iv) $$\text{Fructose-1,6-bisphosphate} \xrightarrow[\text{bisphosphatase}]{\text{C-1 Fructose}} \text{Fructose-6-P} + \text{P}_i$$

$$\text{Fructose-6-P} \xrightarrow[\text{isomerase}]{\text{Hexose phosphate}} \text{Glucose-6-phosphate}$$

These metabolites are determined sequentially. Firstly to the reaction mixture containing the tissue extract and $NADP^+$, glucose-6-phosphate dehydrogenase is added and the amount of NADPH formed gives an estimate of glucose-6-phosphate in the sample. Then phosphoglucomutase is added to the above reaction mixture and enhancement in NADPH production corresponds to the amount of glucose-1-phosphate. This is followed by monitoring the increase in NADPH level in the assay mixture after the addition of hexose phosphate isomerase which provides quantitative information about the presence of fructose-6- phosphate. Finally the difference in the absorbance between this step and on supplementing the reaction mixture with fructose-1,6-bisphosphatase serves as a measure of the amount of fructose-1,6-bisphosphate in the sample preparation. It is evident that to obtain reliable values for these metabolites, the auxiliary enzymes used should be free of other enzymes. For instance, use of phosphoglucomutase containing hexose phosphate isomerase would over estimate glucose-1-phosphate and values obtained for fructose-6-phosphate would be lower than actual content of this metabolite.

Materials and Reagents

1. Refrigerated centrifuge.
2. Spectrophotometer.
3. Water bath.
4. Experimental material: Leaves or any other part of vigorously growing plant such as wheat, pea, lupins etc.
5. Enzymes: Glucose-6-phosphate dehydrogenase, phosphoglucomutase,

hexose phosphate isomerase, C-1 fructose-1,6-bisphosphatase. These enzyme preparations should ideally be free of other auxiliary enzymes. Prepare separate solutions in 0.1 M Tris-HCl (pH 8.5) of these enzymes in such a way that 0.1 ml of the solution contains 1 unit of the enzyme activity.

6. Perchloric acid 20% (*w/v*).
7. 5 M K_2CO_3.
8. 1 M Tris-HCl buffer (pH 8.5).
9. 0.005 M NADP: Prepare in 0.1 M Tris-HCl buffer (pH 8.5).
10. 0.5 M $MgCl_2$ solution.
11. Standard curve for NADPH: Prepare a standard curve of NADPH over a range of 0–0.3 μmoles in final volume of 1.5 ml. Use 0.1 M Tris-HCl (pH 8.5) to dissolve NADPH and also for making up the final volume.

Procedure

1. Weigh 1 g of the plant material and hand homogenize it in cold (0–4°C) in glass pestle and mortar in presence of 5 ml of chilled 20% (*w/v*) $HClO_4$.
2. Centrifuge the tissue homogenate at 7,000 × g for 5 min and decant the supernatant.
3. Carefully titrate the supernatant with 5 M K_2CO_3 to pH 5.5.
4. Remove the precipitated $KClO_4$ by centrifugation for 3 min at 3,000 × g. Use the clear supernatant for the estimation of these metabolites.
5. Prepare assay mixture by pipetting 0.15 ml of 1 M Tris-HCl buffer (pH 8.5), 0.05 ml of 0.005 M NADP, 0.02 ml of 0.5 M $MgCl_2$ and suitable aliquot (0.2-0.5 ml) of the sample and make the final volume to 1.5 ml with water. For blanks, take 5 ml of 20% (*w/v*) $HClO_4$ and process it as described (Steps 3 and 4). Use equivalent volume of this $KClO_4$ solution as the sample extract for the blanks.
6. Note the absorbance of the blank after adjusting the instrument at 340 nm to zero absorbance with 1.5 ml of reaction mixture containing 0.15 ml of 1 M Tris-HCl buffer (pH 8.5), 0.05 ml of 0.005 M NADP and 0.02 ml of 0.5 M $MgCl_2$ and water to make the total volume to 1.5 ml. Let the reading be '*a*' O.D. units.
7. Record the absorbance of the sample containing assay mixture and then add 0.15 ml solution (1.5 units) of glucose-6-phosphate dehydrogenase. Mix the contents properly and note down the absorbance after it has become constant. Let this be '*b*' O.D. units.
8. Remove the cuvette and pipette 0.15 ml solution (1.5 units) of hexosephosphate isomerase. Record the reading after there is no further change in absorbance. Let this be '*c*'.
9. Now add 0.15 ml (1.5 units) of phosphoglucomutase and note the absorbance after it has become steady. Let this be '*d*'.
10. Finally add 0.02 ml of 0.1 M EDTA and 0.15 ml (1.5 units) of C-1 fructose-1,6-bisphosphatase (free of aldolase and transaldolase). Wait till there is no further change in absorbance and then record the reading. Let this be '*e*'.

Calculations

Since in all the reactions, one molecule of NADPH is formed for every molecule of the metabolite, the concentration of these compounds will be equivalent to the amount of NADP reduced.

Suppose, from the standard curve for NADPH,

O.D. of blank (Step 6)	= *a*	nmoles of NADPH
O.D. after addition of glucose-6-phosphate dehydrogenase (Step 7)	= *b*	"
O.D. after addition of hexose-phosphate isomerase	= *c*	"
O.D. after addition of phosphoglucomutase	= *d*	"
O.D. after addition of fructose-1,6-bisphosphatase	= *e*	"
∴ nmoles of glucose-6-P in sample aliquot	$= (b - a)$	
nmoles of fructose-6-phosphate	$= (c - b)$	
nmoles of glucose-1-phosphate	$= (d - c)$	
nmoles of fructose-1,6-bisphosphate	$= (e - d)$	

Calculate the results in form of μmoles of metabolite/g tissue of the sample.

7.8.12 EXPERIMENT: Enzymatic determination of 3-phosphoglyceric acid in the sample

Principle

3-Phosphoglycerate kinase converts 3-phosphoglyceric acid (3-PGA) and ATP to glycerate-1,3-bisphosphate. The enzyme glyceraldehyde-3-phosphate dehydrogenase brings about NADH dependent reduction of glycerate-1,3-bisphosphate to glyceraldehyde-3-phosphate. The amount of NADH utilized is thus a quantitative measure of 3-phosphoglyceric acid. Oxidation of NADH can conveniently be determined from the decrease in absorbance at 340 nm. The enzymic reactions involved in estimation of 3-phosphoglyceric acid are given below:

$$\text{3-Phosphoglyceric acid} + \text{ATP} \xrightarrow[\text{kinase}]{\text{3-Phosphoglycerate}} \text{Glycerate-1,3-bisphosphate} + \text{ADP}$$

$$\text{Glycerate-1,3-bisphosphate} + \text{NADH} \xrightarrow[\text{dehydrogenase}]{\text{Glyceraldehyde-3 -phosphate}} \text{Glyceraldehyde-3-phosphate} + \text{NAD}^+ + \text{P}_i$$

Materials and Reagents

1. Items 1-4, 5, 6 and 10 as listed in EXPERIMENT 7.8.11.
2. Tris-EDTA buffer: Prepare 0.4 M Tris-HCl buffer (pH 7.6) in 0.2 M EDTA.
3. 0.005 M NADH: Prepare in 0.1 M Tris-HCl (pH 7.6).
4. 0.05 M ATP.

5. Enzymes: Prepare solutions, containing 1.5 units of activity per 0.1 ml, of 3-phosphoglycerate kinase and glyceraldehyde-3-phosphate dehydrogenase in 0.1 M Tris-HCl buffer (pH 7.6).
6. Standard curve of NADH: Make a standard curve of NADH over a range of 0–0.3 μmoles in a total volume of 1.5 ml.

Procedure

1. Prepare sample extract as described in Steps 1-4 of EXPERIMENT 7.8.11.
2. Make 1.5 ml of the blank by taking 0.15 ml of Tris-EDTA buffer (pH 7.6), 0.15 ml of ATP solution, 0.03 ml of 0.5 M $MgCl_2$, 0.1 ml of solution of glyceraldehyde-3-phosphate dehydrogenase and suitable aliquot of the sample extract. Make up the volume with water. Using this blank set the instrument to zero absorbance at 340 nm.
3. Now add 0.05 ml of NADH and note the reading.
4. Pipette 0.1 ml of 3-phosphoglycerate kinase, mix the contents properly and immediately place the cuvette in the spectrophotometer and take the reading after there is no further decrease in absorbance.

Calculations

1. Suppose absorbance of the blank (Step 2) $= x$
   Final absorbance of the sample after addition of 3-phosphoglycerate kinase (Step 5) $= y$
   Decrease in absorbance in the sample $= x - y$
2. From the standard curve for NADH find out amount of the coenzyme corresponding to this value of O.D. which would represent the amount of NADH oxidized during the reaction.
3. Since one molecule of NADH is oxidized for each molecule of 3-phosphoglyceric acid, using the value obtained in the preceding step calculate the amount of 3-phosphoglyceric acid/g of the sample.

7.8.13 EXPERIMENT: Estimation of oxaloacetate in the sample using malate dehydrogenase

Principle

Oxaloacetate is a substrate for the enzyme malate dehydrogenase which reduces it to malate utilizing NADH as the reductant as shown in the reaction:

$$\text{Oxaloacetate} + \text{NADH} + \text{H}^+ \xrightarrow[\text{dehydrogenase}]{\text{Malate}} \text{Malate} + \text{NAD}^+$$

As may be observed one molecule of NADH is oxidized to NAD^+ for each molecule of oxaloacetate which is reduced to malate. Hence by measuring the amount of NADH oxidized, oxaloacetate content in the sample aliquot can be determined.

Materials and Reagents

1. All the items at serial nos. 1–4, 5, 6 and 10 under *Materials and Reagents* in EXPERIMENT 7.8.11.
2. 0.15 M Tris-HCl buffer (pH 7.6).
3. 0.39 M EDTA
4. 1.8 mM NADH
5. Malate dehydrogenase: Dilute with 0.1 M Tris-HCl (pH 7.6) so that the final preparation contains 30 units of activity per ml.
6. Standard curve for NADH: Prepare as described in EXPERIMENT 7.8.12.

Procedure

1. Prepare the sample extract as described in EXPERIMENT 7.8.11 (Steps 1-4).
2. Pipette 1.0 ml of 0.15 M Tris-HCl buffer (pH 7.6), 0.05 ml of EDTA solution, 0.05 ml of malate dehydrogenase (1.5 units) and suitable volume of sample extract into a cuvette. Add appropriate amount of water to make the total volume to 1.5 ml. Adjust the instrument to zero absorbance at 340 nm using the above mixture.
3. Add 0.05 ml of NADH solution, and after mixing immediately replace the cuvette in the spectrophotometer and note the initial reading.
4. Let the reaction proceed till there is no further decrease in absorbance. Record the reading.

Calculations

1. Let the initial reading (Step 2) be $= x$ O.D. units
   and the final reading (Step 3) be $= y$ O.D. units
   $\therefore$ Decrease in A_{340} $= x - y$ O.D. units
   $= z$ O.D. units
2. From the standard curve find amount of NADH which corresponds to z O.D. units. This will represent µmoles of NADH oxidized during the reaction.
3. Using the above information, calculate the amount of NADH oxidized/g of sample and this will be equivalent to oxaloacetate content in the sample.

7.8.14 EXPERIMENT: Determination of aspartate content in given sample by enzymatic method

Principle

Aspartate is converted to oxaloacetate by glutamate-oxaloacetate transaminase. Oxaloacetate produced in this reaction is reduced by malate dehydrogenase which is accompanied by oxidation of NADH to NAD. The sequence of these reactions is:

$$\text{Aspartate} + \text{2-oxoglutarate} \xrightarrow[\text{oxaloacetate transaminase}]{\text{Glutamate-}} \text{Oxaloacetate} + \text{Glutamate}$$

$$\text{Oxaloacetate} + \text{NADH} + \text{H}^+ \xrightarrow[\text{dehydrogenase}]{\text{Malate}} \text{Malate} + \text{NAD}^+$$

Utilization of NADH in the second reaction can be monitored quantitatively from the decrease in absorbance at 340 nm and this will eventually be equivalent to the amount of aspartate in the reaction mixture.

Materials and Reagents

1. Equipment and reagents listed at serial nos. 1–4 and 10 under *Materials and Reagents* in EXPERIMENT 7.8.11.
2. 5.7 mM NADH solution.
3. Malate dehydrogenase: Prepare solution of malate dehydrogenase containing 10 units of activity/ml.
4. Glutamate-oxaloacetate transaminase: Make solution having 40 units of activity/ml of this enzyme.
5. 0.1 M 2-oxoglutarate solution.
6. Standard curve for NADH as given in EXPERIMENT 7.8.12 (Reagent No. 6).

Procedure

1. Process the sample for preparation of its extract as given in Steps 1–4 of EXPERIMENT 7.8.11.
2. Prepare a reaction mixture containing 1 ml of 0.15 M tris-HCl buffer (pH 7.6), 0.05 ml of malate dehydrogenase solution, 0.05 ml 2-oxoglutarate and an appropriate amount of the sample extract. Make the final volume to 1.5 ml with water. Adjust the instrument to zero absorbance at 340 nm.
3. Now add 0.05 ml glutamate-oxaloacetate transaminase and 0.05 ml of NADH solution. Mix the contents rapidly and place the cuvette in the spectrophotometer and note the initial reading.
4. Allow the reaction to proceed till the absorbance attains a constant value and record this reading.

Calculations
Calculate the amount of NADH oxidized, which will be equivalent to the quantity of aspartate, as given under *Calculations* in EXPERIMENT 7.8.13.

7.8.15 EXPERIMENT: To immobilize chymotrypsin on agarose gel beads by oxirane method

Principle
Highly strained three membered electrophilic ring systems may be used for the immobilization of enzymes:

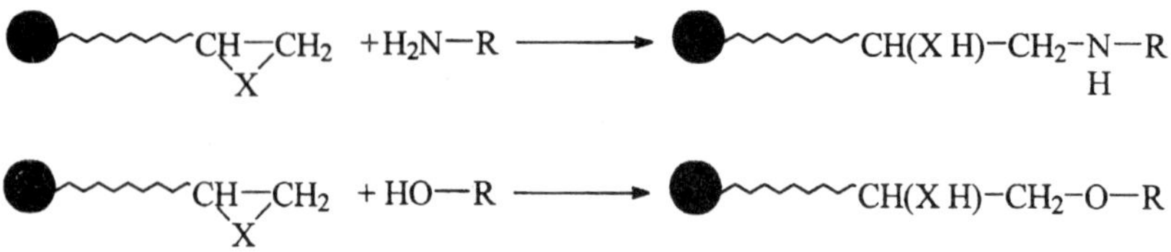

Where X may be NH, O or S. Oxirane groups may be introduced in many hydroxyl polymers by a specific reaction with bisoxiranes, 1-halo-2-hydroxy or 1-hydroxy-2-halo compounds.

The reactivity towards nucleophilic groups in the proteins follows the usual order SH>NH_2>OH. Nucleophilic attack on aliphatic hydroxyl groups takes place at a strongly alkaline pH (~11), while the amino groups react at a lower pH and SH-groups still lower (around neutrality or slightly above). Aromatic hydroxyls, such as those in tyrosyl residues, occupy a position intermediate between aliphatic hydroxyls and amino groups. Guanidino groups and imidazoles can also react.

Materials and Reagents

1. Agarose beads
2. 2.5 M sodium hydroxide
3. Sodium borohydride
4. 1,4-butanediol diglycidyl ether
5. Acetone
6. 0.5 M sodium bicarbonate
7. Chymotrypsin—procure from commercial source
8. 0.1 M sodium borate buffer (pH 8.5)
9. 0.1 M sodium acetate buffer (pH 4.1)
10. 0.01 M sodium acetate buffer (pH 4.1)
11. 0.1 M Tris-HCl buffer (pH 7.8) containing 0.1 M $CaCl_2$
12. N-Benzoyltyrosine ethyl ester (BTEE): Prepare 0.001 M solution of BTEE in 50% methanol by dissolving 15.7 mg of BTEE in 30 ml of methanol (spectral grade) and make the volume to 50 ml with water.

Procedure

1. Take 10 g of washed, suction-dried 2% agarose gel beads.
2. Suspend the beads in 5 ml of 2.5 M sodium hydroxide solution and add 20 mg of sodium borohydride.
3. Add 1 ml of 1,4-butanediol diglycidyl ether drop wise under vigorous stirring. After 6 h at room temperature, wash the gel beads with water on a glass filter until neutrality is reached, then with 50 ml of acetone and finally with water. This step completes the step of introduction of oxirane group into the agarose gel beads. (The content of oxirane function, as determined by sodium thiosulphate titration, is about 100–120 μmoles per g of dry product).
4. For immobilization of chymotrypsin, suspend the oxirane derived gel in 25 ml of 0.5 M sodium bicarbonate solution.

5. Add 50 mg of chymotrypsin and allow the reaction to proceed at ambient temperature for 5 h.
6. Wash the product with the following solutions in the given order:

   0.1 M sodium borate buffer (pH 8.5) containing 1 M NaCl.
   0.1 M sodium acetate buffer (pH 4.1) containing 1 M NaCl and finally with 0.01 M sodium acetate buffer (pH 4.1).

7. Store the conjugate in the last buffer at 4°C. The content of enzyme, as determined by amino acid analysis, is found to be 50 mg of protein per g of dry product.
8. To a cuvette add 1.5 ml of BTEE and 1.5 ml of 0.1 M Tris-HCl buffer containing 0.1 M $CaCl_2$. Adjust the spectrophotometer at zero absorbance at 256 nm with this mixture.
9. In another matched cuvette, take 1.5 ml of BTEE solution, 1.3 ml of 0.1 M Tris-HCl buffer containing 0.1 M $CaCl_2$ and 0.2 ml of the enzyme preparation and record the change in absorbance upto about 5 min.

Calculations

Activity of chymotrypsin is calculated using change in the molar absorbance coefficient at 256 nm which is $\Delta\varepsilon_{256} = 964$.
Use the following formula:

$$\text{Activity (mmoles BTEE hydrolysed/ml of preparation/min)} = \frac{1}{964} \times \frac{\Delta A \times 3}{v}$$

where ΔA = change in absorbance/min
v = ml of enzyme preparation used for assay.

Suggested further reading

Bernhard, S.A. (1968). The Structure and Function of Enzymes, W.A. Benjamin Inc., New York.

Dixon M. and Webb E.C. (1979). Enzymes. Longman Group Ltd., London.

Engel P.C. (1996). Enzymology. Bios Scientific Publishers Ltd., London.

Godfrey T. and West S. (1996). Industrial Enzymology. Macmillan Press Ltd., London.

Lea P.J. (1990). Methods in Plant Biochemistry, Vol. 3, Enzymes of Primary Metabolism. Academic Press, New York.

Price N.C. and Stevens L. (1982). Fundamentals of Enzymology. Oxford University Press, Oxford.

Roberts D.V. (1977). Enzyme Kinetics. Cambridge University Press, London.

Segel I.H. (1975). Enzyme Kinetics: Behaviour and Analysis of Rapid Equilibrium and Steady-state Enzyme Systems. Wiley-Interscience Publication.

Wharton C.W. and Eisenthal R. (1981). Molecular Enzymology. Blackie, Glasgow and London.

Wilsen K. and Walker J. (1996). Practical Biochemistry: Principles and Techniques, 4th ed., Cambridge University Press, London.

Wiseman A. (1995). Handbook of Enzyme Biotechnology, 3rd ed., Ellis Horwood Ltd., UK.

8. Cell Fractionation

8.1 Introduction

Living organisms have been grouped into two main classes: prokaryotes and eukaryotes (*karyon*, a Greek word meaning 'nut' or kernel or nucleus). Eukaryotes are the organisms in which the cellular genetic material is segregated from rest of the protoplasmic melieu by being encased in a prominent double-membraned envelope, the nucleus. Prokaryotes, as the term implies, originated during evolution at a stage prior to acquisition of discrete nuclei and their hereditary material is not enclosed in a limiting membrane. Prokaryotes are thus primitive organisms with relatively simple internal cellular structure. They do not contain nucleus or any other membrane bound compartments in the cell. The only distinguishable structures visualized under an electron microscope are cell wall, plasma membrane, cytosol, nuclear zone (nucleoid), endoplasmic reticulum, ribosomes and storage granules such as granules containing β-hydroxybutyric acid. In contrast, cells of eukaryotic organisms are not only larger in size but have a much more complex internal structural organisation. Their protoplasm contains a variety of well defined membrane bound compartments. Distinct bodies like nucleus, mitochondria, lysosomes, microbodies, golgi bodies, endoplasmic reticulum and ribosomes can readily be identified in the protoplasm of sectioned animal cells under an electron microscope. Such membrane bound inclusions are known as sub-cellular organelles. Plant cells, in addition contain plastids (chloroplasts, proplastids, amyloplasts, etioplasts and chromoplasts) and in place of lysosomes they have prominent vacoules.

A living cell has to carry out a myriad of diverse chemical reactions to fulfil its metabolic needs. Generally, different sets of reactions occur in a sequence leading to synthesis or degradation of specific biomolecules such as various carbohydrates, lipids, amino acids, nucleotides, hormones, alkaloids etc. Such sequences are referred to as metabolic pathways. For example, metabolism of glucose to pyruvate through a sequence of eleven reactions of glycolytic pathway. Pyruvate is converted to acetyl CoA which is then completely oxidised to CO_2 and H_2O via tricarboxylic acid cycle or Krebs cycle which entails a sequence of nine different enzymatic reactions. Similarly, starting with relatively simple compounds viz. phosphoribosyl pyrophosphate, glycine, glutamine and aspartate, the synthesis of AMP is accomplished via a complement of thirteen different enzymic reactions. In a eukaryotic cell each sub-cellular compartment is assigned a well defined biochemical function. For example, in parenchymatous leaf cell, the metabolic task of carrying out photosynthetic reactions is allocated to chloroplasts whereas that of generation of energy via respiratory activity is performed by mitochondria. The endowment of each organelle with a complement of certain set of enzymes results in spatial distribution of specific roles to different sub-cellular compartments. This facilitates towards a more ordered and efficient performance of metabolic activities in eukaryotic cells. The sub-cellular organelles are, however, inter-dependent on each other and intermediates or products generated in a particular compartment may be translocated in a well regulated manner to other parts of the cell. The metabolic activity of a cell is, in fact, the sum total of reactions/pathways operating in different protoplasmic structures. Elucidation of role of each of the sub-cellular structure, therefore, constitutes a very important facet of biochemical investigations for gaining an insight into the relationship between structure and function of a cell and also for unravelling as to how the biochemical events occurring in different parts of the cell are co-ordinated spatially and temporally. One of the most fruitful approaches towards achieving this goal has been to isolate different intracellular structures and then examine the biochemical attributes of each isolated fraction. This chapter deals with the techniques which are generally adopted for fractionation of a eukaryotic cell.

8.2 Cell Fractionation Technique

The procedure for cell fractionation comprises essentially of two distinct steps, viz. breakage of cells followed by separation of different sub-cellular organelles and particles. The important aspects of these two stages of cell fractionation are discussed separately.

8.2.1 Disruption of cells

The first stage in cell fractionation entails breaking of the encasing cell coat or cell wall and the protoplasmic membrane in order to release the intracellular components into the liquid medium. Irrespective of the nature and source of tissue, it should be ensured that the media used is well buffered, isotonic and contains requisite supplementary additives for obtaining active tissue homogenates. Adequately buffered media prevents any large changes in pH due to release of organic acids following disruption of the cell. Isotonic media minimizes the chances of swelling and bursting of the sub-cellular organelles due to osmotic shock. For fractionation purposes, sucrose in the isotonic range (0.15–0.33 M but generally around 0.25 M) in a dilute Tris-HCl or phosphate buffer is used. Quite often, sucrose has been substituted by sorbitol or mannitol in the extraction media. The other supplements include: sulphydryl reagents (dithiothreitol, 2-mercaptoethanol, glutathione or cysteine) for protection of sulphydryl groups of the cellular components against oxidation, EDTA for chelation of heavy metals which might interfere with the biological functioning of the isolated organelles or inhibit activity of the resident enzymes, Mg^{++} (around 10 mM) to maintain integrity of nucleus and ribosomes and also to prevent shrinkage of mitochondria. In the case of plant tissues, it is advisable to include PVP in the media to protect the organelles and enzymes against phenolic compounds.

The choice of method for breaking open the cells depends upon the type of experimental material. Ideally the method employed should be sufficiently forceful to rupture the outer covering of the cells but gentle enough to cause minimum damage to internal structures. Animal cells are bound by a rather thin and delicate cell coat which can be ruptured by relatively mild procedures such as osmotic shock, repeated freezing and thawing or by ultrasonicator. Maceration of finely sliced or minced tissue, e.g. small pieces of liver, kidney, heart, pancreas etc. can also be accomplished with motor-driven (Potter-Elvejham homogenizer) or hand operated (Dounce or Tenbroeck) glass homogenizer using a close fitting glass, leucite or teflon plunger. The tissue, suspended in extraction medium, is taken in the glass tube and the plunger inserted into it. The plunger is rotated (about 600 rpm) by an electric motor for about 0.5–2.0 min. During this process, the static glass tube is rapidly moved up and down (about 20 times). The sample experiences a shearing force and gets disrupted when it comes in the space between walls of the glass tube and fast rotating plunger. In case of hand operated homogenizers, the plunger is forced up and down with rotating motion with hand till the tissue is completely disintegrated. High speed mixters such as Waring blender can also be used and breakage of the cells occurs due to the immense shearing force generated by the fast rotating blades. Unlike the animal cells, the plant and fungal cells are surrounded by a thick and tough cell wall. This necessitates adoption of relatively harsher techniques such as an extended spell of 2–5 sec at maximum speed in Waring blender or homogenization in glass or porcelain pestle and mortar quite often in the presence of sand, alumina or tiny glass beads which act as abrasives. The excessive mechanical force applied does tend to cause substantial damage and breakage of the sub-cellular organelles resulting in rather poor yields. This problem to some extent has been circumvented by using protoplasts rather than intact plant tissues. Protoplasts are prepared by incubating the chopped plant material with cell wall degrading enzymes, commonly referred to as macerase, which is a mixture of hydrolases such as cellulase, hemicellulase and polygalacturonase. The derived protoplasts (cells devoid of cell wall) are then broken by gentler procedures such as sqeezing them through a mesh whose pore size is slightly smaller than the protoplast, forcing them through a hypodermic syringe fitted with a long narrow gauze needle or simply by subjecting them to an osmotic shock by suspending them in a hypotonic medium. The last procedure has been used primarily for the release of intact vacuole which are usually not preserved in other methods.

8.2.2 Isolation of sub-cellular organelles and particles

Once the cell contents are discharged into the extraction media, the separation of various sub-cellular organelles and particles is accomplished by centrifugation in a supporting media which is usually close to isotonic. During centrifugation the particles differing in their size, shape or density sediment at different rates under influence of applied centrifugal force or gravitational force and hence get separated from each other. The salient features of various organelles and particulate entities of a plant cell are presented in Table 8.1.

8.2.2.1 Principles of centrifugation

Before discussing the methods used for separation of sub-cellular components, it will be worthwhile to consider the theoretical aspects of centrifugation. Every particle in suspension (such as in tissue homogenate) has tendency to sediment

Table 8.1 Physical characteristics of sub-cellular organelles and particulate bodies of a typical plant cell

Component	*Shape*	*Size*	*Density (g/ml)*
Nucleus	Spherical	6–8 μm (diameter)	1.32
Chloroplasts	Lens shaped	3–10 μm (long diameter)	<1.24
Proplastids	"	1–1.5 μm (diameter)	1.23
Etioplasts	"		1.26
Mitochondria	Short rods	6 μm*	1.19
Microbodies	Spherical	0.2–1.5 μm	1.25
Rough endoplasmic reticulum	Spherical**	50–150 μm**	1.16
Golgi bodies	Spherical**		1.155
Ribosomes	Roughly spherical		
Cytosolic		26 nm***	
Mitochondrial and chloroplastic	"	22 nm*** "	
Microsomes		50–150 nm (diameter)	

* Distance along the long axis.

** Get fragmented during homogenization and recovered as small vesicles in the microsomal fraction alongwith fragmented vesicles derived from golgi bodies.

*** Largest diameter of the particle.

due to earth's gravitational force. However, the rate of sedimentation of a particle can be accelerated or increased by applying, as during centrifugation, a force greater than the earth's gravitational force. The rate at which a given particle moves downwards is proportional to the centrifugal force (G), which is directed radially outwards and is represented by following relationship:

$$G = \omega^2 r \qquad (1)$$

where G = applied centrifugal force
ω = angular velocity (in radians/sec)
r = radial distance from the axis of rotation (in cm)

Since one revolution of the rotor is equal to the circumference of the circular path, therefore, one revolution is equal to 2π radians. The angular velocity can be expressed in terms of revolutions per min (rpm) which is a common way for denoting the centrifugation speeds. So,

$$\omega = \frac{2\pi \cdot \text{rpm}}{60}$$

$$\omega^2 = \frac{4\pi^2 \cdot (\text{rpm})^2}{3600} \qquad (2)$$

By substituting value of ω^2 from equation (2) in (1)

$$G = \frac{4\pi^2 \cdot (\text{rpm})^2 \cdot r}{3600} \qquad (3)$$

The applied centrifugal force (G) is calculated in terms of multiples of earth's gravitational force (g = 981 cm/sec). This is referred to as Relative Centrifugal Force (RCF) and can be derived from equation (3).

$$RCF = \frac{4\pi^2 \cdot (\text{rpm})^2 \cdot r}{3600 \times 981} \tag{4}$$

Equation (4) on simplification becomes

$$RCF = (1.18 \times 10^{-5}) \cdot (\text{rpm})^2 \cdot r \tag{5}$$

So the RCF (denoted by $\times\, g$ i.e. times the gravitational force) generated during centrifugation in a given rotor of fixed radius at different speeds (rpm) can be readily computed from equation (5). In fact, nomograms have been prepared and are available (Fig. 8.1) from which for the desired g values, one

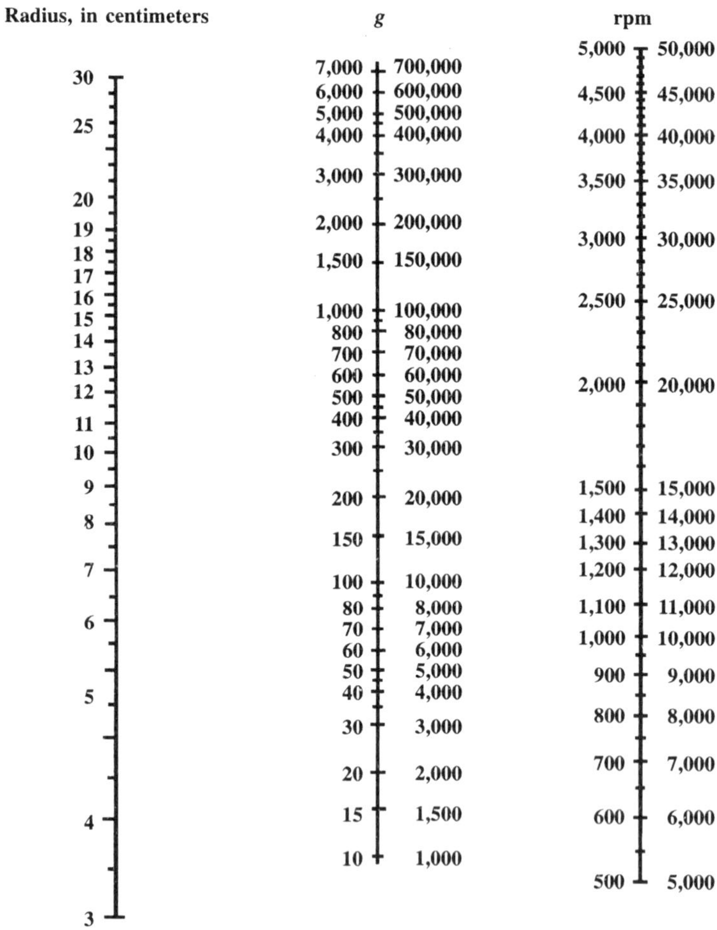

Fig. 8.1 Nomogram for relation between radius of rotor, centrifugation speed (rpm) and Relative Centrifugal Force (g). For finding the value for g keep a ruler at proper position on the left hand bar showing radius and align it with the bar on the right giving values for rpm and the point of intersection on the middle bar will correspond to g.

can directly read out the required centrifugation speed (rpm) for the rotor of a given size.

During centrifugation, the rate of sedimentation of a spherical particle suspended in a liquid medium will depend on its mass (volume × density) and the density and viscosity of the medium. The net outward force (*F*) which it experiences is given by the following equation:

F = Mass of the particle × Difference in density of the particle and media

or
$$F = \frac{4}{3}\pi r_p^3 (\rho_p - \rho_m)\omega^2 r \tag{6}$$

where $\frac{4}{3}\pi r_p^3$ is volume of the spherical particle or radii

ρ_p = density of the particle

ρ_m = density of the medium

r = radial distance (in cm)

ω = angular velocity (radians/sec)

While travelling through the medium, the particle also experiences friction which tends to retard its movement and at its any given velocity, the opposing frictional force will be:

$$F_0 = vf \tag{7}$$

where v = velocity or sedimentation rate of the particle

f = frictional coefficient of the particle in the solution

The magnitude of frictional coefficient of the particle is dependent on its size, shape and hydration and also viscosity of the medium. The frictional force for an unhydrated spherical particle is given by the following relationship:

$$f = 6\pi\eta r_p \tag{8}$$

where f = frictional coefficient

η = viscosity coefficient of the medium

r_p = radius of the particle

Substituting the value of f from equation (8) into (7)

$$F_0 = 6\pi\eta r_p \cdot v \tag{9}$$

An unhydrated spherical particle of fixed volume and density suspended in a medium of uniform density accelerates on application of centrifugal force till the net force of sedimentation equals the frictional force resisting its motion through the medium, i.e.

$$F = F_0$$

Putting the values of F from equation (6) and F_0 from equation (9)

$$\frac{4}{3}\pi r_p^3 (\rho_p - \rho_m)\omega^2 r = 6\pi\eta r_p \tag{10}$$

This balancing of opposing forces is attained quickly and soon the particle sediments at a constant velocity because the frictional resistance increases at higher sedimentation velocity of the particle and under these conditions the net force experienced by particle is zero. The particle does not accelerate any more but achieves a maximum velocity and hence sediments at a constant rate. The rate of sedimentation of particle then is defined by the following equation:

$$\overline{\omega} = \frac{d_r}{d_t} = \frac{2r_p^2 (\rho_p - \rho_m)}{9\eta} \cdot \omega^2 r_p \tag{11}$$

From equation (11) it may be observed that sedimentation rate of a particle is proportional to its size, difference in densities of the particle and the medium and the applied centrifugal field. When $\rho_p = \rho_m$, then $\rho_p - \rho_m = 0$ and under such a situation, its sedimentation velocity also becomes zero or, in other words, it does not move any further. It may also be observed that its sedimentation rate decreases in more viscous media and increases at higher centrifugal force. Equation (11) contains square of particle radius (r_p^2) as one of the terms and it becomes immediately apparent that size of the particle will have a very profound effect on its sedimentation rate. Particles which have almost identical density but differ even slightly in their size sediment at significantly different rates.

The following equation can be derived by integrating equation (11).

$$t = \frac{9\eta}{2\omega^2 r_p^2 (\rho_p - \rho_m)} \ln \frac{r_b}{r_t} \tag{12}$$

where t = sedimentation time in sec
r_t = radial distance from the axis of rotation to liquid miniscus
r_b = radial distance from the axis of rotation to bottom of the tube

This equation relates time for sedimentation of a particle in a centrifuge tube with different variables such as centrifugal force, viscosity of media, particle size, difference in densities of the particles and the media. It is apparent that the mixture containing nearly spherical particle can be separated on the basis of their densities and/or their size by centrifugation of the sample for time required for complete sedimentation of the particle of interest or by the extent of distances moved by different particles after a fixed time. Equations (11) and (12) are of paramount importance and form the basis for separation of sub-cellular organelles by centrifugation.

Two different approaches have been employed for separation of sub-cellular organelles and particulate entities from tissue homogenates. These are:

(i) Differential centrifugation
(ii) Density gradient centrifugation

8.2.2.2 Differential centrifugation

According to this method the preparation of broken cells is poured into a centrifuge tube and is initially centrifuged at low centrifugal force long enough to completely sediment the largest and heaviest sub-cellular component. The supernatant obtained is carefully decanted and it is again centrifuged at a higher centrifugal force for sedimenting the next heavier entity in the extract. This process is continued and, at each ensuing step the centrifugal force as well as the centrifugation time is increased to successively sediment the lighter components and particles. In this method, the separation occurs due to the differential sedimentation rates of the sub-cellular organelles because of the differences in their sizes and densities. As apparent from equation (11), centrifugation will initially sediment the largest particle. Between the particles of the same mass but different density, the ones with highest density (e.g. peroxisomes, ρ = 1.23 g/ml in sucrose solution) will sediment at faster rate than those with lesser density (plasma membranes, ρ = 1.16 g/ml in sucrose solution). For particles having similar density (most sub-cellular organelles, ρ = 1.1 – 1.3 g/ml), effective separation by differential centrifugation can be achieved only if there is about 10 fold difference in their masses. An example of isolation of various sub-cellular components from rat liver cells by differential centrifugation method is shown in Fig. 8.2.

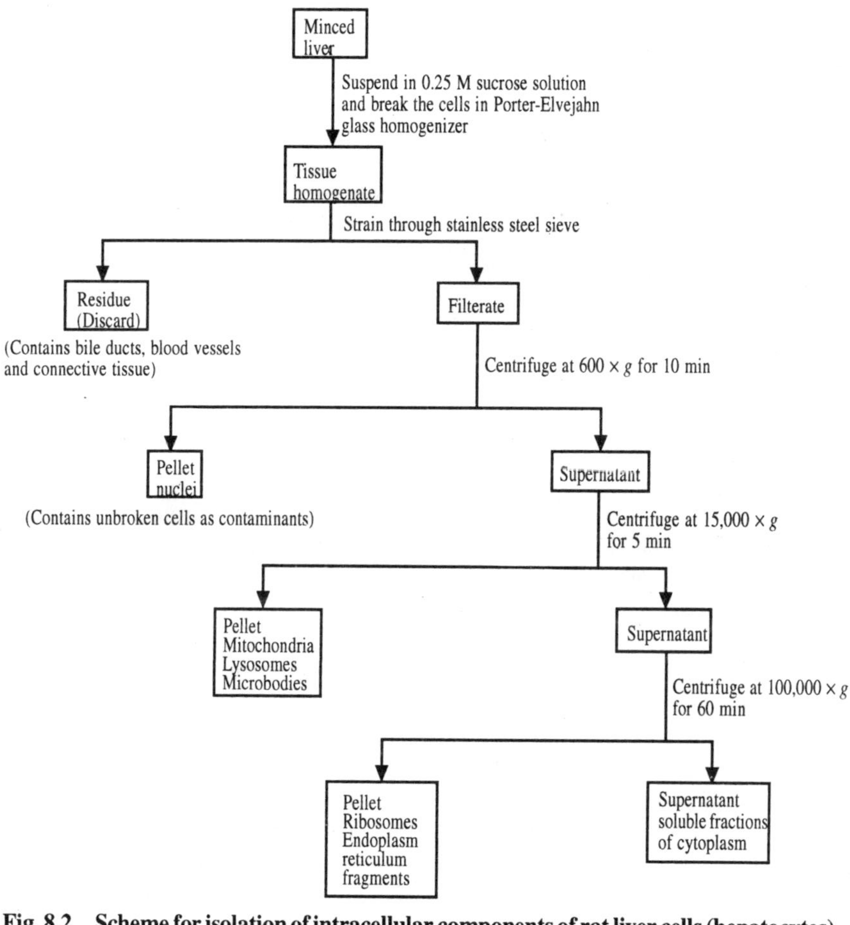

Fig. 8.2 Scheme for isolation of intracellular components of rat liver cells (hepatocytes).

The fractions prepared by differential centrifugation are not pure. Each fraction is enriched with particular component but it inevitably contains substantial amounts of other organelles as well. The principal reason for relatively poor efficiency in obtaining homogenous and pure fractions is that prior to centrifugation all the intracellular particles are evenly dispersed throughout the homogenate in the centrifuge tube. On centrifugation the heaviest particles near bottom of the tube sediment first while longer period is required for the same species of particles which are in the upper region of the centrifuge tube. By the time these get pelleted, lighter particles originally present near the base of the tube also get sedimented within this period. Infact the pellet obtained is enriched with heaviest particles but is likely to contain varying amounts of other particles also. The amount of impurities in the fractions obtained by differential centrifugation technique can be reduced by suspending the pellet in the extraction media and again subjecting it to the same centrifugal force for the same duration. The lighter particles, freed from the sediment and which are towards the top of the tube, will remain in the supernatant. This is repeated twice or thrice and at each step, the quantity of lighter entities diminishes progressively thus reducing the extent of cross-contamination. Alternatively, purification of the pellet may be achieved by subjecting the resuspended preparation of the pellet to density gradient centrifugation. For example, during fractionation of plant cells, the mitochondrial fraction is often contaminated with proplastids and microbodies. The purification of crude mitochondrial preparation may be effected by density gradient centrifugation. The mitochondria-enriched pellet obtained after centrifugation at $10{,}000 \times g$ for 30 min is suspended in 1 M sucrose buffered solution at pH 7.5 and centrifuged at $65{,}000 \times g$ for 3–5 h on a linear sucrose gradient ranging from 30% (*w/w*) to 60% (*w/w*) sucrose at pH 7.5. After centrifugation, mitochondria are recovered as a major band (density = 1.19 g/ml) and microbodies (density = 1.25 g/ml) and proplastids (density = 1.23 g/ml) get separated from mitochondria and also from each other as two distinct bands.

8.2.2.3 Density gradient centrifugation

This method involves layering of the extract of disrupted cells on top of a density gradient with density of the medium progressively increasing from top to bottom of the centrifuge tube. Generally a gradient of 5 to 20 per cent (*w/w*) is used for this purpose. The other materials available for preparation of density gradients include Ficoll (a co-polymer of sucrose and epichlorohydrin), Percoll (a colloidal silica) or non-ionic iodinated aromatic compounds such as metrizamide and nycodenz. The gradients may either be of discontinuous or continuous type. Discontinuous gradients can be prepared by either over-layering or under-layering methods. In the over-layering method a layer of fixed volumes of the most dense media is taken at the bottom of a centrifuge tube followed by gently pipetting a fixed volume of media of decreasing densities on the surface of the preceding layer. In the under-layering method, first fixed volume of the least dense solution of the media is taken at bottom of the centrifuge tube. This is followed by addition of next dense solution below the solution in the tube with a pipette whose tip is positioned at base of the centrifuge tube. This more dense solution displaces upwards the layer of less dense solution already present in the centrifuge tube. Continuous gradients are prepared with a density gradient making device (for a simple gradient maker see Section 9.5.3.3, Fig. 9.7). In these gradients there is a linear and uniform change in density of the media from bottom to top of the tube.

Two different operational modes employed for separation of sub-cellular components on density gradients are:

(i) Rate zonal or band centrifugation
(ii) Isopycnic or equilibrium density centrifugation

(i) Rate zonal or band centrifugation: In this approach the broken cell preparation is applied as a narrow layer on top of the surface of the gradient. The density gradient range used is generally less than that of the particles to be separated. During high speed centrifugation, organelles having different size, shape and densities migrate at different rates as distinct bands through the gradient. Centrifugation is carried out for a period long enough till a good separation of these bands is achieved. After centrifugation, liquid from the tube is collected as fractions of a definite volume from bottom of the tube by piercing a hole at its base. The fractions may also be collected from top of the tube. This is done by inserting a tubing through the gradient in such a way that its mouth lies at the bottom of the centrifuge tube. Then a very dense solution is injected or pumped through this tubing and media originally in the centrifuge tube is pushed upwards by this dense solution. Definite volume of the liquid from top of the tube may be withdrawn with a syringe or may be collected from the outlet tubing of an unloading cone fitted at mouth of the centrifuge tube.

(ii) Isopycnic or equilibrium density centrifugation: In this case also the sample is applied as a thin layer on top of the density gradient and subjected to high centrifugal force for much longer periods till each type of particle occupies a position in the tube where density of the media is equal to density of the particle and hence it comes to rest and does not move further down the gradient. Unlike rate zonal method, isopycnic centrifugation is an equilibrium method in which different types of particles form discrete bands at different zones in the gradient where the density of the media is equal to bouyant density of each of these particles. Thus under carefully controlled conditions quite narrow, distinct and reasonably well separated bands of particles of similar size and shape but differing only slightly in their density can be obtained. After centrifugation the liquid from different zones of the gradient is collected in form of fractions from either top or bottom of the centrifuge tube as described earlier under rate zonal centrifugation.

8.2.3 Non-aqueous technique for isolation of chloroplasts

Fairly purified preparations of chloroplasts can be obtained by non-aqueous method. This technique is particularly useful for studies aimed at examining concentrations of low molecular weight metabolites in chloroplasts and the chloroplastic enzymes also do not leak out. The main shortcomings of the non-aqueous technique are the loss of outer chloroplastic membrane, the preparation is not pure and does contain appreciable levels of mitochondria and microbodies and finally the isolated chloroplasts are not photosynthetically active. The procedure involves freeze drying of the finely chopped leaf material followed by grinding of this material in chilled mixture of cyclohexane and carbon tetrachloride having a density of 1.32 g/ml. The brei is filtered through several layers of muslin cloth to remove the cell debris and unbroken cells. The filtrate is layered above (over-layered) the density gradient prepared from mixtures of cyclohexane and carbon tetrachloride having densities of 1.36 g/ml at the bottom of the tube followed upwards by layers with densities of 1.32, 1.30, 1.29 g/ml. The tubes are subjected to centrifugation at 12,000 × *g* for 15 min in a swinging bucket rotor. Chloroplasts being lighter migrate to surface of the gradient whereas other organelles remain lower down in the gradient. The chloroplast containing band is then removed with a Pasteur pipette.

8.2.4 Marker enzymes

The cell fractionation studies have been extremely useful in examining biochemical attributes of different sub-cellular fractions and in elucidating the relationship between various structural components with functioning of the cell. However, the available fractionation techniques are far from being perfect and do suffer from several inadequacies which must not be overlooked while interpreting the obtained results. The main shortcomings include cross-contamination of the prepared fractions with other sub-cellular organelles, non-specific adsorption of biomolecules and enzymes on the outer surface of isolated organelles and leakage of contents, particularly small water soluble molecules or even enzymes, from damaged and even apparently intact organelles. Quite often a protein, such as bovine serum albumin, is included in the extraction media to suppress non-specific binding of enzymes to organelles. The problem of leakage of water soluble molecules can be minimized by isolating an organelle in a non-aqueous media. This method is not suitable for metabolic studies since the organic solvents cause extensive damage to the membranes, often inactivate the enzymes and destroy the biochemical machinery of isolated organelles.

The extent of presence of other intracellular components is assessed by monitoring the markers in the isolated fractions. Co-sedimentation of different sub-cellular organelles is perhaps inevitable due to several reasons including that particles of the same organelle might be at different stages of development and hence of different sizes and densities which might overlap with those of other organelles. In sub-cellular fraction prepared by differential centrifugation, some other organelles might get entrapped and co-sediment with the main organelle. Two types of markers have been used for determining the purity of isolated fractions: morphological and biochemical. The morphological markers are based on microscopic examination of the separated fractions. Reliable quantification of presence of other organelles as impurities in the fraction of interest is often difficult by this method. However, one of the advantages of this method is that intactness of the isolated sub-cellular organelles can be assessed. Biochemical markers, e.g. measurement of enzyme activities or a specific biomolecule (e.g. DNA for nucleus and chlorophyll for chloroplasts) is a more popular method to evaluate the extent of cross-contamination of sub-cellular fractions. Such enzymes are known as the marker or indicator enzymes. Marker enzymes are uniquely and exclusively located in or associated with a particular intracellular fraction and hence can be used to detect the presence of that specific sub-cellular component. The marker enzymes for various sub-cellular organelles are listed in Table 8.2.

Table 8.2 Commonly used markers for various intracellular components

Sub-cellular fraction or organelle	*Markers*	
	Plant cells	*Animal cells*
Nuclei	DNA NMN adenyltransferase NADP pyrophosphorylase DNA-dependent RNA polymerase I	DNA NMN adenyltransferase NADP pyrophosphorylase DNA-dependent RNA polymerase
Chloroplasts	Ribulose-1,5-bisphosphate carboxylase NADP-triose phosphate dehydrogenase	—
Proplastids	Ribulose-1,5-bisphosphate carboxylase Nitrite reductase Fatty acid synthetase	—
Mitochondria	Cytochrome *c* oxidase Succinate dehydrogenase Succinate: cytochrome *c* reductase Fumarase Antimycin A insensitive NADH-cytochrome *c* reductase	Cytochrome *c* oxidase Succinate dehydrogenase NAD-glutamate dehydrogenase
Microbodies		
Peroxisomes	Catalase D-amino acid oxidase Urate oxidase Hydroxypyruvate reductase	Catalase D-amino acid oxidase 2-hydroxy acid oxidase Urate oxidase
Glyoxysomes	Malate synthetase Isocitrate lyase	—
Vacuoles	Ribonuclease Phosphodiesterase	—
Lysosomes	—	Acid phosphatase Acid deoxyribonuclease β-galactosidase
Endoplasmic reticulum	NADPH-O_2 linked monoxygenase Antimycin insensitive NAD(P)H-cytochrome *c* reductase	Glucose-6-phosphatase NADPH-O_2 linked mixed function oxidases NADPH-linked acyl dehydrogenases
Soluble fraction (cytosol)	Phosphofructokinase	Phosphofructokinase Glucose-6-phosphate dehydrogenase Acetyl CoA carboxylase

From the data on total activity of a particular marker enzyme in the original preparation taken for fractionation and its distribution pattern in different sub-cellular fractions, it is possible to quantitatively determine to what extent are the various fractions contaminated with that particular component. For example, the enzyme catalase is located exclusively in microbodies in plant and animal cells. By monitoring the activity of this marker enzyme in various fractions, such as mitochondrial and chloroplastic, the degree of contamination of each of these fractions with microbodies can be calculated as follows:

$$\text{\% of total microbodies in mitochondrial fraction} = \frac{\text{Catalase activity in mitochondrial fraction}}{\text{Total catalase activity in original preparation}} \times 100$$

In differential centrifugation, the type of sub-cellular organelle recovered in the pellet can be judged from the applied centrifugal force and duration of centrifugation. However, except for chloroplasts which due to chlorophyll have distinct green colouration, all the other sub-cellular particles are colourless. Hence the position of different sub-cellular organelles in a density gradient or the fractions obtained from it can not be determined by mere visual observation. However, by monitoring the activities of marker enzymes for different intracellular components, the predominant sub-cellular organelle in the collected fractions can be established. This is illustrated below in the Fig. 8.3. From the activity profile of the marker enzymes it is apparent that fractions 3–7 contain chloroplast, fractions 11–15 are enriched in mitochondria whereas microbodies are recovered in fractions 16–20. Hence the main purpose of marker enzymes is, firstly to provide the information about the extent of cross-contamination of a prepared sub-cellular fraction with other intracellular components and secondly to facilitate identification of fractions which contain the organelle of interest.

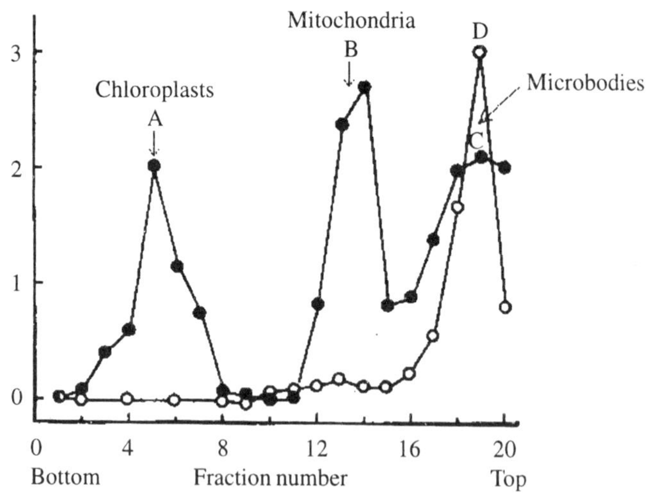

Fig. 8.3 Identification of types of sub-cellular components in different fractions obtained after density gradient centrifugation. Activity peaks of marker enzymes for: Peak A—chloroplasts (NADP-glyceraldehyde-3-phosphate dehydrogenase); Peak B—mitochondria (succinate dehydrogenase); Peak C—microbodies (catalase). Peak D—distribution profile of enzyme under investigation (see Section 8.2.5).

8.2.5 Processing of data and its interpretation

The cell fractionation techniques have been employed for three types of investigations: for studying the morphological and ultrastructural features of an organelle, for examining the biochemical role of each organelle and finally for determination of intracellular location of enzymes. For biochemical studies, purified preparations in fairly large quantities, of the organelle under examination are required. Hence the organelles are separated initially by differential centrifugation which enables obtaining high yields of the intracellular components and, if need be, their further purification is accomplished by density gradient centrifugation. Homogeneity of the preparation is indicated by absence of activities of marker enzymes for the other sub-cellular components.

The degree of purity of isolated fractions can be evaluated as follows: total activity of the marker enzymes as well as protein content of the crude homogenate and also of the different fractions is determined. After this, specific activity (activity/mg protein) of the marker enzyme is calculated. On enrichment of a fraction with a particular organelle, specific activity of the marker enzyme for that organelle increases and the extent of increase in specific activity provides information regarding fold purification of the organelle. For the identification of intracellular site of an enzyme, activities of the enzyme under investigation as well as of marker enzymes in different fractions are determined alongwith the protein content of the individual prepared fractions. For differential centrifugation, the data is usually presented in a tabular form as illustrated in Table 8.3. The sum of enzyme activity and the protein content in different fractions should be more or less equal to the total enzyme activity and protein content in the crude homogenate taken for preparation of the fractions. From the data in Table 8.3 it may be observed that the bulk of activity of the enzyme under investigation is associated with chloroplasts and the specific activity of the enzyme in this organelle is also considerably higher than in the unfractionated crude extract as well as the other

Table 8.3 Distribution pattern of activity of an enzyme in the various fractions obtained by differential centrifugation procedure

Fraction	*Enzyme activity in fractions (units)*	*Activity in fraction as % of total activity*	*Total protein content in fractions (mg)*	*Protein content in fractions as % of total protein*	*Specific activity of enzyme in fractions (units/mg) protein*
Crude homogenate	900.0		1200	—	0.75
Nuclear	20.4	2	60.5	5	0.33
Chloroplastic	690.6	58	580.3	48	1.19
Mitochondrial	163.2	16	302.2	25	0.54
Microbodies	20.0	10	36.2	3	0.55
Microsomal	23.8	4	48.6	4	0.48
Cytosolic	102.0	10	181.1	15	0.56
Total	1020.0	100	1209	100	

fractions. If the marker enzyme for chloroplast also shows a similar quantitative pattern of distribution (i.e. 60% recovery of the total activity in chloroplastic fraction with 1.6 fold specific activity), then the results would strongly suggest that the enzyme under examination is a chloroplastic enzyme.

In case where density gradient centrifugation procedure is used for separation of the sub-cellular organelles, the activities of marker enzymes for individual intracellular components as well as of the enzyme whose location has to be ascertained, are assayed in each fraction and the data presented graphically as in Fig. 8.3 and intracellular site of this enzyme is deduced by examining with which of the marker enzymes does its distribution profile coincide. However, it is not unusual that the enzyme being examined may occur in two different compartments of the cell. In such a situation its distribution profile is likely to qualitatively, but not quantitatively, follow the pattern of the marker enzymes for those compartments.

8.3 Laboratory Exercises

8.3.1 EXPERIMENT: To prepare various sub-cellular fractions of rat liver cells

Principle

Liver is one of the vital organs in an animal body and its function is to process the incoming nutrients from alimentary tract and maintain them at a steady level in the blood. In addition it performs excretory functions by catabolizing the amino acids and the other nitrogen containing compounds via urea cycle and degradation of protoporphyrin IX to metal-devoid open chain tetrapyrroles, the bile pigments, such as bilirubin. It also serves as the principal storage organ for glycogen which may account for about 10% of fresh weight of the liver in normal adult.

Bulk of the liver constitutes of multipurpose soft epithelial cells, the hepatocytes, and it has been used quite extensively for studies on fractionation of animal cells. The method described here is applicable for preparation of sub-cellular organelles from other soft animal tissues as well. The hepatocytes are disrupted by liquid shear technique which involves extrusion of finely chopped tissue suspension through the space between wall of the glass tube and hand operated or motor driven pestle. The brei after being filtered through several layers of cheese cloth is subjected to differential centrifugation (see Section 8.2.2.2) for obtaining various sub-cellular fractions.

Materials and Reagents

1. Experimental animal: A young adult rat
2. Scissors
3. Potter-Elvejhem homogeniser
4. Cheese cloth
5. Wide mouthed cannula (1 mm internal diameter) attached to a syringe
6. Refrigerated high speed centrifuge

7. Pentobarbitone sodium: Prepare a solution containing 6 mg/ml. 1 ml of this solution per 100 g of the rat weight will be required
8. Homogenization media: Make 5 mM Tris-HCl buffer (pH 7.4) containing 0.25 M sucrose to make it iso-osmotic

Procedure

1. Anasthesize the rat by injecting pentobarbitone sodium (6 mg/ml) intraperitoneally using 1 ml for 100 g of the rat weight. Decapitate to kill the rat and let it bleed from neck to drain out the blood.
2. Dissect the rat and open the abdomen with a mid line incision, displace the intestines to right and excise the liver and quickly transfer it to a pre-weighed chilled beaker containing 20 ml of the homogenization media. After weighing the liver, drain out the media and slice it into small pieces with scissors. Add calculated volume of ice cold homogenization media (4 ml/g liver). Preparation of the liver slices and all the subsequent steps should be carried out at 0–4°C.
3. Transfer the liver slices alongwith homogenization media into chilled Potter-Elvejhem homogenizer. Operate the homogenizer at about 500 rpm and push the glass tube up and down 6–8 times to ensure proper breakage of the cells. Filter the homogenate through 3–4 layers of cheese cloth which has been wetted with homogenization media. Stir the homogenate with glass rod to facilitate quicker filtration.
4. Pour the filtrate into 50 ml centrifuge tubes and centrifuge at $1{,}000 \times g$ for 10 min to sediment the heaviest material (P_1). Since the pellet formed is not well packed and is fluffy, remove the supernatant carefully with the help of a wide bore cannula (1 mm internal diameter) attached to a syringe. Centrifuge this supernatant at $3{,}000 \times g$ for 10 min to obtain P_2 which is again a loose packed one. Recover the pellet by withdrawing the supernatant with a syringe as described above.
5. Subject the supernatant from the preceding step to centrifugation at 10,000 $\times g$ for 30 min. Gently recover the pellet and label it as P_3.
6. Finally centrifuge the remaining supernatant from Step 5 at $100{,}000 \times g$ for 40 min to obtain pellet (P_4). Decant the supernatant into a chilled beaker and store it on ice.
7. Suspend the pellets (P_1–P_4) in original volume of the homogenization buffer and store them in cold on ice.

8.3.2 EXPERIMENT: To evaluate the purity of different cellular fractions of rat liver cells prepared by the differential centrifugation method

Principle

The various sub-cellular fractions prepared by differential centrifugation are impure and each fraction contains varying amounts of other intracellular components. Presence of these contaminants can give misleading results regarding intracellular location of an enzyme under investigation if the extent of contamination of each fraction by other components is not taken into account. The purity or homogeneity of the obtained fractions can be assessed by microscopic examination or by monitoring the levels of various biochemical markers. Biochemical markers are the biomolecules or enzymes which are exclusively located in a particular cell fraction. Hence by determining the level of these markers one can obtain a reliable quantitative estimate about the extent of cross-contamination of a given fraction by each of the different types of sub-cellular components. Let us suppose that 30% of the total succinate dehydrogenase activity in crude homogenate is recovered in the nuclear pellet. It would signify that the nuclear fraction contains 30% of the population of mitochondria in the original extract as an impurity. The biochemical markers used for the present experiment are:

Marker	*Sub-cellular fraction*
DNA	Nuclear fraction
Succinate dehydrogenase	Mitochondrial fraction
Aryl sulphatase and acid phosphatase	Lysosomal fraction
Catalase	Peroxisomes
Glucose-6-phosphatase	Endoplasmic reticulum
RNA	Rough endoplasmic reticulum and ribosomes
Pyruvate kinase	Cytosolic fraction

Materials and Reagents

1. All the requirements as listed in EXPERIMENT 8.3.1
2. Water bath
3. Spectrophotometer
4. Spectrofluorimeter
5. Microfuge
6. Reagents for estimation of DNA:

   (i) Diphenylamine reagent: Dissolve 100 g diphenylamine in 100 ml of glacial acetic acid and add 2.75 ml of conc H_2SO_4 to it.
   (ii) Standard DNA solution: Dissolve 500 μg DNA/ml in water.
   (iii) 20% (*w*/*v*) Trichloroacetic acid (TCA).

7. Reagents for assay of succinate dehydrogenase:

   (i) 0.2 M phosphate buffer (pH 7.8).
   (ii) Bovine serum albumin (BSA): Weigh 1 g of BSA, dissolve it in water and make the volume to 100 ml.
   (iii) Succinic acid: Prepare 0.6 M succinic acid solution in water and adjust its pH to 7.8 with NaOH.
   (iv) 2,6-dichlorophenol indophenol solution (DCPIP): Prepare 1.5 mM solution of DCPIP in water.
   (v) Phenazine methosulfate (PMS): Make 9 mM solution of PMS in water.
   (vi) 0.03 M aqueous solution of potassium ferricyanide.

(vii) Potassium cyanide: Prepare 0.03 M solution of potassium cyanide in water (CAUTION: KCN is extremely poisonous and should be handled with care).
(viii) Cytochrome *c*: Weigh 23 mg cytochrome *c* and dissolve it in 20 ml of water to obtain a solution containing 0.009 M of cytochrome *c*.

8. Reagents for catalase:

(i) 50 mM Phosphate buffer (pH 7.0).
(ii) H_2O_2: Take 0.68 ml of 30% H_2O_2 and dilute it to 100 ml with phosphate buffer.

9. Reagents for aryl sulphatase:

(i) 0.1 M sodium acetate buffer (pH 5.6).
(ii) 30 mM lead acetate.
(iii) 5.0 mM 4-methyl umbelliferyl sulfate (4-MUS).
(iv) 0.2 M glycine-carbonate buffer (pH 10.3) containing 1 M EDTA.
(v) 4-methyl umbelliferone: For preparation of standard curve and its concentration range depends on sensitivity of the spectrofluorimeter.

10. Reagents for acid phosphatase:

(i) Substrate solution (3.65 mM): Dissolve 20 mg of *o*-carboxyphenyl phosphate in 25 ml of water.
(ii) 0.15 M sodium acetate buffer (pH 5.0).

11. Reagents for glucose-6-phosphatase:

(i) Assay mixture: Prepare by mixing 0.01 M glucose-6-phosphate (sodium salt), 35 mM histidine buffer (pH 6.5) and 10 mM EDTA in the ratio of 2:5:1.
(ii) 8% (*w*/*v*) Trichloroacetic acid (TCA).
(iii) 2.5% ammonium molybdate in 2.5 M H_2SO_4.
(iv) Reducing solution: 0.5 g of 1-amino-2-naphthol-4-sulfonic acid, 1.0 g anhydrous Na_2SO_3 in 200 ml of 15% $NaHSO_3$.

12. Reagents for estimation of RNA:

(i) Orcinol reagent: Dissolve 0.5 g orcinol and 0.25 g $FeCl_3 \cdot 6H_2O$ in 50 ml conc HCl. This reagent should be prepared fresh and kept at 4°C until use.
(ii) RNA standard solution: Prepare a solution containing 250 μg RNA/ml in water.
(iii) 20% (*w*/*v*) Trichloroacetic acid (TCA) solution.

Procedure

1. Prepare the various sub-cellular fractions as described in EXPERIMENT 8.3.1.
2. Determine the amount of marker molecules or activities of marker enzymes for various sub-cellular components alongwith their protein content in the

various fractions (original unfractionated liver homogenate, P_1–P_4 and S_4) as described in Steps 3–10 that follow.

3. Estimate DNA, the biochemical marker for nucleus, in all the fractions and the crude homogenate. Take a suitable aliquot (about 0.5 ml) of each fraction and remove sucrose by precipitating DNA by adding an equal volume of cold 20% (*w/v*) TCA. Centrifuge at 1,000 × *g* for 20 min and wash the pellet twice with cold 10% TCA to ensure complete removal of sucrose which could interfere with DNA estimation. Add 1 ml of 5% TCA to the pellet and prepare a suspension. At the same time dilute the DNA stock solution to give 0–200 μg DNA in a sample volume of 1 ml in 5% TCA. Keep these standards as well as the fractions in a water bath at 90°C for 20 min to solubilize DNA. To 1 ml samples, add 2 ml of diphenylamine reagent and incubate in water bath at 100°C for 10 min. Allow the tubes to cool down to room temperature and record the optical density at 595 nm using a mixture of 2 ml of diphenylamine reagent and 1 ml of 5% TCA as the blank. Calculate the amount of DNA/ml of the sample.
4. Determine the activity of succinate dehydrogenase in the crude homogenate as well as the various fractions. In a spectrophotometric cuvette take 0.75 ml of phosphate buffer, 0.10 ml potassium cyanide, 0.2 ml succinic acid, 0.10 ml of DCPIP and an appropriate volume of PMS (0–0.4 ml, the activity is dependent on amount of PMS) and add water so that the final volume of the assay mixture is 2.5 ml. Start the reaction with a suitable aliquot of fraction so that change in absorbance is 0.05-0.2 per min. Record change in O.D. at 600 nm against water. Run appropriate blank containing corresponding amount of PMS in the absence of succinate to correct for non-enzymatic reduction of PMS. Calculate the activity in mM of succinate oxidized from the following formula: $0.0476 \times \Delta A_{600}$
5. Assay the activity of catalase, the marker enzyme for peroxisomes, as follows: Take two silica cuvettes (reference and test). In the reference cuvette take 1 ml of phosphate buffer (Reagent 8, i), 0.5 ml of the sample, water to make the total volume to 3 ml and adjust the instrument to zero absorbance at 240 nm with it. In the test cuvette pipette 0.2 ml of the sample, 1.8 ml water and start the reaction by adding 1 ml of H_2O_2 solution (Reagent 8, ii). Mix well and record the decrease in absorbance at 240 nm for 1 min at 15 sec interval. The initial absorbance of this test mixture versus reference should be around 0.82 and the ΔA should not be greater than 0.10 or less than 0.02 per 15 sec. Calculate the amount of H_2O_2 disappearance from the decrease in A_{240} using the following formula:

$$\text{Units/ml} = \frac{\Delta\text{A/min/ml of fraction} \times 1000 \times 3}{43.6}$$

where one unit is equal to 1 μmole of H_2O_2 decomposed per min at 25°C (The molar extinction coefficient of H_2O_2 at 240 nm is 43.6).
6. Aryl sulphatase is one of the marker enzymes for lysosomes. Assay the activity by incubating a mixture of 100 μl acetate buffer, 20 μl lead acetate, 20 μl of 4-MUS, 50 μl water and 10 μl of sample (containing

10 μg protein) at 37°C for 15 min. Terminate the reaction with 1 ml of 0.2 M glycine-carbonate buffer. Measure fluorescence of the solution with an excitation wavelength of 387 nm and emission wavelength of 470 nm. Calibrate the instrument with 4-MU (4-methyl umbelliferone) in glycine-carbonate buffer as a standard. Calculate the enzyme activity in terms of 4-MU produced/min/ml of fraction.

7. Acid phosphatase is another enzyme which can be used as a marker for lysosomes. Prepare the following two reaction mixtures (Reference and test) in two cuvettes: sodium acetate buffer 2.0 ml, 0.3 ml of sample and 0.7 ml water. Adjust the instrument to zero absorbance at 300 nm with this blank. In the test cuvette, instead of 0.7 ml water add 0.5 ml of substrate and 0.3 ml water, mix quickly and record the increase in absorbance at 300 nm due to production of salicylic acid on hydrolysis of *o*-carboxyphenyl phosphate after 1 min intervals. Calculate the activity (μmoles of substrate hydrolysed/min/ml of sample) from the following formula:

$$\text{Enzyme activity} = \frac{\Delta A_{300} \times 1000 \times 3}{3500}$$

where ΔA_{300} is change in absorbance by 1 ml of sample/min.
(Molar extinction coefficient of salicylic acid at 300 nm = 3500)

8. Glucose-6-phosphatase serves as a marker enzyme for endoplasmic reticulum. For determining its activity pipette 0.45 ml of assay solution and 0.05 ml of sample (about 1 mg/ml protein) into a tube and incubate at 37°C for 30 min. For each sample, prepare a blank by omitting substrate and another reagent blank without the sample to correct for non-enzymatic hydrolysis of the substrate. Stop the reaction by adding 2.5 ml of cold 8% (*w*/*v*) TCA and keep the tubes on ice for 20 min. Centrifuge at 700 × *g* for 20 min, take 1 ml aliquot of the supernatant, add 3.65 ml of water followed by 0.25 ml of ammonium molybdate solution and 0.1 ml of $NaHSO_3$ solution. Measure the absorbance after 10 min at 660 nm. Make correction for the blanks and calculate the activity (amount of P_i produced/min/ml of fraction) either in terms of O.D. units or by determining the amount of inorganic phosphate released from a standard curve prepared over a range of 0–1 μmoles P_i using K_2HPO_4.

9. Since bulk of RNA in particulate fraction of cell is associated with rough endoplasmic reticulum and ribosomes, it has been used as a marker for these components. Take in different test tubes 0.5 ml of the fractions and upto 0.5 ml of appropriately diluted solutions of standard RNA to cover a range of 0–250 μg RNA. Make the final volume to 0.5 ml in each tube. Add TCA to a final conc of 5% (*w*/*v*) and incubate for 20 min at 90°C. Centrifuge in microfuge, decant supernatant and add equal volume of orcinol reagent. Keep in boiling water bath for 20 min, let the solutions cool down and measure absorbance at 660 nm. Prepare a blank in the same way as the sample except that take 0.5 ml of water in place of the sample. Calculate the amount of RNA/ml of the sample.

10. Pyruvate kinase is an enzyme of glycolytic pathway and is located in

cytosol. Hence it can be used as an indicator enzyme for the soluble fraction. In a cuvette take 0.2 ml of Tris-HCl buffer, 0.3 ml of $MgSO_4$, 0.1 ml of ADP, 0.1 ml of lactate dehydrogenase. Suitable aliquot of the fraction and water to make the final volume to 3 ml. Adjust the instrument to zero absorbance at 340 nm with this mixture. Now add 0.1 ml of NADH and record the initial A_{340} and then follow the decrease in A_{340} at intervals of 15 sec upto 3 min. This serves as a blank. Start the reaction by adding 0.1 ml of phosphoenol pyruvate solution, mix and immediately start noting the decrease in A_{340} after 15 sec intervals upto 3 min. From the linear part of the curve determine the rate of A_{340} for sample and blank. Deduct the value of blank from that of the sample. Using a molar extinction coefficient value of 6.22×10^3 for NADH, calculate the enzyme activity in terms of μmoles of pyruvate formed (which will be equivalent to μmoles of NADH oxidized)/min. Apply the following formula:

μmoles of pyruvate produced/ml sample/min

$$= \frac{\Delta A_{340}/\text{ml sample/min} \times 3 \times 1000}{6220}$$

11. Finally determine protein content in a suitable aliquots of the various fractions by Lowry's method as described in EXPERIMENT 4.4.2 and calculate the amount of protein per ml of the fraction.

Calculations and interpretations

1. Calculate the total activities and specific activities of various marker enzymes in different fractions as shown in Table 8.3. While calculating the total activity in each fraction remember to take into account of all the dilution factors, e.g. volume of fraction used for assay, total volume of the fraction etc. and identify the main sub-cellular component in each fraction from the activities of various markers.
2. Calculate the extent of cross contamination of each fraction by other sub-cellular components using the following formula, e.g. for calculating proportion mitochondria as impurity in nuclear fraction (P_1)

$$= \frac{\text{Total activity of mitochondrial enzyme in } P_1}{\text{Total activity of the enzyme crude homogenate}} \times 100$$

Note

1. The various fractions obtained are likely to contain the following intracellular components:

P_1 = Predominantly nuclei and unbroken cells. Contaminants plasma membrane sheets and heavy mitochondria.

P_2 = Heavy mitochondria. Likely impurities include plasma membrane fragments.

P_3 = Lysozymes, mitochondria, peroxisomes, golgi membranes.

P_4 = Microsomes.

S_4 = All soluble components (cytosol).

2. The fractions recovered by this procedure are impure and do contain other intracellular components as contaminants. The particulate fractions (P_1–P_4) may be partially purified by removing the more slowly moving impurities. For this, resuspend the pellet in the homogenisation media and repellet it using originally adopted centrifugation conditions (× *g* and time).

8.3.3 EXPERIMENT: To determine intracellular location of glutamate dehydrogenase and glucose-6-phosphate dehydrogenase in liver cells fractionated by differential centrifugation procedure

Principle

Whenever an animal is on a protein rich diet part of the amino acids are deaminated in liver and the organic acids so produced are oxidized via Tricarboxylic acid cycle to generate energy. Glutamate dehydrogenase is a key enzyme in deamination of amino acids in liver and it carries out the following reaction:

$$\text{Glutamate} + \text{NAD} \xrightleftharpoons{\text{Glutamate dehydrogenase}} \alpha\text{-Ketoglutarate} + NH_3 + \text{NADH} + H^+$$

The enzyme catalyses a reversible reaction and its activity can conveniently be determined from the rate of NAD reduction in the forward reaction or oxidation of NADH to NAD in the reverse direction by monitoring the change in absorbance at 340 nm.

About two-thirds of the glucose entering liver is phosphorylated to glucose-6-phosphate by hexokinase and approximately 50% of glucose undergoing degradation in liver is metabolised via oxidative pentose phosphate pathway which provides NADPH for the biosynthesis of fatty acids. Glucose-6-phosphate dehydrogenase is the first enzyme of this pathway and it catalyses the following reaction:

$$\text{Glucose-6-phosphate} + \text{NADP} \xrightleftharpoons{\text{Glucose-6-phosphate dehydrogenase}} \text{6-phosphogluconate} + \text{NADPH} + H^+$$

The activity of this enzyme can readily be determined from the increase in absorbance at 340 nm due to production of reduced NADPH during the reaction.

Materials and Reagents

1. All the requirements listed in EXPERIMENT 8.3.1 for preparation of different sub-cellular fractions from rat liver.
2. All the materials and reagents given in EXPERIMENT 8.3.2 for determining activities of marker enzymes and quantitative estimation of other biochemical markers (DNA and RNA).
3. Reagents for measurement of activity of glutamate dehydrogenase:

    (i) 0.1 M Tris-HCl buffer (pH 7.5)

(ii) 0.6 M sodium glutamate
(iii) 20 mM NAD solution

4. Reagents for assaying activity of glucose-6-phosphate dehydrogenase:

    (i) 50 mM Tris-HCl (pH 7.8) containing 3.3 mM $MgCl_2$
    (ii) 100 mM solution of glucose-6-phosphate
    (iii) 6 mM NADP solution

Procedure

1. Prepare various sub-cellular fractions of the freshly dissected liver from rats as described in EXPERIMENT 8.3.1.
2. Determine the activities of various marker enzymes or other biochemical markers and protein content in all the fractions as outlined in EXPERIMENT 8.3.2.
3. Assay each fraction for activity of glutamate dehydrogenase as follows: Pipette 2 ml of Tris-HCl buffer (pH 7.5), 0.1 ml of sodium glutamate and appropriate volume of the sample and add water to make the total volume to 2.9 ml. Adjust the spectrophotometer to zero absorbance at 340 nm with it. Now add 0.1 ml NAD solution, mix well and start recording the increase in absorbance at 340 nm at 30 sec intervals upto 5 min. Calculate the activity as μmoles of NADH produced/min from the molar extinction coefficient value of NADH as given in Step 10 of EXPERIMENT 8.3.2.
4. For determining the activity of glucose-6-phosphate dehydrogenase take 2.6 ml of 50 mM Tris-HCl buffer (pH 7.8) containing 3.3 mM of $MgCl_2$ in a silica cuvette and add 0.1 ml solution of NADP and 0.3 mM aliquot of the various fractions. Equilibrate contents of the cuvette at 30°C. Using this mixture adjust the absorbance of the spectrophotometer at 340 nm to zero. Follow the change in A_{340} for 2-3 min at intervals of 15 sec. Now start the reaction by adding 0.1 ml of glucose-6-phosphate and again monitor the increase in A_{340} upto 2-3 min after every 15 sec. Calculate the activity in terms of μmoles of NADPH produced/min from the molar extinction coefficient value of 6.22×10^3 for NADPH at 340 nm by applying the same formula as for NADH which is given in Step 10 of EXPERIMENT 8.3.2.

Calculations and interpretation of results

1. Prepare a Table for total enzyme activity and specific activity of various marker enzymes and glutamate dehydrogenase as given in Table 8.3.
2. Examine which of the fractions contains highest activity of glutamate dehydrogenase and whether its specific activity is higher than that of the crude homogenate and other fractions.
3. Compare the distribution pattern of total and specific activity of glutamate dehydrogenase with that of the biochemical marker (enzyme or metabolite)

for the sub-cellular fraction which contains the highest activity of glutamate dehydrogenase.

4. If trend of total and specific activities of glutamate dehydrogenase and the marker enzyme are more or less similar, then glutamate dehydrogenase is likely to be associated with that particular sub-cellular component.
5. Follow the same procedure to identify possible intracellular location of glucose-6-phosphate dehydrogenase.

8.3.4 EXPERIMENT: To purify lysosomes obtained by differential centrifugation of rat liver homogenates by sucrose density gradient centrifugation

Principle

The main advantage of differential centrifugation technique is that various sub-cellular components can be prepared in large quantities. However, as indicated earlier, the method suffers from the drawback that the fractions obtained are impure and do contain other cellular components in appreciable amounts. Presence of these impurities may lead to erroneous inferences regarding the type of biochemical processes or reactions occurring in that particular organelle. One of the methods for further purification of the recovered fractions is to subject them to density gradient centrifugation. This experiment deals with purification of lysosomal pellet so that the investigations carried out with the purified preparation can be interpreted with greater confidence.

Materials and Reagents

1. All the materials listed in EXPERIMENT 8.3.1.
2. Gradient maker.
3. Sucrose solutions: Prepare 0.3, 1.1, 1.4, 1.7 and 2.1 M sucrose solutions in 5 mM Tris-HCl (pH 7.0) containing 1 mM EDTA.

Procedure

1. Prepare two 10 ml linear gradients of 1.1–2.1 M sucrose in 14 ml polycarbonate tubes with help of a gradient maker. Alternatively, make a discontinuous gradient by layering 2.5 ml of 2.1, 1.7, 1.4 and 1.1 M sucrose solution from bottom upwards in the centrifuge. Allow it to stand for 16 h at 4°C and during this period a linear gradient is formed due to diffusion of sucrose.
2. Proceed stepwise as described in EXPERIMENT 8.3.1 till the pellet after centrifugation of the supernatant at 10,000 × *g* for 20 min is obtained. Remove the supernatant alongwith some of the lighter material, with a wide bore metal cannula attached to 20 ml disposable syringe.
3. Carefully add 10 ml of 0.3 M sucrose solution along the side of the centrifuge tube and lightly swirl the contents so that the dark brown coloured region of the pellet at bottom of the tube, which contains lysosomes, is not disturbed.

4. Remove the solution and repeat this step. Add 4 ml of 0.3M sucrose solution, disoldge the brown coloured material with a glass rod or spatula and suspend it by 2–3 strokes of a loose fitting pestle of a homogenizer.
5. Layer 2 ml of this suspension on surface of each of the two gradients and gently stir the sample/gradient junction to reduce discontinuity in density.
6. Centrifuge at 95,000 × *g* for 2–5 h at 5°C in a swing-out rotor.
7. Pierce bottom of the tubes and collect fractions of 0.5 ml each.
8. Determine the protein content and activity of marker enzymes for lysosomes mitochondria, peroxisomes and rough endoplasmic reticulum in each fraction as described in EXPERIMENT 8.3.2.
9. Identify the lysosome containing fractions and calculate the specific activity of its marker enzymes and also calculate the extent of other sub-cellular contaminants in lysosome containing fractions.

Note

Lysosomes form a band at density range of 1.21–1.26 g/ml whereas the band of mitochondria is in the density range of 1.17–1.21 g/ml but this method is not effective in separating lysosomes from peroxisomes because both of them have almost identical densities.

8.3.5 EXPERIMENT: Preparation of protoplasts from pea leaves

Principle

The plant cells are enclosed in a rigid and tough cell wall and consequently high shearing force is required to disrupt them. Usually the cells are broken by homogenization in pestle and mortar or by a short burst of a few secs at full speed in Waring blender. Despite all the precautions the intensity of force applied results in a considerable damage and breakage of the internal sub-cellular components, which in addition to poor yield, poses the risk of contamination by soluble enzymes. This problem is overcome by preparation of protoplasts i.e. cells devoid of cell wall followed by their very gentle disruption by either forcing them through a nylon net (10–20 μm mesh) with pores slightly smaller than size of the proplastids. Alternatively, the intracellular components may be released by subjecting the protoplasts to an osmotic shock and this method is quite commonly used for release of intact vacuoles. Lysed protoplasts generally yield a high proportion of intact sub-cellular organelles. For preparation of protoplasts the plant material is treated with cell wall degrading enzymes which are available commercially by names such as cellulysin, macerase, onozuka RS cellulase. It needs to be emphasized that conditions required to obtain good preparation of protoplasts vary for different plant species and the most appropriate conditions for each species or even organ may have to be standardized.

Materials and Reagents

1. Experimental material: 3 week old pea plants raised under natural conditions in a net house.
2. Crystallization dishes (shallow).

3. Carborendum (powdered, 380 mesh).
4. Forceps.
5. Coarse nylon cloth (pore size 100 μm).
6. Table centrifuge.
7. Graduated glass centrifuge tubes (conical, 15 ml capacity).
8. Pasteur pipette.
9. Solution A: Prepare a solution containing 0.4 M sorbitol, 1 mM potassium phosphate, 5 mM $MgCl_2$ and adjust its pH to 5.5.
10. Solution B: Same as Solution A except add 1.5% cellulysin and 0.5% macerase to it.
11. Solution C: Same as Solution A except adjust its pH to 6.6.
12. Solution D: Make a solution containing 0.5 M sucrose, 5 mM $MgCl_2$, 50 mM Tricine (pH 7.5).
13. Solution E: Prepare solution containing 0.5 M sucrose, 0.1 M sorbitol, 5 mM $MgCl_2$ in 50 mM Tricine (pH 7.5).
14. Solution F: This solution comprises of 0.5 M sorbitol in 50 mM Tricine (pH 7.5) and 5 mM $MgCl_2$.

Procedure

1. Take fully expanded leaves and abrade their lower surface with carborendum powder by gently rubbing it in circular motion till the leaf surface appears shiny dark green in colour. Conversely, the leaf surface may be rubbed very softly with a tooth brush having soft bristles.
2. Weigh about 3 g of abraded leaves and transfer them to solution A.
3. After 5 min take out the leaves and float them, with their abraded surface down, on a sorbitol buffer Solution B containing 1.5% cellulysin and 0.5% macerase in shallow crystallization dishes and keep the dishes at room temperature, without shaking, for 3 h.
4. Then remove the leaves with forceps and place them in cold Solution C at 0°C and very slowly shake the contents to release the protoplasts.
5. Filter the protoplast containing solution through coarse nylon and centrifuge it at 200 × *g* for 2 min to collect the protoplasts.
6. Suspend the harvested protoplasts in minimum volume of Solution D and transfer them to 12 ml graduated conical centrifuge tubes. Then carefully add 2 ml of the Solution E along the side of the tube so that it forms a distinct layer. On the surface of this solution, overlay 2 ml of solution F.
7. Centrifuge at 200 × *g* for 4 min. With help of a Pasteur pipette withdraw the intact protoplasts which move to upper interface of the liquid in the centrifuge tube.

8.3.6 EXPERIMENT: To examine the intracellular distribution of NADP-isocitrate dehydrogenase in pea leaves

Principle

Intracellular distribution of an enzyme in leaves can be determined by assaying its activity in the various isolated sub-cellular organelles. The sub-cellular organelles may be derived from either differential or density gradient centrifugation of leaf homogenates prepared by maceration of leaves or by separation of the organelles from protoplast lysates.

NADP-isocitrate dehydrogenase catalyses the following reaction and its activity can be monitored by following the increase in absorbance at 340 nm.

$$\text{Isocitrate} + \text{NADP} \rightleftharpoons \alpha\text{-Ketoglutrate} + \text{NADPH} + \text{H}^+$$

Activity of NADP-isocitrate dehydrogenase is assayed in the various fractions and at the same time the activities of the following markers are determined in these fractions: Chlorophyll and Phosphoglycolate phosphatase (for chloroplasts); Fumarase (mitochondria), 3-Phosphoglycerate phosphatase (cytosolic). A graph of activities of these markers as well as NADP-isocitrate dehydrogenase *vs* fraction number (from bottom to top of the gradient) is drawn as shown in Fig. 8.3. The graph is examined to ascertain with which of the markers does the activity of NADP-isocitrate dehydrogenase coincides. The overlapping peak with that marker enzyme would indicate the intracellular location of NADP-isocitrate dehydrogenase. If the peaks for this enzyme coincide with two or more markers it could suggest multiple location in more than one cellular compartment.

Materials and Reagents

1. All the materials and reagents listed in EXPERIMENT 8.3.5.
2. Syringe with a nylon net (10-20 μm mesh) firmly fitted at its end with polypropylene tubing.
3. Gradient maker.
4. Preparative ultracentrifuge (Spinco, rotor SW-23).
5. Sucrose solutions: Prepare 60%, 42%, 30%, and 20% (*w*/*w*) solutions of sucrose in 50 mM HEPES buffer (pH 7.4).
6. Acetone.
7. Assay reagents for NADP-isocitrate dehydrogenase:

   (i) 0.1 M Tris-HCl buffer (pH 7.4), 1 mM EDTA, 0.3 mM dithiothreitol.
   (ii) 20 mM $MnSO_4$
   (iii) 1.5 mM NADP
   (iv) 80 mM isocitrate

8. Assay reagents for 3-phosphoglycerate phosphatase:

   (i) 0.1 M solution of 3-phosphoglycerate
   (ii) 0.4 M sodium cacodylate buffer (pH 5.9)
   (iii) 10% (*w*/*v*) Trichloroacetic acid (TCA)
   (iv) Reagents for estimation of phosphate as listed in EXPERIMENT 6.3.2.4 (Reagents 1, 2 and 5)

9. Assay reagents for phosphoglycolate phosphatase:

   (i) 0.1 M solution of phosphoglycolate
   (ii) 0.4 M sodium cacodylate buffer (pH 6.3)
   (iii) 75 mM $MgCl_2$ solution

(iv) Reagents for estimation of phosphate as indicated earlier for 3-phosphoglycerate phosphatase

10. Assay reagents for succinate dehydrogenase: As in EXPERIMENT 8.3.2.
11. Assay reagents for catalase: As listed in EXPERIMENT 8.3.2.
12. Assay reagents for pyruvate kinase: As given in EXPERIMENT 8.3.2.

Procedure

1. Prepare protoplasts from mature fully expanded pea leaves as described in EXPERIMENT 8.3.5.
2. Carry out all the operations at 0–4°C. Lyse the protoplasts by extruding them thrice through a syringe with a nylon net (10–20 μm mesh) fitted firmly at its end.
3. Using a gradient maker, prepare a gradient consisting of 10 ml of 60 to 42% (*w*/*w*) sucrose, 4 ml of 42% (*w*/*w*) sucrose, 14 ml of 42 to 30% (*w*/*w*) sucrose and finally 2 ml of 30% (*w*/*w*) sucrose at the top.
4. Layer the extract of broken protoplasts on surface of the gradient and centrifuge in preparative ultracentrifuge (using SW 23 rotor) at 4,000 rpm for 5 min and then 10,000 rpm for 10 min to separate the organelles by rate zonal centrifugation. Allow the rotor to decelerate to 5000 rpm before turning on the brakes.
5. Remove the tubes handling them very gently. Pierce a hole at the bottom of the centrifuge tube and collect fractions of 1.5 ml each. Store these fractions in cold at 0–4°C.
6. For measuring activity of NADP-isocitrate dehydrogenase, pipette in a cuvette 1 ml of 0.1 M Tris-HCl containing 1 mM EDTA, and 0.3 mM dithiothreitol, 0.2 ml of 20 mM $MnSO_4$, 0.2 ml of 1.5 mM NADP, suitable aliquot of fraction (0.1 ml) and 1.45 ml of water. Set the spectrophotometer to zero absorbance at 340 nm with the above solution. Pre-incubate this mixture at 25°C for 5 min. Start the reaction by adding 0.05 ml of 80 mM isocitrate solution and note the increase in A_{340} after every 15 sec upto 3–5 min. Calculate activity of the enzyme (μmoles of NADP reduced/ml fraction/min) using the following formula:

$$\text{Activity} = \frac{A_{340} \times 3}{6.22 \times v}$$

where A_{340} is change in absorbance/min
v = volume of fraction used for assay

7. Assay the activity of 3-phosphoglycerate phosphatase from the amount of phosphate produced as follows: Take the required number of test tubes and add 0.25 ml of 0.4 M sodium cacodylate buffer (pH 5.9), 0.1 ml of 0.1M 3-phosphoglycerate and appropriate amount of water so that the final volume of the assay mixture is 0.75 ml after the addition of the sample preparation (fraction). Incubate the tubes at 30°C for 2 min and start the reaction with the fraction aliquot. After 15 min, stop the reaction with 0.25 ml of 10% TCA. In zero min blanks, add TCA before the

fraction aliquot. Centrifuge at 3000 rpm for 20 min to remove the precipitate. Take 0.5 ml of the aliquot and add 3 ml of water to it. Determine the amount of phosphate by pipetting 1 ml of sulphomolybdate reagent followed by 0.5 ml of freshly prepared stannous chloride. Record the absorbance of 660 nm within 20 min. Prepare a standard curve of phosphate using K_2HPO_4 over a range of 0–1.0 μmole, as described earlier.

8. Assay the activity of phosphoglycolate phosphatase by following the procedure as described in Step 7, except that use sodium cacodylate buffer of pH 6.3, include 0.1 ml of 75 mM $MgCl_2$ and replace 3-phosphoglycerate with 0.1 ml of 0.1 M phosphoglycolate in the reaction mixture. The final volume of the assay mixture should be 0.75 ml.
9. Measure the activities of succinate dehydrogenase, catalase and pyruvate kinase as described in Steps 4, 5 and 10 of EXPERIMENT 8.3.2, respectively.
10. For determination of total chlorophyll content, take 0.5 ml of the fraction and add 2.0 ml of acetone to it. Shake well and note the absorbance at 645 and 663 nm. Calculate the amount of total chlorophyll as follows:

$$\mu\text{g chlorophyll (ml of acetone extract)} = 20.2\ A_{645} + 8.02\ A_{663}$$

11. Prepare graph by plotting activities of various markers vs fraction number (from bottom towards top of the centrifuge tube) as well as the activity of NADP-isocitrate dehydrogenase and from it suggest the likely intracellular site of this enzyme.

8.3.7 EXPERIMENT: To fractionate leaf cell by differential centrifugation technique

Principle

The general principle for isolation of sub-cellular components by differential centrifugation has been enumerated in EXPERIMENT 8.3.1. However, in case of plant cells, the fractionation by differential centrifugation may be done either by using tissue homogenates or with the extracts of lysed protoplasts as the starting material. Another notable difference is the presence of chloroplasts in leaf cells whereas animal cell do not contain this organelle. Hence leaf cells can be processed to give five distinct sub-cellular fractions viz. nuclei, chloroplasts, mitochondria, peroxisomes, microsomes and an impure soluble cytosolic fraction which is quite heavily contaminated with material released from the other broken sub-cellular organelles. As mentioned earlier, generally the fractions isolated by differential centrifugation are not pure but the contaminants can be removed by repeated recentrifugation of the suspended pellet or by density gradient centrifugation. This experiment describes the technique for fractionation of leaf extracts by differential centrifugation. It is relevant to point out that there is a considerable variation in the composition of homogenization media used by different workers and there is no ideal or universal extraction media for obtaining most active preparations of sub-cellular organelles.

Materials and Reagents

1. Experimental material: Freshly harvested young leaves of spinach or pea
2. Pestle and mortar (or Waring blender/high speed mixer)
3. Muslin cloth
4. Razor
5. High speed refrigerated centrifuge
6. Homogenization media: Prepare 0.05M HEPES buffer (pH 7.8) containing 0.33M sorbitol, 10 mM cysteine, 5 mM $MgCl_2$, 2% dextran and 1% bovine serum albumin
7. Reagents for determining activities of marker enzymes: All reagents listed at Serial Nos. 8-12 in EXPERIMENT 8.3.6
8. Reagents for measuring activity of NADP-glyceraldehyde-3-phosphate dehydrogenase:

   (i) 0.1 M Tris-HCl (pH 8.5)
   (ii) 0.5 M sodium arsenate
   (iii) 0.6 M sodium fluoride
   (iv) 4.8 mM NADP
   (v) 60 mM EDTA
   (vi) 60 mM 2-mercaptoethanol

Procedure

1. Take the freshly harvested leaves and slice them into small pieces with a sharp razor.
2. Unless otherwise indicated, carry out all the operations in cold at 0–4°C. Weigh 5 g of the leaf material and macerate it in presence of 20 ml of the homogenization media in chilled pestle and mortar, applying only moderate pressure, for 2–3 min. Conversely they may be homogenized in Waring blender for a short burst of 2 sec at full speed.
3. Filter the leaf homogenate through 3–4 layers of muslin cloth. Transfer the filtrate into centrifuge tubes and centrifuge at 600 × *g* for 5 min. Carefully decant the supernatant and retain the pellet which contains nuclei alongwith cell debris.
4. Centrifuge the supernatant from the preceding step (Step 3) at 2,500 × *g* for 3 min to sediment the chloroplasts. Gently pour out the supernatant and subject it to centrifugation at 10,000 × *g* for 10 min. Mitochondria alongwith microbodies are recovered in the pellet.
5. Take the supernatant from Step 4 and centrifuge at 1,50,000 × *g* for 1 h. The soluble components remain in the supernatant and the pellet contains microsomes and ribosomes.
6. Suspend the pellets from each fraction in the homogenization media. For removing the impurities, recentrifuge the pellets employing the same centrifugation conditions under which they were originally obtained. Repeat this twice or thrice and finally suspend the pellets in media lacking sorbitol.
7. Assay each fraction for activities of marker enzymes for mitochondria, microbodies and cytosol as described in EXPERIMENT 8.3.2. Determine

chlorophyll according to the procedure given in EXPERIMENT 8.3.6 as a marker for chloroplasts. In addition, measure the activity of NADP-glyceraldehyde-3-phosphate dehydrogenase, which is a marker enzyme for chloroplasts, as described in the next step.

8. Prepare assay mixtures in two cuvettes, the reference cuvette and sample cuvette. To each cuvette add 1 ml of 0.1 M Tris-HCl (pH 8.5) and 0.1 ml each sodium arsenate, sodium fluoride, EDTA, 2-mercaptoethanol and NADP solutions. Pipette 0.2 ml of the fraction and 0.2 ml of water. Incubate the cuvettes at 25°C for 7 min. Then add 0.1 ml of 0.1 M Tris-HCl buffer (pH 8.5) to the reference cuvette and use it to set the spectrophotometer to zero at 340 nm. Start the reaction in the sample cuvette with 0.1 ml of glyceraldehyde-3-phosphate solution and note the change in absorbance at 15 sec intervals upto 4–5 min. Calculate the enzyme activity in terms of μmoles NADP reduced/ml extract/min using the following formula:

$$\text{Enzyme activity} = \frac{\Delta A_{340} \times 3}{6.22 \times v}$$

where ΔA_{340} = change in absorbance/min
v = volume of the fraction used for assay

9. Determine the activity of 3-phosphoglycerate phosphatase in the various fractions according to the procedure described in EXPERIMENT 8.3.6 (Step 7).

Calculations and interpretation

1. Present the data for the individual marker enzymes as well as for 3-phosphoglycerate as suggested in EXPERIMENT 8.3.3.
2. From the results obtained can any conclusion be drawn about intracellular location of 3-phosphoglycerate phosphatase in the plant material studied by you?

8.3.8 EXPERIMENT: To isolate mitochondria from rat liver

Principle

Availability of intact, purified and metabolically active preparations in fairly large amounts is one of the pre-requisites for the investigations aimed at studying the metabolic attributes of a sub-cellular organelle. The methods described earlier (EXPERIMENTS 8.3.1 and 8.3.7) are not suitable since the quantity of each sub-cellular fraction recovered is low. Specific methods have been developed to selectively obtain a particular organelle in purified and large amounts. This experiment describes such a method for isolation of mitochondria from rat liver and involves centrifugation of liver homogenate directly on a discontinuous gradient of 19, 31, 42 and 52% (v/v) Percoll. The mitochondrial band around the 42%/52% interface, which contains nearly 60% of mitochondria, is quite pure though it may contain lysosomes and peroxisomes as minor impurities.

Materials and Reagents

1. Experimental animals: As in EXPERIMENT 8.3.1.
2. Scissors.
3. High speed refrigerated centrifuge.
4. Homogenization media: Prepare 10 mM HEPES-NaOH (pH 7.4) containing 210 mM mannitol, 60 mM sucrose and 10 mM KCl.
5. Percoll solutions: Dilute Percoll in homogenization media to obtain 19, 31, 42 and 52% (*v*/*v*) solutions of Percoll.

Procedure

1. Dissect the liver from rats as described in EXPERIMENT 8.3.1 and prepare liver homogenate in the homogenization media (Reagent No. 4) according to the method given in EXPERIMENT 8.3.1 (Steps 1–4).
2. Prepare a gradient by pipetting 3 ml of 52% Percoll at bottom of a 15 ml centrifuge tube followed by layering on top of this 2 ml each of 42% followed by 31% and finally 19% (*v*/*v*) Percoll.
3. Carefully layer 2 ml of the liver homogenate on surface of the gradient. Place the centrifuge tube in a fixed angle rotor and centrifuge at 5°C at 37,000 × *g* for 30 sec with an acceleration time of 95 sec and a braking time of 180 sec.
4. Bands are formed at all the four interfaces alongwith a pellet. Using Pasteur pipette withdraw the band at the interface of 42%/52% Percoll in which about 60% of the mitochondria originally present in the crude homogenate are recovered.

8.3.9 EXPERIMENT: To isolate mitochondria from plant leaves

Principle

As in the case of rat liver (EXPERIMENT 8.3.8), intact and active mitochondria can be prepared in fairly large amounts by discontinuous density gradient centrifugation on Percoll gradient. The integrity and intactness of mitochondrial preparation can be evaluated from degree of respiratory control and P/O ratio, which should be close to the theoretical value of 3. The intactness of outer mitochondrial membrane may be assessed by measuring the rate of succinate: cytochrome *c* reductase activity before and after an osmotic shock with 0.3 M sucrose. The undamaged mitochondria exhibit very tight respiratory control and unlike broken mitochondria, do not respire in absence of exogenously added ADP. Similarly, defective mitochondrial preparations give very low P/O ratios due to uncoupling of oxidative phosphorylation. Mitochondria with intact outer membrane show negligible succinate: cytochrome *c* reductase activity. Hence these criteria can be employed for judging quality of the mitochondrial preparation.

Materials and Reagents

1. Experimental material: Freshly harvested shoots of 5 days old maize seedlings.

2. Pestle and mortar.
3. Muslin cloth.
4. Spectrophotometer.
5. High speed refrigerated centrifuge (Beckmann L5-75 Ultracentrifuge).
6. Clark type oxygen electrode.
7. Extraction buffer: Prepare 0.1 M HEPES-KOH buffer (pH 7.5) containing 0.4 M mannitol, 1 mM EDTA, 0.1% (*w*/*v*) bovine serum albumin, 0.6% (*w*/*v*) insoluble polyvinylpyrrolidone (PVP).
8. Suspending buffer: Prepare 20 mM HEPES-KOH buffer (pH 7.5) containing 0.3 M mannitol, 0.1% (*w*/*v*) fat-free bovine serum albumin.
9. Percoll: Dilute stock solution of Percoll to obtain 62, 45, 28 and 5% (*v*/*v*) Percoll solutions each containing 0.25 M sucrose, 20 mM HEPES-KOH (pH 7.5) and 0.1% (*w*/*v*) fat-free bovine serum albumin.
10. Assay mixture for measurement of oxygen consumption: Make 10 mM sodium phosphate buffer (pH 7.2) which contains 0.3M mannitol, 5 mM $MgCl_2$, 10 mM KCl, 0.1% fat-free bovine serum albumin.
11. Succinate: Prepare 0.3 M solution of succinate and adjust its pH to 7.2.
12. ADP: Make 0.3 mM solution of ADP in water.
13. 1.2 M sucrose solution in 20 mM phosphate buffer (pH 7.2).
14. 20 mM phosphate buffer (pH 7.2).
15. 1.5 mM Cytchrome *c* solution.
16. KCN: Prepare a 30 mM solution of KCN (CAUTION—it is extremely toxic and should be handled with care).

Procedure

1. Weigh 50 g of maize leaves, wash them with distilled water and dry by dabbing with filter paper.
2. After Step 1 perform all the operations at 0–4°C. Cut the leaves into small pieces and homogenize them in chilled pestle and mortar in 250 ml of the extraction buffer (Reagent 7).
3. Filter the homogenate through four layers of muslin cloth and centrifuge at 1,000 × *g* for 10 min to remove cell debris, chloroplasts, starch granules and aggregated organelles.
4. Decant the supernatant and centrifuge it at 12,000 × *g* for 10 min to sediment mitochondria.
5. Prepare a discontinuous gradient of Percoll by adding 3 ml of 60% (*v*/*v*) Percoll at the bottom of centrifuge tube then 4 ml of 45% (*v*/*v*) Percoll, 4 ml of 28% (*v*/*v*) Percoll and finally 4 ml of 5% (*v*/*v*) Percoll.
6. To the pellet obtained in Step 4, add 1 ml of suspending buffer and disperse it to obtain a homogeneous suspension.
7. Layer the above mitochondrial suspension very gently on top of the gradient and centrifuge at 30,000 × *g* for 30 min using SW 27–1 rotor (Beckmann ultracentrifuge or equivalent).
8. The mitochondria gather at interface between 45 and 28% Percoll. Remove them using a Pasteur pipette.
9. Dilute the collected mitochondria ten folds with suspending buffer and

centrifuge at 10,000 × *g* for 10 min. This step is carried out to remove Percoll from the preparation. Suspend the sedimented mitochondria in the suspending buffer and recentrifuge at 10,000 × *g* for 10 min to get rid of residual Percoll.

10. Suspend Percoll-free sedimented mitochondria in 0.5 ml of suspension buffer.
11. Determine the intactness of mitochondria by examining them for respiratory control by monitoring the rate of oxygen consumption in absence and on addition of ADP using Clark type oxygen electrode. Pipette 0.90 ml of the reaction mixture (Reagent 10) and 0.05 ml of 10 fold diluted solution of succinate (Reagent 11) in the chamber and set the instrument to 100% oxygen saturation. Now add 0.05 ml of the mitochondrial preparation and record the rate of decrease in oxygen saturation at every 15 sec intervals for 5 min. (The rate of oxygen become will be low or negligible under these conditions.) Inject 0.05 ml of ADP solution with a syringe into the solution through the groove of electrode holder and again record the rate of oxygen consumption upto 5 min at 15 sec intervals. If the chart recorder facility is available, the tracings of oxygen consumption can be obtained directly, otherwise, the readings can be noted manually.
12. The intactness of outer mitochondrial membrane can be tested by determining the activity of succinate: cytochrome *c* reductase spectrophotometrically from the increase in absorbance at 551 nm due to reduction of cytochrome *c*. In a cuvette, pipette 0.75 ml of 20 mM phosphate buffer (pH 7.2), 0.1 ml of cytochrome *c* solution, 0.1 ml KCN, 1.7 ml of water and 0.02 ml of mitochondrial preparation. Mix the contents and set the spectrophotometer to zero absorbance at 551 nm. Note any change in absorbance upto 3 min at 15 sec intervals. Start the reaction by adding 0.1 ml of 0.3 M succinate solution and again record the change in absorbance at 15 sec intervals.
13. Repeat the same assay for succinate: cytochrome *c* reductase (Step 12) except use 0.75 ml of 20 mM phosphate buffer containing 1.2 M sucrose, so that the concentration of sucrose in the assay mixture is 0.3 M.

Calculations and conclusions

1. Determine the rate of oxygen consumption on the basis of decrease in % oxygen saturation of the solution in the absence and presence of ADP. For example:

$$\begin{aligned} \text{Initial \% saturation} &= 100\% \\ \text{\% saturation after 30 sec} &= 80\% \\ \text{Decrease in \% saturation per min} &= (100 - 80) \times 60/30 = 40\% \end{aligned}$$

2. Compare the rates of oxygen consumption in presence and absence of ADP. If the rate increases significantly, say 4–5 fold, it would indicate a good quality of preparation in which the coupling mechanism of the respiratory chain is functional and that the mitochondria are not uncoupled.

3. Calculate the activity of succinate: cytochrome *c* reductase in terms of arbitrary units of change in O.D. per min in presence and absence of 0.3 M sucrose. In the absence of sucrose the outer membrane gets ruptured. The proportion of mitochondria in the preparation with damaged outer membrane can be determined from the activity in the absence of 0.3 M sucrose as % of that in its presence as:

   Activity (ΔA_{551}/min) in presence of sucrose = 0.02

   Activity (ΔA_{551}/min) in absence of sucrose = 0.10

   $$\% \text{ mitochondria with damaged outer membrane} = \frac{0.02}{0.10} \times 100 = 20$$

8.3.10 EXPERIMENT: Isolation of chloroplasts from plant leaves

Principle

For elucidating the type of biochemical processes and activities performed by chloroplasts, it is essential that structurally intact, functionally active and pure preparations be available in a reasonable quantity. Depending upon the conditions employed for isolation, chloroplasts of different qualities may be obtained. Type A are those which are structurally and functionally intact, Type B those in which the outer membrane gets ruptured but resealed again during isolation and Type C refers to broken chloroplasts. The intactness of chloroplast preparation is generally assessed from either the rate of potassium ferricyanide reduction spectrophotometrically at 430 nm or more commonly from potassium ferricyanide-dependent oxygen evolution in presence of NH_4Cl as an uncoupler before and after lysis by osmotic shock. Osmotic lysis of chloroplasts is accomplished by suspending them in buffer of molar conc < 25 mM in presence of $MgCl_2$. Since potassium ferricyanide is unable to penetrate into intact chloroplasts, the rate of ferricyanide reduction or of ferricyanide-dependent O_2 evolution by illuminated chloroplasts is very low. Both these reactions occur at fairly rapid rates in broken chloroplasts. The functional integrity can be monitored by assaying activities of stromal enzymes such as NADP-glyceraldehyde-3-phosphate dehydrogenase or nitrite reductase before and after osmotic shock. These enzyme assays also enable to distinguish between Types A and B of chloroplasts. Preparations with high degree of functional integrity give photosynthetic activity of 50–200 μmoles of CO_2 fixed/mg chlorophyll/h or similar rates of CO_2-dependent O_2 evolution. In this experiment a rapid procedure for isolation of intact chloroplasts from pea leaves is described.

Materials and Reagents

1. Experimental material: 9–10 days old pea seedlings grown in soil under natural conditions.
2. Waring blender.
3. Muslin cloth.
4. Mira cloth.

5. Refrigerated centrifuge with swinging bucket rotor (HB-4, Sorvall, RC-2B).
6. Clark type oxygen electrode (Biological Oxygen Monitor, Yellow Springs Inc., Ohio).
7. Slide projector.
8. Grinding medium: Prepare medium containing 0.33 M sorbitol, 50 mM Tricine-KOH (pH 7.9), 2 mM EDTA, 1 mM $MgCl_2$ and 0.1% bovine serum albumin.
9. Percoll media: Make a solution containing 0.33 M sorbitol, 50 mM Tricine-KOH (pH 7.9), 0.1% bovine serum albumin and 40% (*v*/*v*) Percoll.
10. Assay mixture for determination of ferricyanide-dependent O_2 evolution: Prepare a solution containing 0.33 M sorbitol, 50 mM HEPES, 2 mM EDTA, 1 mM $MgCl_2$ and 1 mM $MnCl_2$, 1.5 mM potassium ferricyanide and 10 mM glyceraldehyde. Adjut pH of this solution to 7.6.
11. 50 mM NH_4Cl solution.

Procedure

1. Weigh 20 g of leaves and cut them into small pieces and put in the chilled grinding medium (Reagent 8) in the ratio of 4 ml of media per g of the leaf material.
2. Homogenize the tissues in cold (0–4°C) for 3 sec in Waring blender at full speed.
3. Filter the brei through 2 layers of muslin and 2 layers of Mira cloth.
4. Add 30 ml of the homogenate in each centrifuge tube (50 ml capacity). With the help of pipette with its tip at the bottom of the tube, add 14 ml of ice-cold Percoll media (a layer of about 3.5 cm in height).
5. Centrifuge the tubes in a swinging bucket rotor (HB-4, Sorvall) at 2500 $\times g$ for 1 min.
6. Remove the supernatant and then the Percoll layer by aspiration and suspend the intact chloroplasts, which are recovered in the pellet, in the grinding media.
7. Suspend a small proportion of the pellet in 25 mM Tricine buffer (pH 7.9) containing 10 mM $MgCl_2$ to break the chloroplasts by subjecting them to osmotic shock.
8. For measuring the rate of ferricyanide-dependent oxygen evolution, pipette 1.95 ml of Reagent 10 in the vessel of the oxygen monitor. Set the instrument to 100% saturation of air. Turn the knob to per cent oxygen saturation and it should give reading of about 21%. Cover the vessel with black paper and add 0.05 ml of chloroplast preparation (containing about 200 μg chlorophyll) and record the dark rate of change in % oxygen saturation for 3 min. Remove the black paper and focus beam of light from a slide projector (placed at a distance of about 15 cm from the vessel) and again note the change in % oxygen saturation. Using a reaction mixture as prepared for determination of oxygen evolution, monitor the rate of ferricyanide reduction spectrophotometerically by illuminated chloroplasts at 430 nm. For this irradiate the mixture for 5 min with the slide projector

and transfer the solution to cuvette of spectrophotometer and record the absorbance at 430 nm. Determine the activity of NADP-dependent glyceraldehyde-3-phosphate dehydrogenase in the suspension containing intact chloroplasts as described in EXPERIMENT 8.3.7, Step 8.

9. Prepare lysed chloroplasts by adding 0.2 ml of the chloroplast suspension to 0.2 ml of distilled water, stir for 1 min and then add 0.4 ml of double strength of grinding media to restore isotonic solution. Measure the rate of O_2 evolution by 0.2 ml of lysed preparation and after 2 min inject 0.1 ml of NH_4Cl into the mixture and continue to record the change in O_2 evolution.

Conclusions

1. Negligible rates of oxygen evolution in presence of ferricyanide (or reduction of ferricyanide if spectrophotometric method is used) indicate intactness of the chloroplast preparation. In lysed chloroplasts the rate of these reactions is quite high. If the rates with intact preparation are 10% of osmotically shocked chloroplasts, it would indicate that 90% of the chloroplasts in the preparation are intact.
2. The rate of these activities increases, both in intact and lysed preparations, on addition of NH_4Cl due to uncoupling of the photosynthetic electron transport chain. However, enhancement in the rate is about 2 fold higher in lysed or broken chloroplasts than in intact chloroplasts.

Suggested further reading

Goodwin T.W. and Mercer E.L. (1983). Introduction to Plant Biochemistry, Pergamon Press, Oxford.

Graham J. (1984). Isolation of sub-cellular organelles and membranes. *In* Centrifugation: A Practical Approach, (Rickwood D., ed), 2nd ed, IRL Press, Washington D.C., pp 161–182.

Griffiths A. (1996). Centrifugation Techniques, *In* Practical Biochemistry: Principles and Techniques, (Wilson K and Walker J., eds), Cambridge University Press, Cambridge, pp 275–323.

Linskens H.F. and Jackson J.F. (eds), (1985). Cell components, *In* Modern Methods of Plant Analysis (New Series) Vol. 1, Springer-Verlag, Berlin.

Quail H.P. (1979). Fractionation of Plant Cells. *Ann. Rev. Plant Physiol.* **30:** 425–484.

9. Chromatography

9.1 Introduction

The term "Chromatography" was coined by a Russian botanist Tswett in 1906. It includes all those techniques in which various components in a mixture get separated during their passage through a porous media due to the differences in their migration rates. In chromatography, the sample is applied at one end of a porous support which holds the stationary phase (which may be a solid or a liquid) and the mobile phase (a liquid or a gas) is then made to flow over it. The various constituents in a sample get separated essentially due to differences in their partition or distribution behaviour between the stationary phase and the mobile phase. The partition or distribution coefficient (K_d), is defined as the ratio of concentration of a compound in two phases (say A and B) at equilibrium. Hence, a situation where both the phases are liquids

$$K_d = \frac{\text{Concentration of the compound in solvent A}}{\text{Concentration of the compound in solvent B}}$$

Different compounds vary in their distribution coefficients in a given biphasic system and this constitutes the underlying principle for their separation in the different types of chromatographic techniques. In simple terms, movement of those substances in the sample which have greater affinity for stationary phase is retarded since they tend to spend more time in this phase. On the other hand, the compounds with relatively higher affinity for mobile phase travel at a much faster rate as these are carried alongwith the flowing mobile phase. This eventually results in the separation of various components in the mixture.

9.2 Classification of Chromatographic Techniques

Chromatographic techniques may broadly be classified on the basis of nature of support used for holding the stationary phase. These include:

(i) Paper chromatography: In this system a filter paper sheet is used as a support for stationary phase.

(ii) Thin layer chromatography (TLC): In this technique a glass plate, plastic sheet or a piece of metal foil serves as a support for the stationary phase which is applied in the form of a thin layer on these materials.

(iii) Column chromatography: Here stationary phase is packed into a tubular glass, polypropylene or metal columns.

The various chromatographic techniques have also been classified into the following types depending upon the forces or the interacting phenomenon between the solute molecules and the stationary phase:

(i) Partition chromatography
(ii) Adsorption chromatography
(iii) Ion-exchange chromatography
(iv) Molecular sieve chromatography
(v) Affinity chromatography

In the following sections individual chromatographic techniques are discussed in detail.

9.3 Paper Chromatography

9.3.1 Theory

In paper chromatography, the sample to be analyzed is applied in the form of a spot near one of the edges of a Whatman filter paper. This filter paper is then kept in an atmosphere already saturated with water vapors to form a thin film of water around cellulose fibres of the paper which then act as a stationary phase. An appropriate solvent system, which functions as a mobile phase, is then allowed to flow over the sample spot. On coming in contact with mobile phase, the various components of the sample get partitioned between the stationary and the mobile phases. Those constituents having a higher affinity for the stationary phase move less rapidly as compared to those having higher affinity for the mobile phase. Evidently the components will get well separated from each other if their K_d values are sufficiently different. Satisfactory separation of components of interest can hence be achieved by judicious selection of the mobile phase.

Depending upon the direction of flow of mobile phase, the two commonly used systems are: ascending and descending paper chromatography. In the ascending method, the solvent is kept at the base of the chamber and the edge of the paper where the sample has been applied is immersed in the solvent taking care that the sample spots do not get dipped in the solvent but remain just above the surface of the solvent (Fig. 9.1a). The solvent moves up or ascends the paper by capillary action and the separation of different components occurs on the basis of differences in their partition coefficients. In the descending method, the paper is hung in such a way that the side where the sample has been spotted dips in a trough which is fitted at the top of the chamber and contains the mobile phase and this solvent travels down the paper under the force of gravity (Fig. 9.1b).

9.3.2 Component detection

After development, the filter paper is dried in air. The location of the compounds under investigation is carried out by making use of their specific chemical, physical or biological property. For instance, if these components form coloured complex or product with a particular reagent, they can then be conveniently located by spraying the chromatogram with that reagent. In case the compounds absorb UV light or show UV fluorescence, the paper can be examined under strong UV light in darkness. The UV absorbing compounds would appear as dark spots while UV

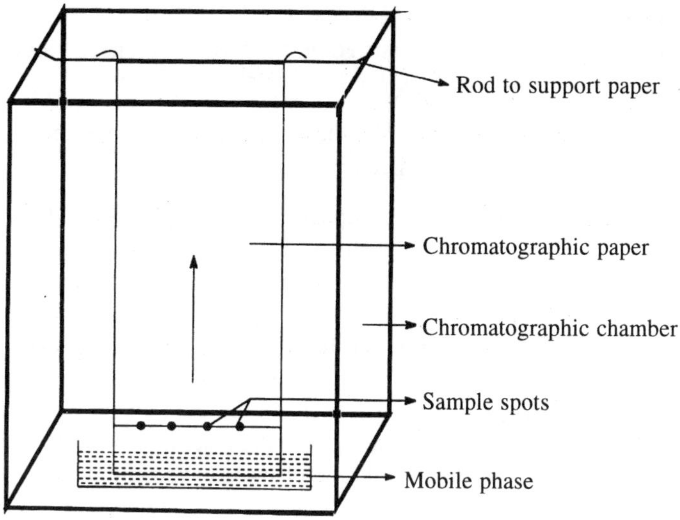

(a)

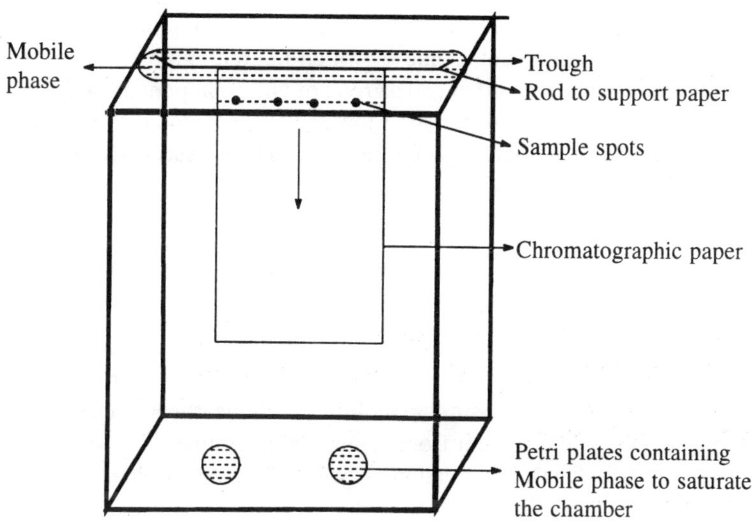

(b)

Fig. 9.1 A typical set-up of paper chromatography. (a) Ascending and (b) Descending, arrow indicates the direction of flow of mobile phase.

fluorescent compounds would show a characteristic fluorescence under UV light. If in metabolic studies, a radioactive precursor has been used, then the products derived from it can be detected from the radioactive zones or spots on the chromatogram.

9.3.3 Identification

The separated compounds are identified on the basis of their R_f values, which denotes relative to front. The R_f value is calculated as follows:

$$R_f = \frac{\text{Distance travelled by the component from base line}}{\text{Distance travelled by the solvent from base line}}$$

Since a number of factors such as composition of solvent system, temperature, pH, mode of development, grade of filter paper and flow rate influence the R_f value, it is always advisable to run authentic standards alongwith the sample.

9.3.4 Different modifications for development of chromatograms

Quite often a single run with one solvent system in a single dimension is not sufficient to obtain satisfactory separation of the components. A number of modifications has been employed to achieve better resolution of such overlapping or very closely located compounds. Some of these have been described below:

(*i*) *Multiple development:* In this procedure, the paper is developed with the same solvent system several times in the same direction. The chromatogram is air dried between the successive developments. This mode of development of chromatogram can be used advantageously when the sample contains mixture of components some of which migrate quite fast and get well separated from each other and a group of other components which remain clustered near origin due to their low and closely similar R_f values. These slow moving components get further apart with each successive run resulting in their better separation.

(*ii*) *Continuous development:* It is also known as over-run chromatography. In this case the chromatogram is developed in the same solvent system in the same direction continuously for a long time even after the solvent has started dripping down from the other end of paper. Care, however, must be taken that the fastest moving compound does not get eluted out of the paper. This approach is useful for achieving better separation of compounds having very low and similar mobilities. Evidently R_f value cannot be calculated as the solvent front cannot be determined due to overflow of the solvent. The identification is hence done on the basis of R_g value which is calculated as follows:

$$R_g = \frac{\text{Distance moved by unknown}}{\text{Distance moved by a standard compound}}$$

(*iii*) *Two dimensional (bidimensional) chromatography:* In this method, the chromatogram is first developed in one solvent system in one direction. After air drying, the paper is again developed in a second solvent system in a direction perpendicular to the previous run. The compounds having similar R_f values in first solvent system might have different mobilities in the second solvent system and hence they get separated. However, the limitation of bidimensional chromatography is that only one sample can be spotted or applied for analysis on each filter paper.

(*iv*) *Reverse phase chromatography:* In all the above cases the aqueous phase of the solvent system is used to saturate the internal atmosphere of the chamber and water absorbed by the cellulose fibres of the paper serves as the stationary phase. Thus the stationary phase is hydrophilic, whereas the mobile phase is hydrophobic in nature. For separation of hydrophobic substances like fatty acids, the filter paper is treated with lipophilic compounds like silicon grease, kerosene oil, paraffin, vaseline etc while the mobile phase is hydrophilic in nature. Since the characteristics of stationary and mobile phases are converse of that in the conventional paper chromatography, this modified form is referred to as "Reverse Phase Chromatography".

9.4 Thin Layer Chromatography (TLC)

9.4.1 Theory

This technique is similar to paper chromatography but is more convenient and less time consuming. Here, instead of paper, the supporting material is either a glass plate, a plastic sheet or a piece of metal foil. A thin layer of the stationary phase is laid over this inert support. The layer may be as thin as 250 μm for analytical separations and as thick as 2–5 mm for preparative separations. A binding agent such as calcium sulfate or gypsum may be incorporated into the chromatographic media to facilitate firm adhesion of adsorbent to the plate. Most widely used adsorbent in TLC is Silica Gel 'G' Merck which contains 13% $CaSO_4$. Another commercially available media is Silica Gel 'H' Merck which does not contain $CaSO_4$. Since Silica Gel 'G' has a binder, its slurry should not be kept for a long time and it should be spread immediately after preparation of its suspension.

TLC has a number of advantages over paper chromatography. Some of these are:

(i) In paper chromatography, the separation is based on partitioning only while in TLC, depending upon the nature of the chromatographic media used, the separation can be achieved either by partitioning, adsorption, ion exchange or molecular sieving phenomenon.

(ii) This method is extremely rapid as the separation can be achieved within an hour or so while in paper chromatography it may take upto about 8–24 h.

(iii) It is relatively more sensitive since lower concentrations of compounds in the mixture can be successfully separated and detected.

(iv) Corrosive agents like H_2SO_4 and high temperatures can be used to locate the separated compounds which is not possible in paper chromatography.

9.4.2 Development, location and identification of components

For development, the plate on which the sample spots have been applied is placed in an air tight glass jar containing the solvent. The location and identification of separated components is carried out in the same manner as in the case of paper chromatography. Also like paper chromatography, TLC is amenable to two-dimensional mode of development.

9.5 Column Chromatography

Paper and thin layer chromatography described in preceding Sections use a flat bed or surface for the separation of molecules. In column chromatography, separation is achieved by passage of the sample through vertically fixed tubular glass or polypropylene column which is packed with an appropriate chromatography media. Usually commercially available columns have a porous sintered plate fused at their base which prevents the stationary phase from flowing out of the column. This sintered base is positioned as near the base as possible in order to minimize the dead space to reduce the chances of post column mixing of the separated compounds. Alternatively, a simple glass burette with a plug of glass wool at the base can be used as a column. At the base there is a small capillary tubing through which the effluent from the column flows into the test tubes in which fractions are collected. At the top of the column a solvent reservoir with a delivery system is fitted (Fig. 9.2).

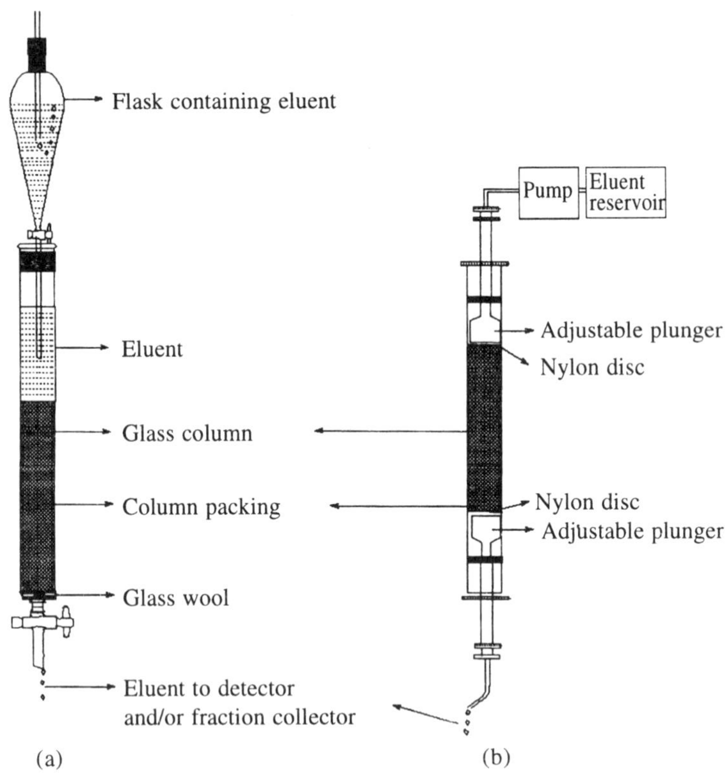

Fig. 9.2 Simple (a) and more sophisticated (b) version columns for column chromatography.

The stationary phases used in column chromatography are water insoluble, porous, solid particles and the resolution of sample components occurs depending upon the principle underlying the separation phenomenon. The flow rate and resolution characteristics are influenced by the size and shape of the stationary phase. Large and coarse particles have higher flow rate but give comparatively poor resolution while finer particles with large surface to volume ratio have slower flow rate but greater resolution efficiency. Generally, particles of 100–200 mesh are used for routine analysis but for high resolution the smaller particles of 200–400 mesh are preferred.

For fractionation and separation of components, the sample is loaded on top of the column and eluted with an appropriate buffer. The effluent emerging from base of the column is collected in the form of fractions of fixed volume or fixed time in individual test tubes either using an automatic fraction collector or manually. The collected fractions are then analyzed for the presence of the desired substance. The detection technique depends on physical, chemical or biological property of the compound. Presence of coloured compounds can be identified simply from visual observation but for colourless compounds, alternative methods of detection are employed. They can be either the colour reactions or may be based on its unique physical property such as ultraviolet absorption, fluorescence, refractive index etc. or biological activity such as enzyme activity.

Column chromatographic techniques have been classified on the basis of the nature of the interactions occurring between solutes and the stationary phase which ultimately results in their separation. Various types of column chromatography are described below.

9.5.1 Adsorption chromatography

Adsorption is a phenomenon in which compounds are held onto the surface of a solid adsorbent, having specific adsorption sites, through weak non-ionic interactions such as van der Waal's forces and hydrogen bonding. Different compounds bind with varying strengths and hence can selectively be desorbed. For good resolution, selection of right type of the adsorbent and the eluent or mobile phase is essential. Some of the commonly used adsorbents include charcoal, silica, alumina, hydroxyapatite etc. Eluent influences quality of separation since polarity of the mobile phase influences the adsorption considerably. Non-polar solvents favour maximum adsorption which decreases with increase in polarity of the solvent. In general, the polar solvents are preferred for the substances having polar or hydrophilic groups and nonpolar solvents for substances having hydrophobic or nonpolar groups. For example, alcoholic solvents for OH group containing substances; acetone or ether for substances with carbonyl groups and hydrocarbons such as toluene or hexane for non-polar substances. For gradient elution, mixture of the polar and non-polar solvents of different ratios can be used to obtain eluent of varying polarities.

9.5.2 Gel filtration or Exclusion column chromatography

9.5.2.1 Theory

The chromatographic media used in this technique are porous, polymeric organic compounds with molecular sieving properties. These are cross-linked polymers which swell considerably in water forming a gel of a three dimensional net work of pores. The size of pore is determined by degree of cross-linking of polymeric chains. Different solutes in a mixture get separated on the basis of their molecular size and shape during their passage through a column packed with the swollen gel particles. The terms 'exclusion chromatography', 'gel filtration' and 'molecular sieve' chromatography are used for this separation process. The large molecules in sample are unable to penetrate through the pores into the gel and thus remain excluded while the small molecules enter into the gel beads. Obviously the volume of solvent accessible to large molecules is very much less (V_o), whereas small molecules, which can freely penetrate into the gel have access to solvent inside (V_i) as well as outside (V_o) the spherical beads (Fig. 9.3).

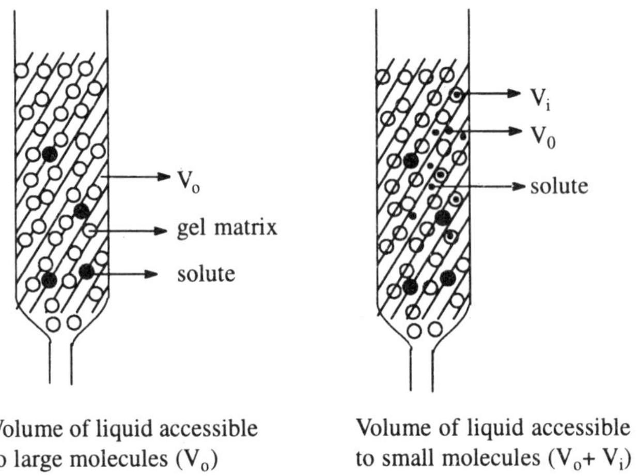

Fig. 9.3 Diagramatic representation of spatial accessibility to small molecules and excluded molecules during gel filtration. V_o represents volume of liquid present outside the gel particles ; V_i denotes volume of liquid present inside the gel particles.

Separation process of components of a sample during gel filtration is illustrated in Fig. 9.4. A sample containing a mixture of large and small molecular weight substances is applied onto the column (Fig 9.4a). Solvent, usually a buffer is used as an eluent. As components of the sample travel down the column, the compounds whose molecular size exceeds the fractionation range of the gel, are unable to enter into gel particles and hence remain completely excluded from beads and travel through the interstitial spaces. Molecules of smaller compounds, however, diffuse into the gel matrix through the pores and get distributed between the mobile phase inside as well as outside the gel particles (Fig. 9.4b). These thus follow a longer path than the larger molecules and hence their movement down the column is retarded. Consequently different components in the sample get separated from each other with larger molecules getting eluted first followed by smaller molecules (Fig 9.4c).

For a given type of gel, the distribution coefficient, (K_d) which represents fraction of the liquid within the gel particles accessible to molecules of a given

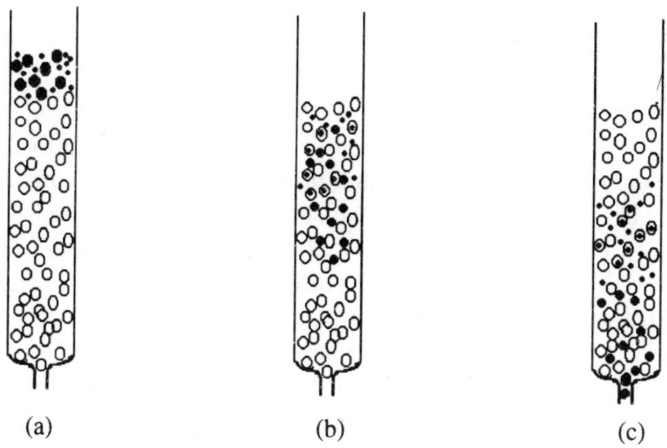

Fig. 9.4 Separation of solutes during gel filtration due to molecular sieving: ○, gel matrix; ●, large molecules; •, small molecules.

substance will predominately depend upon molecular size of that compound. Very large molecules, due to their exclusion, have no access to the mobile phase within the gel hence have a K_d value of 0, whereas smaller molecules for which the inner mobile phase is completely accessible have K_d value of 1. For the molecules of the intermediate size, the K_d value varies within this range (0–1). This difference in K_d value of various compounds in the sample accounts for their separation during gel filtration. The relationship of K_d with elution volume (V_e) is given by Equation 9.1.

$$V_e = V_o + K_d V_i \quad (9.1)$$

where V_e is the elution volume and represents volume of the mobile phase required to elute the compound from the column.

V_o is void volume or the volume of the mobile phase outside the gel particles

V_i is the volume of the mobile phase present inside the gel particles and can be calculated from the relationship: $V_i = a \cdot W_r$, where a = dry weight of the gel and W_r = water regain. Values of W_r for different types of gels are given in Appendix (Table 12b).

K_d is the fraction of V_i accessible to a particular compound

Transformation of Equation 9.1 gives

$$K_d = \frac{V_e - V_o}{V_i}$$

As can be noted, for very large molecules which are completely excluded from the gel $V_e = V_o$ and so $K_d = 0$, but for small molecules which can freely diffuse into and out of the gel beads $V_e = V_o + V_i$ and so $K_d = 1$.

Materials commonly used for gel filtration chromatography alongwith their fractionation range are listed in the Appendix (Table 12a).

9.5.2.2 Applications of gel filtration

Gel filtration phenomenon has been used for several purposes which are briefly described as follows:

(*i*) *Desalting or group separation:* In a number of investigations, as during preparation of nucleic acids and protein purification, inorganic salts or other low molecular weight reagents are extensively used which often have to be removed from the final preparation. This can effectively be achieved by passing preparation through a column of Sephadex G-10 or G-25. The high molecular weight macromolecules like proteins and nucleic acids remain excluded from the gel particles and are recovered from the column immediately after void volume, while the movement of low molecular weight compounds, such as salts, is considerably impeded due to their entry into the beads. This method of desalting is relatively faster than the process of dialysis and hence is particularly suitable for desalting of labile compounds.

(*ii*) *Concentration of dilute solution:* Dry Sepahadex (G-25, coarse) when added to a dilute solution of high molecular weight substance absorbs water during swelling. The high molecular weight substances, however, remain in solution due to their complete exclusion. Because of reduced volume, the concentration of the higher molecular weight substances in solution increases without any effect on pH and ionic strength. The added beads are then removed from the solution by centrifugation.

(*iii*) *Fractionation or purification:* The availability of a large variety of gel media (dextran, polyacrylamide, agarose) with varying fractionation range has extended the use of this technique for separation as well as for purification of biological materials ranging from low molecular weight compounds like sugars, polymers like proteins and nucleic acids to superstructures such as viruses. Generally, for purification or fractionation purposes, the gel selected is such that the compound to be purified falls within the fractionation range of the gel.

(*iv*) *Molecular weight determination:* As mentioned earlier, separation of various components in the sample is based on the differences in their molecular size. Larger the molecular weight of a compound, lesser will be its elution volume. In fact, this method has been exploited for determination of molecular weight of a solute, particularly proteins, from their elution characteristics (V_e or K_d value). A plot of log molecular weight vs elution volume or K_d gives a linear curve (Fig. 9.5). Hence by using proteins of known molecular weights, a calibration curve can first be prepared and from the elution volume or K_d value, the molecular weight of protein of interest can readily be obtained. This method of molecular weight determination is simple, inexpensive, and does not require homogeneously purified preparation of the protein or compound of interest and is non-destructive.

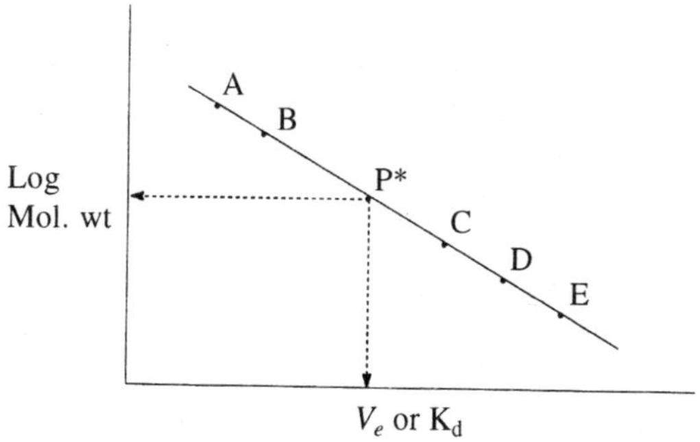

Fig. 9.5 A graph showing relation between molecular weight and elution volume during gel filtration. A-E are standard proteins of known molecular weight and P* is the protein whose molecular weight is being determined.

9.5.3 Ion exchange chromatography

9.5.3.1 Introduction

Ion exchange chromatography is a type of adsorption chromatography in which retention of a solute occurs due to its reversible electrostatic interaction with the oppositely charged groups on an ion exchanger. Hence this technique is useful for separation of compounds which bear a net electric charge such as proteins, amino acids and nucleic acids etc. Ion exchangers are prepared from either certain synthetic resins which are insoluble porous organic molecules or naturally occurring biopolymers such as cellulose to which various groups known as *fixed ions* are covalently attached. These fixed ions are balanced by equal and oppositely charged ions from the solution referred to as *counter ions*. Depending upon the nature of the counter ions, these ion exchangers are of two types: cation exchangers in which the counter ions are cationic or positively charged ions and anion exchangers which have negatively charged counter ions. Counter ions are mobile and can be easily exchanged by other similarly charged molecules in the sample. Nature of the resin matrix remains unchanged during this exchange process. List of different types of ion exchangers and their chemical nature is given in the Appendix (Table 11).

Generally resin based ion exchangers are used for separation of low molecular weight biomolecules and cellulosic ion exchangers are more suitable for isolation of macromolecules such as proteins and nucleic acids.

9.5.3.2 Principle

Ion exchangers have to be precycled to get an appropriate charge and their complete swelling. In case of anion exchanger, it is treated first with alkali and then with acid and finally washed with water till it is neutral. On the other hand, cation exchanger is first given the acid treatment and then alkali treatment and finally washed with water till neutral. After precycling, the ion exchanger is packed into a column and is equilibrated with the counter ion by passing 2–3 bed volumes of the starting buffer of a required pH. In case of anion exchanger, Tris-HCl buffer and for cation exchanger K^+ or Na^+ buffer is used. The process of ion exchange takes place in different steps as depicted in Fig. 9.6. In the example illustrated below, the ion exchanger has positive fixed ions which are balanced by negatively charged counter ions (Fig 9.6a). When sample containing mixture of compounds is applied onto the top of the exchanger bed (Fig 9.6b), the solute molecules having charge similar to that of the counter ions (in present case –ve) get exchanged with the counter ions and bind reversibly but strongly to the ion exchanger (Fig 9.6c). Different compounds are bound with different strengths. Strength of binding depends upon the degree of the charge and the charge density (amount of charge per unit volume of the molecule) of the solute. The other solute particles (+vely charged and neutral) have no affinity for the stationary phase and are washed down alongwith the starting buffer. The bound solute molecules are then released in succession by altering pH or ionic strength of the elution buffer. This type of elution is known as gradient elution (see Fig. 9.6).

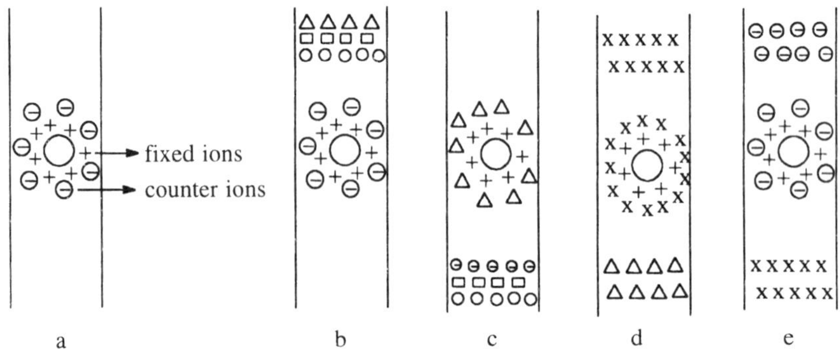

Fig. 9.6 Separation process of solutes during ion exchange chromatography. Δ is a negatively charged solute and ○ and □ are positively charged and neutral solutes.

Stepwise (discontinuous) or continuous gradient may be applied but continuous gradient is preferred because it gives better resolution and less tailing. In general with anion exchanger a buffer with increasing pH and increasing ionic strength while with cation exchanger a buffer with decreasing pH and increasing ionic strength gradient is employed for desorption of bound substances (Fig. 9.6d). Increasing ionic strength in both the cases displaces the bound substances while change in pH on either side would bring the pH of the bound substance to isoelectric point where the substance will have no net charge and so it will get detached from the ion exchanger and will be eluted. The ion exchanger can be regenerated by again passing the starting buffer continuously (Fig. 9.6e). The ion exchanger is ready for reuse.

9.5.3.3 Gradient elution

Gradient elution is of two types:

(*i*) *Discontinuous gradient:* It is also known as stepwise gradient. In this, a fixed volume of the mobile phase of a particular pH, ionic strength or polarity is passed through the column. This is then followed by elution with fixed volume of mobile phase of different composition and this process of developing the column with definite volumes of mobile phase of changing composition is continued. Here, the polarity, the pH or the ionic strength of the eluent is changed in a stepwise manner and at every step, the change is rather abrupt.

(*ii*) *Continuous gradient:* In this method, the composition of the eluent is changed at a steady and constant rate. This can be done by using a gradient maker. The lätter constitutes of two vessels of equal diameter with an outlet tube at their base. These are connected by a teflon tube having an adjustable screwtype pinch cock (Fig. 9.7). The recepient chamber is connected to the column via teflon tubing having a pinch cock. In the beginning, both the pinch cocks are tightened and the two eluents of different pH, ionic strength or polarity are poured into each of the two vessels. The recepient chamber generally contains the starting buffer. The solution in the recepient chamber is continuously stirred using a magnetic stirrer. Both the pinch cocks are opened. As the eluent from the recepient vessel starts flowing into the column, the same volume of the eluent from the donor chamber enters into the recepient chamber where it is mixed by stirrer. In this way, the pH, the ionic strength or polarity of the buffer in the recepient chamber changes continuously at a constant rate depending upon the composition of the eluent in the donor chamber.

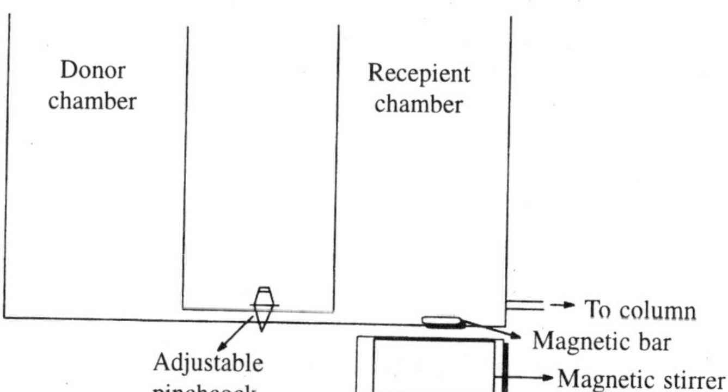

Fig. 9.7 Diagram of a gradient maker.

9.5.4 Affinity chromatography

Selective separation of a particular group of biopolymers from a complex mixture has undergone a major transformation due to introduction of affinity chromatography. Purification by affinity chromatography is different from all other forms of chromatography in the sense that this technique does not make use of the differences in the physical properties (like solubility, adsorption, molecular weight and ionic properties) of the molecules to be separated. Rather it exploits one of the unique and fundamental properties of biopolymers, i.e. the specificity of their interaction with other biomolecules. In other words, affinity chromatography is a type of adsorption chromatography in which the substance to be isolated is specifically and reversibly bound to a complementary binding substance (ligand) immobilized on an insoluble chromatographic bed material (matrix). The other substances in the mixture remain unbound and are, therefore, washed away while the substance of interest (the bound substance) is subsequently recovered by displacement from the ligand either by specific (affinity) elution or by non-specific (change in pH or ionic concentration) elution. The general principle of affinity chromatography is illustrated in the Fig. 9.8.

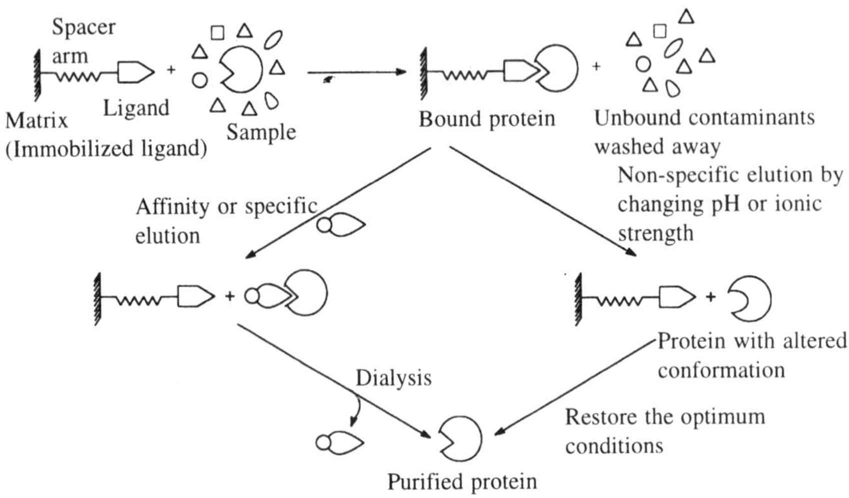

Fig. 9.8 Principle of affinity chromatography showing use of this technique for purification of a protein.

Purification by affinity chromatography is often of the order of several thousand fold and recoveries of the material are generally very high. Before the method can successfully be employed for the purification, one must have the following information:

(i) The knowledge about the structure and biospecificity of the compound to be isolated.
(ii) Various biospecific ligands which can be covalently attached to the chromatographic bed material (matrix).
(iii) Whether the ligand retains its biospecific binding affinity for the substance of interest after its attachment to the matrix.
(iv) Different methods for selectively desorbing the bound substance in an active form after washing away the unbound substances.

A large number of biospecific ligands can be employed. In practice, it is possible to select a ligand which displays absolute specificity and binds exclusively with one particular compound only. It is also possible to select a ligand which displays group specificity e.g. for the purification of enzymes a substrate analogue, or an inhibitor, a cofactor or an activator can be used as ligand while a specific receptor or a carrier protein can be used for the purification of vitamins and hormones. Antigen-antibody interactions can be exploited for the purification of either of these. Similarly, for nucleic acid purification a complementary base sequence or histones and for lectins, cell surface receptors or polysaccharides can be successfully employed as ligands. But before selecting a ligand, two attributes must be kept in mind: (a) that the ligand must exhibit specific and reversible binding with the substance to be purified and (b) it must have chemically reactive functional groups which allow it to be attached to the matrix without destroying its binding activity with the substance of interest. If several functional groups are available, the ligand should be coupled with the matrix via the group least likely to be involved in its specific interaction with the molecule to be isolated. The most common such groups are $—NH_2$, —COOH, —SH and —OH.

An ideal chromatographic bed material (matrix) to which the ligand is covalently bound must possess these attributes:

(i) It must possess suitable groups to which ligand can be covalently coupled. Many groups may be introduced into matrix in order to couple ligand. They may be nucleophilic as NH_2, SH, OH or electrophilic such as activated acid chlorides, carbonyls activated by carbodiimide, isothiocyanate, or diazonium salts.
(ii) It must remain unchanged under the conditions of attachment of ligand.
(iii) During the binding of the macromolecule and its subsequent displacement from ligand, it must retain its physical and chemical stability.
(iv) It must not exhibit nonspecific adsorption.
(v) It should have an open pore structure.

The most commonly used matrices are cross-linked dextrans (e.g. Sephacryl), agarose (e.g. Sepharose), polyacrylamide gel (Bio gel P), polystyrene, cellulose, porous glass and silica. Sepharose is a bead-form of agarose gel which displays virtually all features required of a successful matrix for immobilizing biologically active molecules. The hydroxyl groups on the sugar residues can be easily derivatized for covalent attachment of a ligand. The different matrices which are in common use for affinity chromatography alongwith their physical properties are given in Appendix (Table 13).

9.5.4.2 Spacer arm

Generally, it is observed that the active site of the biological substance (e.g. enzymes) is located deep within the molecule and the attachment of the ligand directly to the matrix, sometimes, interferes with its ability to bind the macromolecule due to steric hinderances between the matrix and the substance to be bound to the ligand. This interference can be prevented by interposing a 'spacer arm' between the matrix and the ligand to facilitate effective binding (Fig. 9.9).

The length of the spacer arm is critical. If it is too short, the arm is ineffective and the ligand fails to bind the substance in sample. If it is too long, non specific effects become pronounced and reduce the selectivity of separation. The optimum length of the spacer arm is generally six to ten C-atoms or their equivalent. Chemical nature of the spacer arm (e.g. whether hydrophobic or hydrophilic) is also critical for the success of separation.

There are two approaches which are generally followed for coupling of the ligand. In one case, spacer arm is first linked to the matrix followed by the coupling of the ligand, whereas the second approach involves the binding of the spacer arm to the ligand which is then linked to the matrix. It is the first approach which is more convenient and is preferred over the second one.

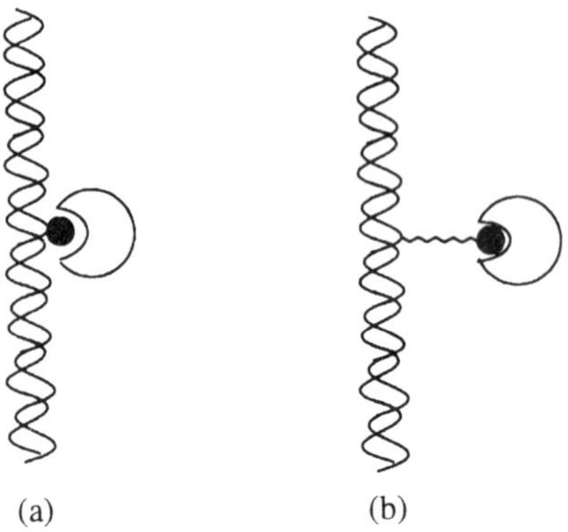

Fig. 9.9 Role of spacer arm in proper binding of the macromolecule.

For attachment of the ligand with the matrix, the matrix is given preliminary treatment with cyanogen bromide at pH 11. This causes activation of the matrix and the molecules containing primary amino groups could then easily be coupled to CNBr activated matrices. In 1972, Pharmacia Fine Chemical Co. developed CNBr—activated Sephrose 4B, a stable ready to use medium for immobilizing ligands. Table 13 (Appendix) gives a list of coupling gels which allow the convenient immobilization of ligands (containing particular functional groups) without the need of complex chemical synthesis. Different spacer arms including 1,6-diamino hexane, 6-amino hexanoic acid, and 1,4-bis(epoxy-propoxy) butane have been used to which the ligand can be attached by conventional organo synthetic procedures involving the use of succinic anhydride and a water soluble carbodiimide. The list in Table 13 (Appendix) also includes a number of supports of agarose, dextran and polyacrylamide type which are commercially available with a variety of spacer arms and ligands preattached.

9.5.4.3 Applications

Affinity chromatography occupies a unique place in separation technology since it is the only technique which enables purification of almost any biomolecule on the basis of its biological function. The first application of the affinity chromatography was the selective adsorption of amylase onto insoluble starch in 1910. However, the technique could not be generally established because of the complex organic chemistry involved in the synthesis of reliable matrix and in the covalent attachment of the ligand. However, in 1967, Axen and coworkers could successfully couple the molecules containing primary—NH_2 group to polysaccharide matrix activated by CNBr. Since then the CNBr activation has been the most generally used method for coupling ligands. Although the technique became routinely available little more than ten years ago, its application has developed so rapidly that it is now used in almost every laboratory concerned with purification of biological substances. The widespread use of affinity chromatography in such a short time reflects its success in achieving rapid separations which are time consuming, difficult or even impossible using conventional techniques.

The principle of affinity chromatography has been extended to purify a large number of enzymes, other proteins including immunoglobulins and receptor proteins and nucleic acids and so has contributed considerably to recent developments in the field of molecular biology. Poly (U) Sepharose 4B, poly(A) Sepharose 4B and immobilized single stranded DNA have been successfully used to isolate mRNA, viral RNA and complementary RNA and DNA, respectively. For the purification of proteins involved in nucleic acid metabolism, immobilized nucleotides are quite useful. The technique of affinity chromatography has also been successfully employed for the separation of a mixture of cells into homogenous populations where it relies either on the antigenic properties of the cell surface or on the chemical nature of exposed carbohydrate residue on the cell surface or on a specific membrane receptor-ligand interactions. Useful modifications or methods of affinity chromatography technique have also been developed. For example, a number of proteins including interferon, plasminogen and restriction endonucleases have been purified by *dye ligand chromatography*. In this technique, immobilized cibacron blue, one of triazine dyes, is used as ligand to purify proteins. The ionic groups and the conjugated ring system of the ligand bind with catalytic or effector site of some proteins and so can be exploited to purify such proteins. In *metal affinity chromatography* metals such as Zn^{2+}, Cu^{2+}, Cd^{2+}, Hg^{2+}, Co^{2+} and Ni^{2+} are immobilized by chelation using either iminodiacetic acid or tris (carboxymethyl) ethylene diamine. Binding of proteins with such metals is pH dependent. The proteins having similar molecular weights and isoelectric points can be best separated by this technique if they have the differential binding affinity for metal ions. The bound proteins can be eluted either by changing the pH or ionic strength of the buffer. Proteins such as fibrinogen, superoxide dismutase and non-histone nuclear proteins have been purified by this method. Another approach is the *covalent chromatography.* In this technique a covalent bond such as disulfide bond is formed between the immobilized ligand and the protein to be isolated. Ligands like thiopropyl Sepharose and thiol Sepharose are used to purify proteins having large number of thiol groups such as papain and urease. The bound protein is then eluted either by dithiothreitol or cysteine. The larger the number of thiol groups on the protein to be isolated, better is the separation.

9.5.5 Gas Chromatography

9.5.5.1 Principle

Gas chromatography (GC) as the name suggests, is particularly suited for the separation of gases and volatile liquids or solids in their gaseous state. The compounds of low polarity are best separated by gas chromatography. The technique is highly sensitive, reproducible and has high speed of resolution. When the stationary phase is an active solid such as silica, the method is referred to as Gas Solid Chromatography (GSC). However, if the stationary phase is a liquid such as polymers of silicone coated onto the surface of an inert granular solid then the technique is known as Gas-Liquid Chromatography (GLC). The stationary phase, whether a solid or a liquid coated as a thin film on surface of a solid support, is packed in a glass or stainless steel column which is narrow, coiled, 1 to 3 m long and with 2–4 mm internal diameter. An inert carrier gas (mobile phase) such as nitrogen, helium or argon is made to flow through the column. The temperature of the column is maintained high in an oven to keep the compounds to be separated in their volatile state. These volatilized compounds get partitioned between the liquid or solid stationary phase and the gaseous mobile phase and hence get separated because of differences in their partition coefficients. After leaving the column, the separated compounds pass through the detector, sensed and recorded by the recorder. The essential components of a gas chromatographic equipment are presented in Fig. 9.10.

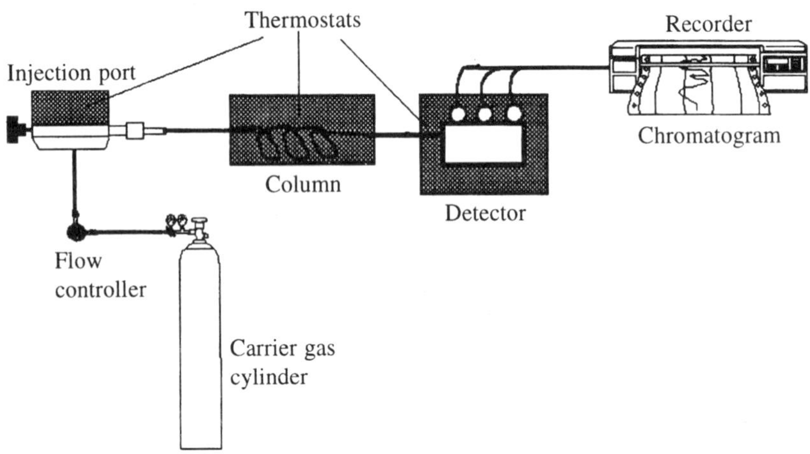

Fig. 9.10 Schematic drawing of a gas chromatographic system.

Capillary columns of internal diameter 0.03 to 1 mm and length upto 100 m made of glass or steel are used for performing gas chromatography. Two types of capillary columns viz. wall coated open tubular columns (WCOT) and support coated open tubular columns (SCOT) are available for this purpose. In WCOT, as the name suggests, walls of the capillary column are coated with the stationary phase. Since the stationary phase is a liquid and is directly coated on the walls of the capillary column, only a small amount of the stationary phase is present in this system. Accordingly, only a small amount of the sample can be applied on to the WCOT column. In SCOT, the stationary phase is in the form of a thin layer on surface of a solid support which in turn is packed into the capillary column. Hence, the capacity of SCOT system is much higher than that of WCOT and a larger amount of sample can be applied.

9.5.5.2 Solid support and stationary phase

The purpose of the solid support is to provide a large uniform, inert surface area for holding a thin layer of the liquid stationary phase. The support should be inert, and should have high mechanical strength, large surface area, regular shape and uniform size. The most commonly used support is celite, the OH groups of which are modified by silanization with hexamethyl disilazane to minimize interaction with the sample.

The correct choice of the stationary phase is perhaps the most important parameter in GC. Ideally the stationary phase must be nonvolatile and thermostable at temperature used for analysis. It should be chemically inert towards the solutes of interest at the column temperature. The high boiling point organic compounds such as polyethylene glycols, methyl-phenyl and methyl vinyl silicon gums, esters of adipic, succinic and phthalic acid, polyesters, polyethylene glycols are used as liquid stationary phases. The operating temperature must be compatible with the phase chosen. At very high temperature, the phase may get volatilized and cause excessive column bleeding which may contaminate the detector.

9.5.5.3 Column packing

The columns are generally dry packed under a slight positive gaseous pressure. Pre-packed columns are also available but are costly. After packing, the column is kept in an oven for 24–48 h at temperature near the upper working limit. This is done to condition the column. While conditioning, the carrier gas is passed through the column at normal flow rates but the column is not connected to the detector, otherwise the detector may get contaminated.

9.5.5.4 Sample preparation and its application

The sample should be prepared in such a way that it does not get retained on the column for excessive period of time. This will lead to poor resolution and peak tailing. The polar groups such as —NH_2, —COOH, —OH are derivatized by methylation, silanisation and trifluoromethylsilanisation to increase the volatile character and distribution coefficients of the compounds.

Solvents such as ether, heptane or methanol are used to dissolve the sample which is then injected with the help of a microsyringe onto the column through a rubber septum in the injection port. The temperature of the injection port is generally maintained higher than the temperature of the column to ensure rapid and complete volatilization of the sample. As a thumb rule, the temperature of the injection port should be 50°C higher than the boiling point of the sample. Too high temperature of the injection port may decompose the sample. Therefore, the temperature of the injection port should be such that it causes rapid vaporization of the sample without decomposing it.

9.5.5.5 Carrier gas

The primary function of the carrier gas is to carry the volatile components through the column. The gas used should be inert and should not react either with sample or with stationary phase. Its secondary purpose is to carry the separated components to the detector so it should be suitable for detector use. It should be readily available in extra pure form and should be inexpensive.

Normally, nitrogen, helium and argon are the three most commonly used carrier gases. The column temperature must be high enough so that analysis can be accomplished in a reasonable length of time. The retention time doubles with every 30°C decrease in column temperature. Lower the temperature better is the resolution and longer analysis time. Therefore, a balance has to be struck between the peak retention time and resolution. Chromatographic separation can be achieved isothermically where a constant temperature is employed or by temperature programming where the temperature is increased gradually.

9.5.5.6 Detectors

Characteristics such as selectivity, sensitivity, response, noise, minimum detectable quantity and linear range, should be given consideration while making choice of the detector. The detector should be simple to operate, inexpensive and as far as possible insensitive to changes in flow rate and temperature. Some of the commonly used detectors are:

(*i*) *Flame ionization detector* (*FID*): Since it responds to almost all organic compounds, it is the most widely used detector. It has a wide linear response range (10^6) and can detect as low a concentration as 1 ng. The detector consists of two electrodes. One of the electrodes is the jet of the flame which is produced by introducing a mixture of hydrogen and air into the detector, while the other electrode is made of brass or platinum wire which is mounted near the tip of the flame. When the carrier gas carrying the sample components emerges from the column, the sample components are ionized in the flame, sensed by the detector giving the signal which is recorded by the recorder. An FID has an upper temperature limit of 400°C and the minimum quantity which can be detected by this detector is of the order of 5×10^{-12} g. A typical FID is shown diagramatically in Fig. 9.11.

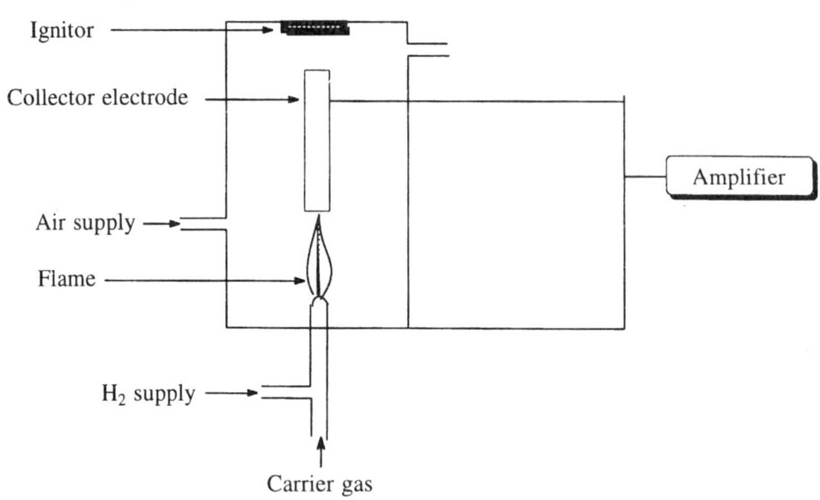

Fig. 9.11 Flame ionization detector.

(*ii*) *Nitrogen phosphorus detector* (*NPD*): As the name suggests, NPD responds efficiently to detect the compounds containing N and/or P. It shows poor response for the compounds which possess neither of these elements. This detector has an upper temperature limit of 300°C, narrow linear response range of 10^4 and detection limits of 10^{-11} g sec^{-1}. The principle of operation is same as that of FID but the NPD has the sodium salt fused onto the electrode system. NPD is widely used in the analysis of organophosphorus pesticides.

(*iii*) *Electron capture detector* (*ECD*): The compounds which have the capacity to capture the electrons are best detected by this detector. Here a radioactive source (^{63}Ni) ionizes the column gas and produces electrons which give a current across the electrodes to which suitable voltage is applied. When the carrier gas carrying the electron capturing substance emerges from the column, it captures the ionized electrons. This results in the drop of the current which is traced on a chart paper by the recorder. The detector has an upper temperature limit of 300°C, high detection sensitivity (10^{-12} g sec^{-1}) but much lower linear range (10^2–10^4). The detector is best suited for the halogen containing compounds and is widely used for the analysis of polychlorinated compounds such as pesticides DDT, dieldrin and aldrin.

9.5.5.7 Amplifiers and recorders

When components leave the column and pass through the detectors discussed above, small and weak electrical signals are produced which are amplified by an amplifier before they are fed to the recorder. Recorders generally consist of two basic parts viz. a chart paper and a pen which moves on the chart paper and traces the signals being activated from the amplifier in form of peaks.

9.5.6 High performance (pressure) liquid chromatography

9.5.6.1 Theory

As in other forms of liquid chromatography, separation in High Performance Liquid Chramotography or High Pressure Liquid Chromatography (HPLC) is also obtained on the basis of partitioning, adsorption, ion exchange or molecular sieving phenomena. The conventional column chromatography suffers from two major drawbacks; namely it is generally a time consuming process and quality of resolution is poor. This is mainly because of the fact that in conventional column chromatography the mobile phase percolates through the column under the force of gravity or by small pressure applied by peristaltic pump. This accounts for the slow flow rate which in addition to extending the time required for elution of the sample creates the problem of peak broadening through diffusion phenomenon resulting in poor resolution. In general, resolution of individual components can be improved by decreasing the particle size of stationary phase. However, in conventional column chromatogrpahy, this is not feasible because the use of finer gel material will further lower the permeability of the column contributing to decreasing flow rate thereby providing greater time for band broadening. The resistance to flow of mobile phase can be overcome by use of high pressure. In recent years, stationary phases of smaller particle size which can withstand high pressures, have been developed, which has facilitated the development of a new chromatographic technique called HPLC, which gives faster and superior resolution with sharp and compact peaks.

9.5.6.2 Equipment of HPLC

The basic HPLC equipment consists of the following components: (i) solvent reservoir, (ii) pump, (iii) damping device, (iv) pressure gauge, (v) sampling device, (vi) column, (vii) detector, (viii) fraction collector and (ix) recorder. The essential features of HPLC equipment are shown diagramatically in Fig. 9.12.

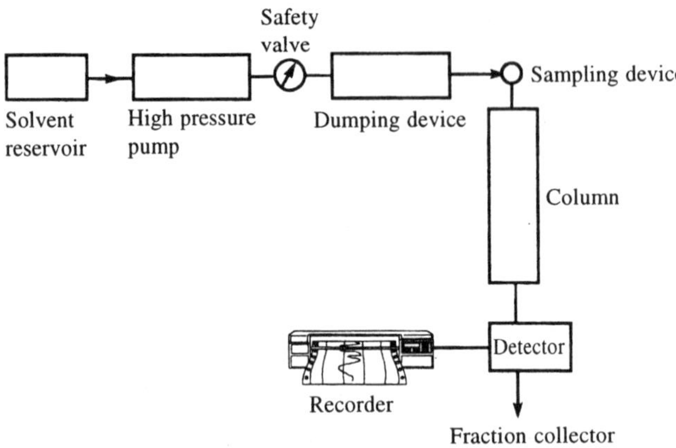

Fig. 9.12 Diagram of essential components of HPLC.

The pump delivers the solvent from the reservoir at a constant flow which is smoothed out by means of damping device. At a place where maximum pressure developes (normally after the pump), a safety valve is incorporated. From the damping unit the mobile phase flows via sample injector into the column. The inlet pressure of the column is monitored with a manometer. After leaving the column, the sample components are monitored by the detector and their tracings drawn by the recorder.

Solvent reservoir: Solvents, as supplied, contain substantial amount of dissolved air. Formation of air bubbles can seriously interfere with satisfactory separation by HPLC because the air bubbles affect the column efficiency and also the solute detection. Thus some of the conventional solvent reservoirs are equipped with degassifiers. In these the solvent reservoir is provided with a heater, temperature regulated magnetic stirrer or a condenser. Alternatively, solvent can also be degassed (by heating, stirring, subjecting it to vacuum, ultrasonic vibrations or bubbling helium gas) before pouring into the reservoir. The solvent used in HPLC must be of high purity as any traces of impurities or suspended material can seriously affect the column efficiency and can interfere with the detection system. Highly pure solvents labelled as "For HPLC" are available commercially, but even with such solvents also it is advisible to introduce a 1-5 μm microfilter prior to the pump to prevent any particulate impurities from entering into the column.

Pumps: In HPLC, the column is quite narrow and is packed with superfine particles. There is a high resistance to flow of solvent and high pressures are, therefore, required to achieve satisfactory and constant flow rates. Therefore, a good pumping system which delivers pulse free solvent flow upto 20 ml/min at pressures upto 300–400 atmospheres is one of the most important features in HPLC. All materials in the pump should be chemically resistant to all the solvents used in HPLC. Various pumping systems operate on the principle of either constant pressure or constant displacement.

Constant pressure pumps facilitate delivery of the solvent at a constant pressure. A gas at high pressure is introduced into the pump which forces, in turn, the solvent from pump into the column. Constant pressure is maintained through out, which causes a decrease in permeability of the column with time which in turn results in decreased flow rates. Such pumps do not compensate for this decrease in flow rate and so provide uniform and pulse less solvent flow.

The second type of pumps are the constant displacement pump. Such pumps displace a constant amount of the solvent from the pump into the column and so maintain a constant flow rate irrespective of the changing conditions within the column. These pumps produce small pulses of flow between two displacements and so pulse dampners are usually introduced between the pump and the column to smoothen the flow and to minimize the pulsing effect. Two commonly used constant displacement pumps are: (i) motor driven syringe type pump and (ii) reciprocating pump which deliver a fixed and constant volume of the solvent onto the column at each stroke.

Sampling device: Sample can be introduced into the column either by a syringe injection through a septum of an injection port into eluent stream or by a sample loop from which it is swept into the column by the eluent. The sample is loaded directly on top of the column to avoid appreciable mixing of the sample with the eluent. In the syringe injection mode, the sample is injected with the help of a microsyringe (which can withstand high pressures) directly onto the column bed. While loading the sample, the system should not be under pressure. Hence, before applying the sample, the pump is turned off and when the pressure is dropped near atmospheric pressure, the sample is introduced. After the sample has been injected, the pump is switched on again. This procedure is known as stop flow injection.

The second type of system is loop injection. Here the sample is introduced with the help of a metal loop of fixed small volume.The loop is filled with the sample and by the appropriately adjusting the sample valve, the solvent from the pump is channelled through the loop. The sample is thus flushed by the solvent from the loop whose outlet opens directly at top of the column bed.

Columns: Since glass tubing cannot withstand pressures in excess of 70 atm, stainless steel precision bored columns with an internal mirror finish for efficient packing, are normally used. These straight columns of 15–50 cm length and 1–4 mm diameter can withstand very high pressures of upto 5.5×10^7 Pa and are relatively corrosion resistant. At the end of the column, homogenously porous plugs of stainless steel or teflon are used to retain the packing material and to ensure the uniform flow of the solvent through the column. At times, repeated application of impure samples such as urine, blood or crude cell extracts results in clogging and the loss of resolving power of the column. To prevent this, a short column of length 1–2 cm and internal diameter equal to that of analytical column is generally introduced between the injector and the analytical column. This short column is called guard column and is packed with material with which analytical column is packed. The guard column retains the solid particles in the sample before sample enters the main column. The guard columns can be replaced at regular intervals.

Matrices and Stationary phases: One of the basic requirements for HPLC is that the packing material which serves as stationary phase or support for stationary phase should be pressure stable, i.e. it must be designed to withstand the operating pressure applied during separation. Three forms of column packing materials are available based on nature of the rigid solid structure:

(i) Totally porous materials or microporous supports: In these supports the micropores ramify through particles which are generally 5–10 μm in diameter.
(ii) Porous layer beads or pellicular supports: These are superficially porous supports where a thin, porous, active layer is coated onto a solid core such as impervious glass beads. The thickness of the porous layer is generally 1–3 μm. The size of glass beads used is between 25–50 μm.
(iii) Bonded pliases: The stationary phase is chemically bonded to an inert support such as silica. The type of particular stationary phase will depend on the separation principle. Some of these stationary phases, with their commercial names, which can be used for different types of chromatographic separations are listed in Table 9.1.

Table 9.1 Examples of HPLC Stationary Phases

Chromatographic separation principle	*Commercial name*	*Nature of stationary phase*	*Type of support*
Partition	ULTRA Pak TSK ODS	Octadecylsilane	Porous
	ULTRA Pak TSK NH_2	Alkylamine	Porous
	Bondapak-C_{18}/Corasil	Octadecylsilane	Pellicular
	μ Bondapak-NH_2	Alkylamine	Porous
Adsorption	Corasil	Silica	Pellicular
	Partisil C_8	Octylsilane	Porous
	Pellumina	Alumina	Pellicular
	Micropak	Alumina	Microporous
Exclusion	Superose	Agarose	Soft gel
	Fractogel TSK	Polyvinylchloride	Semi-rigid gel
	Bio-Glas	Glass	Rigid solid
	Styragel	Polystyrene divinyl benzene	Semi-rigid gel
Ion exchange	Perisorb-KAT	Strong acid	Pellicular
	Partisil-SAX	Strong base	Porous
	Micropak-NH_2	Weak base	Porous
	Partisil-SCX	Strong acid	Porous

Detectors: Detectors are the devices which continuously monitor changes in the composition of the eluent coming out of the column. Most commonly used detectors are refractive index detector, UV detector, electrochemical and fluorometric detectors.

(a) Refractive index detector (RID): Refractive index (RI) of dilute solutions changes proportionally with solute concentration. This relationship is exploited for quantitative detection of solutes in the column eluate. The relationship between change of RI and solute concentration is only moderately dependent on the type of solute, making this a quite universal, yet not very sensitive detection principle. RID can, therefore, be applied to general purpose. This detector suffers from many defects including low sensitivity, tendency to be affected by temperature or flow speeds and incompatibility for being used in gradient elution unless chosen solvents are of identical RI. It measures the bulk RI of sample eluent system. Hence any substance whose RI differs sufficiently from that of eluent can be detected. To attain adequate sensitivity, the temperature of the eluent and measuring cell is held constant to $\pm 0.001°C$. Variations in flow rates also interfere with response of differential refractometer. Hence very good damping is essential for the pumps producing pulsating flow.

(b) UV-VIS-absorption detectors: The basis of quantitative absorbance photometry is Lambert-Beer's law, i.e. the absorbance of a solution is proportional to the concentration of the absorbing solute, the light path length and the extinction coefficient. Fixed wavelength detectors utilize lamps which emit light of a few discrete wavelengths. The most common of these lamps is the low pressure mercury lamp emitting over 90 per cent of its light at 254 nm. Lower wavelength lamps are available such as zinc lamps (214 nm) and cadmium lamps (229 nm). The combination of the lamp and a filter determines the fixed operating wavelength of the detector. Variable wavelength (VW) detectors use a light source with a continuous emission spectrum and a continuously adjustable (narrow) band filter, called monochromator. The most common light source for these detectors is the deuterium lamp whose usable emission spectrum ranges from about 190 nm to about 350 nm, with an intensity maximum between 220 and 240 nm. Above 300 nm, the output intensity is low, therefore, some VW detectors have an additional or optional tungsten lamp, which can be used at wavelength above 350 nm. In recent years, specially designed VW detectors for HPLC have been introduced which allow automatic rapid change of wavelength setting within 1 to 2 seconds or less across their entire wavelength span, which typically ranges from about 190 nm to about 600 nm.

(c) Electrochemical detectors: Methods used in HPLC based on electroanalysis can be classified as bulk property and solute property electrochemical detectors. Bulk property electrochemical detectors respond to a change in an electrochemical property of the bulk liquid flowing through the measuring cell, whereas the solute property electrochemical detectors respond to a change in voltage (potentiometry) or current (voltammetry or coulometry) when an analyte passes through the cell. The potentiometric detectors have not been commercialized. Voltammetric and coulometric detectors on the contrary are offered by various manufacturers.

(d) Fluorescence detectors: The quantity of fluorescent light emitted from excited molecules in dilute solutions is proportional to intensity of excitation source, illuminated volume of the sample solution, quantum efficiency of fluorescence of sample and the concentration of the solute to be detected. Ideally, fluorescence radiation, as a result of suitable excitation of the sample molecules, is measured against a dark background. Therefore, the main source of detector noise is dark current noise of the photodetector, which is mainly determined by temperature. As a rule for every 10°C increase in temperature the dark current noise of the photodiode gets doubled.

9.6 Laboratory Exercises

In this Section a few selected laboratory exercises have been described to familiarize the students with techniques of chromatography.

9.6.1 EXPERIMENT: Separation and Identification of amino acids by descending paper chromatography

Principle

Amino acids in a given mixture or sample aliquot are separated on the basis of differences in their solubilities and hence differential partitioning coefficients in a binary solvent system. The amino acids with higher solubilities in stationary phase move slowly as compared to those with higher solubilities in the mobile phase. The separated amino acids are detected by spraying the air dried chromatogram with ninhydrin reagent. All amino acids give purple or bluish purple colour on reaction with ninhydrin except proline and hydroxyproline which give a yellow coloured product. The reactions leading to the formation of purple complexes are given below:

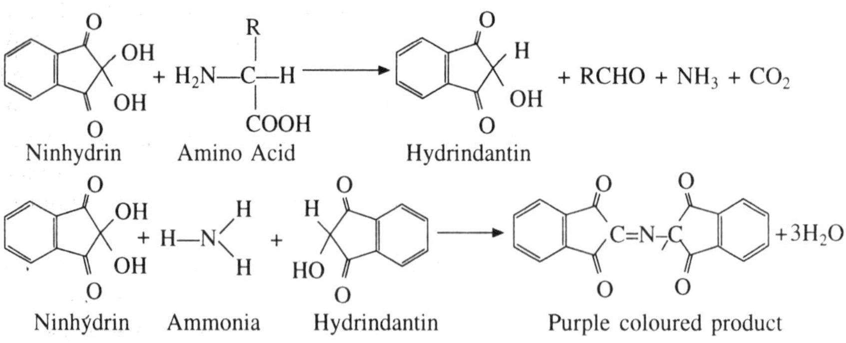

Materials and Reagents

1. Whatman No. 1 filter paper sheet.
2. Micropipette/microsyringe.

3. Hair drier.
4. Sprayer.
5. Oven set at 105°C.
6. Chromatographic chamber saturated with water vapours.
7. Developing solvent: Take butanol, acetic acid and water in the ratio of 4:1:5 in a separating funnel and mix it thoroughly. Allow the phases to separate out completely. Use the lower aqueous phase for saturating the chamber. The upper organic phase is used as mobile phase.
8. Ninhydrin spray reagent: Prepare fresh by dissolving 0.2 g ninhydrin in 100 ml acetone.
9. Standard amino acids: Prepare solutions of authentic samples of amino acids such as methionine, tryptophan, alanine, glycine, threonine etc. (1 mg/ml of 10% *iso*-propanol).
10. A sample containing mixture of unknown amino acids.

Procedure

1. Take Whatman No. 1 filter paper and lay it on a rough filter paper. Throughout the experiment care should be taken not to handle the filter paper with naked hands and for this purpose either gloves should be used or it should be handled with the help of folded piece of rough filter paper.
2. Fold the Whatman No. 1 filter paper about 2.0–2.5 cm from one edge. Reverse the paper and again fold it 2 cm further down from the first fold.
3. Draw a line across the filter paper with a lead pencil at a distance of about 2 cm from the second fold. Put circular marks along this line at a distance of 2.5 cm from each other.
4. With the help of a micropipette or microsyringe apply 20 μl of solution of each standard amino acid on a separate mark. Also apply spot of the sample or mixture to be analyzed, preferably on the mark at center of this base line. The size of the spot should be as small as possible so that the developed spots are compact and do not overlap. If necessary, the wet sample spot should be dried with hair drier before applying additional aliquot.
5. Hang the filter paper in a chromatographic chamber which has previously been saturated with aqueous phase of the solvent system. This is done by keeping Petri plates containing the aqueous phase at bottom of the chamber. The paper is hung from the trough/tray and a glass rod is kept to hold it in place. Care should be taken to ensure that the base line should not get submerged when the mobile phase is added to the trough otherwise the spotted material would get dissolved in the solvent.
6. Close the chamber firmly so that it is airtight. Allow sufficient time for cellulose fibres of the paper to get fully hydrated.
7. Pour the mobile phase through the holes provided on the lid of the chamber into the trough. Replace the rubber bungs in the hole and allow the mobile phase to run down the paper till the solvent front reaches about 5 cm from the opposite edge.

8. Remove the paper and mark the solvent front with lead pencil and let it dry at room temperature.
9. Spray the filter paper (chromatogram) with ninhydrin reagent and after drying it at room temperature, transfer it to an oven at 105°C for 5–10 min.
10. Blue or purple coloured spots would appear on the paper. Mark the boundary of each spot with lead pencil.
11. Measure the distance between the center of the spots and also the distance of the solvent front from the base line.
12. Calculate the R_f value of standard amino acids as well as those in the given mixture or sample as follows:

$$R_f = \frac{\text{Distance travelled by unknown amino acid}}{\text{Distance travelled by the solvent system}}$$

13. Identify the amino acids in the mixture or sample by comparing their R_f values with those of the reference standards.

Note: It is advisable to carry out chromatography in three different solvent systems before the identity of amino acid in the mixture or sample can be established with any degree of certainty.

Observations

Distance travelled by the solvent front from base line = x cm
Distance travelled by glycine from base line = a cm
Distance travelled by alanine from base line = b cm
Distance travelled by threonine from base line = c cm
Distance travelled by methionine from base line = d cm
Distance travelled by spot no.1 in sample from base line = a cm

Calculations

R_f value of glycine = a/x
R_f value of alanine = b/x
R_f value of threonine = c/x
R_f value of methionine = d/x
R_f value of spot no.1 = a/x

Conclusion

The sample contains glycine since R_f value of spot no. 1 is identical to that of authentic glycine standard.

Precautions

1. Do not touch the paper with naked hands because sweat on hands contains significant amount of amino acids.
2. The spots of the applied sample should be as compact as possible. Larger the spot, poorer will be the resolution.

3. At the time of fixing paper in chromatography chamber, it should be ensured that the base line on which the sample has been applied does not dip into the solvent otherwise the sample might get washed away in the solvent.
4. Allow sufficient time for filter paper to absorb sufficient water (which will act as a stationary phase) before pouring the solvent into the trough/ tank. Inadequate conditioning or equilibration will result in improper or poor quality resolution.
5. Solvent front should advance in a straight line and should not be zig-zag or sloping but should be parallel to the base line.
6. Dry the paper thoroughly before spraying with the detection reagent. Wet paper may interfere with the appearance of evenly shaped compact spots.
7. The chromatography should be carried out in temperature controlled room because any fluctuation in the temperature would cause the uneven flow of the solvent and may alter the R_f value.

9.6.2 EXPERIMENT: Separation and identification of amino acids by ascending paper chromatography

Principle
Same as in EXPERIMENT 9.6.1.

Materials and Reagents

Same as given in EXPERIMENT 9.6.1 except that cylindrical chromatography chambers are needed for this experiment.

Procedure

1. Take Whatman No. 1 filter paper sheet of appropriate size so that it can be rolled into a cylinder and can be accommodated in the cylindrical chromatographic jar.
2. Draw a base line 2 cm from one of the breadthwise edges of the paper. Put small circular marks along the base line in such a way that the distance from the edge of the paper and the first spot and the distance between the adjacent spots is not less than 2.5 cm.
3. Apply 20 μl aliquots of the standard amino acids and of the sample on different spots. Diameter of the spotted material should be as small as possible and, if required, the applied solution may be dried prior to loading additional volume.
4. Roll the paper into a cylinder, fasten its edges with a paper clip. Pour sufficient volume of the mobile phase into the chromatographic jar which has been earlier saturated with water vapours by lining the tank with filter paper saturated with aqueous phase of the solvent system.
5. Gently place the rolled filter paper upright in the jar ensuring that it does not touch the sides of the chamber and at the same time taking care that the base line where the spots have been applied does not dip into the solvent.

6. Close the tank with an airtight lid or a glass plate to which sufficient amount of silicon grease has been applied.
7. Leave the set up undisturbed and allow the solvent to move up till it reaches about 5 cm from the upper edge.
8. Remove the chromatogram from the chamber and air dry it.
9. Spray the paper with ninhydrin reagent and let it dry again at room temperature prior to transferring it to an oven at 105°C for 5–10 min. Locate the position of amino acids from the bluish or purple coloured spots on the chromatogram.
10. Calculate the R_f values of the standard amino acids and those in the sample or mixture as described in EXPERIMENT 9.6.1.
11. Identify the amino acids in the mixture or sample by comparing R_f values with those of applied standard amino acids.

Precautions

Observe all the precautions mentioned in EXPERIMENT 9.6.1.

9.6.3 EXPERIMENT: To separate and identify the amino acids in a given mixture by two dimensional paper chromatography

Principle

Amino acids having very close R_f values in a particular solvent system may appear as a single or overlapping spots in a single dimensional chromatography and may be mistaken as one component. They can be separated into individual components by developing the chromatogram again in a direction perpendicular to the first run in a second solvent system in which they have different R_f values. The main limitation of this method is that only one spot either of the sample or of a standard amino acid can be applied on each filter paper sheet necessitating running of a large number of chromatograms for the standard amino acids.

Materials and Reagents

1. Same as listed in EXPERIMENT 9.6.1 except that an additional chromatographic chamber for the second solvent system is required.
2. Solvent system No. 2: Phenol:water (80:20, *w*/*v*) is used as second solvent system. Add 125 ml of water to 500 g of phenol. Add a few drops of ammonia (0.88%) to this mixture just before use (CAUTION—Phenol is corrosive and can cause burns on the skin).
3. Standard amino acids: Prepare 1% solution of standard amino acids such as aspartate, glycine, serine, arginine in 10% *iso*-propanol (*v*/*v*).

Procedure

1. Lay the chromatographic paper sheet flat on the rough filter paper using gloves.
2. Draw a base line 5 cm from one of the edges of the paper.
3. Draw another line perpendicular to the first line again 5 cm away from the adjacent edge.

4. Apply 60 μl of the sample solution or given mixture containing unknown amino acids at the point of intersection of these two lines. The sample should be applied in small volumes at a time with the help of a micropipette with intermittent drying to ensure that the zone of applied solution is as small as possible.
5. Repeat the same procedure for a mixture of three standard amino acids using a separate chromatographic sheet for each mixture. The composition of mixture of standard amino acids should be such that each amino acid is present in at least two different mixtures so that its identity can be established.
6. Hang the paper in the chromatographic tank whose interior has previously been saturated with aqueous phase of solvent system No. 1 (Butanol:Acetic acid:Water mixture is 4:1:5).
7. After allowing an equilibration period of half an hour, pour the solvent No. 1 into the trough of the chamber and let it run till it is about 10 cm from the opposite edge of the paper.
8. Take the paper out, air dry it, turn it at 90° angle and now develope the paper in the second chromatographic chamber using the solvent system No. 2 (Phenol:Water).
9. Remove it when the solvent has travelled upto about 10 cm from the opposite end.
10. Dry it at room temperature and spray it with ninhydrin reagent. After air drying it, keep the chromatogram in an oven at 105°C for 10 min. Mark the blue and/or purple coloured zones which appear on the paper.

Observations

Calculate the R_f value of the standard amino acids and those in given mixture as given in EXPERIMENT 9.6.1 in both the solvents. From these values identify the amino acids in the given mixture.

9.6.4 EXPERIMENT: Separation and identification of sugars by adsorption TLC

Principle

Sugars get separated on the basis of differential adsorption onto silica gel. The sugars which have higher affinity for stationary phase are adsorbed more strongly and hence they migrate slowly when mobile phase moves over them. On the other hand, those having lower affinity for stationary phase, are weakly adsorbed and are more easily carried by the mobile phase. The separated sugars are then located as coloured zones by spraying TLC plates with aniline diphenylamine reagent.

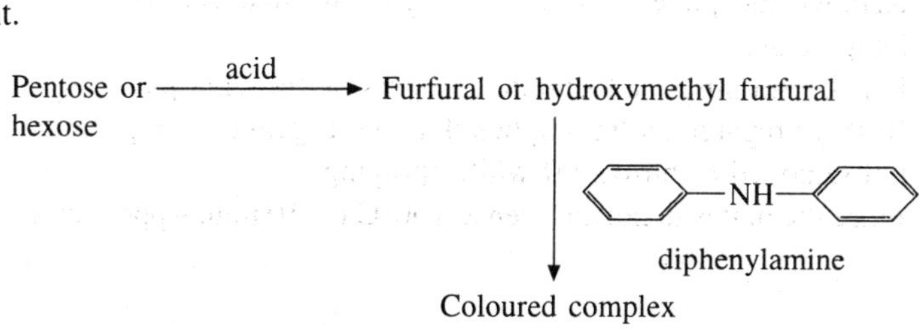

Materials and Reagents

1. TLC chromatographic tank
2. Glass plates (20 × 20 cm)
3. Spreader
4. Micropipettes/microsyringes
5. Oven maintained at 105°C
6. Hair drier
7. Sprayer or atomizer
8. Solvent System: Prepare a mixture of ethyl acetate: *iso*-propanol:water: pyridine (26:14:7:2, *v*/*v*)
9. Standard sugar solutions: Prepare 1% solution of standard sugars such as glucose, ribose, lactose etc. in 10% *iso*-propanol (*v*/*v*). For mixture in which the sugars have to be identified, mix the sugar solutions in equal proportions
10. Aniline diphenylamine reagent: Mix 5 volumes of 1% aniline and 5 volumes of 1% diphenylamine in acetone with 1 volume of 85% phosphoric acid

Procedure

1. Place thoroughly cleaned and dried glass plates (20 × 20 cm) on a flat plastic tray side by side with no gap between the two adjacent plates.
2. Prepare a slurry of the stationary phase (Silica Gel G) free of clumps in water or in an appropriate buffer.
3. Spread a uniform layer of 250 μm thickness with the help of a spreader or an applicator by moving it from one end of the tray to its other end.
4. Activate the plates by keeping them at 105°C for 30 min. Allow the plates to cool in a desiccator before use.
5. Gently put marks in a straight line with the help of a pin at a distance of about 2 cm from one edge of the plate. The adjacent marks should be about 1.5–2.0 cm apart from each other. Extreme care should be taken that silica does not get scratched off while putting these marks.
6. Carefully apply the solution of individual standard sugars and the mixture or alcoholic extract of the sample on the separate marked spots.
7. Gently put marks or draw a line 1 cm from the opposite edge.
8. Place the plate in chromatographic tank which has already been equilibrated with the solvent taking care that base line on which samples have been applied does not dip into the solvent
9. Close the chromatographic tank with air tight lid and allow the solvent to ascend along the plate by capillary action till the solvent reaches the marked line on the upper side of the plate. This may take about 90 min.
10. Remove the plate from chromatographic tank and let it dry at room temperature.
11. For determining the location of sugars on the TLC plates, spray it with freshly prepared aniline-diphenylamine reagent ensuring that silica gel is not removed or blown off while spraying.
12. Place the plates in hot air oven at 100°C for 10 min. Appearance of bluish

spots on the white background indicates presence of sugars at that region of the plate.

13. Measure the distance from the base line to the center of the colored spot and calculate the R_f value of each sugar as described in EXPERIMENT 9.6.1.
14. Identify the sugars in the given mixture or sample by comparing their R_f values with those of sugar standards.

Precautions

A number of precautions should be observed while performing TLC. Some of these are:

1. Thoroughly cleaned glass plates free of any greasy spots or finger marks should be used.
2. Thickness of the layer should be uniform throughout the length of the plate.
3. The slurry of the chromatographic media should be free of any clumps. This can be ensured by vigorously shaking it in an Erlenmeyer flask or by gently preparing the slurry in pestle and mortar to ensure uniform mixing.
4. The TLC plates should be activated at recommended temperature and duration. Poor resolution of components occurs on over or under activated plates.
5. The layer of chromatographic media should not get scraped off at the time of putting marks of application of samples.
6. Size of the applied spot should be as small as possible. If large volume of the sample has to be spotted, then it should be done in small aliquots with an intermittent drying. Overloading of the sample should be avoided.
7. The chromatographic tank should be airtight and chromatography should be performed under temperature controlled conditions.

9.6.5 EXPERIMENT: To identify lipids in a given sample by thin layer chromatography

Principle

In biological materials, lipids are found as lipoprotein complexes and these have to be extracted. Lipids, being soluble in non-polar organic solvents and proteins being soluble in polar aqueous solvents, the efficient lipid extraction can be achieved only when an aqueous solvent like ethanol or methanol is included in the non-polar organic solvent like chloroform and diethyl ether. This would help in breaking the lipoprotein complexes. Extracted lipid components can be separated on TLC based on their differential mobility along the porous stationary phase such as silica gel and these can be located by spraying the plates with either 2′,7′-dichlorofluorescein or 50% sulfuric acid.

Materials and Reagents

1. TLC Tank

2. Glass plates (20 × 20 cm) for TLC
3. Oven set at 110°C
4. Ultraviolet lamp
5. Solvent system: Petroleum ether (b.p. 60–70°C) or hexane:diethyl ether: glacial acetic acid (80:20:1, *v*/*v*)
6. Lipid standards: Various lipids such as cholesterol acetate, vitamin A palmitate, triacyl glycerol (e.g. trioleate, tripalmitate, tristearate)
7. Sulfuric acid (50%, *v*/*v*)
8. 2′,7′-dichlorofluorescein: Prepare 0.2% solution of 2′,7′-dichlorofluorescein in 95% (*v*/*v*) ethanol

Procedure

1. Extraction of lipids from sample: Grind 1 g of the tissue in the extraction solvent [either ethyl ether: ethanol (3:1) or chloroform:methanol (2:1)] in pestle and mortar. Transfer the homogenate to a separating funnel. Shake the contents vigorously and allow it to stand till the two phases have completely separated. Drain out the lower organic layer which contains the lipids. Evaporate the solvent under vacuum and keep the concentrated lipid extract protected from light under N_2 atmosphere.
2. Prepare the TLC plates using Silica Gel G as the adsorbent as described in Experiment 9.6.4.
3. Activate the TLC plates at 110°C for 30 min, cool them in a desiccator and spot the lipid samples, standards as well as unknown, as outlined in Experiment 9.6.4.
4. Develop the plates in the solvent system consisting of petroleum ether (b.p. 60–70°C) or hexane:ethyl ether:glacial acetic acid (80:20:1) till the solvent has travelled upto 1cm from the opposite side of the plate.
5. Remove the plate and allow it to air dry.
6. Locate the lipid spots by either of the following methods:

   (i) Spray the plate with 2′,7′-dichlorofluorescein and examine it under UV light. Lipids show up as green fluorescent regions against the dark background.
   (ii) Spray the plate carefully with 50% H_2SO_4 and heat it in an oven at 110°C for 10 min. Areas containing lipids get charred and appear as black spots.

7. Calculate the R_f value of the lipid components in the sample as described in Experiment 9.6.1 and identify them by comparing their R_f values with lipid standards.

Precautions
As given in Experiment 9.6.4.

9.6.6 EXPERIMENT: Separation of pigments from leaves or flowers by adsorption column chromatography

Principle

Different pigments get adsorbed to alumina to different extents. They can be selectively desorbed by using mobile phase of increasing polarity in a stepwise manner.

Materials and Reagents

1. Leaves or flowers
2. Pestle and mortar
3. Glass column or a burette
4. Whatman No. 1 filter paper
5. Alumina or icing sugar
6. Benzene:methanol (2:1)
7. Sodium sulfate (anhydrous)
8. Acetone

Procedure

(a) Preparation of extract:

1. Homogenize the leaves or flowers (5 g wet weight) in a pestle and mortar, using sand as an abrasive in 20 ml of benzene:methanol (2:1) adding a small amount of this extractant at a time.
2. Filter the extract through Whatman No. 1 filter paper and transfer the filtrate to a separating funnel.
3. Add 10 ml of water to the filtrate and after shaking the contents and allowing the phases to separate out, drain out the lower aqueous methanol layer. Repeat this step. Avoid very vigorous shaking.
4. Collect the benzene layer in a beaker and add small amount of solid anhydrous Na_2SO_4 to remove the traces of moisture.
5. Decant the clear benzene layer to another beaker and concentrate the extract by evaporating the solvent on a boiling water bath.

(b) Preparation of column:

1. Mount a burette or a glass column vertically on a burette stand with the help of clamps.
2. Place lightly a plug of glass wool at the base of the burette and close the stopcock or outlet at the bottom of the column.
3. Take 5 g of alumina or icing sugar (adsorbent) previously dried at 120°C for 8 h and prepare its slurry in benzene. Pour the slurry carefully into the column or burette, by gentle tapping of the column so that no air bubbles get trapped in the adsorbent.
4. Allow the adsorbent to settle by opening the outlet. After the adsorbent has completely settled, add 20 ml of benzene and let it pass out of the column. Care should be taken not to let the adsorbent to get dried.

When about 1 cm layer of the solvent remains at the top of the chromatographic bed, close the outlet.

(c) Sample application:

1. Allow the solvent at the surface of the column to drain out slowly and transfer the leaf or flower extract with the help of pipette without disturbing the surface of the column adsorbent. Let it enter into the column and then add a few drops of benzene to wash the traces of the extract sticking to the wall of the column.
2. Add 20 ml more benzene to wash out the column of any unadsorbed material.

(d) Column development:

1. For desorption of the adsorbed substances change the polarity of the solvent in a stepwise manner. After 20 ml of benzene has passed through the column add 10 ml of 5% acetone (*v*/*v*) in benzene and let it percolate through the column and collect 1–2 ml fractions of the effluent from the outlet. Continue increasing the concentration of acetone in benzene at every succeeding step. Finally pass pure acetone through the column.

Observations

Note the change in the colour of the collected fractions. In case of the leaf extract, the initial fractions are colourless followed by yellow coloured and then by the green coloured ones. The colourless fractions do not contain any pigments but it is quite possible that these fractions may contain some UV absorbing materials.

9.6.7 EXPERIMENT: Separation of amino acids by ion exchange column chromatography using cation exchanger

Principle

Ion exchange chromatography can be used for separation of substances which possess a net electrical charge. Anion exchangers reversibly bind negatively charged compounds through electrostatic forces whereas positively charged molecules interact with cation exchangers. Different compounds are held by ion exchangers with varying strengths depending upon charge. Amino acids are amphoteric substances and have a net charge of zero at isoelectric pH. At pH below the isoelectric point, the amino acids are positively charged. At very low pH of 1.0 almost all amino acids (including acidic amino acids) exist as cation and hence these can be separated on a cation exchanger.

Materials and Reagents

1. Chromatographic column (2.5 × 25 cm)
2. Colorimeter
3. Dowex—50 resin in 0.05 M citrate buffer pH 3.0
4. Citrate buffer 0.05 M, pH 3.0

5. Citrate buffer 0.05 M, pH 6.0
6. Citrate buffer 0.05 M, pH 9.0
7. Amino acids (aspartic acid, alanine, lysine and histidine; 2 mg of each amino acid/ml of 0.1 M HCl, pH 1.0
8. 4 N HCl
9. Ninhydrin reagent

Procedure

(a) Preparation of ion exchanger: Dowex-50 is a cation exchanger. Before packing it into the column, it must be fully saturated with H^+ first and this H^+ form can be converted to Na^+ form so that the resin can function as cation exchanger.

1. Suspend 10 g of Dowex-50 into sufficient volume of 4 N HCl for 15 min to ensure that the resin is saturated with H^+ ions.
2. Filter the suspension and wash it repeatedly with distilled or deionized water till the pH of the filtrate is neutral.
3. Transfer the resin into 2N NaOH for 15 min to get Na^+ form. Wash it till the pH of the filtrate is neutral.

(b) Equilibration of the resin:

1. Suspend the resin in citrate buffer (pH 3.0) and allow it to stand for 1 h.
2. Mount the column upright and pour the suspension with the help of a glass rod while tapping the column gently.
3. Allow the suspension to settle down, open the outlet and pass two to three bed volumes of the citrate buffer (0.05 M, pH 3.0) through the column. This will fully equilibrate the resin to pH 3.0. When only a thin layer of the buffer remains at the top of the resin, stop the flow by closing column outlet.

(c) Sample loading:

1. Open the outlet and let the buffer at the top drain into the column surface. Close the stopcock.
2. Load the column with 1 ml of mixture of amino acids in 0.1 M HCl having a pH value of 1.0.
3. Add a small amount of buffer to wash the traces of the sample from inside walls of the column and when the level just reaches the surface, close the stopcock.

(d) Development of column:
All the amino acids would be in cationic form at pH 1.0 and so would be bound to the cation exchanger. Gradient elution using increasing pH and ionic strength facilitates sequential elution of the bound amino acids.

9.6.8 EXPERIMENT: Desalting of protein sample by gel filtration

Principle

Separation is based on the fact that proteins are macromolecules whereas the salts are low molecular weight substances. When the sample is passed through a column packed with Sephadex G-10 or G-25, proteins remain totally excluded from the gel and move with the void volume while salts enter into the gel particles and take longer time to get eluted.

Materials and Reagents

1. Glass column
2. Sephadex G-25
3. Bovine serum albumin
4. Sodium phosphate
5. 0.1 M Tris HCl buffer (pH 7.0)

Procedure

1. Suspend 5 g of Sephadex G-25 (coarse) in 0.1 M Tris-HCl buffer (pH 7.0) and swell it by keeping it for 3–4 h at room temperature with intermittent stirring.
2. Decant the excess of buffer alongwith any suspended fine particles to obtain a slurry of reasonable thickness.
3. Fix the column upright on a burette stand with the help of clamps.
4. Keep the outlet of the column closed, place a plug of glass wool at the base of the column and pour a small volume of the buffer or water into the column.
5. Pour the slurry gradually into it along the inner surface of its wall and, if necessary, with gentle tapping to expel any air bubbles.
6. Allow the chromatographic media to settle down evenly and then open the outlet to drain excess liquid from the column.
7. Place a filter paper disc or a nylon gauze on the surface of the packed bed to prevent disturbance of the upper layer while loading the sample or feeding the eluent into the column.
8. Prepare a mixture of 10 mg of bovine serum albumin and 40 mg of sodium phosphate in 2 ml Tris-HCl buffer (pH 7.0).
9. Apply it to chromatography column by any one of the two methods:

   (i) The mobile phase at the top of the packing is drained out till the bed surface gets exposed. Close the outlet and gently apply the sample uniformly over the bed surface with pipette and the loaded sample is then allowed to just enter into the column by opening the outlet. A small amount of mobile phase (or buffer) is added to wash the traces of the sample into the column.

   (ii) In the second method, sucrose or glycerol, upto the concentration of 1% is added in the sample to increase its density. This sample is applied just above the surface of bed directly through the layer of the buffer in the column bed. Since the sample has higher density, it automatically settles on surface of the gel. Then open the outlet to facilitate entry of the sample into the column. When using this procedure, it is advisable to ensure that addition of glycerol or sucrose does not interfere with the separation and subsequent analysis of the separated compounds.

10. Add sufficient amount of buffer on top of the column and connect it to the buffer reservoir.
11. Collect fraction (2 ml each) either manually or using an automatic fraction collector. Determine the protein content either by monitoring absorbance at 280 nm or by Lowry's method (EXPERIMENT 4.4.2) and phosphate ions (method described in Step 8 of EXPERIMENT 3.4.2) in each of the fractions.
12. Plot a graph of concentration of protein and phosphate vs fraction number or elution volume (Fig. 9.13).

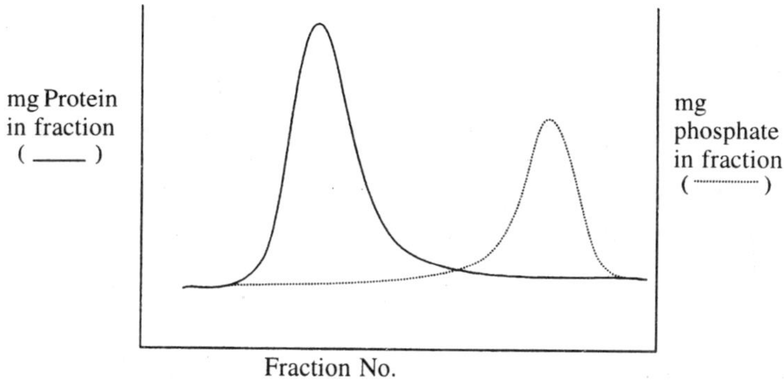

Fig. 9.13 Elution profile of protein and phosphate on Sephadex G-25 column.

9.6.9 EXPERIMENT: Concentration of dilute protein solutions using Sephadex G-25

Principle

When the dry beads of Sephadex G-25 are added to a dilute protein solution, they start swelling and in the process absorb water. The proteins being macromolecules are excluded from the swollen gel and hence remain in the solution. Due to absorption of water, volume of the solution decreases without affecting amount of high molecular weight solutes such as proteins in it resulting in concentration of the solution.

Materials and Reagents

1. Bovine serum albumin
2. NaCl (1 mM)
3. Sephadex G-25

Procedure

1. Dissolve 20 mg bovine serum albumin in 100 ml of 1 mM NaCl. Retain 1 ml of this solution for protein estimation either by spectrophotometric or by Lowry's method (see EXPERIMENT 4.4.2).
2. Add 5 g of dry Sephadex G-25 (coarse) to the remaining protein solution.
3. Let it stand at room temperature for 30 min to swell and then centrifuge it at 3000 × *g* for 10 min. Carefully decant the supernatant into the measuring cylinder and note its volume. Again retain 1 ml for protein estimation.
4. Subject the supernatant to the same treatment as described in Steps 2 and 3 twice or thrice every time recording the volume of the supernatant and keeping 1 ml aside for protein estimation. Decrease the amount of added Sephadex G-25 progressively (say from 5 g in first step to 1 g in the final step) at each step.
5. Determine the concentration of protein in 0.5 ml of the supernatant obtained at each step and express the concentration of protein in terms of amount of protein/ml of the solution.

Observations

It will be noted that at each step the volume of the protein solution decreases but the concentration of protein (amount of protein/ml of solution) increases.

9.6.10 EXPERIMENT: Determination of molecular weight of a given protein by gel filtration

Principle

During gel filtration, solutes are separated primarily on the basis of their molecular size. Due to molecular sieving effect, the large molecules are eluted from the column first followed by compounds of smaller molecular mass. A plot between K_d or elution volume *vs* $\log_{10}$ molecular weight gives a straight line. Molecular weight of a given protein can be established from its elution volume through gel filtration column which has previously been calibrated with standard marker proteins of known molecular weight.

Materials and Reagents

1. Glass column (2.5 × 70 cm)
2. Sephadex G-100
3. HEPES—NaOH buffer (20 mM; pH 8.0)
4. $MgCl_2$
5. DTT
6. Glycerol
7. Standard protein markers:
   (i) β-amylase (200 Kd)
   (ii) Alcohol dehydrogenase (150 KD)
   (iii) Bovine serum albumin (66 KD)
   (iv) Carbonic anhydrase (29 KD)

(v) Cytochrome *c* (12.4 Kd)
(vi) Blue dextran

Procedure

1. Suspend 15 g of Sephadex G-100 in 20 mM HEPES-NaOH (pH 8.0) buffer containing 5 mM $MgCl_2$ and 5 mM DTT for 5 h in boiling water bath.
2. Allow it to cool and pack it into the glass column taking all suggested precautions for avoiding entrapment of any air bubbles in the gel bed. (See EXPERIMENT 9.6.8 Steps 3–7.)
3. Equilibrate the column by passing buffer equivalent to 2 to 3 volumes of the bed volume.
4. Find out the void volume (V_o) of the column by determining elution volume of blue dextran solution (2 mg/ml) through the column. Again pass 2 bed volumes of the starting buffer.
5. Apply the mixture of the standard marker proteins of known molecular weight and elute the column with the buffer.
6. Collect fractions of 2 ml each and determine the protein content in each of these fractions either by Lowry's method or by recording absorbance at 280 nm (see EXPERIMENT 9.6.8).
7. Determine the elution volume of the standard proteins and prepare a graph of $\log_{10}$ molecular weight vs V_e or $\left(K_d = \frac{V_e - V_o}{V_i}\right)$. Again pass 2 volumes of the starting buffer.
8. Layer the sample containing protein whose molecular weight has to be determined. Elute it and collect the fractions of 2 ml each. Test each fraction for the presence of protein and determine its V_e.
9. Determine the molecular weight of the given protein from the calibration curve prepared in Step 7 as shown in Fig. 9.14.

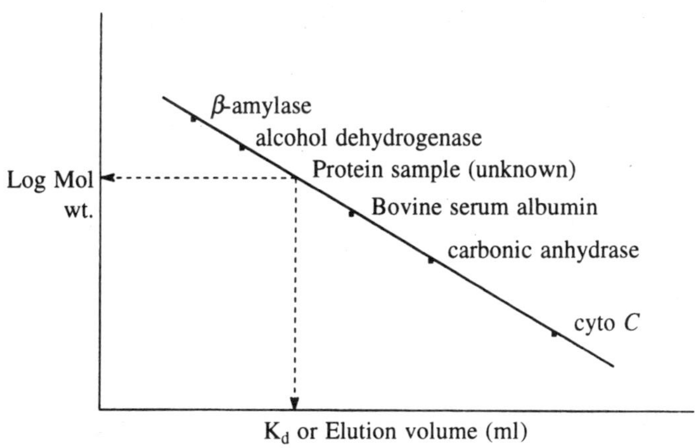

Fig. 9.14 A curve showing relationship between elution volume or K_d vs log molecular weight of proteins.

Suggested Further Reading

Boyer R.F. (1986). Modern Experimental Biochemistry, Addison Wesley Publishing Company.

Clark J.M. and Switzer R.L. (1977). Experimental Biochemistry, Freeman and Company, San Francisco.

Clement R.E. (1990). Gas Chromatography—Biochemical, Biomedical and Clinical Applications, Wiley Interscience, Toronto.

Hearn M.T.W. (1991). HPLC of Proteins, Peptides and Polynucleotides, VCH Publishers Inc., New York.

Hunt B.J. and Holding S.R. (1989). Size Exclusion Chromatography, Blackie, New York.

Robyt J.F. and White B.J. (1987). Biochemical Techniques: Theory and Practice. Brooks/ Cole Publishing Company, Monterey, California.

Wilson K. and Walker J. (1996). Practical Biochemistry: Principles and Techniques, Cambridge University Press, Cambridge.

Wood E.J. (1989). Practical Biochemistry for Colleges, Pergamon Press, New York.

10. Electrophoresis

10.1 Introduction

Electrophoresis is the study of movement of charged particles in an electric field. The earliest technique, pioneered by A.Tiselius in 1937, was a *moving boundary electrophoresis* wherein electophoresis was carried out in buffer solution and had a drawback of sample diffusion and mixing of sample components due to convection currents. These limitations have been overcome in *zone electrophoresis* in which a buffer saturated solid matrix is used as an electrophoretic medium in place of a buffer solution. The sample is applied onto the medium as a narrow zone or band and the molecules with different mobilities travel as distinct zones which gradually separate from each other. The technique is widely used for separation and analysis of a large number of biomolecules such as amino acids, peptides, proteins, nucleotides and nucleic acids.

10.2 Principle of Electrophoresis

Any charged ion or molecule migrates when placed in an electric field. The rate of migration of a compound depends on its net charge, size, shape and the applied current. This can be represented by the following equation:

$$v = \frac{E \cdot q}{f}$$

where v = velocity of migration of the molecule
E = electric field in volts/cm
q = the net electrical charge on the molecule
f = the frictional coefficient which is a function of the mass and shape of the molecule

The movement of a charged molecule in an electric field is often expressed in terms of electrophoretic mobility (μ) which is defined as the velocity per unit of electric field.

$$\mu = \frac{v}{E} = \frac{E \cdot q}{f \cdot E} \left(\text{since } v = \frac{E \cdot q}{f}\right)$$

$$= \frac{q}{f}$$

For molecules with similar conformation, f varies with size but not with the shape. Thus electrophoretic mobility (μ) of a molecule is directly proportional to the charge density (charge/mass ratio). Molecules with different charge/mass ratio migrate under the electric field at different rates and hence get separated. This is the underlying basic principle for all the electrophoretic techniques.

10.3 Types of Electrophoresis

Depending upon the nature of support medium, zone electrophoresis is of different types such as paper, starch, polyacrylamide and agarose gel electrophoresis.

10.3.1 Paper electrophoresis

In this form of electrophoresis, the sample is applied as a circular point or spot on a strip of Whatman filter paper or cellulose acetate paper moistened with the buffer solution. The ends of the paper are immersed in separate reservoirs containing buffer and in which the electrodes are fitted. Upon passing electric current, the ions in the sample migrate towards oppositely charged electrodes at characteristic rates. This method is suitable for separation of low molecular weight compounds such as amino acids, small peptides and nucleotides.

10.3.2 Starch gel electrophoresis

Starch gels are prepared by heating and then cooling a solution of partially hydrolyzed starch in an appropriate buffer. The separation is accomplished due to differences in charge as well as in size of the molecules because of molecular sieving effect of the gel. Assuming that the charge on molecule of two compounds is identical, the compound with smaller molecular mass will travel at a faster rate and *vice versa*. An important application of this form of electrophoresis is in the analysis of isozymes.

10.3.3 Polyacrylamide gel electrophoresis

Electrophoresis in polyacrylamide gel is the most widely used technique for analysis and characterization of proteins and nucleic acids. Use of polyacrylamide gels has several advantages such as: it is chemically inert, gives superior resolution, is amenable to preparation of gels with wide range of pore sizes and that the gels are stable over a wide range of pH, temperature and ionic strength.

Polyacrylamide gel is prepared by polymerization of acrylamide monomers into long chains and cross-linking these by bifunctional reagents such as N, N′-methylene bisacrylamide (Fig. 10.1).

Polymerization is initiated by ammonium persulfate or riboflavin. N, N, N′, N′-tetramethylethylene diamine (TEMED) catalyses formation of free radicals from persulfate which in turn initiate polymerization. In the case of riboflavin-TEMED system, light causes photodecomposition of riboflavin and production of necessary free radicals. Oxygen inhibits polymerization. Gel solutions have, therefore, to be degassed prior to the polymerization step.

The pore size of polyacrylamide gels can be varied by changing total concentration of acrylamide and bis-acrylamide monomers in a fixed volume of gelation solution. Gels ranging from 3 to 30% acrylamide concentration can be made and these can be used for separation of molecules of size upto 1×10^6

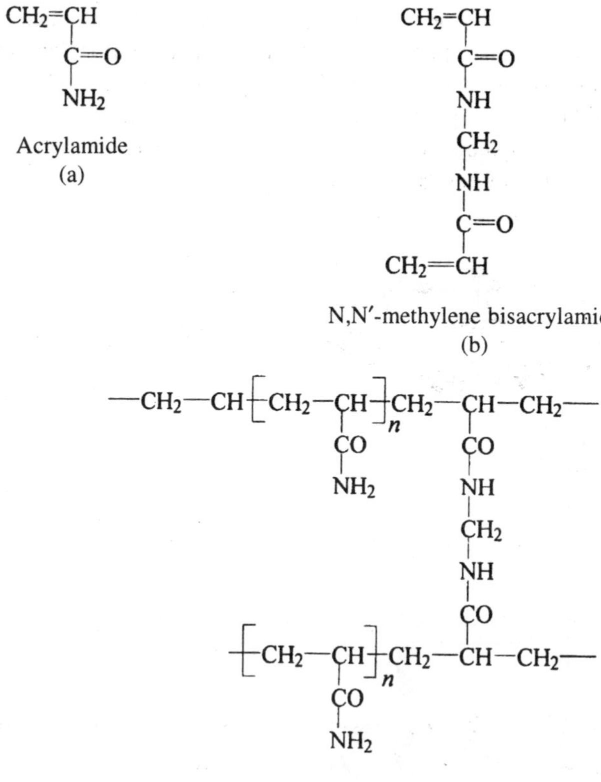

Fig. 10.1 Structures of acrylamide (a), bis-acrylamide (b) and their polymerization and cross-linking to yield porous polyacrylamide gels (c).

daltons. A gel with low percentage has larger pore size and is suitable for separation of high molecular weight compounds and a high percentage gel has smaller pore size and is used for separation of relatively low molecular weight compounds. Alternatively, gels of different pore sizes may be prepared by varying the relative concentrations of acrylamide and bis-acrylamide. Gradient gels with linear gradients of increasing acrylamide concentration give better resolution and are quite commonly used.

10.3.3.1 Apparatus

It consists of a power pack and an electrophoresis unit. The power pack supplies a stabilized direct current at controlled or required voltage and current output. The electrophoresis unit contains the electrodes, buffer reservoirs and gel casting assembly. Initially, electrophoresis in polyacrylamide gels was carried out in cylindrical rods of gel in glass tubes (~7 mm in diameter and 10 cm in length) which were then fixed in a specially designed apparatus (Fig. 10.2). However, now slab gel, (0.75–1.5 mm thick), in which the gel is in the form of a rectangular or square flat slab, is more commonly used. Two types of slab gel apparatuses are available—horizontal and vertical, the latter is usually more preferred and is shown in Fig. 10.3. Slab gels have advantage over rod gels since a number of samples can be co-electrophoressed under identical conditions on a single gel so that band patterns can be directly compared. On the other hand only one sample can be loaded onto each gel rod. Also, the heat produced during electrophoresis is more easily dissipated by thin slab gels as compared to thick rod gels thus reducing distortion of bands due to heating.

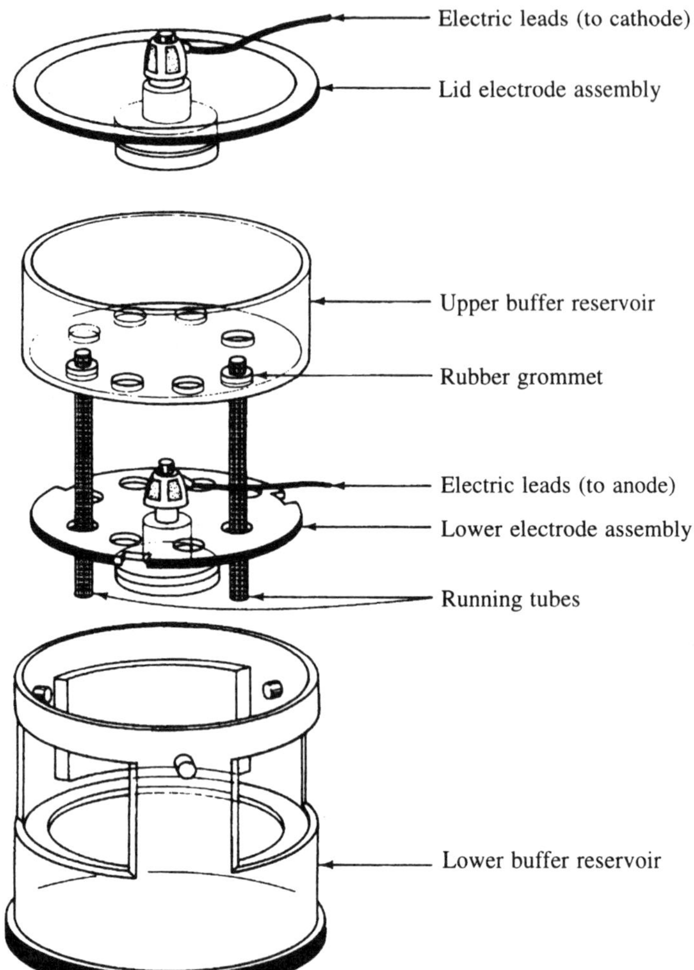

Fig. 10.2 A typical rod gel electrophoresis apparatus.

10.3.3.2 Buffer systems

Gel electrophoresis can be carried out in buffers ranging from pH 3–10. Ionic strength of the buffers is generally kept low (0.01–0.1 M) to reduce heat production during electrophoresis to a minimum. Depending upon the types of buffer systems used, polyacrylamide gel electrophoresis may be of the following kinds:

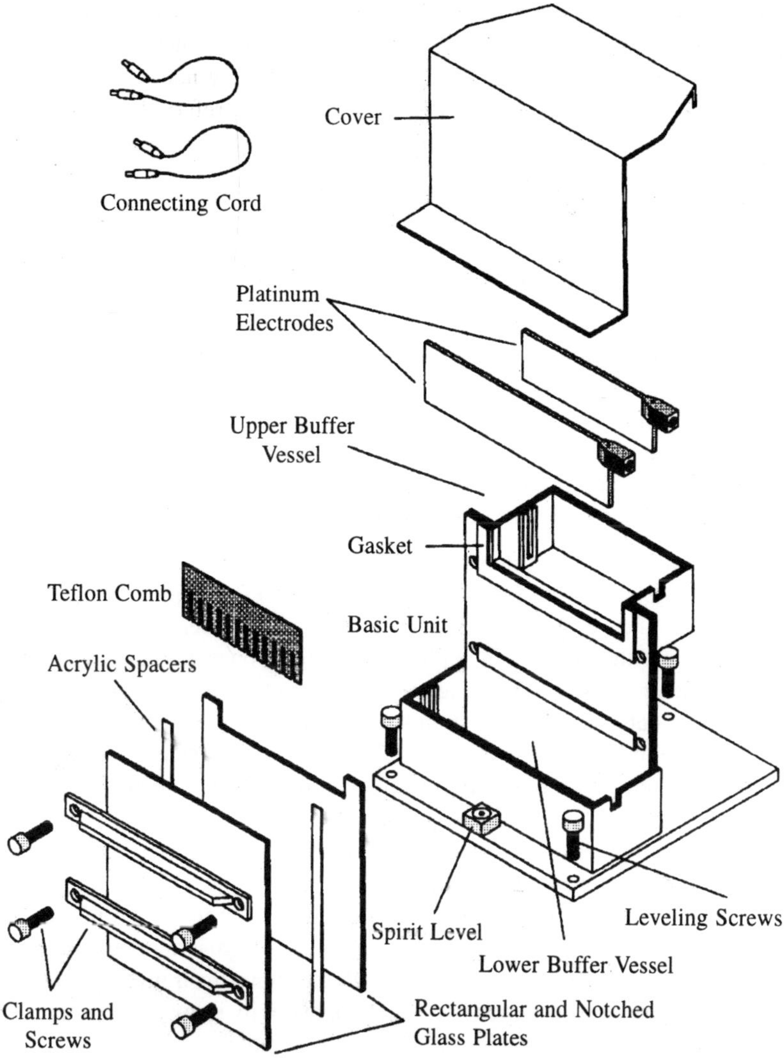

Fig. 10.3 Vertical slab gel electrophoretic apparatus

(a) Native continuous polyacrylamide gel electrophoresis
(b) Native discontinuous polyacrylamide gel electrophoresis or Native disc gel electrophoresis
(c) SDS—polyacrylamide gel electrophoresis (reducing)
(d) SDS—polyacrylamide gel electrophoresis (non-reducing)

The following Section deals with details of these different types of gel electrophoresis alongwith appropriate laboratory exercises.

10.3.3.2(a) Native continuous polyacrylamide gel electrophoresis (native PAGE)

In native continuous polyacrylamide gel electrophoresis, same buffer ions are present throughout the sample, gel and electrode vessel reservoirs at constant pH such that proteins retain their native conformation and biological activity and hence it is known as native continuous PAGE or simply native PAGE. In this system, protein sample is loaded directly onto the gel, referred to as resolving gel, in which separation of the sample components occurs.

10.3.3.2(b) Native discontinuous polyacrylamide gel electrophoresis (Disc-gel electrophoresis)

In contrast to continuous system, discontinuous or disc gel electrophoresis embodies the following features:

(a) contains different buffer ions in the gel than those in the electrode reservoirs
(b) consists of two types of gels—a large pore stacking gel which is polymerized on top of small pore 'resolving gel'. The buffers used to prepare these two gels are of different ionic strength and pH
(c) most discontinuous buffer systems have discontinuities of both buffer composition and pH (Table 10.1)

Discontinuous buffer systems have an advantage over continuous systems as relatively large volumes of dilute solutions of protein samples can be applied onto the gels and yet good resolution of sample components can be obtained. This is because the proteins in applied sample get concentrated into extremely narrow zone or get stacked during migration through large pored stacking gel prior to their entry and separation in small pored resolving gel. The process of sample concentration in the gel is as follows. Let us consider a discontinuous system where electrode buffer is Tris-glycine, pH 8.3, sample and stacking gel contain Tris-HCl buffer of pH 6.8 and the resolving gel contains Tris-HCl buffer of pH 8.8. At the pH of the sample and stacking gel (pH 6.8) glycine is very poorly ionized so that its effective mobility is very low, chloride ions have a much higher mobility at this pH while mobilities of proteins are intermediate between that of chloride and glycine. The moment voltage is applied, the chloride ions (leading ions) migrate away from the glycine ions (the trailing ions) leaving behind a zone of lower conductivity. Since conductivity is inversely proportional to field strength, this zone attains a higher voltage gradient which now accelerates glycine molecules so that it keeps up with the chloride ions. A steady state is established where the products of mobility and voltage gradient for glycine and for chloride are equal and these charged species now move at the same velocity with a sharp boundary between them. As this glycine/chloride boundary moves through the sample and then the stacking gel, a low voltage gradient moves before the moving boundary and a high voltage gradient behind it. Any proteins in front of the moving boundary are rapidly overtaken since they have lower velocity than the chloride ions. Behind the moving boundary, in the higher voltage gradient, the proteins have a higher velocity than glycine. Thus the moving boundary sweeps up the proteins so that they become concentrated into very thin zones or 'stacks'.

When the moving boundary reaches the interface of the stacking and resolving gels, the pH of the gel increases markedly and this leads to a large increase in the

Table 10.1. Buffers for native discontinuous polyacrylamide gel electrophoretic systems

Type of buffer	*Preparation of buffer*	
(a) *High pH buffer system*		
(i) Stacking gel buffer (Tris-HCl, pH 6.8)	Tris	6.0 g
	1 M HCl	48.0 ml
	Adjust pH to 6.8 and make final volume to 100 ml with water	
(ii) Resolving gel buffer (Tris-HCl, pH 8.8)	Tris	36.3 g
	1 M HCl	48.0 ml
	Adjust pH to 8.8 and make volume to 100 ml with water	
(iii) Reservoir buffer (Tris-glycine, pH 8.3)	Tris	3.0 g
	Glycine	14.4 g
	Adjust pH to 8.3 and make volume to 1 L with water	
(b) *Neutral pH buffer system*		
(i) Stacking gel buffer (Tris-phosphate, pH 5.5)	Tris	4.95 g
	Water	40.0 ml
	Adjust pH to 5.5 with 1 M orthophosphoric acid and make final volume to 100 ml with water	
(ii) Resolving gel buffer (Tris-HCl, pH 7.5)	Tris	6.85 g
	Water	40.0 ml
	Adjust pH to 7.5 with 1 M HCl and make volume to 100 ml	
(iii) Reservoir buffer (Tris ethyl barbiturate, pH 7.0)	Diethylbarbituric acid	55.2 g
	Tris	10.0 g
	Add water to make 1 L	
(c) *Low pH buffer*		
(i) Stacking gel buffer (Acetic acid—KOH, pH 6.8)	1 M KOH	48.0 ml
	Glacial acetic acid	2.9 ml
	Adjust pH to 6.8 and make volume to 100 ml with water	
(ii) Resolving gel buffer (Acetic acid—KOH, pH 4.3)	1 M KOH	48.0 ml
	Glacial acetic acid	17.2 ml
	Adjust pH to 4.3 and make final volume to 100 ml	
(iii) Reservoir buffer (Acetic-acid-B-alanine, pH 4.5)	β-alanine	31.2 g
	acetic acid	8.0 ml
	Add water to make 1 L	

degree of dissociation of glycine. Therefore, the effective mobility of glycine increases so that glycine overtakes the proteins and now migrates directly behind the chloride ions. At the same time, the gel pore size decreases thereby retarding the migration of the proteins because of molecular sieving. These two effects cause the proteins to be unstacked. The proteins now move in the zone of uniform voltage gradient and pH value and are separated according to their intrinsic charge and size.

10.3.3.2(c) SDS-polyacrylamide gel electrophoresis (non-reducing) (SDS-PAGE)

In this system, proteins are dissociated into their constituent subunits using anionic detergent such as sodium dodecyl sulfate (SDS) or sodium lauryl sulfate. The protein mixture is denatured by heating the sample at 100°C in the presence of excess SDS. Under these conditions most polypeptides bind SDS in a constant weight ratio (1.4 g SDS per g of polypeptide). The intrinsic charges of the polypeptides are insignificant compared to negative charges provided by the bound detergent. Consequently SDS polypeptide complexes have essentially identical charge densities (charge/mass ratio) and these migrate in polyacrylamide gels of an appropriate porosity strictly according to polypeptide size or their molecular mass. Thus, in addition to the analysis of polypeptide composition of the sample, the molecular weight of the polypeptides of the sample protein can also be determined by using standard molecular weight markers.

SDS-PAGE can be carried out using either continuous or discontinuous buffer systems (Table 10.2). Resolution is, of course, superior in a discontinuous system.

Table 10.2 Buffers for SDS-polyacrylamide gel electrophoresis

(*a*) *Continuous system*		
	Resolving gel buffer	
		0.1 M sodium phosphate, pH 7.2
	Reservoir buffer	
		0.1 M sodium phosphate, pH 7.2
		0.1% SDS
(*b*) *Discontinuous system*		
	Stacking gel buffer	
		0.125 M Tris-HCl, pH 6.8
	Resolving gel buffer	
		0.375 M Tris-HCl, pH 8.8
	Reservoir buffer	
		0.025 M Tris
		0.192 M glycine
		0.1% SDS
		pH 8.3

10.3.3.2(d) SDS—polyacrylamide gel electrophoresis (reducing)

It is similar to SDS-PAGE described above except that in addition to SDS, a thiol reagent such as β-mercaptoethanol is included which cleaves the disulfide bonds. Hence, polypeptides in a native protein which still remain associated even in the presence of SDS get separated. So in this method a protein sample is boiled in presence of excess SDS and β-mercaptoethanol to denature proteins to their individual polypeptides. SDS polypeptide complexes (1.4 SDS: 1 protein) have net negative charge and migrate towards anode at rates based solely on size or molecular weight of the polypeptides.

10.3.3.3 Staining of proteins in polyacrylamide gels

After electrophoresis, protein bands in gels can be visualized by staining with dyes such as Amido Black 10B, Coomassie brilliant blue R-250 or silver staining.

Coomassie brilliant blue R-250 is the most commonly used stain though silver stain is about 100 times more sensitive than it and can detect upto 1 ng protein/band.

10.3.3.4 Protein (Western) blotting

The transfer of proteins from an acrylamide gel onto a more stable and immobilizing support such as nitrocellulose filter is called as protein blotting or Western blotting. It is so named in view of the previously used nomenclature for nucleic acid blotting procedures. Blotting of protein bands allows them to become more accessible for detection and identification using a variety of methods such as immunological detection, autoradiography, analysis of DNA binding proteins, glycoproteins etc. and also the filters can be stored for much longer periods. A number of support matrices such as nitrocellulose, diazobenzyloxymethyl cellulose, nylon membrane, Hybond membrane etc. with high protein binding capacity are available, but nitrocellulose which is available in a range of pore sizes (0.2–0.45 μm), is sufficiently effective for most purposes and is commonly used. The transfer of proteins from gels can be achieved by any of the methods such as simple capillary action, application of vacuum or electrophoretically. Thus based on the technique employed Western blotting is of different types namely, capillary blotting, vacuum blotting and electroblotting, the latter being the most efficient and has become a standard method.

10.3.4 Agarose gel electrophoresis

Agarose gels are more porous and have a larger pore size as compared to polyacrylamide gels and are, therefore, used to fractionate large macromolecules such as nucleic acids that cannot readily penetrate into and move through other types of supporting materials.

Agarose is a linear polymer of D-galactose and 3,6-anhydro-L-galactose. Agarose gels are cast by boiling agarose in presence of a buffer, then poured into a mold and allowed to harden to form a matrix. Porosity of the gel is determined by concentration of agarose. Higher the agarose concentration, smaller the pore size and lower the agarose concentration, larger is the pore size. When an electric field is applied across the gel, DNA molecules that are negatively charged at neutral pH, migrate towards oppositely charged electrode at rates determined by their molecular size and conformation. Since charge/mass ratio in nucleic acids is one, rate of migration of DNA molecules is inversely proportional to $\log_{10}$ of their molecular weights, i.e. smaller DNA molecules will travel faster as compared to the larger ones. Further, DNA molecules of the same size but with different conformations travel at different rates. The order of migration velocity in the increasing order of various forms of DNA is: supercoiled DNA > linear double stranded DNA > open circular DNA.

10.3.4.1 Apparatus

Both vertical as well as horizontal types of apparatuses for casting of agarose gels are available. However, horizontal submarine or submerged gels (so named because the gel is immersed in buffer) are more commonly employed (Fig 10.5). Horizontal slab gels have advantages over vertical ones, in that low agarose concentration can be used as the entire gel is supported from beneath and also the processes of loading, pouring and handling of gels are more convenient.

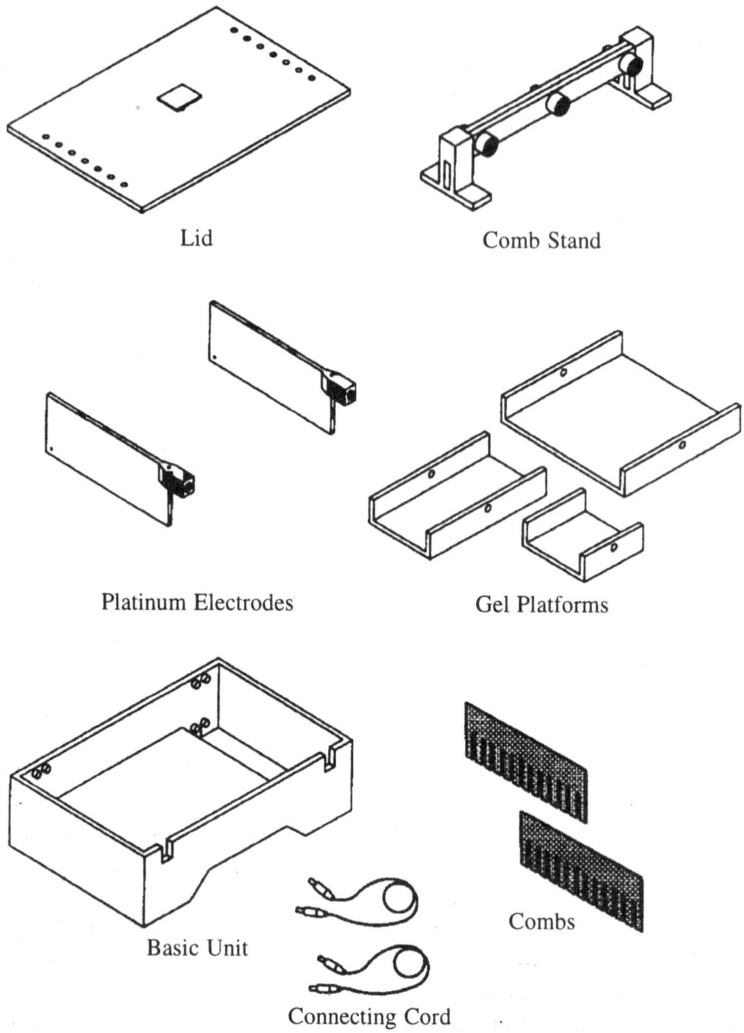

Fig. 10.4 Submarine gel electrophoretic apparatus.

10.3.4.2 Buffers

Commonly used buffers include Tris-acetate (TAE), Tris-borate (TBE) or Tris-phosphate (TPE) at a concentration of approximately 50 mM, over a pH range of 7.5–8.0. EDTA is almost invariably incorporated in these buffers to suppress the activity of DNase.

10.3.5 Other types of electrophoresis

10.3.5.1 Isoelectric focussing (IEF)

It is an electrophoretic method for separation of proteins which differ in their isoelectric points. The proteins are made to migrate in a pH gradient maintained by a mixture of low molecular weight ampholytes. Proteins migrate depending upon their charge till they reach a region where the pH corresponds to respective isoelectric point at which pH proteins possess no net charge and hence get focussed. The method has high resolution and separation of proteins that differ in isoelectric points by as little as 0.01 pH units can be achieved.

10.3.5.2 Two dimensional gel electrophoresis

The technique was developed by O'farrell (1975) and combines the resolving power of isoelectric focussing (IEF) and SDS-polyacrylamide gel electrophoresis (SDS-PAGE). Proteins in the first dimension are separated according to their charge by IEF and in second dimension SDS-PAGE facilitates their separation on basis of their size. It is a standard method for judging purity of a protein preparation.

10.3.5.3 Immunoelectrophoresis

It is a type of agar/agarose gel electrophoresis which exploits the separation of proteins on the basis of their charge to mass ratio and their antigenicity. Protein mixtures are first separated by electrophoresis on agarose gel and are then allowed to interact with a specific antibody preparation placed in a trough cut in the center of the gel. The antibody diffuses through the gel and forms a visible precipitate with the electrophoresed proteins for which the antibody has affinity. The technique allows the identification of the number of antigens in a sample and also indicates the relative amount of imunoreactive material.

A modification of the above technique is rocket immunoelectrophoresis so called because the precipitin patterns that form on the gel after electrophoresis look like rockets. This technique is used in the determination of antigen concentration. Here varying antigen concentrations are placed in the sample wells and electrophoresed into the agarose gel which contains antibody. As the antigen moves into the gel, it encounters antibody and precipitates when antigen-antibody concentrations are equivalent resulting in the precipitate spreading as a rocket from the antigen well. The area of the peak of the rocket serves as a measure of antigen concentration.

10.3.5.4 Capillary electrophoresis (CE)

It is a more recent technique that has a very high resolving power and speed, is versatile and automated. CE involves electrophoresis of samples in small narrow capillary tubes and a high voltage of 10–50 kV is applied. The technique can be used to separate a wide spectrum of biological materials including amino acids, peptides, proteins, oligonucleotides and nucleic acids as well as small organic molecules such as drugs or even metal ions. Very small amounts of samples (5–10 nl) are required for analysis.

10.3.5.5 Pulse-field gel electrophoresis (PFGE)

The technique was developed by Schwartz and Cantor (1984). In this technique, electric field is changed repeatedly (pulsed) in direction and strength during the separation such that the DNA molecules are forced to continuously orient and reorient themselves. Larger is the DNA molecule, longer is the time required for its reorientation. Very large DNA molecules of size upto 5000 kb can be separated by this technique.

10.4 Laboratory Exercises

10.4.1 EXPERIMENT: Native Disc-gel electrophoresis of proteins

Principle

The separation of proteins in native polyacrylamide gels is based on both charge density and size of the molecules and has been discussed in detail in Section 10.3.3.2(b).

Materials and Reagents

The various components/accessories for rod gel electrophoresis required for this experiment include:

1. Glass tubes (10 cm long × 7 mm diameter).
2. Stand for glass tubes.
3. Upper and lower troughs.
4. Constant voltage power supply (0–500 Volts).
5. Electrophoresis leads.
6. Sucrose (40%, *w/v*): Weigh 40 g of sucrose and dissolve in 100 ml of water.
7. 0.01% Bromophenol blue solution: Take 10 mg of bromophenol blue and dissolve in 100 ml of water.
8. Acrylamide—bisacrylamide stock solution:

Acrylamide	30.0 g
Bisarcylamide	0.8 g

   Dissolve in water and make final volume to 100 ml. Filter the solution through Whatman No. 1 filter paper and store in brown bottle at 0–4°C. This solution is stable for one month.
9. Stacking gel buffer stock (Tris-HCl, pH 6.8):

Tris	6.0 g
I M HCl	48.0 ml

   Adjust pH to 6.8 and make its final volume to 100 ml with water. Filter through Whatman No. 1 filter paper and store at 0–4°C.
10. Resolving gel buffer stock (Tris-HCl, pH 8.8):

Tris	36.3 g
I M HCl	48.0 ml

   Adjust its pH to 8.8 and then make the final volume to 100 ml. Filter through Whatman No. 1 filter paper and store at 0–4°C.
11. 1.5% (*w/v*) Ammonium persulfate (APS) in water: Prepare by dissolving 0.15 g of APS in 10 ml water. This reagent should be prepared fresh just before use.

12. N, N, N′, N′-Tetramethyl ethylene diamine (TEMED): Use as supplied by manufacturer.
13. Reservoir buffer or electrode buffer (Tris-glycine, pH 8.3):

Tris	3.0 g
Glycine	14.4 g

Adjust its pH to 8.3 and make the final volume to 1 L with distilled water.

14. Staining solution:

Coomassie brilliant blue R-250	1.25 g
Methanol	200 ml
Glacial acetic acid	35 ml

Make final volume to 500 ml with distilled water. Filter to remove any undissolved material and store at room temperature.

15. Destaining solution:

Glacial acetic acid	75 ml
Methanol	50 ml

Mix the above components and add water to make its final volume to 1 L.

16. Composition of stacking and resolving gels is given in the following table:

Reagent	*Stacking gel (2.5%)*	*Resolving gel (7.5%)*
	ml of solutions for preparation	
(a) Acrylamide-bis-acrylamide stock solution	2.50	7.5
(b) Stacking gel buffer stock solution (Tris-HCl pH 6.8)	5.00	—
(c) Resolving gel buffer stock solution (Tris-HCl pH 8.8)	—	3.75
(d) 1.5% APS	1.50	1.50
(e) Water	11.00	17.25
(f) TEMED	0.015	0.15
Total volume	20.00	30.00

Procedure

(a) Sample preparation:
To 0.2 ml of solution of the sample preparation containing about 3–4 mg protein/ml add 0.1 ml sucrose (40%, *w*/*v*) and 0.1 ml bromophenol blue (0.01%, *w*/*v*) solution just prior to loading of the sample. In place of sucrose, glycerol at a final concentration of 10% (*v*/*v*) may be used. Add one drop of glycerol with a pipette.

(b) Preparation of the gel rods:

1. Take clean glass tubes and place them vertically in the rubber septum in the rack or tube stand so that the tubes fit snuggly in it.
2. Mix all the components of the resolving gel, except TEMED and APS as indicated in the above table. Degas the mixture for 1 min using a water pump and then add APS and TEMED in the required quantities, mix gently but quickly.
3. Without any delay, fill the resolving gel in each tube to the mark (if tubes with permanent mark are not available, the positions of resolving and stacking gels on all tubes should be marked with the help of a glass marker) using Pasteur pipette to avoid trapping of any air bubbles.
4. Overlay the gel with a drop of water as gently as possible to ensure a flat meniscus. Water also avoids contact with oxygen which inhibits polymerization.
5. Allow the gel to stand for 20–30 min for polymerization.
6. When the gel has polymerized, remove water from the top of the gel and rinse surface of the gel with stacking gel buffer.
7. Mix various components of stacking gel as specified in the Table except APS and TEMED. Degas the solution as above and add indicated amounts of APS and TEMED. Pour the stacking gel upto the mark which should be of about 1 cm in height on top of the resolving or running gel and allow it to polymerize for 10-15 min. When completely polymerized, rinse its surface with reservoir buffer.
8. Remove rubber septum from the base of the glass tubes containing rod gels. Label the tubes and place them in rubber grommets of the electrophoretic system with stacking gel towards the upper side. The lower end of each tube must dip in the buffer in lower chamber. Now pour sufficient amount of the electrode buffer into the top unit so that upper ends of all the tubes are submerged in buffer. Remove any air bubbles from the tubes which might have got entrapped while adding the buffer.
9. Load 0.1 ml of the protein sample with a syringe or a micropipette on the top of glass tubes. Only one sample should be loaded onto each of the tube.
10. Carry out electrophoresis in cold (0–4°C). Initially supply a current of 2 mA/tube for 10 min and then increase it to 3 mA/tube and continue electrophoresis until the tracking dye migrates to about 1 cm from the bottom end of the tubes. This may take 1 to 2 h. Turn off the power supply and disconnect the electrophoretic apparatus from power pack and take out the glass tubes. Remove the gels from the tubes by squirting water with a syringe along the sides between the wall of glass tube and gel while constantly rotating the tubes.
11. Place the gels in petri plates or screw capped tubes. Add sufficient amount of the staining solution to the petri plates or tubes so as to completely cover the gels and allow them to stand for 3–4 h for staining.

12. For destaining, transfer the gels to destaining solution. Change the destaining solution frequently until background of the gels is clear.
13. Measure the distance travelled by the tracking dye and also by the different protein bands and calculate relative mobility (R_m) as given below:

$$R_m = \frac{\text{Distance travelled by the protein band}}{\text{Distance travelled by the tracking dye}}$$

14. Preserve the gels in 7% acetic acid till these can be photographed for permanent record.

General precautions

1. All the chemicals and distilled water should be of high quality.
2. Glassware should be cleaned scrupulously with chromic acid and thoroughly rinsed with distilled water.
3. Acrylamide (monomer) is highly neurotoxic and should be handled with extreme care.
4. The gel solutions should be properly deaerated as oxygen inhibits polymerization.
5. Do not touch the electrophoretic chamber or wires while the electrophoresis is in progress.
6. Avoid mouth pipetting of all the solutions.
7. Unless otherwise mentioned, all the solutions should be stored at 0–4°C.
8. Pour gel mixtures gently so that no air bubbles get entrapped. Air bubbles may cause non-uniform flow of current across the gel and cause disturbances in migration of sample components.
9. Air bubbles, if any, from beneath the gel in the lower reservoir buffer must also be removed.

10.4.2 EXPERIMENT: Separation of isoenzymes of malate dehydrogenase (MDH) by starch gel electrophoresis

Principle

Different isoenzymic forms of malate dehydrogenase (MDH) get separated due to difference in their charge density and size. Separation of isoenzymes may be carried out by starch slab gel or native polyacrylamide gel electrophoresis in rods or slab gels. Activity of the enzyme on the gel is detected from the formation of colored formazan due to NADH linked reduction of nitroblue tetrazolium, shown as follows:

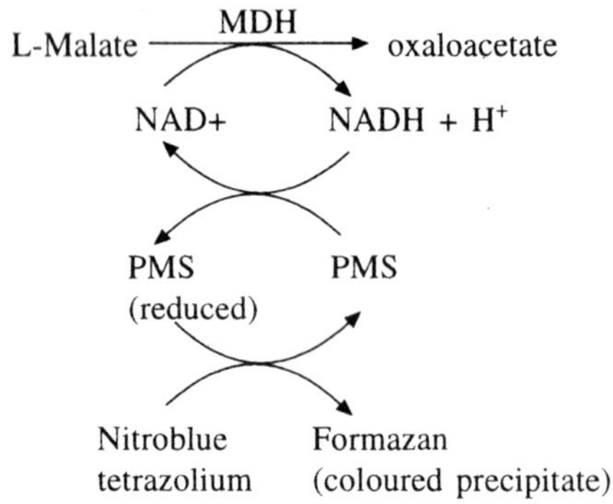

Materials and Reagents

1. Glass plates: Two glass plates (22 × 12 × 0.4 cm) and glass strips of 22 and 14 cm in length, 5 mm wide and 1 and 3 mm thick.
2. Electrophoretic tank and power supply.
3. Gel buffer (pH 8.6):

Tris-HCl	9.2 g
Citric Acid	1.05 g

Dissolve in distilled water, adjust pH to 8.6 and make final volume to 1 L with distilled water.

4. Hydrolyzed starch (electrophoresis grade). *Note*—Hydrolyzed starch, suitable for electrophoresis, can also be prepared by warming potato starch in acidified acetone.
5. Electrode buffer (pH 7.9):

Boric acid	18.6 g
Sodium hydroxide	2.0 g

Dissolve in distilled water, adjust pH to 7.9 and then make final volume to 1 L with distilled water.

6. Enzyme staining solution:

Tris-HCl, pH 8.0	0.1 M
L-Malate	0.036 M
NAD	0.3 mg/ml
Nitroblue tetrazolium	0.8 mg/ml
Phenazine methosulfate	0.14 mg/ml

Adjust pH of the above mixture to 8.0.

7. Bromophenol blue.
8. Enzyme extract: Grind 1g young leaves in 5 ml of 0.05 M Tris-HCl buffer, (pH 7.4) in a chilled pestle and mortar. Centrifuge the extract at 10,000 × *g* for 30 min at 0–4°C. Use the supernatant as the enzyme preparation.

Procedure

1. Prepare the gel mould by sticking 1 mm thick glass edge strips to all sides of one of the glass plates to give a shallow tray of 21 cm length, 14 cm width and 1mm depth.
2. Prepare starch gel (10%, *w/v*) by adding 4 g of dry starch to 40 ml of gel buffer in a conical flask and boil the suspension till it becomes viscous translucent gel.
3. Pour the hot starch solution into the gel mould taking care that no air bubbles are trapped. Insert a comb in the gel at a distance of about 6 cm from one of the ends and allow the gel to set for 10–15 min.
4. Remove the comb, load 15–20 μl of the leaf extract onto a piece of 3 MM Whatman paper and insert it in the slot created by the comb. Similarly load the dye (0.1% bromophenol blue) in one of the slots.
5. Place the gel plate on the platform of electrophoretic tank and connect wicks to both ends of the gel. The wicks are prepared by taking several layers of filter paper cut to the width of the gel, soaked in electrode buffer and then these are placed in such a way that they overlap 1–2 cm of the gel surface on both ends and dip into the electrode buffer in the buffer chambers of the apparatus.
6. Place a glass plate on the top of gel to hold the wicks in place as well as to reduce evaporation of buffer from the gel.
7. Run the gel at 5–20 V/cm in cold room (0–4°C).
8. After the electrophoresis is complete, switch 'OFF' and disconnect the power supply, remove the glass plate at the top as well as the wicks. Lift the gel plate from electrophoresis apparatus and rinse it in cold 0.1 M Tris-HCl buffer (pH 8.0). Now place the gel in enzyme staining solution and incubate at 37°C for 30 min. Observe for the appearance of dark brown coloured bands on the gel.
9. Measure the distance travelled by the dye and the various bands of the enzyme and calculate the R_m values of the individual enzyme bands as follows:

$$R_m = \frac{\text{Distance travelled by the protein band}}{\text{Distance travelled by the tracking dye}}$$

Precautions

1. Since most of the enzymes get denatured at high temperature, the electrophoresis should be carried out at 0–4°C.
2. Prior to use, make certain that the gel does not contain any undissolved starch or air bubbles.
3. Follow all precautions mentioned in EXPERIMENT 10.4.1.

10.4.3 EXPERIMENT: SDS-Polyacrylamide slab gel electrophoresis of proteins under reducing conditions (SDS-PAGE)

Principle

As outlined in Section 10.3.3.2(d) under various types of electrophoresis. As discussed therein, this system of electrophoresis is used for determination of molecular weight of the protein(s) under investigation.

Materials and Reagents

1. Vertical slab gel electrophoresis apparatus: The various items include
   (a) Two glass plates, with one of the glass plates being notched and having vertical glass spacers of 1.5 mm in thickness on its sides
   (b) Gel casting assembly
   (c) Tygon tubing, comb with teeth
   (d) Electrophoretic tank with upper and lower buffer reservoirs
   (e) Electrophoretic leads
2. Power pack (0-500 Volts).
3. Reagents 3–11: These reagents are the same as required for native gel electrophoresis (see EXPERIMENT 10.4.1, Reagents 8–15) except for the reservoir buffer which is different and has the following composition:
   Reservoir buffer (Tris-glycine, pH 8.3):

Tris	3.0 g
Glycine	14.4 g
SDS	1.0 g

   Adjust pH to 8.3 and add water to make final volume to 1 L.
12. SDS (10%, *w/v*): Dissolve 1 g SDS in 10 ml of distilled water. Store the reagent at room temperature.
13. Sample buffer 2X: The sample preparation buffer which contains a two-fold concentration of various components is prepared as given below:

1 M Tris-HCl, pH 6.8	12.5 ml
SDS	4.0 g
β-Mercaptoethanol	10.0 ml
Glycerol	20.0 ml
1% Bromophenol blue	4.0 ml

   Add water to make final volume to 100 ml.
14. Standard molecular weight marker proteins: The standard molecular weight protein kits, having different molecular weight ranges, are commercially available.

Protein	*Molecular weight (kDa)*
β-Lactoglobulin	18.4
Trypsinogen	24.0
Carbonic anhydrase	29.0
Pepsin	34.7
Egg albumin	45.0
Bovine serum albumin	66.0

Dissolve 1 mg of each of the above proteins in 1 ml of the sample buffer diluted with water in ratio of 1:1. About 20 μl of this mixture should be loaded into one of the wells.

15. Stacking and resolving gels: Stacking and resolving gels for SDS-PAGE are prepared according to the particulars given in the following table:

Stock solution	*Stacking gel (2.5%)*	*Resolving gel (12.5%)*
	ml of the solution	
*1. Acrylamide-bis-acrylamide (30:0.8)	2.50	12.50
*2. Stacking gel buffer stock solution (Tris-HCl, pH 6.8)	5.00	—
3. Resolving gel buffer stock solution (Tris-HCl, pH 8.8)	—	3.75
4. 10% SDS	0.20	0.30
5. 1.5% APS	1.00	1.50
6. Water	11.30	11.95
7. TEMED	0.015	0.015
Total volume	20.00	30.00

*Refer to EXPERIMENT 10.4.1.

Procedure

(a) Sample preparation:
Mix the protein sample with equal volume of sample buffer. Boil the mixture for 3 min in a boiling water bath and cool to room temperature. If protein sample is too dilute, precipitate the proteins with 10% TCA (incubate in ice for 30 min), centrifuge at 10,000 × *g* for 5 min. Wash the precipitate with ethanol-ether (1:1) to remove TCA. Dissolve the precipitate in the sample buffer diluted with water in 1:1 ratio, boil for 3 min and cool.

(b) Preparation of slab gels:

1. Thoroughly clean and dry the glass plates and assemble them in gel casting assembly. Seal the two glass plates with the help of tygon tubing, clamp them and place the whole assembly in an upright position.
2. Mix various components of resolving gel as indicated in the above table (Reagent 15) except for SDS, APS and TEMED. Degas the solution for 1 min using a water pump and then add the above remaining components of the gel.
3. Mix gently and pour the gel solution into the mould in between the clamped glass plates taking care to avoid entrappment of any air bubbles. Overlay distilled water on the top as gently as possible and leave for 30 min for setting of the gel.
4. When the gel has polymerized, remove the water layer and rinse the gel surface with stacking gel buffer.

5. Mix the stacking gel components in the same way as described above for the resolving gel.
6. Pour the stacking gel and immediately insert the supplied plastic comb in the stacking gel. Care should be taken that no air bubbles are entrapped. Allow the gel to polymerize for about 20 min.
7. After the stacking gel has polymerized, remove the comb without distorting or damaging the shapes of the wells. Clean the wells by flushing with electrode buffer using a syringe.
8. Remove the tygon tubing and install the gel plate assembly into the electrophoretic apparatus. Pour reservoir buffer in the lower and upper chambers. Remove any trapped bubbles at the bottom of the gel.

(c) Electrophoresis of sample:

9. Load 10–20 μl sample (100–200 μg protein) in the sample wells. Also load molecular weight marker proteins in one or two of the wells.
10. Switch 'ON' the current maintaining it at 10–15 mA for initial 10–15 min until the samples have travelled through the stacking gel. Then increase the current to 30 mA until the bromophenol blue dye reaches near the bottom of the gel slab. This may require 3-4 h.
11. After the electrophoresis is complete, turn 'OFF' and disconnect the power supply and carefully remove the gel slab from in between the glass plates.
12. Place the gel in a trough containing staining solution for 3–4 h or it can be kept for staining overnight. Destain the gel with destaining solution (Reagent 15, EXPERIMENT 10.4.1) till a clear background of the gel is obtained.
13. Record the distance travelled by the dye and the various protein bands and calculate R_m values as in EXPERIMENT 10.4.1.
14. Plot a graph between $\log_{10}$ molecular weight versus Rm values of standard molecular weight markers. A straight line would be obtained. From the R_m values of the sample polypeptides determine their molecular weights using the above standard calibration curve.

Precautions

Take all precautions listed in EXPERIMENT 10.4.1. A number of problems may be encountered while using polyacrylamide gel electrophoretic technique. A summary of common problems, alongwith their causes and remedies, are summarized in Table 10.3 at the end of the Chapter.

10.4.4 EXPERIMENT: Electroblotting (Western blotting) of proteins from SDS-polyacrylamide gel

Principle

When an electric field is applied, proteins migrate from cathode to anode and as they come in contact with nitrocellulose sheet, which has the property to bind proteins, the proteins get immobilized onto the sheet. An exact replica of protein bands as in the gel is obtained on the nitrocellulose sheet.

Materials and Reagents

1. Western blot apparatus consisting of gel holder, sponge and transfer tank.
2. Power pack and electrical leads.
3. Slab gel containing separated proteins (use gel obtained after electrophoresis of proteins on SDS-PAGE in EXPERIMENT 10.4.3).
4. Nitrocellulose sheet cut to the size of the gel.
5. Whatman 3 MM paper cut to the size of the gel.
6. Transfer buffer: It consists of

Tris	3.0 g
Glycine	14.7 g
Methanol	200 ml

After mixing the above components, adjust the pH of the solution to 8.3 and make the final volume to 1 L with distilled water.

Procedure

1. Take the gel obtained after electrophoresis in EXPERIMENT 10.4.3. Do not stain it and mark it by cutting of the one corner. Place the gel for 30 min in transfer buffer for equilibration.
2. Take a nitrocellulose sheet, cut to size of the gel and dip it in the transfer buffer by carefully wetting one edge and then slowly lowering the sheet into the buffer. Leave it under the buffer for 30 min.
3. Soak the sponge in transfer buffer and place the wet sponge on the gel holder. Now keep a sheet of Whatman 3 MM paper (presoaked in transfer buffer) on the sponge.
4. Place the equilibrated gel carefully on the filter paper avoiding trapping of any air bubbles.
5. Now lay down carefully the nitrocellulose membrane with shining side towards the gel on top of the gel. Gently roll a sterile 10 ml pipette over the membrane to remove air bubbles for ensuring a good contact between the membrane and the gel.
6. Complete the sandwich by placing wet Whatman 3 MM filter paper over the membrane and a second sponge on the filter paper. Close the gel holder and place it in the transfer tank containing sufficient transfer buffer to completely cover the blot.
7. Connect to the power supply and run for 5 h at 60 V or at 30 V over night.
8. When the transfer is complete, lift the membrane from the gel. Stain and destain it as in EXPERIMENT 10.4.1 using Coomassie brilliant blue R-250 stain. Examine the nitrocellulose sheet for the presence of blue bands of the transferred proteins.

Precautions

1. Do not touch nitrocellulose sheet with hand. Wear gloves while handling the gel and nitrocellulose sheet.
2. Avoid entrapment of air bubbles in the assembly.

10.4.5 EXPERIMENT: Separation of proteins by isoelectric focussing

Principle
It is a method for separation of proteins that differ in their net charge. When a potential difference is applied across the gel, the ampholytes align themselves in order of increasing isoeletric point (pI) from anode to cathode, thus producing a uniform pH gradient. Proteins migrate depending upon their charge till they reach a region where the pH corresponds to their respective isoelectric point (pI) at which proteins possess no net charge and hence get focussed. After focussing, the separated components are detected by staining.

Materials and Reagents

1. Electrofocusssing apparatus: A vertical slab gel electrophoresis apparatus as mentioned in EXPERIMENT 10.4.3 can be used.
2. Power supply (0–500 Volts).
3. Acrylamide—bisacrylamide solution:

Acrylamide	30 g
Bis-acrylamide	1 g
Distilled water	100 ml

4. Ampholyte solution (pH 3.5–10 and pH 4–6). Ampholytes are amphoteric compounds which are supplied as a mixture of molecules with closely spaced isoelectric points. Ampholyte solutions for different pH range are available commercially.
5. 10% Ammonium persulphate (APS): Dissolve 100 mg ammonium persulphate in 1 ml distilled water. Prepare fresh just before use.
6. N, N, N′, N′-Tetramethylethylene diamine (TEMED).
7. Gel mixture (for pH gradient 4–6): It consists of the following:

Acrylamide—bisacrylamide solution (Reagent 3)	2.0 ml
Ampholyte solution pH 3.5–10	48.0 μl
Ampholyte solution pH 4–6	240.0 μl
Distilled water	9.7 ml
10% Ammonium persulphate	50.0 μl
TEMED	20.9 μl

Note: Add APS and TEMED just before pouring the gel.

8. Anolyte solution (10 mM phosphoric acid): Prepare a stock solution of 1 M phosphoric acid. Dilute 100 times before use.
9. Catholyte solution (20 mM NaOH): Prepare stock solution of 1 M NaOH. Dilute 50 times before use.

10. Sample buffer (2X) for pH gradient 4–6:

Glycerol	3 ml
Ampholyte solution (pH 3.5–10)	33 μl
Ampholyte solution (pH 4–6)	167 μl
Distilled water	1.8 ml

11. 10% Trichloroacetic acid solution (TCA).
12. 1% TCA.
13. Staining solution: See Reagent 14, EXPERIMENT 10.4.1.
14. Destaining solution: See Reagent 15, EXPERIMENT 10.4.1.

Procedure

(a) Sample preparation:
Mix protein sample with an equal volume of 2X sample buffer (Reagent 10).

(b) Preparation of slab gel:

1. Thoroughly clean and dry the glass plates and assemble them in gel casting assembly. Seal the two glass plates with the help of tygon tubing, clamp them and place the whole assembly in an upright position.
2. Mix various components of the gel mixture as indicated in Reagent 7 and pour the gel mixture into the moulds between two glass plates taking care to avoid entrapment of air bubbles. Fill the gel plates to the top with gel mixture.
3. Insert comb taking care not to trap air bubbles in the teeth of the comb.
4. Allow gel to polymerize for about 1 h. After gel has polymerized, remove the comb carefully.

(c) Electrophoresis:

5. Attach gel assembly to inner cooling core and insert into electrophoresis tank.
6. Load about 20 μl sample (50–100 μg protein) in the sample wells with the help of a syringe or gel loading pipette.
7. Pour catholyte (20 mM NaOH) into the upper buffer chamber and anolyte (10 mM phosphoric acid) into the lower buffer chamber.
8. Attach electrodes and run electrophoresis at room temperature for 30 min at 150 volts (constant voltage). Then set the voltage at 200 volts for 2.5 h (constant voltage).
9. After electrophoresis is over, turn 'OFF' and disconnect the power supply.
10. Carefully remove the gel slab from in between the glass plates and place in a trough containing 10% TCA for 10 min to allow fixing of the gel. Replace with 1% TCA and keep the gel overnight in this solution to remove ampholytes.
11. Place the gel in staining solution for 3–4 h or it can be kept for staining overnight.
12. Destain the gel with several changes of destaining solution till a clear background of the gel is obtained and protein bands become clearly visible. The gel is photographed for permanent record.

Precautions

Take all precautions listed in EXPERIMENT 10.4.1.

10.4.6 EXPERIMENT: Determination of molecular weight of plasmid DNA by agarose gel electrophoresis

Principle

DNA molecules are negatively charged at neutral or alkaline pH and migrate towards anode when an electric field is applied. Charge/mass ratio in nucleic acids is unity, thus migration occurs largely on the basis of molecular size of the DNA molecules.

Materials and Reagents

1. Minigel horizontal agarose gel electrophoretic unit. The unit comprises of:

   (a) Gel casting plate
   (b) Electrophoretic tank
   (c) Comb
   (d) Electrophoretic leads.

2. Adhesive tape.
3. Power pack (0–500 Volts).
4. UV-transilluminator with camera.
5. Micropipette or syringe.
6. Gloves.
7. Tris-acetate buffer (TAE) stock solution (5X): A 5-fold concentrated TAE buffer stock solution is prepared as follows:

Tris-base	24.2 g
Glacial acetic acid	5.71 ml
0.5 M EDTA	10.0 ml

   Adjust pH of the above to pH 8.0 and add water to make 1 L. Dilute 5 times before use to obtain the working buffer (1 X buffer).
8. 1% agarose in 1 X TAE buffer: Dissolve 0.75 g agarose in 75 ml of 1 X TAE buffer (working TAE buffer) by boiling and maintain it at 50°C till it is to be used.
9. Gel loading solution: 10% glycerol and 0.025% bromophenol blue in water.
10. Ethidium bromide: Dissolve 10 mg of ethidium bromide per ml of the 1 X TAE buffer. Gloves must be worn while preparing this solution.
11. Plasmid DNA preparation: To 20 μl of plasmid DNA preparation add 10 μl of gel loading solution and mix properly.
12. Standard DNA marker: Take 20 μl of the λ Hind III DNA digest, add 10 μl of gel loading solution and mix them well.

Procedure

1. Take a clean dry gel casting plate and make a gel mould using an adhesive tape along the sides of the plate to prevent running off of the material to be poured on the plate.
2. Pour 50 ml of 1% agarose solution kept at 50°C onto the casting plate. Immediately place the supplied comb about 1 cm from one end of the plate ensuring that teeth of the comb do not touch the glass plate. Wait till a firm layer of gel is formed.
3. Remove the comb and the tape surrounding the plate carefully and transfer the gel plate to the electrophoretic tank such that wells are towards the cathode.
4. Pour 1 X TAE buffer into the tank until the gel is completely submerged. Connect the electrodes to the power supply.
5. Load the plasmid DNA preparation and the standard DNA markers into separate wells with the help of a micropipette or a syringe.
6. Turn 'ON' the power supply and run at 100 V (10–15 mA). Monitor the progress of fast running tracking dye (bromophenol blue) during electrophoresis.
7. Turn 'OFF' the power supply when the tracking dye has reached near the opposite edge of the gel.
8. Transfer the gel from casting plate onto a UV-transparent thick plastic sheet and place it in a staining tray containing ethidium bromide solution. Stain for 20–30 min.
9. For destaining the gel, place it in water for 15–20 min.
10. Now place the gel alongwith UV-transparent sheet on a UV-transilluminator and view the gel in UV light for presence of orange coloured bands. For a permanent record, the gel may be photographed (CAUTION: UV light is extremely injurious for eyes. Wear UV protectant glasses while viewing gel).
11. Measure the distance moved by each band from edge of the loading well. Draw a graph between $\log_{10}$ of molecular weight of standard DNA markers vs the distance travelled by each of them.
12. From the distance travelled by the supplied plasmid DNA preparation, determine its molecular weight using the calibration curve prepared in Step 11.

Precautions

1. High purity chemicals and water should be used.
2. All glassware, plastic ware and solutions should be sterile.
3. Ethidium bromide is carcinogenic. Gloves must be worn while handling gels in ethidium bromide solution.
4. Wear safety glasses while viewing the gel on UV-transilluminator.
5. Do not touch the electrophoresis chamber or the electrical leads while electrophoresis is in progress.
6. At the time of casting of gel, ensure that no air bubbles are entrapped in the gel.

10.4.7 EXPERIMENT: Electrophoretic evaluation of quality of high molecular weight DNA isolated from plant material

Principle

Same as in EXPERIMENT 10.4.6

Materials and Reagents

Same as in EXPERIMENT 10.4.6

1. Tris-acetate (TAE) buffer stock solution (5X): Same as in EXPERIMENT 10.4.6. (Reagent 7).
2. 0.8% agarose in 1 X TAE buffer: Dissolve 0.6 g agarose in 75 ml of 1 X TAE buffer by boiling and maintain it at 50°C until used.
3. Gel loading solution: Same as in EXPERIMENT 10.4.6. (Reagent 9).
4. Ethidium bromide stock (10 mg/ml).
5. Isolated plant DNA.
6. Hind III digest as marker.

Procedure

1. Isolation of DNA from plant tissues: Follow the procedure described in EXPERIMENT 14.2.2.
2. Electrophoresis of DNA sample: The procedure for electrophoresis of DNA is outlined in EXPERIMENT 10.4.6.

Results and inference

Presence of a thick band near mouth of the wells will indicate that DNA is of high molecular weight whereas sheared DNA will appear as a smear throughout the gel lane indicating poor quality of the preparation. RNA contamination is indicated by smear near the bottom of the gel.

Precautions

Same as in experiment 10.4.6.

Table 10.3 Artifacts and trouble shooting

Nature of problem	*Cause*	*Remedy*
1. Slow or failure of gel to polymerise	(i) Catalyst absent or solutions old	(a) Check concentration of catalyst. Use freshly prepared solutions
	(ii) Presence of oxygen	(b) Deaerate the solution
2. Failure of sample to enter and settle into wells	Low density of sample	Increase concentration of glycerol or sucrose in the sample
3. Poor resolution of bands	(i) pH of buffer not correct	(a) Check pH of all solutions and buffers
	(ii) Resolving gel	(b) Change concentration of resolving gel or use gradient gels

(Contd)

Nature of problem	*Cause*	*Remedy*
4. Band streaking	(i) In SDS-PAGE protein remains aggregated	(a) Use additional SDS and boil sample properly. Remove insoluble material by centrifugation
	(ii) In native disc-gel electrophoresis, precipitation of native proteins	(b) Load a less concentrated sample and use continuous system
5. Protein bands in all tracks of gel	Contamination of sample buffer	Use fresh sample buffer
6. Bands broaden at the bottom of the gel	Sample density	Load equal volumes of samples in all wells
7. Wavy bands	Excess APS	Use proper APS concentration
8. Detachment of gel from glass plates during electrophoresis	Glass plates not clean	Use glass plates cleaned with swab of acetone or ethanol and then dried
9. Dye migration not even	(i) Insufficient cooling	(a) Improve cooling or run at low current
	(ii) Air bubbles	(b) Remove bubbles before electrophoresis
10. Poor staining	(i) Protein concentration low	(a) Load more protein
	(ii) Amount of dye low or dye is old	(b) Use more concentrated solution of dye or use fresh stock
11. Patchy staining	Insoluble material in dye	Dissolve dye properly and filter the solution
12. High background of staining	(i) Sample proteolysis	(a) Use protease inhibitors during sample preparation
	(ii) Impure grade SDS	(b) Use pure grade SDS and boil sample properly
13. Gel Cracking		
(a) During polymerization	(i) Excessive heat of reaction	(a) Use cold solutions
(b) During electrophoresis	(ii) Excessive current input	(b) Lower the current
14. Long duration required for run	Air bubbles	Remove air bubbles from gels and use deaerated solutions
15. Variations in staining intensity along width of stained band	Unven gel surface	Overlay water gently
16. Non-reproducibility of protein bands or loss of protein bands after staining	(i) Sample proteolysis	(a) Use protease inhibitors during sample preparation
	(ii) Electrophoretic conditions not proper	(b) Check all conditions properly
17. Fuzzy bands on isoelectric focussing	Focussing carried out for too long or too short period	Increase voltage gradient towards end of electrophoretic run

(Contd)

	Nature of problem	*Cause*	*Remedy*
18.	Skewed bands on isoelectric focussing	Faults in pH gradient	Clean electrodes and ensure that they make good contact with the gel
19.	Overlapping spots on isoelectric focussing	Complex protein mixture	Change pH range of isoelectric focussing gel
20.	Protein band streaking on isoeletric focussing	Protein aggregation or sample overloading	Add 8 M urea or 2% Triton X-100 to sample buffer Centrifuge sample before loading

Suggested further reading

Hames B.D. and Rickwood D. (1990). Gel Electrophoresis of Proteins—A Practical Approach, 2nd edn., IRL Press at Oxford University Press, Oxford, London.

Rickwood D. and Hames B.D. (1990). Gel Electrophoresis of Nucleic Acids—A Practical Approach, 2nd edn., IRL Press at Oxford University Press, Oxford, London.

Sambrook J., Fritsch T. and Maniatis J. (1989). Molecular Cloning—A Laboratory Manual, 2nd edn., Cold Spring Harbour Laboratory Press, London.

Shaw C.R. and Prasad R. (1970). Starch gel electrophoresis of enzymes—A compilation of recipies. *Biochem. Genet.* **4:** 297–320.

Vallejos E. (1983). Enzyme activity staining. *In* Plant Genetics and Breeding, (Tanksley SD and Orton TJ., Eds), Part A, pp 469–516, Elsevier, Amsterdam.

Wilson K. and Walker J. (1996). Practical Biochemistry—Principles and Techniques, 4th edn., Cambridge University Press, Cambridge, London.

11. Spectroscopy

11.1 Introduction

Spectroscopy deals with interaction of electromagnetic radiations with matter. Spectroscopic techniques now occupy a key position in biochemical studies and some of them are employed routinely and quite extensively. These techniques are extremely useful in characterization of chemical compounds in a sample on the basis of their spectral properties. In the beginning the spectroscopy was largely confined to use of light (visible radiation) for quantitative estimation and in some cases even for identification of biomolecules. With new developments and refinements in instrumentation the scope of spectroscopic techniques broadened tremendously and now, in addition to visible light, it encompasses examining the behavior of chemical substances upon their irradiation with regions of electromagnetic radiations such as γ-rays, X-rays, ultraviolet rays, infrared light, microwaves and even radio frequencies. Fairly detailed information about the intermolecular bonding, the type of molecular changes occurring during various enzymatic reactions and even in highly complex cellular systems such as mitochondrial electron transfer chain etc. the spectrophotometric and fluorometric methods offer a rapid and convenient means of qualitative identification and quantitative estimations of biomolecules even in relatively impure samples. In addition these methods are the preferred and most commonly used procedures for assaying activities of a vast majority of enzymes, determining their kinetic attributes and for examining their regulatory properties.

Nuclear Magnetic Resonance (NMR) and Electron Spin Resonance (ESR) have been helpful for solving complex structural and mechanistic problems. Mass spectrometry is the highly sensitive technique for molecular analysis and fairly detailed information can be derived about structure of a molecule using sub-milligram or even sub-microgram quantities of the material. Various spectroscopic techniques, their principle and applications are discussed in this chapter.

11.2 Nature of Electromagnetic Radiations

Electromagnetic radiation is a form of energy which does not require any medium for its transmission. It is made up of two components, viz. an electric field (E) and a magnetic field (H) which are continuously oscillating in space perpendicular to each other. The direction of propagation of electromagnetic radiation is in turn perpendicular to both the electrical and magnetic fields (Fig. 11.1).

The classical sinusoidal model adequately accounts for parameters such as wavelength, frequency, velocity and amplitude of the light wave (Fig. 11.2).

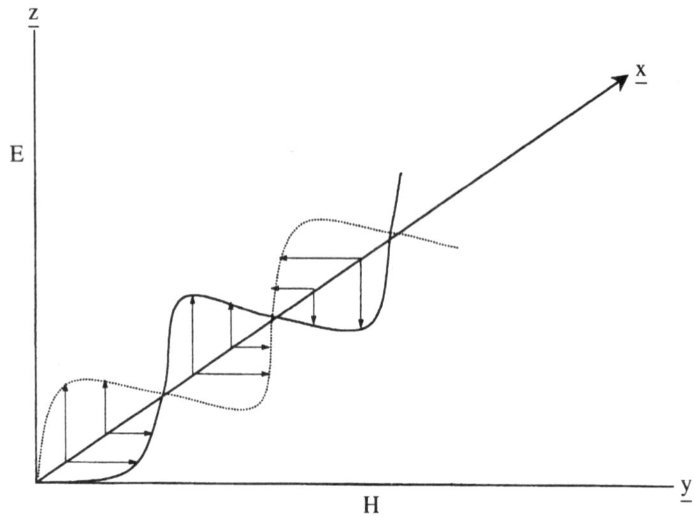

Fig. 11.1 Electromagnetic radiations. E represents electrical field; H magnetic field and electromagnetic radiation is propagated along X-axis. Arrows indicate the planes of electrical field and magnetic fields. Note the change in planes of these fields due to oscillations.

However, the phenomenon like their absorption and emission by matter can not be satisfactorily explained by the sinusoidal wave model alone and particle model has been proposed. According to this model an electromagnetic radiation constitutes of a stream of discrete packets of energy called photons or quanta. The amount of energy in a photon is directly proportional to its frequency (velocity divided by wavelength). Since the velocity of electromagnetic radiation is constant (3×10^8 m/sec) the energy content of a quantum depends on wavelength of the radiation. These two views about the nature of electronic radiations are complementary. Various types or regions of electromagnetic radiations along with their important properties are given in Table 11.1.

11.3 Theory of Spectroscopy

Inorganic as well as organic substances absorb electromagnetic radiations of different frequencies. This absorption of radiation results in transition of an atom or a molecule from its ground state to an excited state of higher energy levels.

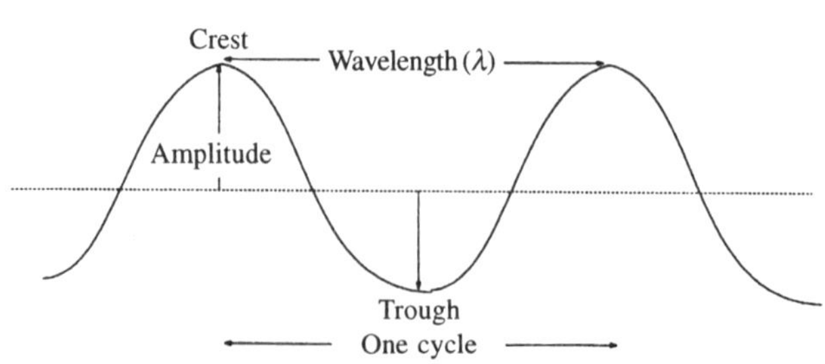

Fig. 11.2 Representation of electromagnetic radiation as a sinusoidal wave.

Table 11.1 Some characteristics of electromagnetic radiations

Region of electromagnetic radiation	*Wavelength (nm)*	*Frequency*	*Wave number (cm^{-1})*	*Energy KJ/mol*	*Phenomenon observed*
γ-rays	.003	1×10^{21}	3×10^{10}	4×10^{8}	Nuclear transitions
	.300	1×10^{19}	3×10^{8}	4×10^{6}	
X-ray	100	3×10^{16}	1×10^{6}	1.2×10^{4}	Inner electronic transisitions $\sigma \rightarrow \sigma^*$
Ultraviolet	300	1×10^{16}	3×10^{5}	4×10^{3}	Outer electronic transitions $\pi \rightarrow \pi^*$ $\pi \rightarrow \pi^*$
	400	8×10^{4}	3×10^{4}	330	
Visible	700	4×10^{14}	1.3×10^{4}	170	
Infra-red	1000	3×10^{14}	1×10^{4}	125	Molecular vibrations
	3×10^{6}		33.3	0.4×10^{-3}	
Microwave	3×10^{8}	1×10^{10}			Molecular rotation, electron spin resonance
	3×10^{12}	3×10^{6}	3×10^{-5}	4×10^{-7}	
Radio frequency	3×10^{14}	3×10^{4}	3×10^{-7}	4×10^{-9}	Nuclear magnetic resonance

The transitions which occur are a quantum phenomenon. In other words, conversion of an atom or a molecule to its excited state occurs only when a discrete amount of energy, which is exactly equivalent to that required for the transition in accordance with the principles of quantum mechanics, is absorbed by it in its ground state. This is referred to as one quantum of energy for an electronic transition. Similarly, when an excited electron reverts to its ground state, it emits radiations of specific wavelength. Spectroscopy, qualitatively and quantitatively, measures such changes in chemical substances. The spectroscopic techniques can thus be divided into two categories: absorption and emission spectroscopy.

The amount of energy involved in transitions of a molecule from its ground state to excited state (following absorption of energy) and from excited to ground state (by emission of radiation) is given by the following equation:

$$\Delta E = E_2 - E_1 = h\nu \qquad (1)$$

where ΔE = radiation energy absorbed or emitted by a molecule

E_1 = Energy of an electron in original level

E_2 = Energy of an electron in the final level

h = Plank's constant (6.63×10^{-34} J sec)

ν = Frequency of radiation in Hertz, i.e.

$$\frac{\text{speed of light } (c)}{\text{wavelength of radiation } (\lambda)}, \text{ or } \frac{3 \times 10^{8} \text{ m sec}^{-1}}{\lambda}$$

Besides elevating the electron to higher energy level, the absorbed light may cause vibration of atoms in a molecule or their rotation about the bond axis giving rise to vibrational and rotational sub-levels. Thus, in case of molecules, ground state and excited state is sub-divided into number of sub-levels (Fig. 11.3). The molecular spectra are composite of these alterations and are, therefore, obtained as *band spectra*. In contrast due to absence of vibrational energy sub-levels in an element, atoms give a sharp *line spectra*.

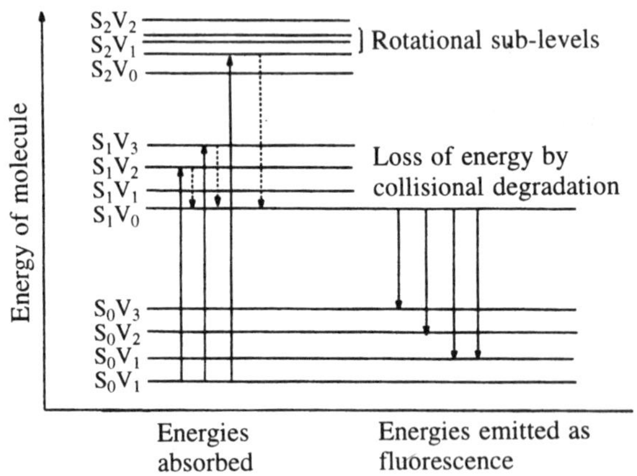

Fig. 11.3 Energy levels and transitions of electrons in a molecule. The rotational sub levels are shown only for vibrational sub-level S_2V_1. The given example is of a fluorescent organic molecule in which part of the absorbed energy is lost as light energy.

The amount of energy involved in these transitions depends on the nature of functionalities and the type of excitations involved. Different types of electromagnetic radiations exert different effects. In the ultraviolet and visible regions the excitations of electrons from lower to higher energy levels take place, the infrared radiations evoke bond vibrations and deformation excitations, microwaves induce molecular vibrations and electron spin resonance whereas radio waves elicit nuclear magnetic responses (Table 11.1)

11.4 Spectroscopic Techniques in Biochemical Investigations

It may be observed from Table 11.1 that molecules interact with radiations of various regions over a wide range of wavelengths. Special type of instruments are required to examine the spectra in different regions. Whereas some of the regions yield highly useful biochemical information with relatively simple instruments, other regions involve use of expensive and sophisticated equipment and require proper expertise for comprehending the obtained spectral information. The various spectroscopic techniques are described in this section and those which are more commonly used in biochemical investigations have been discussed in greater detail.

11.4.1 γ-Ray and X-Ray Spectroscopy

The γ-ray and X-ray spectroscopic techniques have been employed for rather limited purposes in biochemical studies. γ-ray resonance spectroscopy has been used for examining the co-ordination complexes, including those in biomolecules, whereas the application of X-ray spectroscopy is confined to forensic and metal pollutant studies such as for qualitative detection and quantitative estimation of trace metals like lead.

11.4.2 Ultraviolet and Visible Spectroscopy

Perhaps there is no other technique except UV-visible spectroscopy which has been so extensively employed and has such a profound impact on progress in biochemical investigations. UV-visible spectroscopy has undoubtedly become an indispensable tool and is routinely used for identification and quantitative estimation of most of the biomolecules. Besides other applications, most of the enzyme assays and the screening of biomolecules recovered in different fractions by separation processes such as column chromatography, density gradient centrifugation etc. rely heavily on spectro-photometric/ colorimetric techniques.

11.4.2.1 Principle

Use of flame photometry and atomic absorption spectrophotmetry for estimating metals in biological samples has already been discussed in Sections 6.2.1 and 6.2.2. At room temperature, most of the electrons in a molecule are in ground state and when a compound absorbs UV (200–400 nm) or visible (400–700 nm) radiations, both the bonding and non-bonding outer electrons undergo excitation from lower energy level (ground state) to higher energy level (excited state). The most common type of these excitations are of delocalized electrons such as bonding $\pi \rightarrow \pi^*$ electrons which give intense absorption bands. Aromatic and conjugated compounds show these excitations. For saturated compounds low intensity bands are obtained due to transitions of $n \rightarrow \pi^*$ and $n \rightarrow \sigma^*$ type. This technique exploits property of the compound to absorb light of particular wavelength(s) which defines its absorption spectra. The UV-visible spectrum of a compound provides clues regarding the ground state and the next higher excited state because the actual transitions occurring determine wavelength of absorbed light. Hence the absorption peaks can be related to known molecular sub-structures.

The greatest utility of this technique is that it enables quantitative determination of vast range of compounds of biological interest. The relationship between concentration of a compound and the extent of absorption of light is based on two simple laws viz. Lambert's Law and Beer's Law. When a beam of a monochromatic light (light of only one wavelength) passes through a homogenous solution, it may be transmitted as such or a fraction of it may be absorbed (Fig. 11.4).

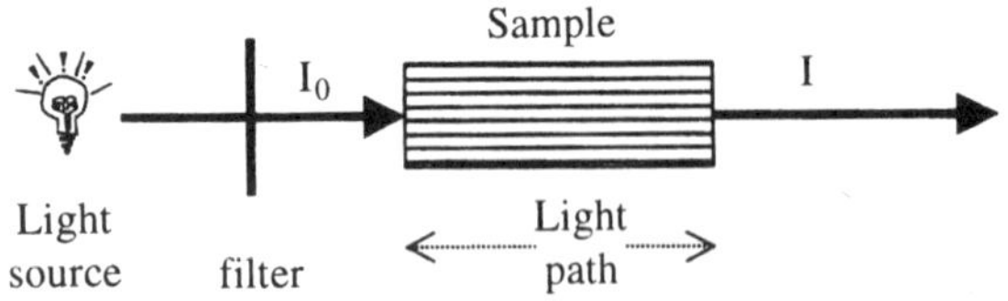

Fig. 11.4 Transmission of light through a solution.

I_0 and I represent intensities of incident and transmitted light, respectively. The proportion of transmitted light can be represented by the ratio of intensity of incident light and the intensity of transmitted light after traversing through the solution. Thus,

$$T = \frac{I}{I_0} \tag{2}$$

where T = Transmitted light or transmittance
I_0 = Intensity of incident light (photons cm^{-2} sec^{-1})
I = Intensity of transmitted light

From the above equation, per cent transmittance can be expressed as:

$$\% \, T = \frac{I}{I_0} \times 100 \tag{3}$$

For a completely non-absorbing transparent solution I will be equal to I_0 and hence T will be 100%. Conversely for an opaque solution, which does not allow any light to pass through it, value of I will be zero and hence T will be equal to 0% . Thus transmittance can vary over a range of 0–100%. Absorbance (A) of light through the solution is inversely proportional to $\log_{10}$ of % T. Thus,

$$A = \log_{10}\left(\frac{1}{T}\right) = \log\left(\frac{I_0}{I}\right) \tag{4}$$

Absorbance has no units and varies from 0 (100% transmittance, hence $\log_{10} (I/I_0) = \log 1 = 0$) to ∞ (0% transmittance, $\log_{10} (I_0/I) = \log 0 = \infty$).

The quantitative determination of compounds by spectrophotometric technique is based on two simple laws viz. Beer's law and Lambert's law. The Lambert's Law states that light absorbed by a solution is directly proportional to the length of light path (Fig. 11.4) through the solution. Hence,

$$A = \log_{10}\left(\frac{I_0}{I}\right) = \varepsilon \cdot l \tag{5}$$

where A = Absorbance
I_0 = Intensity of incident light

I = Intensity of transmitted light after passing through the solution
ε = Molar absorbance coefficient
l = Light path or width of liquid in cell or cuvette in cm

According to Beer's Law the amount of light absorbed is directly proportional to the concentration of absorbing solute in the solution. Thus

$$A = \log_{10}\left(\frac{I_0}{I}\right) = \varepsilon \cdot c \tag{6}$$

where ε = Molar absorbance coefficient
c = concentration of the solute in moles/L

Combining equations (5) and (6), we get

$$A = \log_{10}\left(\frac{I_0}{I}\right) = \varepsilon \cdot c \cdot l \tag{7}$$

If a standard cuvette with light path of 1 cm is used, which in fact is the general practice, then

$$A = \log_{10}\left(\frac{I_0}{I}\right) = \varepsilon \cdot c \tag{8}$$

The above equation indicates that absorption of light by a solution is dependent on the molar absorbance coefficient and concentration (moles/L) of the absorbing solute. Molar absorbance coefficient or molar extinction coefficient, as it is quite often referred to, is the absorbance of light by one molar solution of the solute using a standard cuvette of 1 cm light path. It is generally denoted $\varepsilon_\lambda = y$. For example $\varepsilon_{340} = 6.21 \times 10^3$ would imply that 1 M solution of a particular compound has an absorbance of 6.21×10^3 O.D. units at 340 nm. If the value of ε of the compound is known, then its amount can directly be calculated by determining absorbance of the solution against an appropriate blank. Even if the value for ε is not available a reference or calibration curve can be prepared by taking known amounts of authentic sample of the compound and, if necessary, after suitably processing (such as development of colour etc.) its absorbance at an appropriate wavelength is recorded. A graph of absorbance values against concentrations of the sample is plotted. The amount of solute in the experimental sample can be determined using the standard curve.

11.4.2.2 Instrumentation

Colorimeters and spectrophotometers are used for measuring absorption of light by sample solutions. Colorimeters are relatively simple instruments which are designed to function only in the visible range (400–700 nm). On the other hand, spectrophotometers are equipped to operate both in the visible and UV range since they are fitted with deuterium or hydrogen (for UV light) and tungsten (visible range) lamps as the light sources. Several models of spectrophotometers of varying degree of sophistication are available. These include single beam, double beam, recording and multibeam instruments. In dual beam spectrophotometers the incident light is split into two beams of equal intensity, one of which passes through the reference cuvette and the other through the sample cuvette. The instrument records the change in absorbance in the sample with respect to the reference. The multibeam instrument is designed to simultaneously record absorbance changes at two or more pre-determined wavelengths. Recording spectrophotometers can both scan the absorption spectrum of the sample in the desired range and also record the change in absorbance with time at a fixed wavelength. Reflectance spectrophotometers are used for determining spectra of pastes and suspensions such that of micro-organisms. The latter type of spectrophotometers enable measurement of the radiation absorbed when the light beam is reflected by a sample, which is too opaque to allow transmission of light.

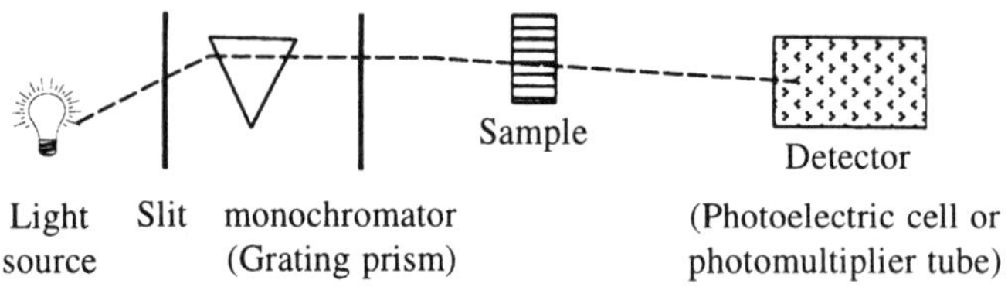

Fig. 11.5 A typical single beam UV-visible spectrophotometer.

11.4.2.3 Applications

(*i*) *Quantitative estimation of biomolecules:* Spectrophotometric methods are the commonly used procedures for quantitative analysis of a vast variety of compounds in purified preparations or impure biological samples. For example, the presence of colourless substances such as proteins, nucleic acids, NAD(P)H etc. can rapidly be examined by measuring the absorbance at a chosen wavelength which is characteristic and unique for that particular substance. During enzyme purification, the presence as well as amount of proteins in different fractions can be monitored simply by measuring their absorbance at 280 nm. This evidently is a much quicker, convenient and non-destructive method of establishing elution profile of such measurable substances during column chromatography or density gradient centrifugation. Most of the analytical methods for quantitative estimation of chromogenic biomolecules such as chlorophylls, carotenoids, anthocyanins, hemoglobin etc. are colorimetric or spectrophotometric and are based on measurement of absorbance at a specific wavelength. The absorption maxima (λ_{max}) of a few compounds of biological interest are listed in Table 11.2.

Even for the colourless substances which do not exhibit absorption of either UV or visible light, such as sugars, amino acids, metal ions etc, their coloured derivatives can be obtained by reacting them with suitable reagents and these can then be quantitatively determined from their absorbance. Numerous examples of such colorimeteric methods are given in Chapters 1–6 of this book.

(*b*) *Identification of compounds:* Spectrophotometry, in conjunction with other methods, is helpful in establishing presence of a particular compound and also in identifying certain types of biomolecules in a sample preparation. This is generally done by examining its absorption spectrum and then comparing it with that of the authentic samples of known compound. Most of the compounds have

Table 11.2 Absorption maxima of some biomolecules

Compounds	*Absorption maxima (nm)*
Phenylalanine	260
Tyrosine	275
Tryptophan	280
Uracil	259
Adenine	260
Cytosine	267
Guanine	246
Chlorophyll-*a*	663,420
Chlorophyll-*b*	650,450
Ergosterol	282
Cytochrome *c* (reduced)	550,521,415*
Cytochrome *c* (oxidized)	415
Flavin nucleotides (oxidized)	370–450
Flavin nucleotides (reduced form)	370
Nicotinamide adenine nucleotides (oxidized)	260
Nicotinamide adenine nucleotides (reduced)	260,340
Proteins	280
Nucleic acids	260

*α, β and γ peaks of the cytochrome, respectively.

a characteristic absorption spectrum which is simply a plot of absorbance of light by a compound at different wavelengths. Absorption spectra of oxidized and reduced forms of cytochrome *c* is presented in Fig. 11.6. For instance, presence of cytochromes, haem or flavins as prosthetic group of a number of enzymes was largely deduced from spectral studies. This technique has also been extremely useful in indicating involvement of certain compounds in various complex processes. A notable example of this is in establishing the role of quinones, flavins and various cytochromes in the mitochondrial electron transport chain as well as in the photosynthetic electron transfer chain. Spectral investigations revealed that,

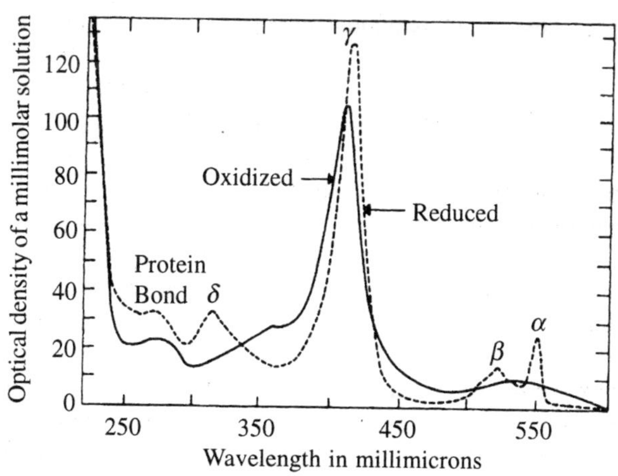

Fig. 11.6 UV-vis spectrum of cytochrome *c*.

under anaerobic conditions, addition of NADH to mitochondrial preparations results in reduction of cytochrome *c*. It undergoes rapid oxidation on introduction of oxygen thereby suggesting that it acts as an intermediate carrier of electrons during their transport from NADH to oxygen via the mitochondrial electron transfer system. Such studies are generally carried out by examining the difference spectra because the various states of a compounds exhibit qualitatively alteration in their spectral characteristics. For example the oxidized and reduced forms of quinones, flavins and cytochromes have distinctive spectral difference. As the name denotes, the difference spectra is a graphical representation of difference in absorbance of light at different wavelengths by two forms of a compound. The absorption spectra and difference spectra of ubiquinone and ubiquinol is shown in Fig. 11.7. Similarly pyridoxal phosphate and pyridoxal phosphate engaged in formation of a Schiff's base show qualitativėly different spectral features.

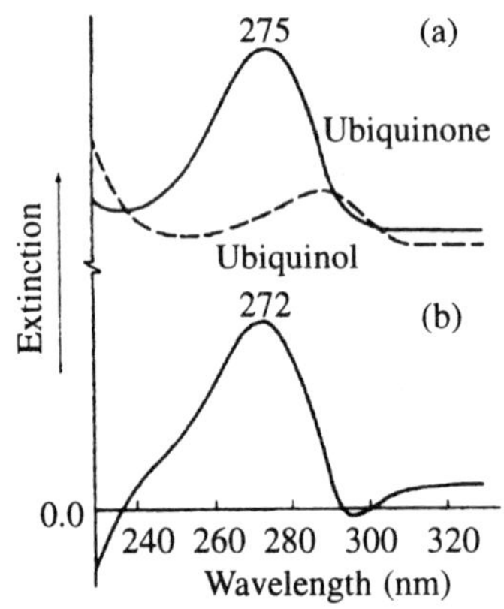

Fig. 11.7 Absorption (a) and difference (b) spectra of ubiquinone and ubiquinol.

Spectrophotometric analysis has also been helpful in identification of the chromophores in light-mediated responses and processes. First the rate of the light dependent response or process is determined at varying wavelengths. A plot of rate of the process (such as photosynthetic oxygen evolution) or at various wavelengths is prepared. Such a plot is known as an action spectrum. Then attempt is made to isolate the compound from the tissue which has an absorption spectrum superimposable on the action spectrum. This approach has successfully been employed in establishing the central role of chlorophylls as the primary light harvesting pigments in photosynthesis and of phytochrome in red-far red induced responses in plants.

(*c*) *Measurement of enzyme activities:* The most wide spread application of spectrophotometric technique is for continuous assay of enzyme activities. The method can be employed for any enzymatic reaction in which the product, substrate or cofactor shows absorbance at a unique wavelength. Then by simply monitoring the change in absorbance at that particular wavelength, the rate utilization of the substrate, formation of the product or changes in concentration of the coenzyme can be followed spectrophotometrically. The most striking example of such assays is the one in which nicotinamide adenine nucleotides undergo reduction or oxidation during the enzymatic reaction. Though both forms of these nucleotides show intense absorbance at 260 nm but, only the reduced form exhibits specific absorbance at 340 nm. This provides a very convenient system for monitoring activities of NAD(P)-linked oxido-reductases, several examples of which are given in Chapter 7. Thus for reactions involving oxidation of NAD(P)H, one needs to simply record the rate of decrease in absorbance at 340 nm and vice versa. Besides oxido-reductases activities of several other enzymes can be determined directly. For example, activity of fumarase is determined directly from the decrease in absorbance at 290 nm because the substrate fumarate, but not malate which is product of the reaction, shows absorbance at this wavelength. Similarly in the spectrophotometric assay of catalase the rate of decomposition of the substrate H_2O_2, which has absorbance at 240 nm, can be measured from decrease in absorbance at 240 nm. Spectrophotometric assay procedures are often preferred because they are convenient, rapid and the information regarding reaction rates with time becomes available continuously and instantaneously which is extremely helpful in ensuring that initial velocity of the reaction is being measured.

Amount of the substrate consumed or the product formed can be calculated from their molar absorption or extinction coefficient. This is illustrated below by taking an example of dehydrogenase:

$$A + NADH + H^+ \rightleftharpoons AH_2 + NAD^+$$

The activity of such an enzyme is measured by following the decrease in absorbance at 340 nm (A_{340}). Suppose the ΔA_{340} is 0.4 O.D. units per min. The molar extinction coefficient of $NADH_2$ is 6.2×10^3. Hence,

$$6.2 \times 10^3 \text{ O.D. units} = 1 \text{ moles L}^{-1}$$

or

$$1 \text{ O.D. unit} = 1/6.2 \times 10^3 \text{ moles L}^{-1}$$

$$\therefore \quad 0.4 \text{ O.D. unit} = \frac{1 \times 0.4}{6.2 \times 10^3} \text{ moles/L} = \frac{1 \times 0.4 \times 10^6}{6.2 \times 10^3} \mu\text{moles/L}$$

$$= 64.5 \ \mu\text{moles/L}$$

or

$$= 64.5 \text{ nmoles/ml}$$

11.4.3 Infrared spectroscopy

The infrared part of the electromagnetic radiations is sub-divided into near far-red (1–2 μm), infrared (2–25 μm) and far-red (25–250 μm) regions. Out of these the infrared region between 2.5-20 μm is the most useful for analytical purposes.

11.4.3.1 Principle

The absorption of electromagnetic radiations in the infrared region results in vibrational and rotational transitions of the bonding atoms in the molecule leading to asymmetrical charge distribution (Fig. 11.8). Non-linear molecules may undergo vibrational motions like stretching and deforming. The energy requirement of stretching vibrations (symmetrical and asymmetrical) are higher and occur at higher frequencies than deforming or bending (in-plane—scissoring and rocking; out-of-plane—wagging and twisting) vibrations.

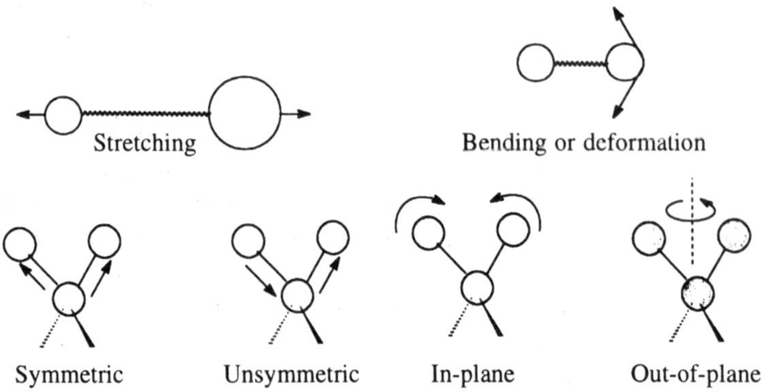

Fig. 11.8 Changes in vibrational and rotational transitions caused by absorption of infrared radiation.

Thus absorption of infrared radiations does not cause excitation of electrons but induces vibrations of bonding atoms, and the vibrations which bring about alteration in dipole moment or displacement of charge can be detected. The other types of vibrations are studied by Raman Spectroscopy.

The fundamental vibrations are those which correspond to $V_0 - V_1$ transitions. The intensity of a particular absorption depends on the dipole moment of the molecule in the ground state and the vibrational excited state. Higher is the difference in dipole moments, greater will be the intensity of absorption. The infra-red spectra are observed only in the heteronuclear molecules since homonuclear molecules have no dipole moment. For obtaining an infrared spectrum, electromagnetic radiation of increasing wavelength is passed through the sample and per cent transmittance or absorbance is measured. The energy of vibration is quantized. The infrared spectrum of the functional groups in a molecules is highly specific and this characteristic spectrum or finger print facilitates identification of the compound. In infrared spectroscopy it is usual to work in frequency units (Hz) rather, than wavelength.

11.4.3.2 Instrumentation

Spectrum of the sample is recorded using an infrared spectrophotometer. Double beam spectrophotometers are preferred since CO_2 and H_2O show significant absorbance of infrared light and in these instruments their interference is compensated automatically as half of the splitted beam of light passes through the reference cell and the other through the sample. The source of light is generally an incandescent lamp having an output maxima at around 2 cm, however, their energy shows a sharp decline (15% of maximum at 15 cm). Nernest-Glober and nichrome alloy wire lamps have also been used for this purpose. The detector is a thermopile which comprises of stack of thermocouple junctions in an evacuated can having an infrared window. Golay cells which consist of sealed container filled with a gas or a liquid which expand due to rise in temperature as consequence of the impinging infrared radiations. The design and components of a double beam infrared spectrophotometer are illustrated in Fig. 11.9.

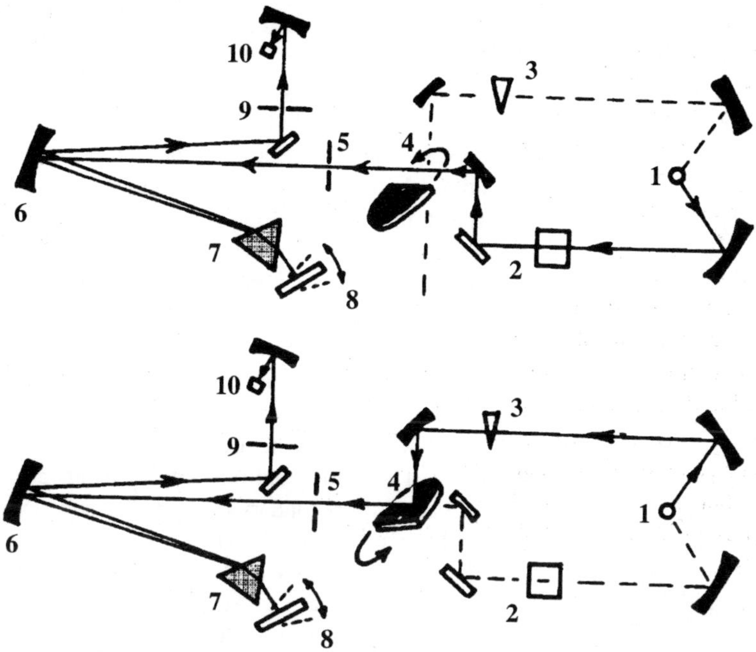

Fig. 11.9 **Schematic representation of the parts of a double beam infrared spectrophotometer showing the use of rotating mirror for sample as well as reference beams. 1, source of light; 2, sample; 3, attenuator wedge (to rccorder); 4, rotating mirror; 5, 9, slits; 6, paraboloid optic; 7, prism; 8, mirror to control frequency of IR radiation; 10, detector.**

Preparation of the sample depends on its physical state. Liquid samples are run either as neat or as solutions. Solid samples are used in form of solutions dissolved in an appropriate solvent such as CCl_4 or $CHCl_3$ or are prepared in mulls or as solids dispersed in KBr. Finely ground mixture is pressed into KBr discs under pressure of 4–8 tons. Mulls are prepared by suspending finely ground sample in nujol. The sample is taken between two highly polished salt plates of NaCl since non-covalent materials are transparent to infrared light, and hence are used for sample containment.

11.4.3.3 Applications

The fundamental frequencies obtained are characteristic of functional groups and these are absolutely specific. The fundamental frequencies emanating from vibrational motion in a molecule of acetaldehyde are as follows: for CH_3 group at 1460 and 1365 cm^{-1} due to the bending vibration, —C—C— and —C=O groups give stretching vibrations at 1165 and 1730 cm^{-1} respectively, —C—H in

CH_3 group exhibit absorption frequencies at 2960 and 2870 cm^{-1} whereas for C—H (in CHO) the observed frequency is at 2720 cm^{-1}. With the increased number of functional groups in a complex molecule, difficulties arise in interpreting the absorption bands. However, in such cases group frequencies obtained are helpful as these can be assigned to a particular type of molecular groupings. A particular group, however, may not always absorb at the same frequency because molecular environment can have a considerable influence on it. Nonetheless these are quite useful in structural analysis as they enable to distinguish between C—H vibrations in methylene (—CH_2) and methyl (—CH_3). When the bonding force between two atoms increases, such as double bond formation, the stretching frequency increases. For a situation where the reduced mass (which is equivalent to $M_a\, M_b/M_a + M_b$ where M is the atomic weight of the atoms *a* and *b*) increases, the vibrational frequency decreases and hence the wavelength of the absorbed light increases. For any given bond,the bending vibrations occur at much lower frequency than the stretching frequency.

The main application of infrared spectroscopy in biochemical investigations is in elucidation of structure of purified biological molecules of intermediate size such as small peptides, metabolic intermediates, substrates, drugs etc. This technique has been employed for examining the secondary structure of proteins. Polypeptides give characteristic infrared spectra. The nature and intensity of bands at 3290, 1650, and 1535 cm^{-1} provides clues about the secondary structure. These absorption bands are attributable to stretching mode of N—H and C=O bonds and deformation mode of N—H admixed with C—N stretching mode respectively. The bands at 1650 and 1535 cm^{-1} are referred to as Amide I band and Amide II band, respectively. On transitions of randomly coiled molecule to an ordered structure, each of these bands gives two maxima due to vibrational interaction among the peptide groups. In α-helix the Amide I band splits into two absorption maxima of 1650 cm^{-1} of a strong intensity and 1652 cm^{-1} of medium intensity, whereas the Amide II band gives a weak intensity band at 1516 cm^{-1} and an intense band 1546 cm^{-1}. For parallel pleated sheets Amide I gives weak absorbance at 1645 cm^{-1} and an intense one at 1630 cm^{-1} but for antiparallel pleated sheets the spectra shows a weak signal at 1685 cm^{-1} and an intense band at 1632 cm^{-1}. The Amide II band at 1535 cm^{-1} in random coil gets bifurcated giving intense band at 1530 cm^{-1} and weak absorbance at 1550 cm^{-1} and a single pronounced spectral band at 1530 cm^{-1}. Thus from the infrared absorption characteristics of the bands associated with Amide I and Amide II bands, preliminary information can be derived about the secondary structure of peptides.

Another important use of infrared spectroscopy is in measurement of different gases such as CO_2 for studies on photosynthesis and respiration and of acetylene for biological nitrogen fixation.

11.4.4 Spectrofluorometry

11.4.4.1 Principle

This technique is based on quantitative measurement of fluorescence. Upon absorption of radiation energy, electrons of a molecule are elevated to their excited state. If the excitation of the molecule does not result in chemical reactions such as dissociation or rearrangement , or energy transfer by collisions, then the molecule returns to its ground state and the excessive energy is dissipated as light, i.e. photons of energy $h\nu$ in form of luminescence by two possible pathways: fluorescence and phosphorescence. Whereas fluorescence ceases as soon as the excitation source is extinguished, phosphorescence is characterized by much longer decay periods of about 10^{-3} sec (as compared to 10^{-8} sec in fluorescence) and usually persists even when the excitation energy source is removed.

In fluorescence the absorbed or excitation light is of a smaller wavelength (hence more energy) and wavelength of the emitted light is longer (less energy). This effect is on account of loss of part of the energy in radiationless transitions before reversion to the ground state. The increase in wavelength of emitted light than the excitation light is referred to as Stokes shift. Those compounds which display significantly large and readily measurable Stokes shift can thus be examined fluorometrically using a spectrofluorometer.

The phenomenon of fluorescence is shown in Fig. 11.3. At room temperature most of the electrons are at their ground state S_0V_0. These electrons may get elevated in less than 10^{-15} sec to higher energy levels (S_1, S_2, S_3) on absorption of photons. The radiant energy may be rapidly discharged as heat due to collision gradation resulting in minimal vibrational energy in the lowest excited state S_1V_0. The electrons in this state further lose energy in less than 10^{-8} sec as radiations thus resulting in fluorescence.

Although organic compounds, absorb UV or visible light, only a few of them fluoresce. Aliphatic molecules which are generally flexible, undergo photo-dissociation whereas aromatic compounds with delocalized electrons sometimes fluoresce. Several organic compounds of biological interest incidentally do show fluorescence and this technique can, therefore, be employed for their quantitative estimation. The quantum efficiency involved during fluorescence is given by:

$$Q = \frac{\text{Quanta fluouresced}}{\text{Quanta absorbed}} \qquad (9)$$

The quantum efficiency is generally independent of exciting wavelength. The maximum value of quantum yield or efficiency can be significantly less than that of the excited molecules which are deactivated by other processes as well. In dilute solutions, the intensity of fluorescence (I_f) radiation is dependent on intensity of the incident radiation (I_0), and the quantitative relationship between them is

$$I_f = 2.3\ I_0\ \varepsilon\ c\ d\ Q \qquad (10)$$

where c is molar concentration of the fluorescing solution, d is the light path in cm of fluorescing solution, ε is the molar extinction coefficient of the fluorescent compound at wavelength (dm^3 mol^{-1} m^{-1}). Under given experimental conditions all the terms (I_0, ε, d, Q) except c in the above equation are constant. Hence

$$I_f \propto c \qquad (11)$$

In spectrophotometric technique the most accurate measurements are generally obtained at very low concentrations and the technique is relatively much more sensitive since amplification of the detector signal can be adjusted over a large range. The major implication encountered in spectrofluorometry is that of quenching. This arises from loss of emitted radiant energy to other molecules by colloisional interactions. This interference accounts for greater accuracy of the spectrofluormetric technique at low concentrations or dilute solutions because of lower collisions due to lesser number of molecules. The extent of quenching can, however be, corrected by using an internal standard. This entails measurement of fluorescence of the sample and then adding a known amount of the compound being determined and again recording the fluorescence.

11.4.4.2 Instrumentation

Spectrofluorometer is used for measurement of the intensity of fluorescence radiation by fluorescent compounds. The instrument constitutes of the following components. A source of light, monochromator for selection of light of a particular wavelength for irradiation of the sample, a sample holder, another monochromator for isolating light of a desired wavelength which reaches the detector or a photocell, an amplifier to magnify the signal and finally a recorder (Fig. 11.10).

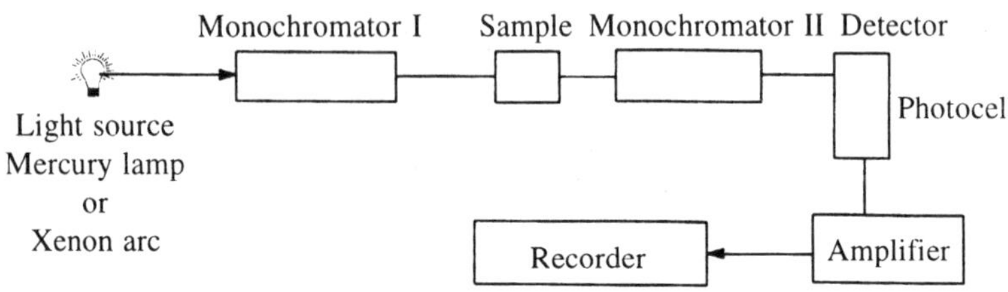

Fig. 11.10 Basic components of spectrofluorometer.

The radiation source is either a mercury lamp or a xenon arc. The excitation wavelengths chosen are frequently in UV region and the emission wavelengths in the visible region. Facility for control of temperature is also important since the intensity of fluorescence may differ by 10–50% for a 10°C decrease in temperature from 30–20°C.

11.4.4.3 Applications

Fluorometric techniques, besides providing information about the electronic structure of molecules in their excited state, have also been used quite extensively for a variety of purposes in biochemical analysis including qualitative and quantitative determination of biomolecules, conformational changes in protein molecules resulting from binding of ligands, investigations on membrane structures, immunofluorscent based techniques for intracellular localization of enzymes (Section 13.11.1) the fluorescence-activated cell sorter (Section 13.11.2) for isolating particular type of cells from heterogeneous cell population. It has also been exploited for quantitative and more efficient measurement of radioactive compounds such as H^3, C^{14}, P^{32} and S^{35} by liquid scintillation counting (Section 12.7.2). Some of these are discussed in brief as follows.

Presence or absence of a particular compound of interest in a sample can evidently be established by monitoring the fluorescence at specific wavelengths for the excitation and light emission which are characteristic for that particular substance. Since the intensity of emitted light is directly proportional to its concentration, it offers a highly sensitive method for quantitative estimation of compounds such as vitamins, hormones, NADH, drugs, pesticides, cholesterol etc. At times this technique can also be extended for estimation of suitable inherently non-fluorescent compounds by attaching a suitable fluorescent probe. A few such examples include detection and estimation of amino acids and peptides by coupling their amino groups with dansyl chloride or phthalaldehyde. The basis for use of phthalaldehyde for spectrophotometric estimation of amino acids is as follows:

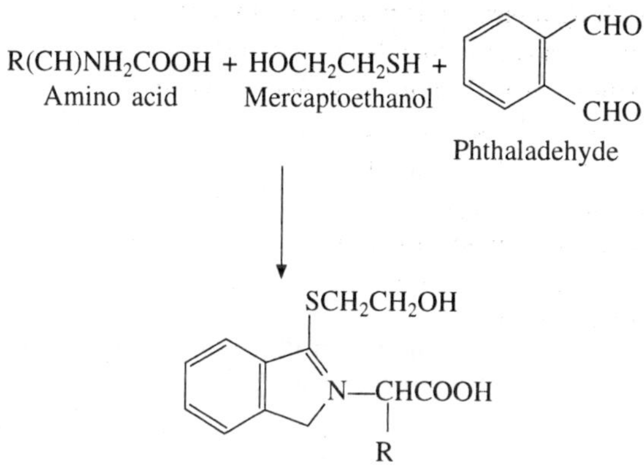

Indole derivative
(fluorescent)

The indole derivatized amino acid exhibits intense blue fluorescence. And enables measurement of as low as 10 pmoles of the amino acids. Similarly, acridine orange is used to establish whether a polynucleotide chain is single or double stranded by virtue of the difference in Stokes shifts of conjugates of the single and double stranded polynucleotides which fluoresce in red and green regions of the spectrum, respectively. Some other such fluorescent probes include ethidium bromide, fluorescein, 4-methylumbelliferone, 1-anilinonaphthalene-8-sulphonate (ANS) and 12-(9-anthranoyl) stearic acid.

Valuable information can also be obtained about conformational changes and aggregation of protein molecules. For example, FAD and tryptophan are fluorescent and binding of ligands such as substrates, inhibitors, modulators close to these residues alters the fluorescence intensity of a protein molecule. The absence of intrinsic fluorescent moiety can be overcome by using appropriate external fluors such as anilinonaphthalene-8-sulphonate (ANS) and 2-*p*-toluidinylnaphthalene-6-sulphonate (TNS) which interact with protein molecules. Both these fluorescent labels are non-fluorescent in aqueous solution but show intense fluorescence on binding to the hydrophobic regions in the protein molecule. Change in fluorescence quantum yield or intensity denotes structural alterations in the environment where these intrinsic or extrinsic fluors are located. Other fluorescent probes such as dansyl chloride or derivatives of fluorescein or rhodamine can also be used for such studies.

Some other applications of fluors such as in liquid scintilation counting (Section 12.7.2), intracellular localization of enzymes (Section 13.11.1) and fluorescence activated cell sorter (Section 13.11.2) are described elsewhere.

11.4.5 Optical rotatory dispersion (ORD)

As mentioned earlier (Section 11.2) light is electromagnetic radiation having two components, viz. an electrical field (E) and a magnetic filed (H). Both these components oscillate in space with the same wavelength and frequency and the propagated light is perpendicular to the planes of both the electrical and magnetic components. Hence light is a transverse wave motion. In case of an ordinary light, the planes of E and H components change continuously and randomly. Consequently the generated light waves also keep on oscillating in all the directions. When an unpolarized light is made to traverse through a polarizer, the light which emerges travels along a particular plane thereby implying that vibrations of the electrical component and consequently those of magnetic field, which is oscillating perpendicular to the electrical component, get fixed in distinct and definite planes. Such a light is known as plane polarized light or linearly polarized light. Materials such as Nicol prism and Polaroid are commonly used for this purpose. Nicol prism is constructed from a crystal of calcite ($CaCO_3$) whereas Polaroid is trade name of transparent films made up of thin sheets of cellulose nitrate packed with ultra-microscopic double refracting crystals arranged in such a way that their optical axes are paralleled.

11.4.5.1 Principle

Optical isomers possess optical activity and exhibit the property of rotating plane of the polarized light. In contrast, ordinary compounds do not show this effect. This is illustrated in Fig. 11.11.

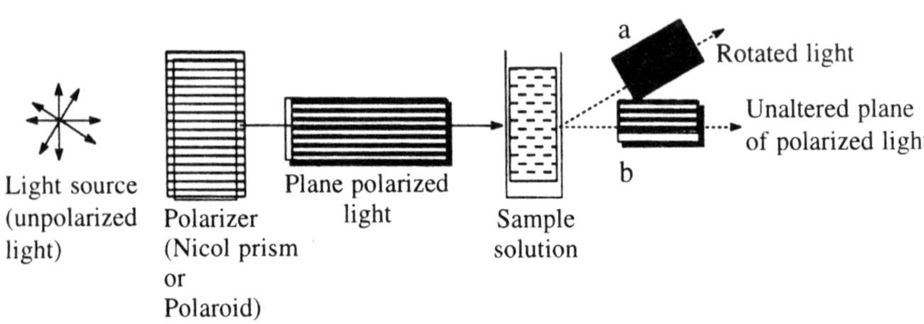

Fig. 11.11 Generation of plane polarized light and effect of sample solution on the plane of the polarized light. (a) Solution of optically active compound, (b) Solution of an ordinary compound.

This phenomenon can be explained as follows: A plane polarized light can be divided into two vibrating vectors E_L and E_R of equal amplitude and frequency which correspond to left and right of the circularly polarized waves. Both E_L and E_R rotate at the same speed when the plane polarized light passes through an ordinary media. However, the speed of rotation of E_L and E_R is differentially affected when the plane polarized light traverses through a solution of an optically active compound. In case of optical isomers (isomers whose mirror images are non-superimposable), being optically active, the refractive index (rates of velocity of light in vacuum to that through a medium) is different for the left and the right circularly polarized light and this causes differences in the speed of rotation. This uneven speed of rotation results in displacement in the angle of plane of the polarized light. Thus rotatory power is a direct consequence of molecular symmetry. The angle of rotation (α) is given by:

$$\alpha = \frac{1800}{\lambda}(n_L - n_R) \tag{12}$$

where λ is wavelength of the polarized light and n_L and n_R are refraction indices for E_L and E_R. This optical activity can be further elaborated in terms of specific rotation:

$$[\alpha]_\lambda^T = \frac{\alpha}{1c} \times 100 \tag{13}$$

$[\alpha]_\lambda^T$ is specific rotation, α is the change in angle in degrees 1 is the light path in decimeters, c is the concentration of solute in g per 100 ml of solution, T is the temperature in degrees (Celsius) and λ is the wavelength. The media which rotates the plane of polarized light in a clockwise direction (right hand direction) is called dextrorotatory and is denoted by symbol (+) and if the rotation is in anticlockwise direction (left hand direction), it is known as levorotatory and designated as (–). This in fact, forms the underlying principle for simple polarimetric determinations in which the measurements are made at a single fixed wavelength (usually D lines of sodium 5890 and 5896 Å).

As can be noted from equation (12) the refractive index and hence rotation of plane polarized light of a media varies with wavelength of light. A simple plot of refractive index versus wavelength is referred to as a *dispersion curve*. The optical rotatory dispersion spectroscopy (ORD) is based on measurements of rotation (α) of the incident plane polarized light over a range of wavelengths and the plot obtained is known as optical rotartory dispersion curve. Thus optical rotatory dispersion is simply based on observing the variation in plane of polarization of monochromatic light of different wavelenghts. Optical dispersion curve of a typical protein is shown in Fig. 11.12. Thus a sharp change in the magnitude of rotation (at about 234 nm) is known as "Cotton effect" and in most of the proteins this effect is observed at 225–235 nm.

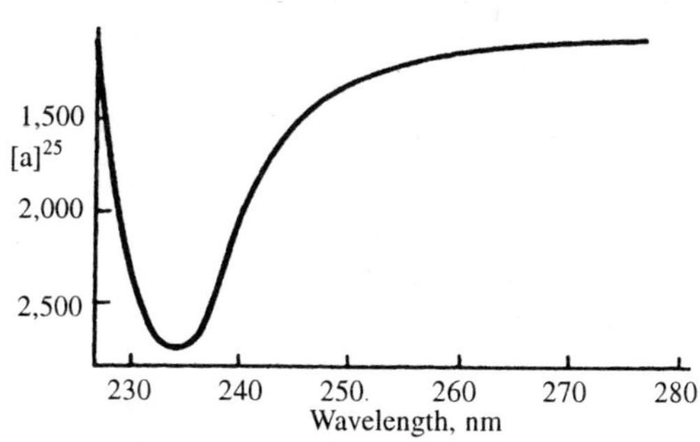

Fig. 11.12 ORD curve of a protein showing Cotton effect.

11.4.5.2 Applications

Optical rotatory dispersion is a direct consequence of molecular asymmetry and has been employed quite extensively for studying secondary structure, particularly in estimating the helical content is influenced by other factors, of optically active polypeptides. However, the rotatory power of polypeptides besides their helical structure depend on the amino acid content because the constituent amino acids are themselves asymmetric. Moffit and Yang equation is used for evaluation of the proportion of α-helical segments in a polypeptide chain by assigning optical activity beyond that due to optically active residues. This additional contribution to total optical activity of the polymer is due to dissymmetry of the helix. By analysing the change in optical rotation over much of the UV and visible spectrum, the percent helix of synthetic peptides and proteins can thus be deduced.

Measurements of optical rotatory dispersion examined with light of single wavelength (D-lines of sodium at 5890 and 5896 Å) show marked differences depending on secondary structure. For example in case of synthetic peptides such as poly-γ-benzylglutamate, polyglutamic acid and natural protein BSA, notable variations occur on conversion of these polypeptides from their helical to random coil structures. Equations have been proposed to quantitatively relate these changes with the fraction of residues in the helix. Drude equation gives the relationship between optical rotation and wavelength for rather a very simple system. According to this,

$$[\alpha]_\lambda = \frac{K}{\lambda^2 - \lambda_0^2} \tag{14}$$

where $[\alpha]_\lambda$ = specific rotation defined by $[\alpha]_\lambda = \dfrac{\alpha}{cd}$

α = observed rotation at specific wavelength

c = concentration of rotatory substance in g per 100 ml

d = length of light path in decimeter

This simple equation is not adequate to account for optical rotation with wavelength and a two-term equation of Moffit and Yang gives satisfactory correlation between these parameters for most of the natural and synthetic polypeptides. According to this equation:

$$[R]_\lambda = a_0 \frac{\lambda^2}{(\lambda^2 - \lambda_0^2)} + b_0 \frac{\lambda^4}{(\lambda^2 - \lambda_0^2)} \tag{15}$$

where a_0, b_0 and λ_0 are all constants for a given system. The values for a_0 depends on temperature and nature of residues of b_0 on the level of organization, e.g. value of $b_0 = -630$ for a perfect α-helix and 0 for random coil. In the above equation $[R]_\lambda$ represents reduced mean ratio and it is equal to:

$$[R]_\lambda = \frac{3}{\eta^2 + 2} \cdot \frac{\text{MRW}}{100} [a]_\lambda \tag{16}$$

where η is the index of refraction. MRW, the Mean Residue Weight, i.e. molecular weight of protein divided by total number of amino acid residues and is usually taken as 115 for a protein of unknown composition.

The parameters a_0 and b_0 are determined experimentally and the chosen value of λ_0 which gives the best fit to Moffit-Yang equation is 212 nm. The values for a_0 and b_0 are determined from a plot of

$$[R]_\lambda \frac{(\lambda^2 - \lambda_0^2)}{\lambda_0^2} \text{ against } \lambda_0^2 (\lambda^2 - \lambda_0^2)^{-1}$$

which, as illustrated in Fig. 11.13, gives a linear curve with a_0 as the intercept on the ordinate and b_0 is the slope.

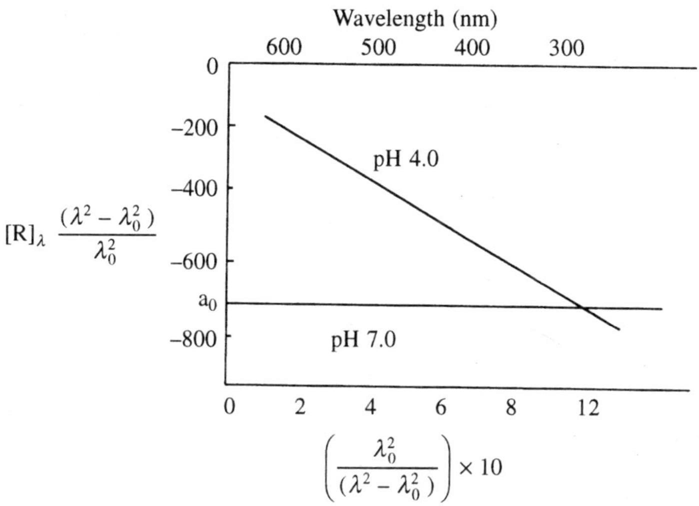

Fig. 11.13 A plot of optical rotatory dispersion of a synthetic polypeptide of tyrosine and L-glutamic acid.

The above plot has been prepared by taking λ_0 as 212 nm for a synthetic co-polymer of 5% tyrosine and L-glutamic acid. This polypeptide exists as an α-helix at pH 4.0 and gets disrupted and assumes a random coil structure at pH 7.0. Investigations with several other synthetic polypeptides such as of L-glutamic acid in non-polar solvents, where they exist in α-helical form, also give a value of –630 but in polar solvents these assume random coil structures and the obtained value for $b_0 = 0$. The polypeptides containing intermediate α-helical content give intermediate values. For instance for a co-polymer of L-glutamic acid and lysine in molar ratios of 10:0, 6:4, 4:6 and 0:10, the reported experimental b_0 values are – 625, – 439, – 310 and – 131 which correspond to 100, 70, 25 and 0 α-helical structure on the basis of the assumption that b_0 value of – 630 represents a fully helical conformation. These results were very similar to those obtained for α-helical content of these polypeptides, using the hard to exchange amide hydrogen (HEAE) technique. Evidently by using Moffit and Yang plot, a rough approximation of the extent of α-helical content in polypeptide chains can be deduced.

Optical rotatory dispersion technique can also be employed for detecting subtle conformational changes in a protein molecule from the changes in Cotton effect.

As indicated earlier, most of the proteins exhibit cotton effect around 225–235 nm and these are highly sensitive to even small structural changes. Hence any alterations in their conformation due to their interactions with ligands such as substrate, modulators, inhibitor etc. can readily be detected as they considerably alter the inflexion points of the Cotton effect.

11.4.6 Circular dichroism (CD) spectroscopy

Circular dichroism technique is also based on ascertaining the deviations in characteristics of a polarized light arising from its passage through a solution of an optically active substance. However, in place of linearly polarised light, a circularly polarized light is used. Circularly polarized light is generated by passing plane polarized light through a quarter-wave plate. Such a light can be resolved into two waves lying in planes at right angles to each other, are of equal amplitude and frequency but these differ in phase by one quarter of a wavelength. The electric vectors of these light waves, at any point in the path of the wave, describes circle about the direction of propagation of light as the axis. Just as the plane polarized light, the circularly polarized light is either E_L (left handed) or E_R (right handed). The relative positions of the peaks of the two plane polarized waves determines whether the circularly polarized light is of *L* or *R* type.

11.4.6.1 Principle

The inherent asymmetry in the structure of chiral molecules or centres results in differential absorption of these two components (E_L and E_R) of the circularly polarized light and this phenomenon is referred to as *circular dichroism*. Due to this difference in molar absorptivity, E_L and E_R will not be equal and will differ in their amplitudes, and as a consequence will oscillate, not in form of a circle, but in elliptical path (Fig. 11.14).

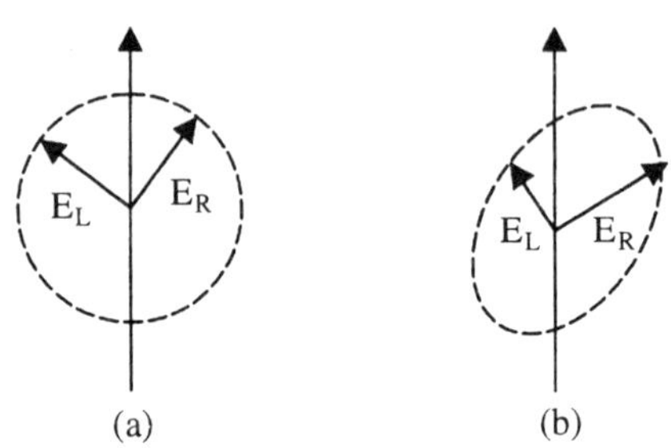

Fig. 11.14 Illustration showing effect of ordinary compound (a) and optically active compound (b) on amplitude and oscillations of E_L and E_R vectors of circularly polarized light.

Circular dichroism spectroscopy is based on the measurements in elipticity on passing a circularly polarized light through the solution under examination. The difference in absorptivity is calculated from the equation:

$$\theta = 2.303\ \Delta A = 33\ \Delta A \text{ degrees}$$

where ΔA is the difference in absorption between E_R and E_L components. Thus a CD spectrum of a solution is a plot of elipicity versus wavelength.

11.4.6.2 Instrumentation

The essential components of a CD spectrometer are depicted in Fig. 11.15.

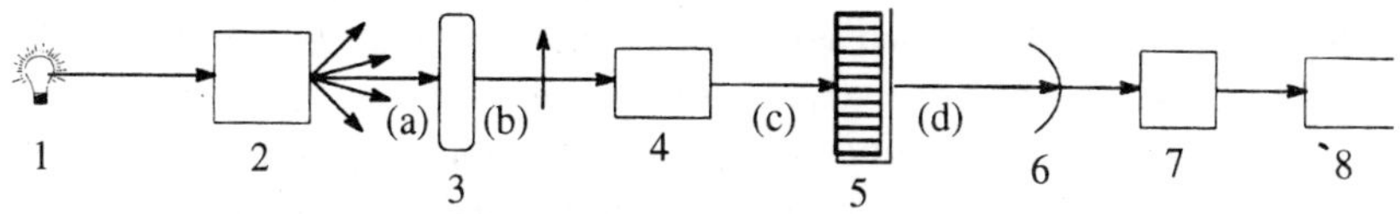

Fig 11.15 Components of CD spectrometer. 1. Light source; 2. monochromator; 3. polarizer; 4. electro-optic modulator; 5. sample; 6. detector; 7. amplifier; 8. recorder. The alphabets represents nature of radiaion, (a) unpolarized monochromatic light; (b) plane polarized light; (c) alternating E_R and E_L circularly polarized radiation; (d) polarized elliptical radiations.

The radiations from the light source are directed to a monochromator, which permits waves of specific wavelength to pass through. The monochromatic beam then is circularly polarized by plane polarizer. The electro-optic modulator is a device which facilitates alternate production of E_L and E_R components of circularly polarized light. It is in the form of a crystal, which when subjected to alternating currents, transmits either E_R or E_L components depending upon the polarity of the electric field applied to it. The detector is usually a photomultiplier which generates a voltage which is proportional to the elipicity of the beam impinging upon it from the sample.

11.4.6.3 Applications

The circular dichroism analysis of polypeptides and proteins solutions provides information about the proportions of various types of secondary structures, i.e. α-and β-helix and random coils. For this purpose synthetic polypeptides of known compositions and structures are used as standards. However, the more important use of circular dichroism technique is for detecting conformational changes in the protein molecule due to the interaction with various ligands. The conformational change is revealed in the form of quantitative and qualitative deviations in CD spectrum of the protein molecule after its interaction with the ligand.

Another application of circular dichroism spectroscopy is in investigations of nucleic acids. Nucleotides show chiral properties and their CD increases markedly when the polynucleotide exists as α-helix. Therefore, this technique has been employed to investigate structural changes in nucleic acids, alterations in the structure due to transitions between single and double stranded DNA, interaction of DNA with histones and interrelating compounds and loss of helicity of single stranded DNA as influenced by temperature, pH or on their interaction with metal ions etc.

11.4.7 Electron spin resonance (ESR)

This technique is based on examining the paramagnetic characteristics of molecules.

It is used for detecting the presence of transition metals in compounds and radicals and is particularly valuable for analysis of paramagnetic species which may be generated *in situ* such as during oxidation-reduction reactions.

14.4.7.1 Principle

The spin of an electron forms the basis for electron spin resonance technique. A spinning charged particle, such as electron, generates a magnetic field and hence acts as a small magnet. However, most of the substances do not show magnetic moment because they contain electrons in pairs with opposite spins and this results in cancellation of the generated magnetic fields. This phenomenon is referred to as diamagnetism and is possessed by all substances because all of them contain miniscule magnets, i.e. electrons. The magnetic susceptibility (ratio of magnetization produced in a substance to the intensity of magnetic field to which it is subjected) of such an atom is of order of $-10^{-6}\ g^{-1}$. However, transition metals like iron, cobalt, manganese, copper, nickel, molybdenum etc. contain unpaired electrons and they exhibit magnetic moment due to the unbalanced electron-electron spins. The magnetic susceptibility of such atoms is of the order of $+10^{-3}$ to 10^{-4}. In such cases diamagnetism is very small and the free electrons give rise to paramagnetism. For this reason the technique is also referred to as electron paramagnetic spectroscopy.

In presence of an external magnetic field these unpaired electrons can exist in two spin levels or energy states. A low energy state if they are aligned parallel with the external magnetic field (i.e. the field generated by the spinning charged particle is parallel to the external field) and the high energy state if they are antiparallel to it. The transition of unpaired electron from low energy to high energy state may be induced on absorption of electromagnetic radiation which provides exactly one quantum of energy thus resulting in attainment of a resonance state. For this, the following relationship must be completely satisfied:

$$\Delta E = E_2 - E_1 = h\nu \qquad (1)$$

$$h\nu = g\ \beta\ H \qquad (17)$$

where h = Planck's constant
ν = Frequency of applied radiation
g = A constant called spectroscopic splitting factor
β = Magnetic moment of an electron expressed in Bohr Magneton
H = Applied magnetic field

In the above equation, h, g and β are constants and the only two variables are ν (frequency of radiation) and H (applied magnetic field). Hence the resonance state can be achieved by either keeping the applied magnetic field constant and varying the frequency of radiation or by varying the magnetic field at fixed frequency. The frequencies required depend on the strength of the magnetic field. Generally a magnetic field of about 1 tesla (10^4 gauss) provides the appropriate quantum of energy from the radiations in the microwave range of electromagnetic radiations. From equation (17) it can also be seen that the frequency of microwave radiation absorbed will depend on the function of paramagnetic species (β) and the applied magnetic field. In electron spin resonance technique, the sample is usually irradiated with a constant microwave frequency appropriate to the species under examination and the applied magnetic field is varied until resonance occurs giving a peak for absorption of microwave frequency. The peak in ESR spectrum corresponds to the paramagnetic species and the peak area reflects the concentration of that species which may be quantitatively determined by comparing with a standard containing known concentration of unpaired electrons. For delocalized or free electrons the g value is 2.0023 and in case of radicals this factor is normally very close to that for free electrons and lies between 1.99–2.01. However, for localized electrons especially those in transition metals, values of g vary in the range of 1.4–3.0.

The energy supplied by microwave radiation also gets transferred from the spin system to the surroundings. This is known as spin-lattice relaxation. When resonance occurs the absorption peak is broadened due to spin-lattice interaction. This interaction of the unpaired electrons with rest of the molecule provides further information about the nature of bonding in the environment of these unpaired electrons within the molecule. In addition, the interactions between electron and nuclear spin magnetic moments lead to the appearance of fine structure in ESR spectrum. These hyperfine splitting of ESR peak yields information about the spatial orientation of the atoms within the molecule.

14.4.7.2 Instrumentation

A typical ESR spectrometer is equipped with source of electromagnetic radiations, a sample chamber situated in a resonant cavity between poles of a magnet, a detector and recorder system (Fig. 11.16).

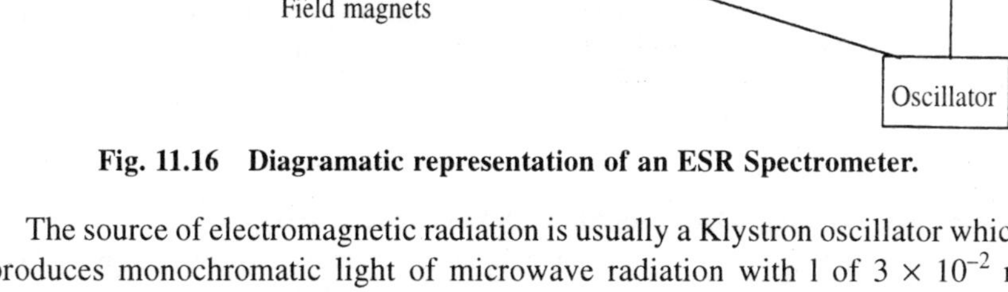

Fig. 11.16 Diagramatic representation of an ESR Spectrometer.

The source of electromagnetic radiation is usually a Klystron oscillator which produces monochromatic light of microwave radiation with l of 3×10^{-2} m (9000 MHz). Electromagnets generating fields of 50–500 milli tesla are provided. Most experiments are carried out at 330 milli tesla with an auxiliary sweep of 10–100 milli tesla. As for analysis the sample must be in solid state, the biological samples used are frozen in liquid nitrogen. The data are plotted in form of dA/dH rather than absorbance (A) versus the applied magnetic field (H) shown as follows:

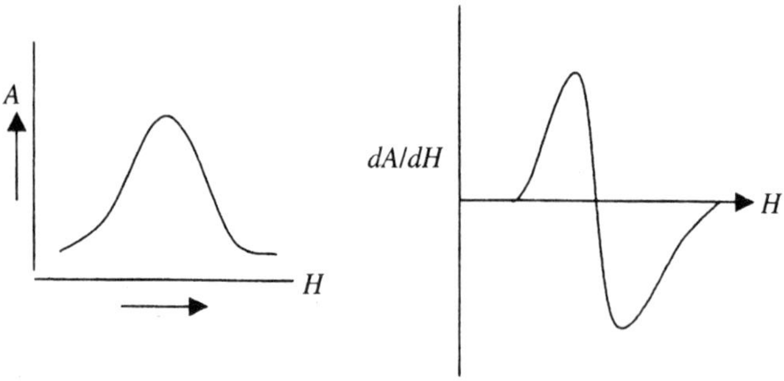

In a plot of *dA/dH* against *H* instead of symmetrical peaks, non-symmetrical peaks adjacent to non-symmetrical troughs are obtained.

11.4.7.3 Applications

Several transition metals constitute prosthetic groups of a number of enzymes and other proteins. Since the transition metals show intense paramagnetism, this technique is particularly useful in studies on metalloproteins such as those containing copper (e.g. cytochrome oxidase), iron (cytochromes, ferredoxin, iron-sulphur proteins, nitrite reductase etc.) molybdenum (aldehyde oxidase, xanthine oxidase, nitrate reductase, nitrogenase, sulfite oxidase etc.). These metal ions possess ESR peaks in one of their oxidation states. Thus by monitoring their ESR signals their role in activity of such metal-containing isolated enzymes can be studied. ESR spectroscopy has also provided useful insight into the functioning of complex multienzymic systems such as mitochondrial and chloroplastic electron transport systems. The ESR spectroscopic data provides information about the environment around the metal component which contributes towards understanding structure of the molecule.

Since free radicals can be detected by ESR spectroscopy, this technique can also be extended to studies on macromolecules which do not contain unpaired electrons by using spin-labels. In spin-labelling, a stable and unreactive free radical is attached to the biomolecule. For example, the lateral movement and the "Flip" rates of glycerophosphatides in the lipid bilayers of biomembranes has been examined by spin-labelling glycerophosphatides with nitroxide free radicals. ESR spectroscopy has also been employed quite extensively for investigations on generation of free radicals following irradiation of biological materials.

11.4.8 Nuclear magnetic resonance (NMR) spectroscopy

The underlying principle of nuclear magnetic resonance spectroscopy is very similar to that for ESR. The main difference is that in ESR the magnetic moment is due to spinning electron, while in NMR it originates from the positively charged protons in the nucleus. Though the neutrons are uncharged particles they do exert a considerable influence on NMR responses and the miniscule magnets are in fact nucleons (protons + neutrons). Atoms having odd number of neutrons such as ^{1}H atom (1 proton) and ^{13}C (6 proton + 7 neutrons) give NMR signal whereas those with even number of neutrons, e.g. deuterium (1 proton + 1 neutron) or ^{12}C

(six protons + six neutrons) do not exhibit this property. ^{15}N and ^{31}P isotopes have also been used in biochemical investigations. Most of the studies involve the use of ^{1}H-NMR and the technique is also referred to as proton magnetic resonance (PMR).

11.4.8.1 Principle

The nuclear magnetic resonance spectroscopy is based on transitions between the spin states of the spinning nuclei which produces magnetic fields along the spin axis of the nucleus. The energies of spin states of a nucleus are degenerated in the absence of magnetic field. This degeneracy of the spin does not occur under the influence of an external magnetic field. When the nucleus is placed in a magnetic field of strength H, the nucleus assumes $(2I + 1)$ spin orientation with respect to the direction of the applied magnetic field where I is the spin quantum number of the nucleus and may have values of 0, 1/2, 1, 3/2, ... ($I = 0$ denotes no spin).

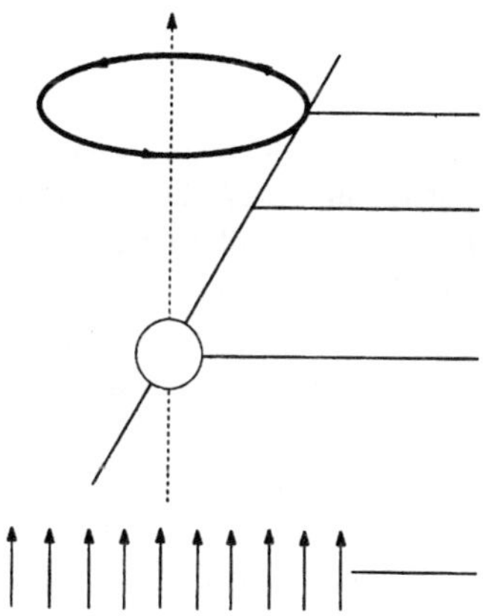

Fig. 11.17 Nuclear spinning in external magnetic field.

The resonance condition is satisfied in an external magnetic field of several hundred milli tesla on absorption of an exact quantum of energy which is supplied as electromagnetic radiations in the radiowave region. The absorption of radiowaves elevates the energy state and the difference in energy (ΔE) is directly proportional to the strength of the applied magnetic field and is defined by the following relationship:

$$\Delta E = h\nu = \mu \, \beta_N \, H/I \qquad (18)$$

where h = Planck's constant
ν = frequency of exciting radiation
μ = magnetic moment of the nucleus
β_N = nuclear magneton constant

The nuclei of different atoms vary in their values for μ and I and hence require different frequencies for nuclear spin transitions in the same applied magnetic fields. Radiation energy with frequency of 40 MHz is used for the resonance of ^{1}H

nucleus. The other nuclei which give NMR signal and occur in biomolecules/organic molecules are ^{13}C, ^{15}N, ^{31}P, ^{19}F, have a spin quantum number, i.e. $I = 1/2$ and uniform charge distribution.

The molecular environment of a proton determines the value of the applied external field at which the nucleus resonates. The relationship between the applied field (H_0) and effective magnetic field (H) on a nucleus is given by the expression

$$H = H_0 (1 - \sigma) \tag{19}$$

where σ is the shielding or screening constant. In case the effective magnetic field is larger than the applied field, the value of σ is negative and the nucleus is said to be deshielded. The shielding is responsible for the shift of resonance frequency to lower values of the applied field i.e. downfield. The chemical shift due to the applied field induces secondary fields (15×10^{-4} to 20×10^{-4} tesla) at the proton by interacting with adjacent bonding electrons. The magnitude of this shift is known as chemical shift (τ) and it is measured relative to an internal standard tetramethyl silane [$(CH_3)_4Si$] which contains 12 identical protons. The magnetic field axis may be calibrated in units on a scale from 0–10 and the value of chemical shift for TMS is taken as zero. The type of proton may thus be identified from the position of the absorption peak and the peak area is proportional to the number of such protons in a particular group. For example the NMR spectrum of ethyl alcohol (CH_3CH_2OH) at low resolution gives three distinct peaks corresponding to protons in methyl group, in methylene group and in alcohol group with the chemical shift for CH_3 protons in group being the lowest followed by those of CH_2 group and then of OH group. The area under peaks of each type of these protons is 3:2:1. High resolution NMR spectrum gives hyperfine splitting which arises due to spin-spin splitting or coupling as a result of interaction of bonding electrons with like or different spins and may extend to nuclei four or five bonds apart. In the spectrum this appears as further splitting of the peaks already separated by chemical shifts, e.g. the high resolution spectra of ethyl alcohol gives three peaks or lines for protons of methyl group, four for methylene protons and one for proton in OH group. Thus high resolution NMR yields further structural information. The general rules for interpreting splitting of signals for protons are: firstly the non-equivalent neighbouring or vicinal protons cause splitting of proton signal, secondly the equivalent protons at the same carbon atom do not cause splitting and, finally the number of peaks obtained by proton signal splitting is equal to or more than the number of vicinal protons. The intensities of the multiple peaks, i.e. doublet, triplet, quartet etc. are not exactly the same. The magnitude of splitting effect as measured from the distance between the peaks of a multiplet is called coupling constant, denoted by J, and is expressed in terms of frequency (Hz). Unlike the chemical shift, the value of J is independent of applied magnetic field strength but is dependent only on the structure of molecules. Thus the hyperfine splitting gives information about the near-neighbour environment of a nucleus.

11.4.8.2 Instrumentation

The layout of NMR instrument is very similar to that of ESR except that

Klystron oscillator is replaced by two sets of coils, a transmitter and a receiver to detect the absorption of energy for generation of radiowaves and reception, respectively. Solution of the sample are taken in sealed tubes which are rotated rapidly in the sample cavity to eliminate irregularities and a uniform signal is reflected to the receiver. The sample tube is placed between the two poles of the magnet operating at 14,000 gauss which for protons corresponds to a frequency of 6×10^7 Hz. The sample compartment is also called 'probe' (Fig. 11.18). The signal is processed and then recorded. For obtaining NMR spectra the frequency of radiation is kept constant and the magnetic field is varied until the resonance state is attained.

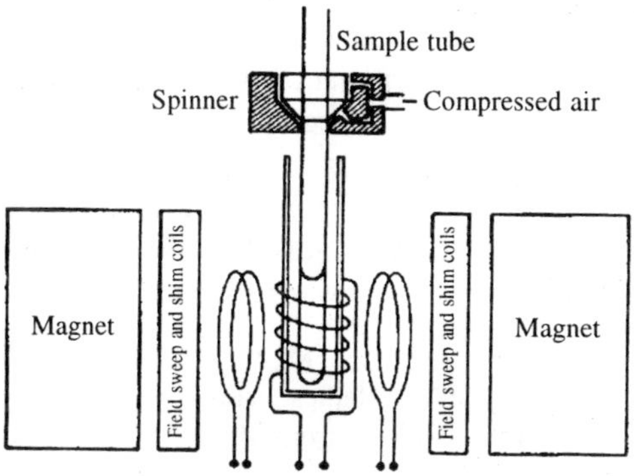

Fig. 11.18 The 'probe' in NMR spectrometer.

11.4.8.3 Applications

NMR is mainly used for studying the structure of molecules, conformational changes in macromolecules, quantitative and qualitative analysis and at times for kinetic investigations. The rates of dynamic processes in proteins and protein-ligand complexes can be measured by NMR spectroscopy. These processes may be related to side chain bond rotation, exchange of some amide protons buried within protein molecule, aromatic ring flipping, breakage of hydrogen bonds and the dissociation rates of protein-ligand complexes. This technique has been useful in direct titration of some ionizing side chains in proteins particularly in determination of pK_a of the histidine residues. The signals from the C(2) and C(4) protons are further downfield than bulk of the resonances of the proteins and can readily be resolved in D_2O solution. The chemical shift changes on deuteration and hence the histidine residues may be directly titrated.

Since the protons of different types of functional groups (—OH, =CH—, CH_2, —CH_3) have different and characteristic magnetic resonances, changes in the local environment of such groups can be detected easily. Of particular importance is the use of NMR spectroscopy which senses angular momentum of ^{13}C isotope which is present in small amounts in natural carbon of proteins. This enables alterations in environment of the carbon atoms in peptide bonds, α-carbon atoms and unsaturated and aromatic groups. High resolution NMR studies have provided direct evidence for existence of multiple conformations in protein-ligand complexes in solutions. This technique has also been used in elucidation of structure, metabolism and action of drugs, antibiotics etc. However, there is an upper limit of molecular mass of 20,000 which restricts its utilization for detailed investigations on structural and functional aspects of macromolecules.

11.4.9 Mass spectrometry

11.4.9.1 Principle

Mass spectrometry is an extremely useful technique as it can provide fairly detailed information about the structure of the molecules with very small amounts of the substance (10^{-6} to 10^{-9} g). The technique essentially involves production of ions of the sample material, their separation on the basis of their mass to charge (*m*/*e*) ratio and finally determination of relative abundance of each ion. A mass spectrometer thus has three basic components: a source of ions, an analyser and a detector.

The simplest process occuring in the mass spectrometer is the interaction of electrons with the parent molecule resulting in loss of another electron from the molecule to give a radical cation shown as follows:

M	+	e^-	$\rightarrow$	$M^{\dot{+}}$	+	$2e^-$
Parent compound		Bombarding electron		Radical cation from M		One bombarding electron plus one electron removed

This radical cation formed after loss of single electron from the molecule is known as the molecular ion. If the molecular ion is sufficiently stable to accord it a life time of 10^{-5} sec, it will be fully accelerated and recorded at its corresponding *m*/*e* value. In such a situation a single peak mass spectrum, which is plotted in the form of ion current as a function of mass to charge ratio is obtained. The majority of ions have single positive charge. Doubly or triply charged ions are also present particularly for metal complexes. Doubly charged ion will appear at half the *m*/*e* value of that of singly charged ions. Since the molecular ion has effectively the same mass as parent compound, it is the ion with highest *m*/*e* ratio in the mass spectrum. It is thus possible to determine the molecular weight of the compound from the high *m*/*e* value in the mass spectrum. The doubly or even triply charged ions give peaks between the integral values. The order of energy required for removal of an electron is $\sigma <$ non-conjugated $\pi <$ conjugated $\pi <$ lone pair.

The molecular ions also on bond cleavage give smaller ions which are called fragment ions. The fragment ions are obtained by homolytic or heterolytic bond cleavage. Homolytic bond cleavage is indicated by fish-hook arrow ($\rightharpoonup$) signifying the transfer of a single electron. For heterolytic cleavage, conventional arrow ($\rightarrow$) is used to indicate transfer of a pair of electrons. The examples of these two types of cleavages are:

Homolytic cleavage

$$CH_3—CH_2—\overset{+}{C}H—CH_2 \rightharpoonup \dot{C}H_3 + [\dot{C}H_2—\dot{C}H—CH_2]^{+}$$

Heterolytic cleavage

$$CH_3—CH_2—\overset{+}{Br} \rightarrow CH_3—CH_2^{\dot{+}} + Br$$

Similarly CO_2 is ionized and fragmented to yield CO_2^+, CO^+, O_2^+, O^+, C^+ and gives a mass spectrum with peaks at *m/e* values of 44, 28, 32, 16, 12 respectively. Other minor peaks corresponding to the other natural isotopes of, e.g. $^{13}C^+$ (*m/e* = 13) and $^{13}CO_2^+$ (*m/e* = 45) are also detectable as minor peaks. When producing ions by electronic bombardment, the degree of fragmentation of a molecule depends on the energy of bombarding electrons. At low energies (1–2 $\times 10^{-18}$ J) only a single electron is removed from the molecule and the resulting charged molecular ion has *m/e* value almost identical with the parent molecule. At the generally used operating beam of energy 10^{-17} J, the molecules undergo cleavage into positively charged fragments of different masses. The fragmentation pattern is characteristic of the compound being examined and is referred to as *cracking phenomenon.* Thus on fragmentation several peaks or lines corresponding to m/e values of positively charged ions are obtained. The height of the peaks gives relative abundance of the molecular ions and fragment ions. Thus the mass spectrum of fragmented compounds provide two types of informations. Firstly the molecular weight of the compound and the mass of various fragments produced from molecular ion. Secondly from the ions detected in a mass spectrum, it is often possible to get information regarding the possible structure of the molecule involved, e.g. for a triatomic molecule RST, four different types of fragment ions may be obtained. If the bond RS undergoes cleavage the $[R]^+$ and ST, R and $[ST]^+$ or if S-T bond breaks then fragments obtained will be $[RS]^+$ and T or RS + $[T]^+$. More than one type of cleavage may occur, e.g. $[RS]^+$ may undergo fragmentation to give $[R]^+$ and S or R and $[S]^+$. So mass spectrum of RST may include some or a few of or all the ions of the molecular ion $[RST]^+$. It may be noted that fragmented $[RT]^+$ is not produced which implies that the structure of the compound is RST and not SRT.

In general following rules are followed for fragmentation: (i) Molecular ion peak appears due to most stable ion, the order of stability of molecular ions is aromatic > conjugated polyenes > alkenes > cycloalkanes > ketones > *n*-alkanes > ethers > esters > carboxylic acids > alcohols > amines.

(ii) The ions are also fragmented to give stable positive ion or stable neutral molecules or free radicals, e.g.

$$[R—C—]^+ \rightarrow R^+ + \dot{C}$$

$$R—C—C^+ \rightarrow R^+ + \quad C{=}C$$

$$[—C—X]^+ \rightarrow —C^+ + X$$

$$[R—C—X]^+ \rightarrow \dot{R} + C^+—X \quad \text{or} \quad C{=}X^+$$

(iii) Resonance stabilised ions are formed preferentially.

There are several fragment peaks in mass spectra which cannot be explained by the simple cleavage of the parent compound. These fragment peaks result due to the rearrangements, which occur during the fragmentation process. The rearrangements may be specific or random. The migration of H-atom from one part of the ion to another during fragmentation gives rise to specific rearrangements. The random rearrangements take place by the scrambling of H- and C-atoms within the cationic species. McLafferty rearrangement is an example of specific rearrangement.

Elimination is generally followed during the rearrangements. The eliminated molecules are HX, H_2O, HOR, HNR_2, X_2. The 1,4-eliminations are most common with long chain alcohols while 1,3-elimination dominates in halides. 1,2-shifts are mostly observed forming more stable species during the fragmentations.

In the mass spectrum a number of peaks are observed at integral *m/e* values but some broad, low intensity peaks at non-integral *m/e* values are obtained due to metastable ions. The rate of decomposition of an ion m_1 formed in the ionization chamber is very fast and before reaching the acceleration region such ions will be decomposed. Only the fragment m_2 will be accelerated and detected in significant quantity. Ion that decomposes at an intermediate rate should show considerable intensity for

$$m_1^+ \rightarrow m_2^+ + \text{neutral fragments}$$

both primary (m_1) and the daughter (m_2) ions. With the intermediate rate of decomposition a part of primary ions will decompose into fragments during the travelling through the accelerating region of the instrument. This kind of ions will be first accelerated as mass m_1 , decompose, loose some kinetic energy to the neutral fragment and then accelerated and deflected as mass m_2. Such kind of ions are recorded as broad peak with low intensity at mass m^* and can be calculated by the following equation.

$$m^* = \frac{(m_2)^2}{m_1} \qquad (20)$$

In general the metastable peaks (m^*) are found at 0.1 to 0.4 mass units higher than the calculated value. Metastable peaks help in the deduction of mechanism of fragmentation.

The mass spectrum of an organic compound is represented graphically by plotting *m/e* versus relative abundance or percentage of base peak. Base peak is the most intense peak of the spectrum. The value of 100 is arbitrarily assigned to the base peak and the other peaks are assigned proportionate values.

11.4.9.2 Instrumentation

The four basic components of a double focussing mass spectrometer are: ion source, electrostatic analyser, magnetic analyser and detector (Fig. 11.19). A single focussing spectrometer has the provision of varying the magnetic field only and not of the electrostatic field.

The ions emerging from the source at first are accelerated by applying a potential of thousands of volts and are then deflected and focussed by an electrostatic analyser. These then pass through a slit and are deflected and refocussed by magnetic analyser, travel through the third slit and are detected by the detector which measures the ion current which is proportional to the number of ions. Several types of ion sources are available and these include electron impact method and chemical ionization methods. For these methods the sample to be analysed is required in gaseous state and hence cannot be used for compounds which are either thermally unstable or non-volatile. Other methods include desorption and fast atom bombardment ionization, plasma desorption, laser ionization methods.

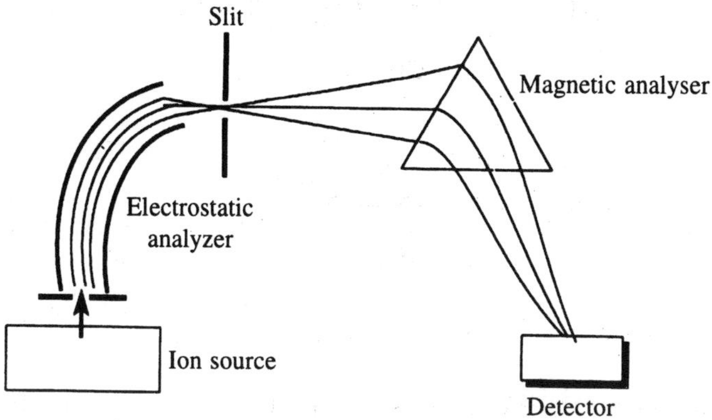

Fig. 11.19 Diagramatic representation of double focussing mass spectrometer.

11.4.9.3 Applications

An example of application of mass spectroscopy for establishing structure of a tripeptide is discussed as follows.

A tripeptide shows glycine, alanine and phenylalanine amino acids by analytical methods. This peptide is acetylated at the N-terminus and methylated with CH_3I at the carboxyl terminal. The N-acetylation removes amide hydrogen bonding and thus imparts increased volatility. This acetylated methyl ester of peptide is used to obtain a mass spectrum (Fig. 11.20) and the *m/e* values are used to determine the exact structure and sequence of amino acids. Modern mass

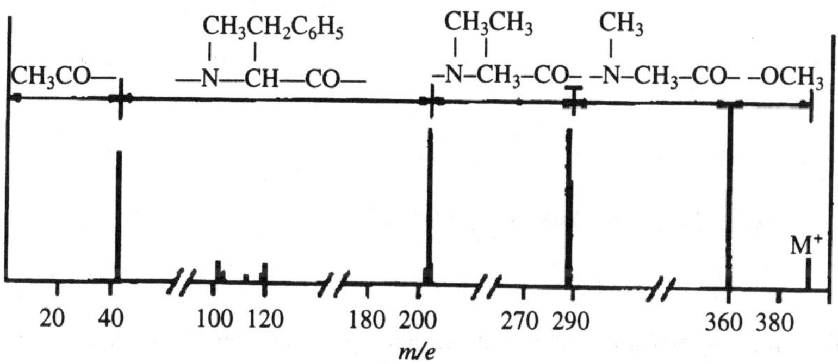

Fig. 11.20 Partial mass spectrum of a tripeptide.

spectrometer with microprocessor or computer interface is very useful in providing the accurate masses of different ions. The molecular mass of different groups which may be formed by ionization under electron impact are calculated as follows.

Amino acid residue		*m/e*
N-methylglycyl	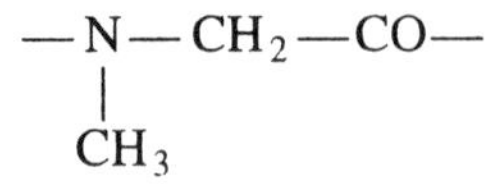	71.04
N-methylalanyl	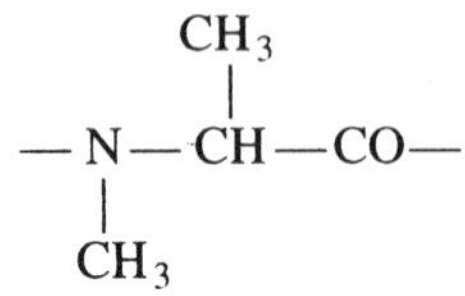	85.05
N-methylphenylalanyl	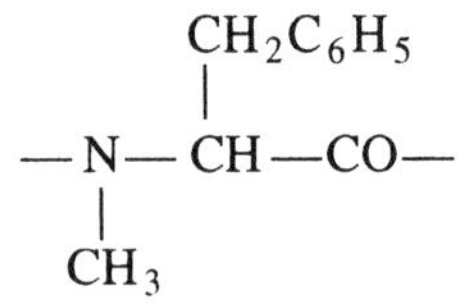	161.08
Acetyl	CH_3CO—	43.02
Methoxyl	CH_3O—	31.02

Beginning at one end of the peptide, with acetyl, added a sum of 43.02 (*m/e* for acetyl), it gives the following *m/e* values for different mass fragments.

Acetyl + N-methylglycyl = 114.06

Acetyl + N-methylalanyl = 128.07

Acetyl + N-methylphenylalanyl = 204.10

The mass spectrum shows a peak at *m/e* 204 (high abundance) and 114 and 128 (both low abundance). This implies that the first amino acid in the sequence of this tripeptide is phenylalanine. Similarly, by adding N-methyl phenylalanine to the two N-methylated amino acid residues, we get,

Acetyl + N-methylphenylalanyl + N-methylglycyl = 275.14

Acetyl + N-methylphenylalanyl + N-methylalanyl = 289.16

The mass spectrum shows a prominent peak at 289.16, so alanine must follow phenylalanine in this tripeptide and the last amino acid would be glycine.

Acetyl+N-methylphenylalanyl+N-methylalanyl+N-methylglycyl +methoxyl

(360.1923) (391.2107)

Ions corresponding to *m/e* 360 and 391 are present in the mass spectrum so the sequence should be Phe-Ala-Gly.

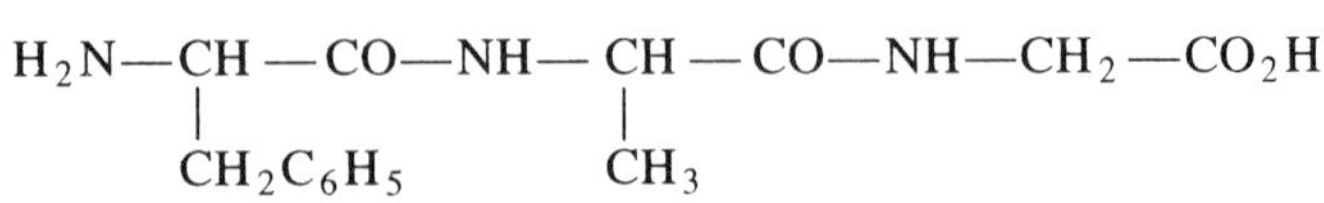

The complete structure of the tripeptide is deduced from the knowledge of the D or L nature of the constituent amino-acid residues, but the mass spectrum has yielded the important sequence information.

It should be noted that the consecutive amino acid residues, Ala-Gly, have the same mass and elemental composition as an asparagine residue after N-methylation. The high-resolution mass spectrometric analysis alone is not sufficient for the structure elucidation. The analysis above relies on amino acid analysis to eliminate certain possibilities, but amino acid analysis does not always resolve all ambiguities as in the case of mixtures of peptides. A simple alternative to amino acid analysis for removing ambiguities is through the use of isotopically labelled derivatives. If, in the example above, iodomethane is replaced with trideuterated iodomethane (CD_3I) for methylation, then there would be an increase of 6 amu (two CD_3 groups) in an Ala-Gly grouping and of 9 amu (three CD_3 groups) in an Asn residue. Thus, differentiation is achieved without the need for high resolution mass spectrometry. The same technique can be applied successfully to most other ambiguous assignments of amino acid residues.

11.5 Laboratory Exercises

11.5.1 EXPERIMENT: To verify the validity of Beer's Law and determine the molar extinction coefficient of NADH

Principle

Quantitative colorimetric estimations are based on two Laws, i.e. Lambert's Law and Beer's Law. While Lambert's Law defines the relationship between the length of light path through the solution, Beer's Law states that the fraction of incident light transmitted is inversely proportional to the concentration of the light absorbing compound in the solution. The amount of transmitted light is inversely related with the absorption of light by the media. Hence as a corollary, the extent of light absorbed by a compound is directly proportional to its concentration. This experiment is designed to verify the validity of Beer's Law.

Reduced forms of nicotinamide adenine nucleotides (NADH and NADPH) show a distinctive absorption peak at 340 nm. These nucleotides are required as cofactors by several oxido-reductases. The spectrophotometric determination of activities of such enzymes is based on measuring the rate of utilization or production of NADH by monitoring the change in absorbance at 340 nm. According to Beer's Law, absorbance of light which often referred to as optical density (O.D. units), should be directly proportional to concentration of NADH in the reaction mixture.

Molar extinction coefficient is defined as absorption of light by 1 M solution of a compound at an optimal wavelength with fixed light path of 1 cm and it is represented as ε with wavelength shown as a subscript (for NADH at 340 nm it is written as $\varepsilon_{340\ nm} = 6.2 \times 10^3$). From this value, concentration of the compound in a sample can directly be calculated without preparing a standard or reference curve.

Materials and Reagents

1. UV-visible spectrophotometer.
2. Standard quartz or silica cuvette of a light path of 1 cm (glass cuvette is not suitable because it absorbs UV-light).
3. NADH solution: Prepare 10 ml of 1 mM solution of NADH in 0.05 M Tris-HCl (pH 7.5).

Procedure

1. Switch 'on' the spectrophotometer. Set the wavelength to 340 nm and after 1–2 min switch 'on' the UV (or deuterium) lamp and allow the instrument to warm up for 15–20 min (follow the instructions given in the 'Operator's Manual' of the instrument supplied by the manufacturer).
2. Take a silica cuvette and add 3 ml of 0.05 M Tris-HCl. Place the cuvette in the cuvette holder and again adjust it to 'zero' absorbance (or 100% transmission) using this blank (buffer alone) cuvette.
3. Check the 'zero' and 100% transmission as described in Step 2 to make sure that the instrument is properly adjusted.
4. In another matched cuvette take 0.1 ml of 1 mM solution of NADH and 2.9 ml of 0.05 M Tris-HCl buffer so that the total volume in cuvette is 3 ml. Mix the contents thoroughly and record the absorbance of this sample solution (it corresponds to 33 μM NADH or 0.1 μmoles in 3 ml).
5. Pour out the solution from the sample cuvette, rinse it with distilled water. Add 0.2 ml of NADH and make the final volume to 2.8 ml with 0.05 M Tris-HCl buffer. Record its absorbance.
6. Repeat step 5 by taking 0.40 ml of NADH solution and 2.6 ml of 0.05 M Tris-HCl buffer and note down its absorbance.
7. Repeat Step 5 by taking 0.6 ml of NADH solution make the volume to 3 ml with 0.05 M Tris-HCl buffer and record the absorbance of this solution.

Calculations

(a) Present the results obtained in form of a table as follows:

Sample No.	*ml of NADH solution*	*Conc. of NADH (μM)*	*Fold conc. of NADH*	A_{340}	*Ratio**
1	0.10	33.3	1.0	—	1.0
2	0.20	66.6	2.0	—	—
3	0.40	133.3	4.0	—	—
4	0.60	200.0	6.0	—	—

*Calculate the ratios of $\frac{A_{340} \text{ of sample 2}}{A_{340} \text{ of sample 1}}$ for samples 2, 3 and 4.

(b) Check whether the calculated values of the ratio for samples 2, 3 and 4 correspond with the folds increase in concentration of NADH in these samples as compared to sample 1. If ratios obtained for absorbance are the same as the fold concentration of NADH in these samples, it indicates that absorbance of light by the sample is directly proportional to concentration of NADH.

(c) Prepare a graph of A_{340} against concentration of NADH in the sample. A straight line will indicate proportional increase in the A_{340} with increasing concentration of NADH in the sample. This graph can be used as a standard curve for NADH. One can determine the concentration of NADH in the sample from its A_{340} by referring to this standard curve.

(d) Calculation of molar extinction coefficient (ε_{340} nm) for NADH: Calculate the molar extinction coefficient of NADH as given below:
Suppose, 133.3 μM solution of NADH gives A_{340} of x
1 μM solution of NADH would correspond to $(x/133.3)$ absorbance
1 M solution of NADH would give A_{340} value of $(x/133.3) \times 10^6$
This is the ε_{340} for NADH. Thus the conc of NADH in unknown sample can directly be calculated from its A_{340} by using the molar extinction coefficient of NADH.

11.5.2 EXPERIMENT: To prepare an absorption spectrum of NADH

Principle

Absorption spectrum refers to absorbance of light of different wavelengths by a compound. Every compound or a group of compounds has some characteristic features in its absorption spectra, which forms the basis for determining its presence and also quantitative estimation of that substance. For example flavin adenine nucleotides have a broad absorption spectra with peaks at 370 and 450 nm. Enzymatic or chemical reduction of flavin nucleotides is accompanied by disappearance of the peak at 450 nm. Thus presence of flavin nucleotides in a sample can be established by examining its absorption spectra before and after its reduction. If peaks are detected at 370 nm and 450 nm and in the oxidized state and on reduction, the latter peak (450 nm) is eliminated, it would be indicative of presence of flavin nucleotides in the test sample. Similarly, oxidized form of cytochrome *c* shows a prominent peak at 415 nm and a broad but less intense, peak at around 550 nm. Reduction with sodium dithionite leads to appearance of a sharp distinct absorbance peak at 554 nm, (α-band) another peak at 524 nm (β-band) and an intense peak at 419 nm. If the sample also shows absorption peaks at these wavelengths and on reduction appearance of a fairly sharp peak 554 nm it would signify the presence of cytochrome *c* in the sample. The amount of cytochrome *c* can be quantitatively estimated from the A_{554} which has a molar extinction coefficient value of 27.7. Absorption spectra also helps in identifying wavelength which is characteristic and unique for particular compound, e.g. absorption peak at 340 nm is unique for NADH and even though the peak at 260 nm is much more intense, it can not be used for detection and quantitative estimation of NADH because NAD and other adenine nucleotides (like ATP, ADP, AMP) also show intense absorption at 260 nm but not 340 nm.

The absorption spectra is obtained by examining the absorbance of light of different wavelengths by the compound. For this a dual beam spectrophotometer having the facility for scanning is used. In these spectrophotometers the wavelength changes in a pre-determined range continuously at a constant rate (say 25 nm per min) and the difference in absrobance between the blank and the sample preparation is automatically recorded on chart paper. However, absorption spectra can also be obtained manually with simple single beam spectrophotometer. At each wavelength the instrument has to be set again to zero absorbance, the absorbance by the sample is then determined and is plotted in form of a graph of absorbance value versus wavelength.

Materials and Reagents

1. Spectrophotometer.
2. NADH solution: Prepare 1 mM solution of NADH in 0.05 M Tris-HCl (pH 7.5).

Procedure

1. Switch 'on' the spectrophotometer and after 1 min switch 'on' the UV lamp.
2. Take two matched silica cuvettes. In the reagent blank cuvette add 3 ml of 0.05 M Tris-HCl buffer (pH 7.5). In the sample cuvette add 0.4 ml of 1 mM NADH solution and 2.6 ml of 0.05 Tris-HCl (pH 7.5).
3. Set the instrument of 200 nm using the reagent blank cuvette (Tris-HCl-Buffer alone) as described in Steps 2 and 3 of EXPERIMENT 11.1 and determine the absorption of sample (containing NADH) against the blank.
4. Now reset the instrument at 210 nm with reagent blank and record reading of the sample. Repeat this Step at intervals of 10 nm each upto 400 nm. For taking readings beyond this wavelength, switch 'off' the deuterium lamp and switch 'on' the tungsten lamp.
5. Draw a graph of absorbance versus the wavelength to obtain the absorption spectrum of NADH.

11.5.3 EXPERIMENT: To determine the difference spectra of oxidized-reduced NAD

Principle

As the name indicates, the difference spectra represents or highlights the difference of absorption of light by two forms of the same compound, e.g. oxidized-reduced states of a compound like NAD and NADH, oxidized form of FAD and its reduced form $FADH_2$ etc. The difference spectra provides useful information as to which is the optimal wavelength to differentiate between the two states of the compound. Infact the selection of the most appropriate wavelength to differentiate and specifically estimate the particular state of the compound from the other state is based on their difference spectra. For obtaining difference spectra an automatic, dual beam scanning spectrophotometer is used where NAD is placed in blank position of cuvette holder and that containing NADH in the sample position. The scanning range (say 210–400 nm) and scanning speed (20 nm per min) adjusted as per requirements of the investigations. The speed of recorder is synchronized (say 1 cm ≡ 20 nm) and scanning speed (20 nm/min) is adjusted and the scanning operation is initiated. Such spectrophotometers will provide a graphical figure of the differences in the absorption spectra of the two states of the compounds which is automatically recorded on the chart paper. If such spectrophotometer is not available then one can determine separately the absorption spectra of these two compounds, calculate the difference in absorption of light at different wavelength and draw a graph of difference in values of light absorbance versus wavelength. At the wavelengths where the two compounds absorb light to the same extent, the value in difference spectra will be zero and only peaks having positive values will appear where the blank (NAD solution) shows lesser absorbance than the sample (reduced NAD) and absorption band of negative value will be obtained where absorption of light by the sample is less than in the blank. The most prominent peak which is unique for the compound of interest is normally chosen for selectively estimating one form of the compound in presence of its other alternate form.

Materials and Reagents

1. Same as in EXPERIMENT 11.2.
2. 1 mM NAD solution in 0.05 M Tris-HCl (pH 7.5).

Procedure

Determine the absorption of light of different wave lengths by taking 0.1 ml of NAD manually as described in Steps 1–4 of EXPERIMENT 11.2 using 0.05M Tris-HCl as the blank. Repeat the experiment using NADH in place of NAD in the sample cuvette.

Calculations

1. Subtract the absorbance values for NAD from those for NADH at each of the wavelengths.
2. Draw a graph of obtained differences in absorbance versus wavelength to obtain the difference spectra of oxidized-reduced NAD.

Suggested Further Reading

Brown S.B. (1980). An Introduction to Spectroscopy for Biochemists. Academic Press, (London).

Burgess C. and Knowles. (1981). Techniques in Visible and Ultraviolet Spectroscopy, Vol. 1, Chapman and Hall, New York.

Burrin D.H. (1975). Spectroscopic Techniques. *In* Principles and Techniques of Practical Biochemistry (William B.L. and Wilson K., eds) Edward Arnold, London, pp 128–169.

Hester R.E. and Girling R.B. (eds) (1991). Spectroscopy of Biomolecules. Royal Society of Chemistry Special Publication, Cambridge.

Ingle J.D. Jr. and Crouch S.R. (1988). Spectrochemical Analysis, Prentice-Hall, New York.

Rose M.E. and Johnstone R.A.W. (1982). Mass Spectrometry for Chemists and Biochemists, Cambridge University Press, Cambridge.

Saunders J.K.M. and Hunter B.K. (1987). Modern NMR Spectroscopy: A guide for Chemists, Oxford University Press, Oxford.

Symons M.C.R. (1978). Chemical and Biochemical Aspects of Electron Spin Resonance Spectroscopy, Van Nostrand Reinhold, New York.

12. Radioisotope Techniques

12.1 Introduction

Radioisotope techniques are one of the most powerful tools in scientific research. Not only can the fate of a given compound in either physical or biological processes be readily traced by the use of radioisotope labelled compound, but the ease and sensitivity of radioactivity assay make possible the detection of extremely minute amount of a given labelled compound in a sample. Before radioisotope methods are used as a research tools, a sound understanding of certain basic concepts involved in radiotracer experiments is essential. One should know the properties of radioisotopes and their radiations. An adequate background of radiotracer methodology is also of great importance in evaluating the research findings. Some of these concepts are discussed below.

12.2 Atomic structure

An atom is composed of a positively charged nucleus and a surrounding cloud of negatively charged electrons. The mass of an atom is concentrated in the nucleus which is composed of protons and neutrons. Protons are positively charged particles with mass approximately 1850 times greater than that of an orbital electron. The number of orbital electrons in an atom is equal to the number of protons present in the nucleus. This number is called the atomic number (Z). Neutrons are uncharged particles with mass equivalent to that of protons. The sum of the number of protons and neutrons in a given nucleus is the mass number (A). Since the number of neutrons in a given nucleus is not related to its atomic number, it does not affect the chemical properties of the atom. Hence, atoms of a given element may not necessarily contain the same number of neutrons. Atoms that contain the same number of protons (same atomic number, Z value) but vary in their number of neutrons are called isotopes, i.e. they are the atoms of the same atomic number but different mass number. A specific nuclear species symbolically is represented by a subscript number for the atomic number and a superscript number representing the mass number, followed by the alphabetical symbol for the element. For example $^{12}C_6$, $^{13}C_6$, $^{14}C_6$. However, in practice, we simply specify the mass number (^{14}C) as the atomic number is the same in every case. The number of known isotopes for each element varies quite widely, e.g. there are three isotopes for hydrogen, seven for carbon ^{10}C to ^{16}C, and 20 or more for some of the elements of high atomic number.

Only a few combinations of neutrons and protons in a given nucleus give stable isotopes. The number again varies widely from element to element. Though the reasons for stability of certain combinations of neutrons and protons are not fully understood, it is apparent that nuclear stability in general is related to the ratio of neutrons to protons, which, in turn is related to the binding within the nucleus. Stable isotopes for elements with low atomic number tend to have equal number of neutrons and protons, whereas, for elements of higher atomic number, stability is associated with neutrons: protons in excess of 1. The unstable isotopes tend to attain stability by attaining this ratio. In this process, they emit radiations and are hence called radioisotopes. These isotopes are generally produced artificially but may occur in nature also.

12.3 Nature of Radioactive Decay

Radioactive nuclides undergo spontaneous nuclear changes leading to more stable conditions. As stated earlier, the condition of stability is related to neutron: proton in the nuclei.

12.3.1 Decay by negative β (negatron) emission

In some cases, nuclides with excess neutrons may gain stability by the conversion of a neutron to proton, resulting in the ejection of a -ve β particles (electron) as follows:

$$n \rightarrow p + \underset{\text{(negatron)}}{\beta^-}$$

It may be noted that this electron originates in the nucleus and is not to be confused with the orbital electron. As a result of this ejection, the nucleus loses a neutron but gains a proton. Therefore Z increases by one and A remains constant decreasing *n/p* ratio. Decay of ^{14}C by negatron emission is

$$^{14}C_6 \rightarrow {}^{14}N_7 + \beta^-$$

Many of the commonly used radioisotopes in biology decay by this mechanism. Important examples include ^3H, ^{14}C, ^{35}S, ^{32}P etc.

12.3.2 Decay by positive β (positron) emission

When the number of protons in a nucleus is in excess, stability is attained by emission of positrons (+vely charged β particles). In this case a nuclear proton is converted to a neutron accompanied by ejection of high speed positron.

$$p \rightarrow n + \underset{\text{(positron)}}{\beta^+}$$

This type of decay does not change the mass number but atomic number (Z) decreases by one. Any excess energy in the nucleus following positron emission is emitted as γ-radiation. ^{22}Na decays by positron emission as:

$$^{22}_{11}Na \rightarrow {}^{22}_{10}Ne + \beta^+$$

12.3.3 Decay by α-particle emission

In elements of higher atomic weight, decay occurs by emission of α-particles (a naked He nucleus composed of 2 neutrons and 2 protons). Accordingly, emission of an α-particle results in considerable lightening of the nucleus as it decreases the atomic number (Z) by 2 units and atomic mass (A) by four units. Alpha emitters are not frequently used in biological work. ^{226}Ra decays by α-emission to ^{222}Rn, which is itself a radioactive thus leading to complex decay series culminating to ^{206}Pb.

$$^{226}_{88}\text{Ra} \rightarrow ^{4}_{2}\text{He}^{++} + ^{222}_{86}\text{Rn} \rightarrow \rightarrow \rightarrow ^{206}_{82}\text{Pb}$$

12.3.4 Decay by electron capture

The number of protons in the nucleus can also be reduced by the process of electron capture. Hence it yields essentially the same result as the positron emission. In this type of radioactive decay, one of the inner orbital electrons is attracted into the nucleus, where it combines with the nuclear proton to form a neutron ($e^- + p^+ = n$), resulting in the loss of one proton and gain of one neutron per nucleus. No particle emission occurs in this case. However, X-rays are emitted as a consequence of rearrangement of the orbital electron energy levels. This process is also known as 'K' capture, as the electron captured by the nucleus is normally from the innermost orbit or 'K' shell. ^{55}Fe follows this mode of decay.

12.3.5 Decay by γ-ray emission

γ-emission involves electromagnetic radiations similar to but with a shorter wavelength of X-rays. γ-rays result from a transformation in the nucleus of an atom and frequently accompany α-and β-particle emission. This does not lead to any change in atomic number or mass. ^{131}I decays by this process.

$$^{131}_{63}\text{I} \rightarrow ^{131}_{64}\text{Xe} + \beta^- + \gamma$$

12.4 Rate of Radioactive Decay

The number of atoms in a radioactive sample that disintegrate during a given time interval decreases exponentially with time (Fig. 12.1).

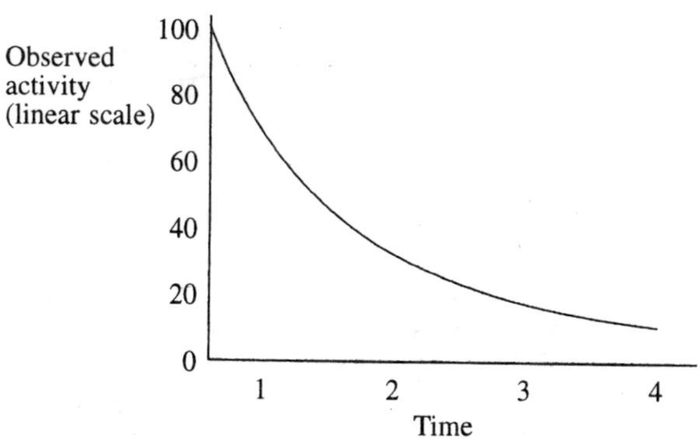

Fig. 12.1 Linear decay curve.

This rate follows always an exponential law and is independent of pressure, temperature, the mass action law, or any other rate limiting factor that commonly affects chemical and physical changes. This decay is characteristic of the source and is related to the degree of instability. Hence the decay rate could be used as a most useful means of identifying a given nuclide. Furthermore, the process is irreversible. The number of atoms disintegrating at any time is proportional to the number of atoms of isotope present (N) at that time (t). The exponential curve (Fig. 12.1) expressed mathematically gives the equation:

$$-dn/dt \propto N$$

or

$$-dn/dt = \lambda N$$

where λ is the decay constant and is defined as the fraction of the isotope decaying in unit time (t^{-1}). On integration, this equation can be converted to a logarithmic form:

$$\ln Nt/N_o = -\lambda t$$

where Nt is the number of radioactive atoms present at time t and N_o is the number of radioactive atoms originally present. It is more convenient to express the decay constant in terms of half life ($t_{1/2}$), which is defined as the time taken for the activity to fall from any value to half of that value (Fig. 12.2). When $Nt = 1/2$ of N_o, then t will be equal to the half life of the isotope. Thus,

$$\ln 1/2 = -\lambda t_{1/2}$$

$$2.303 \log_{10}(1/2) = -\lambda t_{1/2}$$

or

$$t_{1/2} = 0.693/\lambda$$

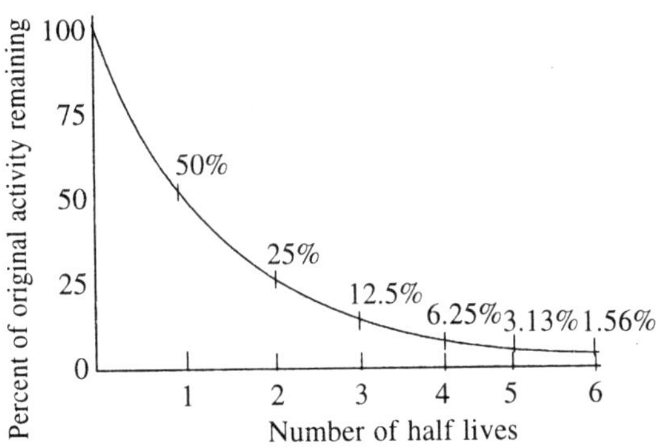

Fig. 12.2 Relation between half-life and radioactivity.

The values of $t_{1/2}$ vary widely from over 10^{19} years for ^{204}Pb to 3×10^{-7} seconds for ^{212}Po. The half lives of some frequently used radioisotopes are given in Table 12.1. In an experiment using radioisotopes, the reduction of radioactivity due to radioactive decay is always taken into consideration before analysis of the results. This is particularly true with fast decaying radioisotopes.

12.5 Units of Radioactivity

The commonly used unit of radioactivity is the Curie (Ci) which is defined as the quantity of radioactive material in which the number of nuclear disintegrations

Table 12.1 Half life values of some selected radioisotopes

Nuclide	$t_{1/2}$	*Nuclide*	$t_{1/2}$
^{42}K	12.42 h	^{35}S	87.200 days
^{24}Na	14.97 h	^{45}Ca	164 days
^{90}Y	64.03 h	^{65}Zn	244.400 days
^{198}Au	2.696 days	^{22}Na	2.580 yr
^{131}I	8.066 days	^{60}Co	5.260 yr
^{32}P	14.221 days	^{3}H	12.262 yr
^{86}Rb	18.660 days	^{90}Sr	28.000 yr
^{51}Cr	27.800 days	^{14}C	5568 yr
^{59}Fe	45.300 days	^{36}CI	3.03×10^5 yr

per second (dps) is the same as that in 1 g of the Ra, namely 3.7×10^{10}. This definition was given in 1950 by the joint commission of the International Union of Pure and Applied Chemistry and the Union of Pure and Applied Physics. Since this unit is too large, several of its subdivisions in common use are millicurie (mCi) which is equal to 1/1000 of a Curie, or 3.70×10^7 dps and microcurie (μCi), which is 1/1,000,000 of a Curie, or 3.70×10^4 dps, or 2.220×10^6 disintegrations per minute (dpm). However, note that the Curie refers to the number of disintegrations actually occurring in a sample, rather than to the disintegrations detected by a radiation counter. which are usually a fraction of the total disintegrations occurring and are referred to as counts, i.e. counts per second (cps). The Systéme International d' unités (SI system) uses Becquerel (Bq) as the unit of radioactivity, which is defined as one disintegration per second (1dps). The bigger units of Bq could be Mega becquerel (MBq), 10^6 Bq or 27.027 μCi, Gegabecquerel (GBq), 10^9 Bq or 27.027 mCi or Terbecquerel (TBq), 10^{12} Bq or 27.027 Ci. While conducting experiments with radioisotopes, a carrier of the stable isotope of the element is added to radionuclide. Under these conditions, it becomes necessary to express the amount of radioisotope present per unit mass. This is achieved by the use of the term specific activity, which could be expressed in number of ways including disintegration rate (dps or dpm), count rate (cps or cpm) or Curies (mCi or μCi) per unit of mass of mixture (units of mass are normally either moles or g). It can also be expressed as the amount of radioactivity per given volume or any other expression of quantity.

12.6 Interaction of Radioactivity with Matter

The most commonly encountered radiations while using radioisotopes as tracers include α-particles, β-particles (+ve or –ve) and γ-rays. These radiations have different physical characteristics and consequently differ in the manner in which they interact with matter. The major characteristics of these radiations are discussed as follows:

12.6.1 α-Particles

The α-particle is a helium nucleus composed of two protons and two neutrons. Being massive particle and doubly +vely charged, these particles exert a considerable electrostatic attraction on the outer orbital electrons of the atom near which they pass. They follow straight path and are normally deflected only by a rare direct collision with a nucleus. These particles have energy between 3 to 8 Mev (one electron volt, ev, is equivalent to 3.85×10^{-20} calories and is the kinetic energy an electron would acquire while moving through a potential gradient of 1 volt) and all the particles from a given isotope have the same amount of energy, e.g. ^{221}Ra emits all its particles at an energy of 6.7 Mev. These particles interact with matter in two different ways. In one case they cause excitation. In this process, orbital electrons take up the energy from the passing particles but are not completely removed from their atoms. Afterwards the excited electrons fall back into their original orbits and emit excess energy as photons in visible or near visible range. In the second mode of interaction, they cause ionisation in which α-particle strips an orbital electron from an interacting atom thereby causing loss of –vely charged electron leaving behind the atom as a +vely charged ion. The electron and the +ve atom constitute an ion pair and the process is known as ionisation. α-emitting isotopes are not commonly used as tracers in biological investigations mainly because most α-emitting radioisotopes are elements of high atomic number (Z above 82), which normally are not metabolically significant in living plants and animals.

12.6.2 β-Particles

Compared with α-particles, β-particles (positrons and negatrons) are very small and their mass is 1/7400 the mass of α-particles. Because of small size, they have higher velocity and much greater penetrating power. Consequently their range is much longer through the matter. Since the negatron is like normal orbital electron, it persists after dissipating all its kinetic energy usually as an orbital electron attached to a positive ion. A positron, on the other hand, has only a transient existence as after expending all its kinetic energy, it interacts with an electron and is annihilated. The mass of these particles is converted to energy in the form of two 0.51 Mev γ-rays. These rays are emitted at angles of 180° to each other. Like α-particles, β-particles also dissipate their energy largely by ionisation or excitation of the atoms with which they interact. However, contrary to α-particles, in β-decay the particles are emitted over a continuous range of kinetic energy upto a maximum value (E_{max}) characteristic of the radioisotope in question. The values of β-E_{max} range from 0.018 Mev for ^{3}H upto 4.81 Mev for ^{38}Cl. This difference in E_{max} affects the penetration of the radiation, e.g. β-particles from ^{3}H travel only few mm in air, whereas, those from ^{32}P can penetrate over 1 m of air. The reason for emission of β-particles over a continuous energy spectrum was given long back by Pauli in 1931. According to Pauli, each event of β-decay actually occurs with a total energy involvement equivalent to E_{max} of β-particle in question. The neutrino shares the total energy with β-particle in varying proportions from disintegration to disintegration. For example, a β-particle emitted from radionuclide ^{32}P (E_{max} = 1.71 Mev) with an energy of 1 Mev would be accompanied by a neutrino equivalent to 0.71 Mev energy, making a total energy emission of 1.71 Mev. Neutrinos have no charge and negligible mass and donot interact with matter.

12.6.3 γ-Rays

In contrast to α- and β-particles, γ-rays are electromagnetic radiations and therefore have no charge or mass. Consequently they penetrate through matter readily with little interaction, leading to greater range in matter compared to α- and β-particles of comparable energy. γ-rays are most frequently emitted immediately following α- or β- particle emission from the nucleus, representing the readjustment of the energy content in the radioisotope from an excited state to a more stable ground state. Their energy falls in the range of 10 Kev to 3 Mev. Since they are uncharged, they cannot bring about ionization of atoms in their path. However, several of their common modes of interaction with matter result in the consequent ejection of electron, which in turn causes ionization in the surrounding matter. They interact with matter in many ways. However, the three most important ways by which they interact with matter, include photoelectric effect, Compton effect and pair production. In photoelectric effect, γ-rays interact with orbital electron, transferring all their energy to the electron, which is then ejected with kinetic energy equivalent to that of γ-rays less the energy needed to overcome the orbital binding force of the electron. Such a photoelectron subsequently behaves as a negatron and will interact with other atoms in its path leading to ionization. This effect is commonly seen with γ-rays with energy below 0.5 Mev interacting with absorbers of high atomic number. In Compton effect, γ-rays of medium energy (0.5–1.0 Mev) collide with loosely bound orbital electron transferring only a portion of the energy to the electron which is ejected. The γ-photon itself is deflected in a new direction with a reduced energy. Again the ejected electron behaves as a negatron. In pair production, γ-rays interact directly with the nuclear force field. In such a case the photon may cease to exist and have all its energy converted into two particles, a positron and an electron. This generally happens with γ-rays having energy equal to or greater than 1.02 Mev. The positron gets annihilated with consequent back to back γ-emission, whereas the ejected electron may cause ionization.

12.7 Measurement of Radioactivity

The three commonly used methods for detecting and quantifying the radioactivity are generally based on the ionization of the gases (proportional counters), or on the excitation of solids or solutions (scintillation counters) or on induction of specific chemical reaction in certain emulsions (autoradiography). The underlying principle for each of these methods and the equipment used are discussed next.

12.7.1 Proportional counters

These operate on the principle of gas ionization. In gas ionization, charged particle while passing through the gas dislodges orbital electron from the atom sufficiently close to its path and causes ionization. The α-particle causes intense ionization (above 10^4–10^5 ion pairs/cm), whereas the β-particles produce rather diffused ionisation (10^2–10^3 ion pairs/cm). γ-rays cause little ionization (about 1–10 ion pairs/cm) and that too by secondary mechanism. Accordingly, α- and β-particles may be detected by gas ionization methods, while these methods are inadequate for γ-radiation. A simple ionization chamber consists of a cylindrical closed metal container filled with air or some other gas and fitted with a central collecting electrode. The chamber wall is connected to the –ve pole of a potential source, making it the cathode; the collecting electrode is connected through a resistor to a +ve side of the power supply so that it serves as the anode. When ionization occurs, a pulse (current) flows, the magnitude of which is related to the applied potential and the number of radiation particles entering the chamber. As radioactive particle produces only one ion pair per collision, the current produced is low and sensitive measuring devices are necessary. Consequently, this method is of little use in quantitative work. However, various electroscopes which operate on this principle, are useful in demonstrating the various properties of radioactivity. At voltage levels higher than that of the simple ionization chambers, electrons produced from ionization, move much faster towards the anode, causing secondary ionization of the gas in the chamber. These secondary ionization electrons cause further ionization. Consequently, a whole torrent of electrons reaches the anode. Thus from a few primary ion pairs, a geometrical increase results in a variable torrent of ions moving towards the chamber electrode. This process is known as gas amplification and the flood of ions produced is termed as 'Townsend Avalanche', after its discoverer. As a consequence of this gas amplification, a very large number of electrons are collected at the anode within a microsecond or less from a single β-particle in the chamber. A strong pulse is thus produced and fed into the external circuit which could be measured directly. However, as the potential gradient is further increased, the number of electrons (mostly secondary) reaching the anode rises sharply and eventually a potential is reached at which chamber undergoes continuous discharge and no longer can be used as a detector. As can be seen from Fig. 12.3, there are three distinct potential regions between the saturation current and continuous discharge. These regions are the proportional region, the limited proportional region and the Geiger-Müeller region. In the proportional region, the number of ion pairs collected is directly proportional to the applied voltage. The counters that are manufactured to operate in the proportional region require a very stable voltage supply because small fluctuations in voltage result in significant changes in amplification. These counters (proportional counters) are particularly useful for detection and quantification of α-emitting isotopes. The next potential region, i.e. the limited proportional region is similar to proportional region in operating characteristics. However, in this region the extent of gas amplification has an upper limit set by the size of the detector chamber, the form of the collecting electrode and the number of gas atoms present. The limited proportional chamber is not normally used for the purposes of detection. In the Geiger-Müeller region, all radiation particles including even weak β-particles or γ-rays are able to initiate sufficient ion pair formation to completely fill the ion space available in the chamber. Consequently the size of the charge collected on the anode is no longer dependent on the number of primary ions produced. Hence various types of radiations cannot be distinguished in this region. Ionization chambers operated in this potential region are generally called Geiger-Müeller detectors. Since maximum gas amplification is realized in this region, the size of the output pulse from the detector remains the same over a considerable voltage range until continuous discharge occurs.

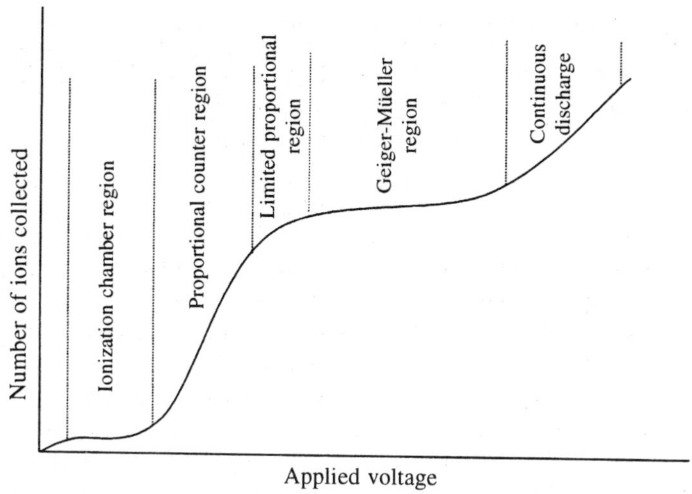

Fig. 12.3 Relation of pulse size to potential gradient in an ionization chamber.

12.7.2 Scintillation counters

Counters based on gas ionization used to be the main instruments employed in quantification of radioisotopes in biological samples. These have now been replaced by scintillation counters. However, small hand held radioactivity monitors based on gas ionization are used for monitoring contamination etc. They can also be used in situations where the presence or absence of radioactivity needs to be established. Scintillation detection is based on the interaction of radiation with substances known as Fluors (solid or liquid), or scintillators. Excitation of the electron in the fluors leads to the subsequent emission of a flash of light (scintillation) which is converted to an electronic pulse by a photomultiplier tube. The magnitude of the pulse is proportional to the energy loss by the incident radiation and the excitation of the fluor. Accordingly, the number of scintillation is proportional to the rate of decay of the sample, i.e. the amount of radioactivity and the intensity of light given out and therefore the signal from the photomultiplier is proportional to the energy of radiation. Hence, the counter could be used both for the qualitative and quantitative purposes.

12.7.2.1 Types of scintillation counting

There are two types of scintillation counting, i.e. solid scintillation counting and liquid scintillation counting. In solid scintillation counting, the sample is placed adjacent to a crystal of a fluorescent material. Crystals of sodium iodide, zinc sulfide and anthracine are the commonly used crystals for γ-isotopes, α-emitters and β-emitters, respectively. The crystals are placed near to a photomultiplier which in turn is connected to a high voltage supply and a scanner. These counters are particularly useful for γ-emitting isotopes because γ-rays are electromagnetic radiations and collide only rarely with neighbouring atoms to cause ionization or excitation. However, they are not good for weak β-emitting isotopes (^3H and ^{14}C) because even the highest energy negatrons emitted by these isotopes hardly have sufficient energy to penetrate the walls of the counting vials in which the samples are placed for counting. In liquid scintillation counting, the radioactive sample and the fluor are mixed intimately in a homogenous medium. However, the detection sensitivity is such that even the lower energy β-emitters (^{3}H, ^{14}C and ^{35}S) can be assayed with quite respectable efficiency.

12.7.2.2 Mechanism of the liquid scintillation detection

Fluor molecules are directly or indirectly excited by ionizing radiation resulting in the emission of photons, which in turn interact with the photocathode of a photomultiplier to yield photoelectrons from the photocathode surface. These photoelectrons pass through a diode series, resulting in the production of a greatly amplified electron pulse at the photomultiplier anode.

In liquid scintillation, the energy of the nuclear radiation is primarily transferred to the more abundant solvent molecules (toluene) which, in turn, get ionized, dissociated or excited. The excitation energy of the solvent can either be emitted as photons (in the UV region) or transferred to the molecules of the primary solute. The excited fluor molecules return to the ground state by emitting photons with a wavelength in visible or near UV region. In case of *para-ter* phenyl in toluene, the fluorescence peak appears at 3500 A°. This may not match the most sensitive range of the photocathode which is used to detect the photons. In such cases, a secondary solute (also a fluor) is used which absorbs energy from primary fluor and re-emits it as light of a longer wavelength. These secondary solutes are therefore called wave shifters. POPOP [1,4-bis-2-(5-phenyl oxazolyl)benzene] is the commonly used secondary solute which greatly improves detection efficiency. Figure 12.4 illustrates the relationship between fluorescence peaks of primary and secondary solutes and the spectral sensitivity of photocathode.

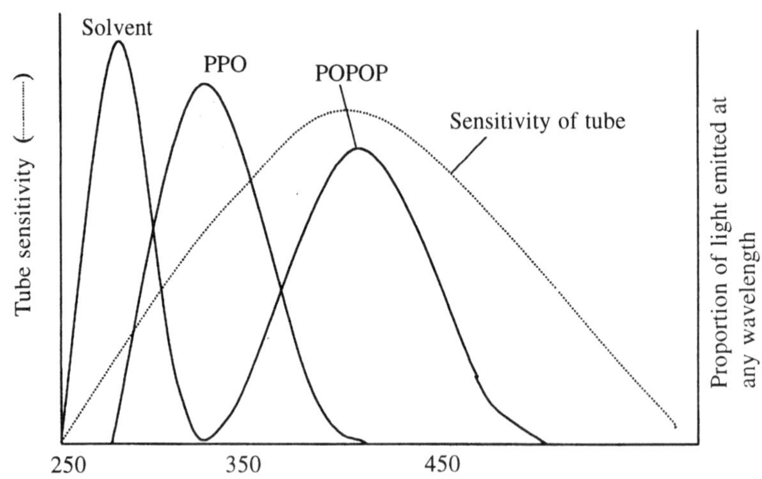

Fig. 12.4 Emission spectra of various fluors in relation to sensitivity of phototubes.

PPO (2,5-diphenyl oxazole) and POPOP were among the most commonly used primary and secondary fluors, respectively. However, a large number of readymade scintillation cocktails are now available in the market which are specifically designed for aqueous and organic samples, separately. Accordingly, it is important to make appropriate selection.

Liquid scintillation counting has several advantages over gas ionization counting. The major advantages include much higher counting efficiencies particularly for low energy β-emitters, ability to accommodate samples of any type including solids, liquids, suspensions and gels, and the ability to count separately different isotopes in the same sample. However, scintillation counters are highly automated and hundreds of samples can be estimated automatically. The built in computer facility can carry out many forms of data analysis, such as efficiency correction, graph plotting, radioimmunoassay etc.

The major problem encountered in use of liquid scintillation counting is that of photomultiplier thermal noise. This is reduced by cooling the photomultipliers. However, some of the latest developments introduced in the instrumentation have taken care of these problems. Another major problem in liquid scintillation detection is the effect known as 'quenching' which generally means reduction of efficiency in the energy transfer process in the scintillation solution. Furthermore, the extent of 'quenching' varies considerably with different sample materials. Thus, in these experiments, a determination of just how much the counting efficiency of every sample has been decreased by quenching is generally determined. Despite the drawbacks mentioned above, the liquid scintillation method of detection is the most effective means of assaying low energy β-emitting samples, especially ^{3}H and ^{14}C. Furthermore, the relative ease of sample preparation, the versatility of sample form possible, and the commercial availability of reliable counting systems with high capacity automatic sample changers have made this detection method the most useful one where large numbers and types of samples are to be assayed.

12.7.2.2(*a*) *Counting of dual-labelled samples:* Since the size of electric pulse produced by the conversion of light energy in the photomultiplier is related directly to energy of the original radioactive event, two different β-emitting isotopes having different energy spectra can be quantified separately in a single sample, e.g. ^{3}H and ^{14}C, ^{3}H and ^{35}S, ^{3}H and ^{32}P, ^{14}C and ^{32}P, ^{35}S and ^{32}P constitute such pairs. The mechanism of estimation is depicted in Fig. 12.5. As may be observed in this Figure the spectra of two isotopes overlap only slightly and by setting a pulse height analyzer to reject all pulses of an energy below X (threshold X) and to reject all pulses of energy above Y (window Y) attributable to radioisotope S and also to reject below a threshold of A and a window of B for isotope T, it is possible to separate the pulses generated by these two isotopes completely. A pulse height analyzer set with a threshold and window for a particular isotope is known as a channel (e.g. a ^{3}H channel). Most of the modern counters have multichannel analyzer and are based on an analogue-2-digital convertor. The electronic signals from a photomultiplier are converted to digital signals stored in a computer and entire energy spectrum is then analysed simultaneously. This type of dual label counting is of great use in certain fields of molecular biology (nucleic acid hybridization and transcription), and metabolism (steroid synthesis and drug development).

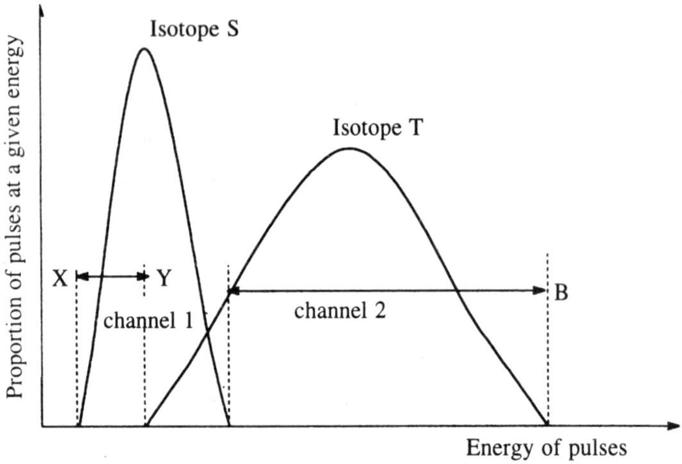

Fig. 12.5 Diagram to illustrate the principle of counting dual labelled samples.

12.7.2.2(*b*) *Determination of counting efficiency:* As stated earlier, a major problem encountered in scintillation counting is that of quenching. This necessitates the determination of the counting efficiency of some, if not all, of the samples in a particular experiment. There are several methods by which counting efficiency can be determined. These are:

(*i*) *Use of internal standard:* It is the most commonly employed method and consists of adding a precisely known amount of a known isotope (an internal standard) in a non-quenching form to a previously counted sample. The resulting mixture is then recounted and the counting efficiency of the sample is calculated as follows:

$$\text{Counting efficiency} = \frac{\begin{pmatrix}\text{Net count rate (in cpm)} \\ \text{of int. std. + sample}\end{pmatrix} - \begin{pmatrix}\text{Net count rate (cpm)} \\ \text{of sample}\end{pmatrix}}{\text{Disintegration rate (in dpm) of int. std.}}$$

Toluene ^{14}C, *n*-hexadecane-^{14}C, benzoic acid ^{14}C, tritiated toluene, tritiated water and tritiated n-hexadecane are some of the commonly used internal standards. However, each standard has its drawbacks. For example, benzoic acid and water are themselves quenching agents and consequently it is necessary to use these compounds at a very low concentration. Similarly, it is difficult to accurately pipette small amount of labelled toluene. Furthermore, in case of heterogeneous samples such as gel preparations, it is important to employ an internal standard that can stay in the same phase as the radioactive compound. The method is time consuming because each sample must be counted twice. Moreover, the accuracy of the determination often depends upon accurate measurement of small volume of the liquid containing the compound used as the standard.

(*ii*) *Channels ratio:* On quenching of the sample in a scintillation counter, the efficiency of scintillation process decreases as less light is produced for a given quantum of radiation energy. Accordingly, the energy spectrum for a quenched sample is lower than for an unquenched sample. Higher the degree of quenching, more pronounced is the effect. Channels ratio method for determining counting efficiency is based on this principle and involves preparation of a calibration curve based on counting in two channels that cover different, but overlapping part of the spectrum (Fig. 12.6). The ratio of counts in each channel varies. For preparing standard curve, a set of quenched standards is counted, the absolute amount of radioactivity is known and the efficiency of counting in each channel is determined. The standard curve (Fig. 12.7) is then drawn by plotting counting efficiency against channel ratio. One standard curve applies only to one set of circumstances (one radioisotope, counter and scintillation fluid). Once the standard curve is drawn, the efficiency of counting experimental samples can be determined. Samples are counted in the two channels, ratio calculated and the efficiency read by putting the ratio in the graph. In practice, all data are stored in the counter computers and the corrected values printed automatically.

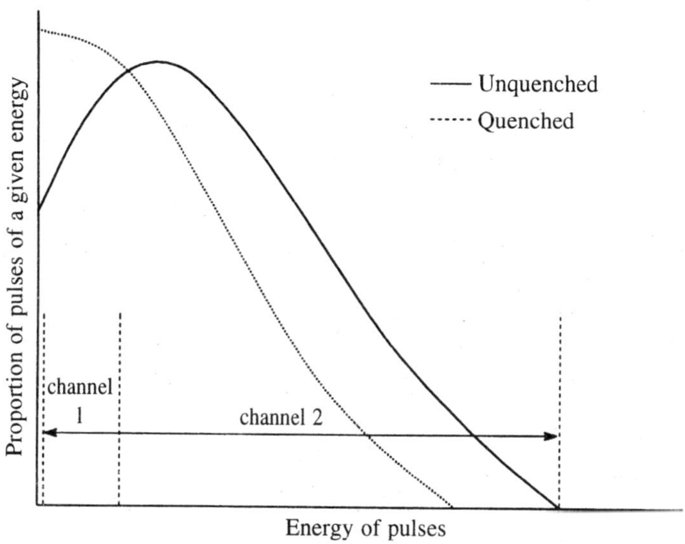

Fig. 12.6 The effect of quenching of β-energy spectrum.

Multichannel scintillation counters are available in the market and they operate on the same principle except that in these instruments the whole shape and position of the spectrum is analyzed. This is given a digital parameter that relates the counting efficiency. For this reason, these systems have better precision. The channels ratio method is suitable for all types and is less time consuming than either internal or external standardization. However, at low count rates and for very highly quenched samples, the method is inaccurate. In more situations the external standardization is most frequently the procedure of choice.

(iii) External standardization: In this method, each sample to be counted is exposed to the external source of γ-emitting external standard built into the counter. The γ-radiations penetrate the vial and excite the scintillation fluid. The resulting spectrum is unique to the source and is significantly different from that produced by the sample in the vial. ^{137}Cs, ^{133}Ba and ^{226}Ra are some of the

commonly used γ-sources. The spectrum obtained is significantly affected by the quenching agents present in the scintillation fluid. The instrument analyses this spectrum and assigns a quench parameter to it. As for the channels ratio method, a standard curve is required. This method is suitable even for samples with low count rates.

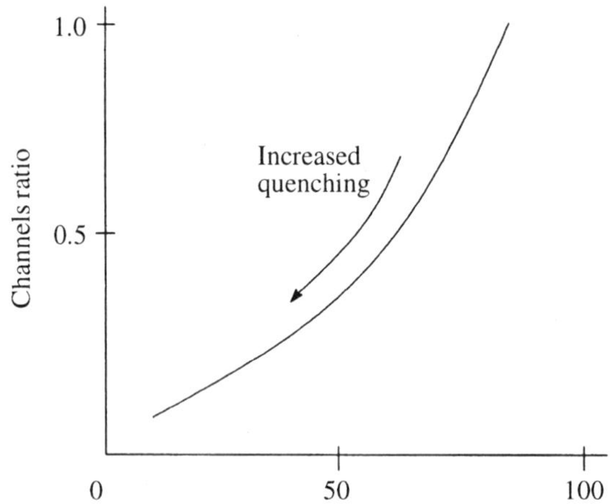

Fig. 12.7 Channels ratio quench correction curve.

The major characteristics of the detectors, i.e. ion chamber with a vibrating-read electrometer, the proportional detector, the Geiger-Müeller detector, and both external sample and internal sample scintillation detector are summarized in Table 12.2.

12.7.2.3 Sample preparation

Biological samples in radiotracer experiments are in diverse forms like blood, urine, water, milk, plant or animal tissue or respiratory gases. Such varied samples can not be assayed directly and must be converted to a suitable form for assay. A sample in the solid form has to be mounted in a uniform and reproducible manner. For liquid scintillation counting, the sample material should be suitably incorporated into the fluor medium. Generally, the choice of the counting sample form is determined by the type and energy of the radiation emitted by the sample radioisotope. In samples containing low energy β-emitters, counting efficiency is poorer because of the problem of self absorption. This necessitates conversion of the sample to a standard and most suitable form. However, sample containing γ-emitting nuclides can be assayed directly with a minimum of pretreatment.

Sample preparation is easy in solid scintillation counting and only involves transferring the sample to a glass or plastic vial compatible with the computer. However, in liquid scintillation counting, sample preparation is more complex. The process starts with the sample vial to be used which could be glass, low potassium glass or polyethylene. Each one of these has its own advantages and disadvantages as polyethylene vials give better light transfer and result in slightly higher counting efficiency but they exhibit more fluorescence than glass vials.

Table 12.2 Summary of Radiation Detector Characteristics

Detector type	*Energy discrimination*	*Detection medium*	*Reproducibility*	*Detector amplification factor*	*Relative electronic amplification required*	*Resolving time (μ sec)*	*Relative background*	*Relative detection efficiency*			*Most commonly used sample form*	*Relative degree of difficulty in preparing sample*
								α	β	γ		
Ion chamber with vibrating read electrometer	Yes	Gas	Excellent	1	Very high	Not appreciable	Low	Good	Good	Poor	Gas	High*
Proportional detector	Yes	Gas	Good	10^2–10^4	High	5–50	Low	Good	Good	Poor	Solid	Medium⁺
Geiger-Müeller detector	No	Gas	Fair	10^7	Low	100–1000	Medium	Good	Good	Poor	Solid	Medium⁺
Solid (external sample) scintillation detector	Yes	Solid	Good	10^6	Low	0.3–>1	High	Fair	Fair	Very good	Solid or liquid	Midium⁺ or low#
Liquid (internal sample) scintillation detector	Yes	Liquid	Fair	10^6	Medium	0.001–<1	Low	Very good	Very good	Fair	Solution	Low

* Usually involves individual combustion of samples to gaseous form.
\+ Usually involves preparation of Planchet-mounted samples.
\# Well-type detector used for liquid samples.

The counters used these days accept many types of vials. If possible, vials should be used to save cost etc.

For scintillation counting, toluene based cocktails are quite good. However, they cannot be used for aqueous samples as toluene and water are immiscible. Cocktails based on 1,4-dioxane and naphthalene are not used due to toxicity. Aqueous samples are mostly counted with emulsifier based cocktails which contain Triton X-100 and can accept upto 50% water. However, readymade cocktails are also available in market which could be used depending upon the precise sample conditions. Sample volume is also one of the factors determining counting efficiency. Accordingly, sample vials in a given series of counts should contain same volume of the sample. In case of colour quenching, sample could be bleached before counting. However, bleaching agents such as hydrogen peroxide may result in chemiluminiscence in some of the cocktails.

Plant and animal tissues (solid samples) can be counted best after solubilization by quaternary amines such as NCS solubilizer or toluene etc. These solutions are highly toxic and care must be taken in handling them. The sample is digested after adding it to the counting vial containing a small amount of the solubilizer. On completion, scintillation cocktail is added and the sample is counted. Chemiluminiscence is again a problem with tissue solubilization. Combustion techniques are good alternatives to bleaching of coloured samples or digestion of tissues. In these methods, the samples are combusted in an atmosphere of oxygen usually in commercially available combustion apparatus. ^{14}C when combusted is converted to $^{14}CO_2$ which is trapped in a solution of sodium hydroxide and then counted. ^{3}H containing sample is converted to $^{3}H_2O$ for counting. However, sample containing β-emitting isotopes could be prepared for counting in liquid scintillation counters by any one of the methods described here. These are some of the important considerations which need to be kept in mind while preparing sample for counting.

12.7.3 Autoradiography

Ionizing radiation acts upon photographic emulsions to produce a latent image much as the visible light does. In photography, light initiates a process in the emulsion which produces electrons leading to the reduction of silver halides to metallic silver. This in turn acts as a catalyst for further reduction of silver halide in its immediate vicinity. Ionizing radiations produce the same end effect. In autoradiography (also known as radioautography), a sample containing radioactive material is placed in close contact with a sensitive emulsion. After exposure, the film is developed and the correct location of the radioactive matter in the sample is then determined from the pattern of darkening of the film. Thus the technique of autoradiography is primarily a means of localization of radioisotope in a given sample. For example, the sites of ^{45}Ca concentrations in growing bone tissue, the relative distribution of ^{32}P in a bean plant, or the localization of ^{3}H-thymidine in DNA of a cell nuclei can be readily demonstrated by this technique. Even radioactive metabolites separated by chromatographic or electrophoretic techniques could be located on the chromatogram or an electrophoretogram.

In general, weak β-emitting isotopes like ^{3}H, ^{14}C and ^{35}S etc. are more suitable for autoradiography, particularly for cell and tissue localization experiments. This is mainly because of the reason that ionizing tracks of these isotopes are short which give rise to a discrete image. This is particularly important when radioactivity associated with some subcellular organelles is being located. ^{3}H is commonly used radioisotope for this purpose. Electron microscopy is then used to detect the radioactivity containing zones in the developed image. For examining distribution pattern within the whole organism or tissue, either ^{14}C or ^{3}H are used. For location of DNA bands in electrophoretic gels, ^{32}P is used. In case of very thin gels, even ^{35}S or ^{14}C can also be detected with high resolution. This precisely is the situation in DNA sequencing gels where ^{35}S is used as the label.

12.7.3.1 Film emulsion

The emulsion is formed by grains of silver halide dispersed in gelatin medium and is backed by sheet of cellulose acetate or glass. The sensitivity and resolution of the emulsion are governed by the size and concentration of silver halide grains. While grain size is directly related to sensitivity, the grain concentration determines the degree of resolution. Various X-ray sensitive films are used for gross autoradiography, whereas nuclear emulsions are used for microscopic autoradiography. In addition, a number of specialized emulsions are also available for specific autoradiographic purposes. Stripping film has high resolution but low sensitivity and is the emulsion of choice for intracellular localization studies. Liquid emulsions are prepared by melting strips of emulsion by heating them to around 60°C and then either the emulsion is poured over the sample or the sample attached to a support is dipped into it. The emulsion is then allowed to set before being dried. The method is known as dipping film method and is recommended when very thin film is required. Accidental exposure to light, chemicals in the sample, natural background radioactivity and even pressure applied during handling and storage can cause a background foam on the developed film. It may cause problems particularly in microscopy and care must be taken to minimize this effect. To decrease background foam, exposure time should always be kept minimum.

12.7.3.2 Determination of exposure time

Exposure time depends upon the isotope, sample type, level of activity, film type and purpose of the experiment. It has been estimated that, 10^{6}–10^{7} β-particles must strike each cm^2 of X-ray film to produce optimum blackening. Accordingly, a rough estimate of exposure time can be made by measuring the activity of the sample per cm^2 with a thin-end-window Geiger Müeller detector.

12.7.3.3 Tissue preparation and artifacts

While preparing gross samples or tissue sections for autoradiography, extreme care must be taken to avoid leaching of radioactive material. The problem of leaching depends upon tissue and type of radioisotope used, e.g. inorganic ^{32}P is easily leached out of many tissues, whereas organically bound ^{32}P is rarely affected. For these reasons, the freeze-drying technique is highly favoured for tissue preparation. Where tissue sections and emulsions are permanently mounted, the subsequent staining may result in a reaction with emulsion. Formation of artifacts constitute a major difficulty in interpreting an autoradiograph. Artifacts of the developed film may result from various sources such as vapors from volatile agents in the samples, mechanical pressure, extraneous light during film processing, dust or debris, finger prints and shrinkage or expansion of either sample or film etc. To evaluate the extent of artifacts, one may run a parallel sample without radioactivity for comparison.

12.7.3.4 Specific autoradiographic techniques

In general, two basic methods of autoradiography are employed in biological investigations. In the first case, there is contact between the emulsion and the sample only during exposure time, after which the emulsion is removed and developed. This method is employed mostly for gross samples. In the second method, there is permanent contact between emulsion and the sample. This method is used exclusively with thin tissue samples. The first method (temporary contact method) is most suitable for use with chromatograms of labelled material, leaf and whole plant tissues, gross bone sections and tissue sections that have well defined outlines. This requires little pretreatment of sample and any subsequent tissue staining does not affect the film. Because of poor contact, the method gives mediocre resolution and cannot normally be used for cellular localization studies. Several unique adaptations of this method have also been proposed. For better resolution, the second method (permanent contact method) is generally preferred. Three modifications of this method in common use include mounting method, coating method and stripping film method. In the mounting method, the tissue section is floated on water and the emulsion, on a glass slide, brought up underneath it so that the tissue lies on the film. Following the exposure, the film is developed and the tissue stained. This is a simple method with reasonably high resolution but suffers from disadvantage of spotty development due to non-uniform penetration of the developer through the sample. To avoid this problem, the tissue section may be mounted on a glass slide and the emulsion applied over it. In coating method, the emulsion is melted and poured onto the tissue sample where it spreads and hardens. This method gives better resolution than the mounting method. In stripping film method, the films used are the ones that allow the emulsion to be stripped off the base and applied directly on to the tissue section by means of water flotation. This gives better resolution than any of the previous methods and is commonly employed for determination of intracellular localization. In more recent techniques, a monolayer of silver halide crystal is applied on ultrathin tissue section embedded in methacrylate. Following development of the emulsion, the tissue section is stained with uranyl or lead stains and examined by means of electron microscope.

12.8 Applications of Radioisotopes in the Biological Sciences

Radioisotopes have wide applications in biochemistry, particularly for tracing metabolic pathways. This generally involves adding a radioactive substrate to a metabolizing tissue, taking samples of the experimental materials at different times, extracting and separating the products chromatographically or by some other appropriate technique. Radioactive detectors could also be attached to GC or HPLC columns etc. In such cases, radioactivity coming out of the column during separation could be monitored. Radioactive products on paper or thin layer chromatograms may be located either by chromatogram scanner or by autoradiography. Most of the metabolic pathways known today largely owe their elucidation to these techniques.

The turn over rates of a particular compound can conveniently be ascertained by radioisotopes. For example, for determining the turn over of proteins in rats, a group of rats is injected with a radioactive amino acid and left for 24 h during which time most of the amino acids get incorporated into proteins. Rats are then sacrificed at suitable time intervals and radioactive-labelled proteins in organs or tissues of interest are determined.

The mechanism and rates of absorption, accumulation and translocation of organic and inorganic compounds can in both plants and animals, be readily studied by radioisotopes. Radioisotopes also have wide applications in metabolic aspects of new drugs etc. This involves determining the site of drug accumulation, the rate of accumulation, the rate of metabolism and the metabolic products etc. For instance, autoradiography on whole sections of experimental animals yields valuable information on the site and rate of drug accumulation. Radioisotopes are also used in enzyme assays. Though these assays are generally expensive but have the advantage of high degree of sensitivity. Radioisotopes are also used in the study of mechanism of enzyme action and in ligand binding studies.

Isotope dilution analysis could be used to assay those compounds which are present in low amounts in living organisms and in mixture of similar compounds. For example, the amount of iron in a protein can be determined by use of a source of ^{59}Fe, which is mixed with the protein and a sample of Fe subsequently isolated and assayed for total iron. These techniques are in common use for the studies on trace elements. Radioimmunoassays developed in recent years are the most important techniques employed for quantitative analysis of hormones, steroids and drugs etc. The technique combines the specificity of the unique reaction with the sensitivity of radioisotope technique and is based on the competition between unlabelled antigen and a finite amount of corresponding radiolabelled antigen for a limited number of antibody binding sites in a fixed amount of antiserum.

Radioisotopes are also used in dating, determining the age of rocks, fossils and sediments. In these techniques, it is assumed that the proportion of an element that is naturally radioactive will remain same throughout. From the time of deposition, there will be decay of the radioactive isotope and by determining the amount of radioisotope left (this could be determined from the half life of the radioisotope), it is possible to date the sample.

Radioisotopes find extensive application in molecular biology particularly in DNA and RNA sequencing, DNA replication, transcription, synthesis of DNA, recombinant DNA technology etc. These techniques are described elsewhere in this book. Radioisotopes are also used in diagnostic tests. ^{133}Xe is routinely used for diagnosis of malfunction of lung ventilation. Similarly ^{133}I-iodohippuric acid is used in diagnosis of kidney infection. ^{131}I is used in the assessing of thyroid functions. Radioisotopes are also used in the study of various aspects of hematology such as blood cell lifetime, blood volumes and blood circulation times. Besides the above, radioisotopes are also employed in ecological studies and food irradiation etc.

Some of the radioisotopes commonly used in biological work with their merits and demerits are given in tabular form (Table 12.3). However, the choice will depend on the precise requirements of the experiment.

Table 12.3 Merits and demerits of commonly used β-emitters

Isotope	*Merits*	*Demerits*
^{3}H	Safety High specific activity possible Wide choice of positions in organic compounds Very high resolution in autoradiography	Low efficiency of detection Isotope exchange with environment Isotope effect
^{35}S	High specific activity Good resolution in autoradiography	Short half-life Relatively long biological half-life
^{14}C	Safety Wide choice of labelling position in organic compounds Good resolution in autoradiography	Low specific activity
^{32}P	Ease of detection High specific activity Short half-life simplifies disposal	Short half-life affects costs and experimental design External radiation hazard Poor resolution in autoradiography

12.9 Laboratory Exercises

12.9.1 EXPERIMENT: Use of labelled carbon dioxide to measure photosynthate partitioning in plants

Principle

Since ^{14}C (radioactive carbon) and ^{12}C (naturally occurring non-radioactive isotope of carbon) have the same atomic number they have the same chemical properties. The living organisms, including plants, are unable to differentiate between these two isotopes of carbon. Thus the plant leaves utilize ^{12}C and ^{14}C carbon dioxide at the same rate. The photosynthetically fixed carbon is then subsequently partitioned between various plant organs, i.e. roots, shoots, seeds, fruits etc. The partitioning of assimilates is influenced by a variety of factors such as plant species, age, stage of development and environmental factors. The partitioning pattern can be quantitatively determined by estimating the radioactivity in different plant parts following exposure of the seedlings to $^{14}CO_2$ in light. The percent distribution of ^{14}C in various parts is then calculated.

Materials and Reagents

1. Syringes
2. Double walled (water jacketed) perspex chamber
3. Air-tight circulation pump
4. Gas proportional or Geiger-Müeller counter or liquid scintillation counter
5. Planchets or scintillation vials
6. 1 N and 2 N HCl
7. NaOH
8. ^{14}C-labelled $NaHCO_3$ (specific activity 50 mCi/mmole)
9. Scintillation fluid: Dissolve 4 g of POP (2,5-diphenyl oxazole) and 100 mg POPOP (1,4-bis-2-(5-phenyl oxazolyl) benzene) in 1 L of toluene
10. Grown up plants of wheat, pea and tomato

Procedure

The experiment is performed in the following two distinct phases (Part A and B) which are described separately.

I Part A: Feeding of plants with $^{14}CO_2$
II Part B: Determination of radioactivity in different plant parts and fractions

Part A: Feeding of plants with $^{14}CO_2$

1. Enclose the plant (e.g. wheat, tomato or pea plant) in a double walled perspex chamber of an appropriate size and place it under natural sunlight.
2. Circulate water between the chamber walls through the water jacket to prevent any rise in temperature in the chamber during the feeding period.
3. Connect outlet of an air tight CO_2 generating tube with a flexible propylene or rubber tubing to the inlet of the air tight circulating pump and outlet of the pump to inlet of chamber. With another tubing connect outlet of the chamber with inlet of CO_2 generating tube.
4. Place 1–2 μmoles of $NaH^{14}CO_3$ (specific activity 50 mCi/mmol) at the bottom of CO_2 generating tube.
5. Release 1N HCl from the stopper into $NaH^{14}CO_3$ containing CO_2 generating tube so that 50 to 100 μCi of labelled $^{14}CO_2$ is liberated.
6. Circulate the evolved $^{14}CO_2$ through the perspex chamber (in which the plant has been enclosed) using a small air tight electrical pump.
7. Allow the plant to fix $^{14}CO_2$ for an appropriate time (5 min to 1 h).
8. Remove excess $^{14}CO_2$ diverting the circulation of gas through NaOH trap which is attached via a 3-way stopcock in between the CO_2 generating tube and the pump.
9. Return the ^{14}C fed plants to natural growth conditions.
10. After 24 h of feeding, harvest the plants and separate the parts, e.g. leaf, stem, roots, seeds, fruits etc.
11. Dry the different parts separately in an oven at 65°C to a constant weight and record their weight.

Part B: Extraction and measurement of radioactivity: Radioactivity can be measured by any one of the following methods after suitably processing the plant sample.

Method 1:

1. Grind the oven dried sample of different plant parts to a fine powder.
2. Place a known weight of the sample in a planchet having some adhesive material on its surface so that the sample gets stuck to the planchet.
3. Count the disintegrations per min (dpm) either by a gas proportional counter or by Geiger-Müeller Counter.

Method 2:

Prepare the ethanol soluble and insoluble factions of the powdered sample and then determine the radioactivity in each of them individually.

1. Take a suitable amount of dried sample in a centrifuge tube.
2. Add 5 ml of 80% ethanol (*v*/*v*) and place in a water bath at 40°C for 1 h. Shake the tube occasionally.
3. Centrifuge at 5000 × *g* for 15 min and decant the supernatant.
4. Extract the residue in the similar manner with 50% (*v*/*v*) ethanol, centrifuge and collect the supernatant.
5. Finally extract the residue with water, centrifuge and collect the supernatant.
6. Pool all the supernatants obtained in Steps 3–5 and this will constitute ethanol soluble fraction.
7. Disperse the residue in 2 N HCl and autoclave it at 15 kg/cm^2 for 1 h.
8. Centrifuge the hydrolyzed material and wash the pellet twice with 2 N HCl.
9. Pool all the acid fractions which will represent the ethanol insoluble fraction.
10. Evaporate both the fractions (ethanol soluble and insoluble) separately to dryness and redissolve each of them in 2 ml distilled water.
11. Pipette 0.2 ml of the above extract on a planchet and dry under infra-red lamp and then record the counts in a gas flow proportional counter.

Method 3: Measurement of radioactivity by liquid scintillation counter.

1. Take a suitable aliquot (1 ml) of the above fractions in scintillation vials and dry it.
2. Add 10 ml of scintillation fluid to each vial.
3. Record the counts on a liquid scintillation counter with an appropriate correction for quenching.

Calculations

1. Determine the total amount of ^{14}C-assimilates in the plant by adding the radioactivity (dpm) in both the fractions of all the plant parts. While doing this take into account weight of the part used for sample preparation, dilution factors (if any in case of fractions) and the total weight of that

plant or organ in the plant. For example, if radioactivity is determined in powdered non-fractionated leaf sample, then:

$$\text{Total } ^{14}\text{C-labelled material} = \frac{\text{dpm} \times \text{total weight of leaves on the plant}}{\text{wt. of powdered sample taken}}$$

For the fractionated sample, take into account the dilution factor as well, i.e. volume of fraction used for determination of radioactivity and the total volume of the fraction.

2. Calculate the total ^{14}C labelled assimilates in each plant part, e.g. *a*, *b*, *c*, *d* in leaves, stem, seeds and roots, respectively from sum of the counts in both the fractions of an organ.
3. Compute the per cent distribution of ^{14}C in different parts as follows:

$$\% \; ^{14}\text{C in leaves} = a/\text{X} \times 100 \text{ dpm}$$
$$\% \; ^{14}\text{C in stem} = b/\text{X} \times 100 \text{ dpm}$$
$$\% \; ^{14}\text{C in seeds} = c/\text{X} \times 100 \text{ dpm}$$
$$\% \; ^{14}\text{C in roots} = d/\text{X} \times 100 \text{ dpm}$$

where,

X is the total ^{14}C-assimilated by the entire plant and *a*, *b*, *c*, *d* represent total ^{14}C-labelled material in each of these plant parts, respectively.

12.9.2 EXPERIMENT: To establish photosynthetic CO_2 assimilation pathway in a given plant species using radioisotopic technique

Principle

The mechanism of photosynthetic CO_2 assimilation in plants varies with species and may occur via either C_3 or C_4 pathway. Whether a C_3 or C_4 mechanism operates in a particular plant can be established by identifying the first product of CO_2 fixation reaction. In case of C_3-pathway, 3-phosphoglycerate (3-PGA) is the first product, whereas in C_4-type of plants the initial product of carboxylation reaction is a C_4 organic acid such as oxaloacetate which is rapidly converted to either malate or aspartate. Hence when plants are exposed to $^{14}CO_2$, the label is incorporated into the primary products. Determination of these primary products (namely, 3-PGA, malate and aspartate) for radioactivity, provides the confirmatory evidence for operation of a particular photosynthetic pathway in the plant species under investigation.

Materials and Reagents

1. Experimental plant.
2. Chamber made of perspex or glass for feeding the plants with $^{14}CO_2$.
3. Gas mixture pump.
4. ^{14}C-labelled sodium bicarbonate ($NaHCO_3$).
5. Chromatographic chamber for descending paper chromatography.
6. Liquid scintillation counter.
7. 50 mM Tricine-KOH buffer (pH 7.5) containing 0.33M sorbitol.

8. Scintillation fluid: Prepare by dissolving 8 g 2,5-diphenyl oxazole (POP) and 100 mg of 1,4-bis-2-(5-phenyl oxazolyl) benzene (POPOP) in 1 L of toluene.
9. Authentic standards: Prepare solution containing 1 mg/ml of authentic sample of 3-PGA, glucose-6-P, aspartate, malate.
10. Solvent system for paper chromatography: Mix *n*-butanol : propionic acid : water in the ratio of 10:5:7 (*v*/*v*) and use as the solvent system.
11. Detection reagents: Prepare various spraying reagents as given in Step 11 in the following Procedure.

Procedure

1. Soak the glass fibre discs in 50 mM tricine-KOH buffer (pH 7.5) containing 0.33 M sorbitol.
2. Quickly transfer 0.1g plant leaf segments on these glass discs to prevent dehydration of plant tissues. Place these glass fibre discs in steel planchets. Prepare five such planchets and place them on a strip having a well in its centre. Add 2 N HCl to this central well and keep the strip in the perpspex chamber (19 × 1.2 × 3.7 cm) with an internal volume of 50 ml.
3. Expose the chamber to direct sunlight for 5 min. Then inject $NaH^{14}CO_3$ through a rubber septum into the acid containing well on the strip such that the concentration of $^{14}CO_2$ in the chamber is 0.1% and circulate the air through the chamber with the help of an air tight gas mixture pump.
4. Remove the samples from the chamber after 20, 40, 60, 120 and 300 sec of exposure to $^{14}CO_2$.
5. After removal, immediately kill the tissue by dropping it in boiling ethanol.
6. Extract the plant material with 80% (*v*/*v*) ethanol, centrifuge the homogenate at 8000 × *g* and collect the supernatant.
7. Extract the residue with distilled water, centrifuge again and carefully decant the supernatant. Pool the obtained supernatants and make to a known final volume of 10 ml with 80% ethanol.
8. Take 1 ml of the extract in a scintillation vial containing 1.5 ml scintillation fluid and evaporate it to dryness. Record the counts in a liquid scintillation counter.
9. Take another aliquot of the extract, evaporate to dryness and extract chlorophyll from the residue with chloroform. Evaporate off the excess chloroform from this preparation and dissolve the residue in water.
10. Apply a suitable aliquot of the above water extract onto Whatman No. 1 filter paper sheet in the form of a spot. Perform descending paper chromatography using n-butanol : propionic acid : water in the ratio of 10:5:7 as the mobile phase (see EXPERIMENT for descending paper chromatography).
11. Remove the chromatogram and dry it at room temperature. Spray the chromatogram for developing colour using the different spraying reagents detailed as follows:

Compound detected	*Spray reagent*
Malate	0.08% (*w/v*) bromocresol green in ethanol
Sucrose	500 mg benzidine + 10 ml glacial acetic acid + 10 ml 40% (*w/v*) TCA and ethanol to make the volume to 100 ml
3-PGA and glucose-6-P	5 ml 60% (*w/v*) $HClO_4$ + 25 ml 4% (*w/v*) ammonium molybdate + 10 ml 1 N HCl + water to make the final volume to 100 ml
Asparatate	0.4% (*w/v*) ninhydrin in 95% (*v/v*) acetone

12. After spraying, dry it at room temperature and then keep it at 80–100°C in an oven for 5 min. Any coloured spots appearing on the chromatogram are due to malate, aspartate and sucrose.
13. For 3-PGA and glucose-6-P, after heating, autoclave the paper for 2 min at 8–10 Psi. The paper becomes blue. Now expose the paper to ammonia vapours. Blue colour of phosphomolybdate due to reaction with these phosphorylated compounds remains while the blue colour of the background would disappear.
14. Identify the compounds by comparing their Rf values with those of authentic standards run alongwith the experimental. Punch the area of paper containing labelled compounds from the developed and sprayed chromatograms and count in scintillation counter. Make sure that necessary corrections for quenching are made for any coloured segments of the chromatogram.

Conclusions

By examining the kinetics of distribution of ^{14}C-label in 3-PGA and C_4 dicarboxylic organic acids, the mode of photosynthetic CO_2 assimilation can be deduced. If at the initial stages, the ^{14}C-label is initially present predominantly in C_4 dicarboxylic acids (oxaloacetate, malate or aspartate) but little in 3-PGA, and this is followed by decrease in ^{14}C-label in C_4 acids accompanied by concomitant increase of ^{14}C in 3-PGA then it is an indication of C_4 pathway of CO_2 fixation. If in contrast, at the early stages the ^{14}C is largely in 3-PGA and ^{14}C-labelled dicarboxylic acids appear at a much later stage then it denotes C_3 mechanism of photosynthetic CO_2 assimilation.

12.9.3 EXPERIMENT: To illustrate the use of ^{14}C as a tracer in microbial metabolism

Principle

This experiment aims at demonstrating application of radioisotopes in investigations on cell metabolism. Microorganisms such as yeast incorporate CO_2 into pyruvate to produce oxaloacetate which is then utilized for biosynthesis of a variety of biomolecules, e.g. amino acids and nitrogenous bases etc. The metabolic fate of the incorporated CO_2 can readily be followed if the CO_2 used is radioactive. The various compounds in the extract are separated, labelled compounds located and identified. This experiment attempts to illustrate utilization of $^{14}CO_2$ for biosynthesis of amino acids in yeast. The cells are fed with $^{14}CO_2$, broken and various compounds in the extract separated by paper chromatography. The amino acids are located on the chromatogram with ninhydrin reagent, presence of radioactivity in these spots determined by a radioactive scanner or autoradiography and the labelled amino acids identified by comparing their R_f values with standard samples of various amino acids.

Materials and Reagents

1. Reaction vessel: This consists of a side flask (25 ml, conical) and a test tube (10 ml capacity) joined to the main flask (125 ml, conical). The mouths of the flasks and the test tube are sealed with a tight fitting rubber cap. The diagram of the reaction vessel is given below:

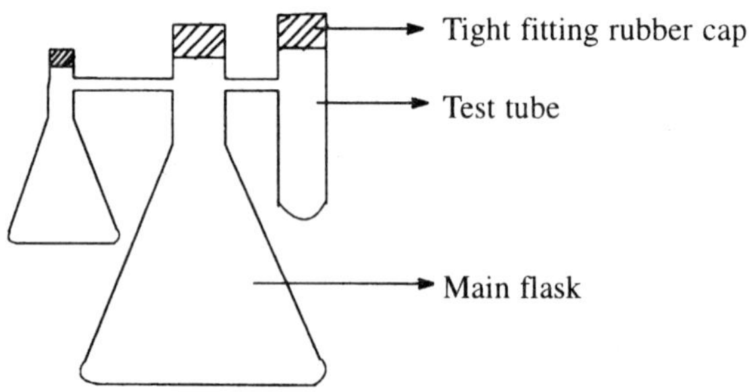

2. Whatman No. 1 filter paper sheets
3. Chamber for paper chromatography
4. Dark room
5. Facilities for development of X-ray films
6. Geiger-Müeller counter
7. Glucose-salt medium: Dissolve 0.2 g $(NH_4)_2HPO_4$, 0.2 g KH_2PO_4, 0.01 g $MgSO_4 \cdot 7H_2O$, 0.05 g yeast extract and 2.0 g glucose in 100 ml water and sterilize by autoclaving at 15 Psi for 30 min
8. ^{14}C-labelled $BaCO_3$ (36 μCi; specific activity 400 dpm/μg)
9. 40% $HClO_4$
10. Chromatographic solvent: Secondary butanol : 3% NH_4OH (3:1)
11. Standard solutions of amino acids: Dissolve various amino acids in 10% *iso*-propanol to give 0.1 μg of α-amino nitrogen per 2.5 μl of solution.
12. Ninhydrin solution: Dissolve 0.625 g of ninhydrin and 2.5 ml of glacial acetic acid in 250 ml of acetone

Procedure

1. Pipette 25 ml of glucose-salt medium into the central flask of the reaction vessel.
2. Add 20 mg of dry commercial yeast into the main flask.
3. Add solid $Ba^{14}CO_3$ containing 36 μCi radioactivity into the side tube. $Ba^{14}CO_3$ should be wetted with water to avoid subsequent blowing during effervescence on addition of the acid.

4. Immediately close mouths of the flasks and the test tube with tight fitting rubber caps.
5. Create partial vacuum in the reaction vessel by connecting tubing of the vacuum pump to a hypodermic needle and pushing it through rubber cap of the main vessel.
6. Inject 1 ml of 40% $HClO_4$ with a syringe through the rubber cap into the side tube to liberate $^{14}CO_2$ into the main flask.
7. Place the reaction vessel on a shaker for 1–1.5 h at room temperature to allow incorporation of $^{14}CO_2$ into various metabolites by the yeast cells.
8. Remove the reaction vessel from the shaker and add 10 ml of 6 N NaOH into the side flask to trap the excess $^{14}CO_2 \cdot NaH^{14}CO_3$ so obtained may be saved for use in another experiment.
9. Inject 10 ml of 6 N HCl into the central flask to inactivate the cells and transfer the contents of the central flask to a centrifuge tube.
10. Centrifuge at 10,000 × *g* for 20 min, discard the supernatant and add 1 ml of 12 N HCl to the sedimented yeast cells. Then quantitatively transfer the suspension into a pyrex test tube (15 ml) and keep the end towards the mouth on a hot flame and by drawing out the glass, seal the tube.
11. Autoclave the sealed tubes at 15 Psi for 6–8 h to hydrolyze the proteins into amino acids. Allow the tube to cool down to room temperature and then break open the sealed end.
12. Transfer the hydrolysate quantitatively to a beaker. Evaporate it to dryness on a steam bath by removing HCl. Repeat this step at least twice and finally redissolve the dried residue in minimum volume of water (0.1 to 0.2 ml).
13. Prepare two sheets of Whatman No. 1 chromatographic paper with coarse serrations at one end of each of these sheets. On the other end of the paper draw a base line, and apply spots of varying amounts of the hydrolysate solution (say 2.5, 5, 7.5 and 10 μl). Simultaneously apply two spots of amino acid standard solutions.
14. After allowing the spots to dry completely, develop the sheets in butanol/ ammonia solvent for about 50 h. The appropriate duration for the development of the paper can be determined by placing a drop of phenol red alongside the hydrolysate spot.
15. Remove the developed sheets from the chromatographic chamber, dry them at room temperature and spray one of the sheets with ninhydrin reagent and save the other sheet for autoradiography or scanning counting. After letting it dry, place the sprayed paper in a warm oven (80°C) to facilitate appearance of violet or purple coloured spots for amino acids.
16. Using a thin end window Geiger-Müeller detector, locate the radioactive spots on the unsprayed chromatogram. From the area of the detector window and ^{14}C-detection efficiency of GM detector end, calculate the exposure time required to produce an adequate image on the film.
17. Working in a dark room, place a 7 in × 7 in sheet of X-ray film in an X-ray film holder. Place chromatogram on it and securely fix it on the film at both ends with small pieces of cellulose tape and then close the holder.

18. After calculated exposure time, develop the film in X-ray developer, wash, fix in X-ray fixer and wash in water for 20 min. Determine the relative amount of radioactive ^{14}C in each spot by comparing the degree of blackening produced on the film. Compare the position of the spots with that of standard amino acids and determine the identity of amino acid at the radioactive spot.
19. Make a table of the amino acids detected on chromatogram of yeast hydrolysate and extent of ^{14}C incorporated in them as judged semi-quantitatively from the autoradiograph.

12.9.4 EXPERIMENT: To study the time course of $^{14}CO_2$ production by rats metabolizing ^{14}C-glucose as a substrate

Principle

Rats kept in a respiratory chamber are fed with labelled glucose. Glucose is catabolized and rate of catabolism is studied by detecting the $^{14}CO_2$ evolved during respiration over an extended period of time.

Materials and Reagents

1. Vibrating-reed electrometer
2. Hormonically-filtered voltage regulating transformer
3. Infra-red CO_2 analyzer
4. Tank containing 0.5% $^{14}CO_2$ in air (specific activity 10 mμCi/L)
5. Pair of strip chart recorders
6. Male albino rat weighing about 350 g
7. ^{14}C labelled glucose with calculated activity
8. Flow system as illustrated in Fig. 12.9

Procedure

1. Assemble the flow system as illustrated in Fig. 12.9. Operate the system as per instruction manual provided by the manufacturer and test it for constancy of air flow and freedom from leakage.
2. Measure the ion chamber background current and adjust the electrometer zero point for the sensitivity range that will cover the anticipated activity levels to be encountered. Adjust the infra red gas analyzer zero point and standardize.
3. Take rat fasted for 24–48 h to reduce the size of endogenous carbohydrate pool.
4. Prepare the glucose dose by mixing calculated amount of ^{14}C-glucose with 1500 mg of unlabelled glucose and dissolving this in sufficient water to yield 50% (*w*/*v*) glucose solution.
5. Administer this glucose solution to the rat by means of a syringe and a stomach tube. Rinse any adhering solution from the tubing into the stomach with a small quantity of water.
6. Weigh the rat and place it in the animal chamber immediately and start

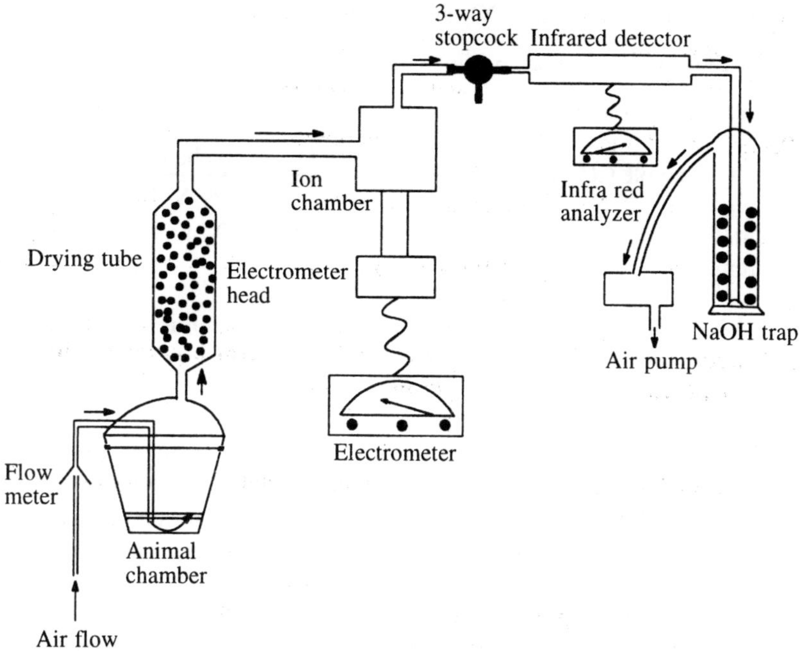

Fig. 12.9 Schematic diagram of continuous $^{14}CO_2/CO_2$ analyzer system.

the flow of air through the system ensuring that a constant flow rate is maintained.

7. Observe the electrometer reading and the meter readings on both the measuring instruments at regular intervals of 10–15 min which are recorded on the chart by the chart recorder.
8. Remove the animal from the flow system at the end of the measurement period of about 6–8 h.
9. Place it in the cage in a fume hood or in a room equipped with an exhaust vent for at least 24 h to prevent $^{14}CO_2$ contamination of the laboratory by the exhaled CO_2.
10. Maintain an air flow through the system until the electrometer reading returns to the initial background level, in order to ensure complete removal of $^{14}CO_2$ and hence prevent the contamination of the apparatus.

Calculations

Record the electrometer readings in milivolts and convert it to $^{14}CO_2$ activity in mμCi as follows:

$$1 \text{ in amperes} = \frac{E \text{ in volts (meter reading)}}{R \text{ in ohms (input resistance)}}$$

$$\text{activity in } \mu\text{Ci} = \frac{1 \text{ in amperes}}{\text{ion chamber constant in amperes}/\mu\text{Ci}}$$

Ion chamber constant in amperes/μCi can be readily determined by the use of the calibration gas mixture ($^{14}CO_2$ in air). Convert activity in μCi to mμCi.

Now convert the recorded CO_2 analyzer meter readings to nmole CO_2.

$$\text{ml } CO_2 \text{ in ion chamber} = \text{Ion chamber vol. in ml} \times \text{Percentage } CO_2$$

Percentage CO_2 in air stream can be determined from meter deflection

$$\text{nmole } CO_2 \text{ in ion chamber} = \frac{\text{ml } CO_2 \text{ in ion chamber}}{22.4 \text{ ml/mmole } CO_2}$$

From the data, calculate specific activity as mμCi $^{14}CO_2$/mmole CO_2 for each measurement and plot against time.

12.9.5 EXPERIMENT: Determination of 2-carboxyarabinitol-1-P in plant leaves by isotope dilution technique

Principle

2-carboxy arabinitol-1-P (CAIP) is a tight binding inhibitor of the enzyme Rubisco that can accumulate to significant levels in chloroplasts of many plant species. Leaf CAIP contents have previously been assayed by measuring the activity of purified Rubisco in the presence or absence of leaf acid extract and calculating the number of inhibited catalytic sites. This measurement of CAIP by Rubisco inhibition gives only approximate values, as there are specific limitations in the use of the inhibition assay to accurately quantify CAIP contents of leaves. Another method involving the use of HPLC has also been developed for the measurement of CAIP, but the method has a relatively low sensitivity to CAIP levels (about I5 nmol), thereby restricting its utility for measuring CAIP contents of leaves. In isotope dilution method described here, the ultimate purification of leaf CAIP ensures assay specificity and allows a considerable concentrating of CAIP from a whole leaf extract. Use of an internal standard [2-^{14}C] CAIP, corrects for any loss of leaf derived CAIP during purification. The isotope dilution assay has a minimum detection threshold of about 0.5 nmol per g fresh weight and an assay sensitivity of ± 0.2 nmol.

Materials and Reagents

1. QAE-Sephadex (Q-25)
2. Dowex 50W (X 8-100)
3. Sephadex G-10
4. Dowex AG1-*x* 8 (200-400 mesh)
5. X-OMAT AR-5 film
6. Scinti Verse II
7. Pyridine
8. Butyl alcohol
9. Chloroform
10. Ribulose 1,5-bisphosphate (Tetrasodium salt)
11. *p*-nitrophenyl phosphate (disodium salt)
12. Malachite Green (Oxalate salt)
13. Cellulose MN 300
14. [^{14}C] NaCN(55.7 Ci mol^{-1})
15. [^{14}C] $NaHCO_3$

16. Acid phosphatase
17. Alkaline phosphatase

Procedure

1. Take freshly harvested leaves and grind them very gently to obtain a coarse leaf macerate. Remove any fibrous unground principal veins.
2. Weigh about 10 g fresh weight of leaf preparation and slowly add to a 400 ml beaker containing 250 ml of 80% (*v*/*v*) boiling ethanol plus about 600,000 dpm of [2-^{14}C) CAIP (specific activity about 1.5 Ci/mol).
3. Boil the contents for another 20 to 30 min until most of ethanol gets evaporated.
4. Centrifuge at 25000 × *g* for 5 min and extract the pellet at 50°C once in 75 ml of 80% ethanol and then three times in 75 ml each of water. Pool the collected supernatants.
5. For partial purification of ^{14}C–CAIP from the above preparation, partition the leaf extract with chloroform and dry the aqueous phase at 47°C in rotary evaporator.
6. Suspend the residue in 25 to 30 ml of H_2O and centrifuge at 25,000 × *g* for 5 min. Pass the supernatant through a 35 ml column of Dowex 50 [H^+] to remove cations and pigments.
7. Concentrate the eluate to about 30 ml and adjust the pH to 10 with 0.1 M KOH and keep the sample at 25°C for 1h to ensure that CAIP is in the free acid form.
8. Load the sample onto a 12.5 ml column (20 × 0.9 cm) of QAE-Sephadex (formate). Elute the sample with 300 ml gradient of 0 to 1.65 N HCOOH (pH 4.0, pyridine) at the flow rate of 12 ml/h. Collect 4 ml fractions. CAIP will be emerged out of the column in eluate volume between 40 to 75 ml which can be monitored by measuring radioactivity of a small aliquot of individual fractions.
9. Pool the fractions containing ^{14}C labelled material and evaporate to dryness in rotary evaporator. Resuspend the residue in about 20 ml water, adjust the pH to 10, dry and resuspend in 5 ml water.
10. Pass through 5 ml column of Dowex 50 (H^+) and concentrate the eluate to about 5 ml and adjust the pH to 10 and let it stand for 1 h.
11. Load the solution onto a 12.5 ml column of Dowex 1 (formate) and elute with 300 ml gradient of 0 to 1.5 N HCOOH (pH 4.0, pyridine), at a flow rate of 12 ml/h and collect fractions of 4 ml each. CAIP will be eluted from the column in the eluate volume between 70 and 90 ml.
12. Pool these radioactive fractions, dry by rotary evaporation twice and resuspend in 300 μl of 25 mM solution of Na_2CO_3 (pH 10) prepared in 10% ethanol.
13. Spot an aliquot containing about 10,000 dpm of the above sample onto a cellulose coated TLC plate and subject the plate to two dimensional thin layer chromatography or thin layer electrophoresis.
14. Locate the spots by autoradiography and remove by scrapping into a

1.5 ml microcentrifuge tube. Resuspend sample in each tube in 0.8 ml of 0.1 M Na_2CO_3 (pH 10.0) and centrifuge at 10,000 × *g* for 5 min.

15. Take 250 μl of the supernatant and assay it for ^{14}C radioactivity. Assay another 225 μl of the aliquot for inorganic phosphate content by using alkaline phosphatase. Each mole of CAIP yields one mole of inorganic phosphate. Calculate sample specific activity (Ci/mole) from the measurements of ^{14}C radioactivity and inorganic phosphate content.

Calculations

nmol CAIP/g fresh wt. = n mol ^{14}C-CAIP added × (Co–C)/C × (1/g fresh wt.)

where,

Co = Specific activity of the added standard.

C = Specific activity of the recovered sample.

12.9.6 EXPERIMENT: To study the incorporation of labelled acetate into leucoplastic fatty acids of developing *Brassica* seeds

Principle

The process of *de novo* fatty acid biosynthesis in all types of plastids is thought to occur by a series of virtually identical enzymatic steps. Starting from acetate, the process involves the enzymes and sequence of repeating reactions well explained invariably in all text books of Biochemistry. Use of labelled acetate in *in vitro* studies on fatty acid biosynthesis in isolated plastids could provide valuable information about the substrate and cofactor requirement of each of the enzymatic reactions involved in the process of fatty acid biosynthesis. The extent of incorporation of radioactivity into fatty acids serves as an indicator of the rate of fatty acid biosynthesis by isolated plastids in presence of various exogenously supplied cofactors and substrates.

Materials and Reagents

1. Developing seeds of *Brassica campestris*.
2. Extraction medium: 50 mM HEPES-NaOH (pH 7.5) containing 0.4 M sorbitol, 0.4 mM EDTA, 1 mM $MgCl_2$, 1 mM DTT, 1% BSA and 1% Ficoll.
3. PBF-Percoll: Prepare a stock solution of Percoll containing PEG-4000 (3% *w/v*), BSA (1% *w/v*) and Ficoll (1% *w/v*).
4. Resuspension buffer: 50 mM HEPES-KOH (pH 7.5) containing 0.4 M sorbitol, 2 mM EDTA and 1 mM $MgCl_2$.
5. Assay buffer: 0.4 M Bis-tris propane buffer (pH 8.5), 0.2 M sorbitol, 8 mM $MgCl_2$, 8 mM ATP, 4 mM $MnCl_2$, 0.8 mM Na (1-^{14}C)-acetate (40.7–51.8 × 10^7 Bq/mmol), 0.6 mM CoA, 20 mM $NaHCO_3$, 1 mM NADH and 1 mM NADPH.
6. KOH (0.65 N).
7. Chloroform : methanol mixture (2:1, *v/v*).
8. KCl (0.5 M).

9. Liquid scintillation mixture: Prepare as described in EXPERIMENT 12.9.2, Reagent No. 8.

Procedure

1. For preparing leucoplasts, hand homogenize 10 g seeds in chilled pestle and mortar using 20 ml of ice cold extraction medium. Filter the homogenate through four layers of cheese cloth and centrifuge the filtrate at 4°C at 500 × *g* for 5 min.
2. Further centrifuge the resulting supernatant in cold at 6000 × *g* for 10 min and resuspend the pellet in 5 ml extraction medium.
3. Layer this preparation onto a discontinuous Percoll gradient which consists of 5 ml of 80% (*v*/*v*) PBF-Percoll, 7.5 ml of 35% PBF-Percoll, 7.5 ml of 22% PBF-Percoll and 7.5 ml of 10% PBF-Percoll, all prepared in the resuspension buffer.
4. Centrifuge in a swing-out rotor at 9200 × *g* for 4 min and collect the band of leucoplasts at 22–35% Percoll interface and dilute it with 20 volumes of the resuspension buffer.
5. Centrifuge at 10,000 × *g* for 10 min and suspend the pellet in 5 ml of resuspension buffer and use it as leucoplast preparation for further investigations.
6. For examining the rate of fatty acid biosynthesis, take appropriate volume of leucoplasts preparation (containing 50–80 μg protein) in 0.2 ml of the assay buffer (Reagent No. 5) containing labelled acetate and cofactors for 2 h at 30°C. Stop the reaction by adding 0.4 ml of 0.65 N KOH.
7. Partition the radiolabelled fatty acids produced by partitioning them into chloroform by adding 750 μl of chloroform/methanol (2:1). Wash the chloroform layer three to four times with 750 μl of 0.5 M KCl.
8. Take the washed chloroform layer into scintillation vial and add 10 ml of scintillation fluid and count using liquid scintillation counter.
9. Calculate the moles of ^{14}C-acetate incorporated from the recorded dpm by applying the value of specific activity of the labelled acetate used for the experiment. Express results in nmol acetate incorporated into fatty acids/mg protein/h.

Suggested further reading

Aronoff S. (1956). Techniques of Radiochemistry, Iowa State College Press, Ames, Iowa.

Baroda E. (1960). Radioactive Isotopes in Biochemistry, Elsevier, Amsterdam.

Chapman J.M. and Avery G. (1981). The Use of Radioisotopes in the Life Sciences. George Allen and Unwin, London.

Chase G.D. and Rabinowitz J.L. (1962). Principles of Radioisotope Methodology, 2nd ed., Burgess Publishing Co., Minneapolis.

Comar C.L. (1955). Radioisotopes in Biology and Agriculture, McGraw-Hill Book Company, New York.

Overman R.T. and Clark H.M. (1960). Radioisotope Techniques, McGraw-Hill Book Company, New York.
Slater R.J. (1991). Radioisotopes in Biology—A practical Approach, IRL Press, Oxford.
Wang C.H. and Willis D.L. (1965). Radiotracer Methodology in Biological Sciences. Prentice-Hall Inc., New Jersey.
Wilson K. and Walker J. (1996). Principles and Techniques of Practical Biochemistry, 4th ed., Cambridge University Press, London.

13. Immunochemical Techniques

13.1 Introduction

Science of immunology originated as a branch of medicine dealing with novel approaches to control infectious diseases of man and animals more than two centuries ago. From the beginning, it was considered an applied science which developed in close association with microbiology following the epoch making discoveries of Edward Jenner and Louis Pasteur. Now immunology has evolved into one of the most challenging and creative branches of modern biology. The understanding of mechanism of immune response and immunological characterization of pathogens has paved the way for development of various approaches to develop effective vaccines against a variety of human and animal diseases. Similarly, detailed knowledge of antigen and antibody interaction under *in vitro* conditions has resulted in the development of a variety of immunological tests used for effective diagnosis of infectious and non-infectious diseases. The principle of antigen and antibody reaction has also been utilised for identification and characterization of microorganisms and their classification into different strains, serotypes and groups, and also for purification and characterization of biomolecules. The very high specificity of antigen-antibody interaction has led to the development of a wide variety of extremely useful and highly sensitive immunochemical techniques. Therefore, the immunochemical tools have become indispensable in clinical microbiology, biomedical, molecular biology and other disciplines of biosciences such as biochemistry, taxonomy, etc. Besides, the commercial potential of immunological assays has also been exploited and a number of commercial kits based on the immunological principles are now available for a variety of investigations.

13.2 Immune System

Resistance of animals to infection by bacteria, viruses, fungi and parasites is called "immunity". In response to natural infection by pathogens or on delibrate inoculation of antigens, the host body immune system gets activated leading to immune response. The immune response against invading pathogens/antigens is of two types: *humoral immune response* characterised by production of immunoglobulins which are called antibodies and *cellular immune response* which is characterised by activation of mainly T lymphocytes. Both types of responses help the individual to fight the pathogenic agents. The activation of immune system does not only cause removal of the antigen but also generates memory. This memory leads to a faster and stronger immune response in the future exposures with the same antigen. Mainly three types of cells (B lymphocytes, T lymphocytes and monocytes) participate in immune response against an antigen.

13.2.1 B lymphocytes (B Cells)

The term B lymphocyte originated from the word 'Bursa' because it was first discovered that antibody secreting cells in birds mature in an organ unique to birds called the 'Bursa of Fabricius'. If the bursa is removed from newly hatched chicks, they are unable to synthesize antibodies. Thus, B lymphocytes originally stood for bursa dependent cells. However, later on it was discovered that in mammals these cells mature in bone marrow (to some extent in foetal liver) and hence B cells now referred to as bone marrow dependent cells. The antigen receptors on B cells are known as antibodies or surface immunoglobulins (SIg) denoting that there are globular proteins on the surface of membrane of an immune cell and have an immunological function. B cells are the only type of cells in an animal body to have surface immunoglobulins. The presence of SIg on cell membrane is, hence, useful in identifying these cells. Each B cell has about 10^5 SIg molecules on its membrane surface. B cells bind to an antigen via the binding sites in the SIg molecules which interact with the complementary epitopes on the antigen.

The antigen binding is an essential requirement for the activation of B cells. For a small population of B cells it is a sufficient activating signal. However, most B cell clones require antigen binding plus additional signals from T cells before they can begin "clonal expansion". The antigens therefore, can be classified as T cell dependent and T cell independent. For example, some bacterial cell walls (*Streptococcus pneumoniae*) contain an antigenic polysaccharide which has a regularly spaced, repeating unit structures and are able to cross link several adjacent molecules of SIg on the surface of B cells. This cross linking is sufficient to trigger B cells to begin the phase of clonal expansion and differentiate into mature antibody secreting plasma cells. B cells may also be activated regardless of their antigenic specificity by certain polyclonal activators. These include powerful mitogens some of which are also components of bacterial cell walls. Another polyclonal activator is Epstein- Barr virus which has been implicated in causing an immune cell cancer (Burkitt's lymphoma). Antigen binding, together with certain activating signals from T cells, initiates the original B cells to proliferate. The expanded clone differentiates into memory cells and plasma cells. The memory B cells play an important role in secondary immune response while the plasma cells synthesize and secrete large amounts of virtually the same type of immunoglobulin molecule which originally participated as surface receptor for interaction with the antigen. The secreted immunoglobulins are antibodies that appear in the blood stream and tissue fluids.

13.2.2 T lymphocytes (T Cells)

This class of lymphocytes is called T lymphocytes because they differentiate and mature in thymus. T cells have three main functions: (i) to kill body cells that have become infected with pathogens and are inaccessible to antibodies, (ii) to maintain an inflammatory response at the site of a persistent infection (for example around the multicellular parasites) and (iii) to regulate many functions of immune system. The T lymphocyte population constitutes of several subsets each with specialized functions. The killing function is carried out by the cytotoxic T cells

(Tc). These cells produce toxic and pore forming molecules to kill the body's own cells which have become infected with intracellular pathogens in much the same way as do "natural killer" (NK) cells. The regulatory functions are carried out by helper T cells (Th), which synthesize and secrete a range of molecules with activating or enhancing effects on other cells involved in the immune response. Suppressor T cells (Ts) have an opposite effect, i.e. they prevent the activation of other immune cells or cause them to reduce their activity or cease functioning altogether. The cells involved in prolonged inflammatory reactions are part of the helper T cell subsets. They are sometimes referred to as a separate group called delayed type hypersensitivity T cells. Cells in each of these subsets have specific marker molecules in their surface membrane that enables them to recognize each other. All these marker molecules are "effectors" for "ligands" on the surface of other cells or for ligands on secreted molecule, and they appear on the T cell surface at a certain stage of cell differentiation. Each of these molecules has been given an internationally agreed CD (cluster of differentiation) number, a term derived from the fact that these molecules are characteristic of a particular type of differentiated lymphocytes.

13.2.3 Monocytes (Macrophages)

Macrophages are the largest of the white cells and have a characteristic horse shoe or kidney shaped nucleus. Less than 7 per cent of the white cells circulating in the blood stream are macrophages. A large number of monocytes are found widely distributed throughout body tissues. Macrophages are long living and highly mobile cells that migrate through connective tissues by squeezing through the intercellular spaces. They are found close to the basement membranes that cover the blood vessels around walls of the gut, in genital tract and abundantly in lungs, lymph nodes and spleen. Specialised macrophages with specific names are found in liver, kidney, brain, bones and in the lymphoid system. From a histological point of view, all mononuclear phagocytes are divided into monocytes of peripheral blood or tissue macrophages (free circulating or fixed macrophages). Circulating and fixed macrophages differ in surface antigens, including the presence of Ia-antigen (an antigen marker which is coded by major histocompatibility gene complex, class II gene, and is located on the surface membrane of antigen presenting macrophages) and in a few other properties. These cells, however, are similar in their ability of phagocytosis and pinocytosis, adhesion to glass or plastic surface, presence of receptors for IgG, IgM monomers as well as for the complement system factors.

Macrophages play an important role in immune responses. They are likely to be the first to capture and digest an antigen in the organism, after which they pass it to the lymphocytes in the form of stronger immunogen as a complex with an Ia-structure that is responsible for interacting with the T cells. Macrophages exert an immunostimulating and immunosuppressing effect when functioning as the effector cells in the immune response to tumors and numerous types of parasites.

13.3 Antigen

An "antigen" (Ag) is defined as any foreign substance which is capable of eliciting an immune response in the host. Lately the term "immunogen" is used more frequently instead of antigen. Macromolecules such as proteins, nucleic acids, carbohydrates, lipids can all act as antigens. The antigenicity of a molecule depends on a number of properties. These alongwith other factors which determine the immunogenic response are described as follows:

13.3.1 Factors affecting immunogenic response to antigen (Ag)

13.3.1.1 Foreignness

Three dimensional structure of the residues in the epitope imparts unique shape and charge profile to the Ag which is rare on the body cells of the responder. Thus, it is only the epitopes which display the property of "foreignness" or *different from self.* Foreignness is a relative term. The epitopes on a bacterial toxin that are recognized as foreign by a mouse may not appear foreign to man or *vice-versa.* So mice and man have antigen receptors of different shapes that bind to different parts of the antigen. Moreover, both the species have different major histocompatibility complex (MHC) molecules (the protein molecules coded by MHC genes which play an important role in the generation of immune response), hence the epitopes that are recognised by the T cells of one individual may differ from those recognised by another.

13.3.1.2 Structure and size of antigen

Most of the large molecular weight substances like proteins and polysaccharides have more than one site capable of provoking an immune response. Such sites on an antigen are called "antigenic determinants" or "epitopes". Protein antigens have been studied more extensively than the other classes of immunogenic macromolecules. There are many B cell clones that recognise non-protein antigens but T cell clones have characteristic receptors for proteins. Hence there is a possibility that non-protein antigens are T cell independent. Certain carbohydrates and lipids can activate a clone of B cell without the help from T cells. Such a response has been ascribed to structural feature of these biomolecules. Many polysaccharide and lipid antigens are polymers with repeating sequences that enable adjacent surface immunoglobulin molecules on B cells to get cross linked by antigens thereby triggering B cell activation. Depending on the number of epitopes on an antigen, the antigen has certain valency. All naturally occurring antigens are multivalent, that is, they have many structurally different epitopes on the same molecule and each of these epitopes may occur more than once (Fig. 13.1).

Epitopes present on the surface and in the interior of an immunogen can be recognised by both B and T cells, though the internal epitopes are mainly recognised by the T cells. Thus it is clear that B and T cells have receptors for different epitopes. Information about approximate size and composition of an antigenic determinant in different types of antigenic molecules is given in Table 13.1.

To be a good antigen, a molecule should have a molecular mass more than 1000 Da. Molecules smaller than this are poor antigens. Haptens are low molecular weight substances which, when bound to a larger carrier molecule, can elicit an immune response.

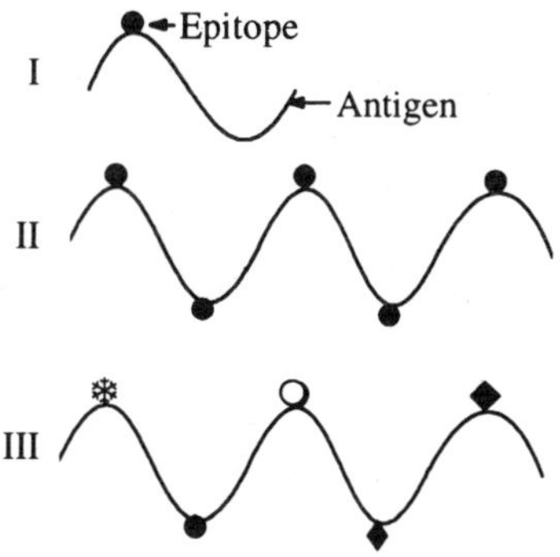

	Description	Example
I	One epitope	Hapten
II	Multiple epitopes of same specificity	Polysaccharides
III	Many epitopes of different specificities	Proteins

Fig. 13.1 Some possible forms of antigenic epitopes. Wavy line represents antigen and symbols denote epitopes.

In case of protein epitopes, some amino acids can contribute to charge and others to hydrophobicity. In addition, some residues might occur more frequently than the others. For example, amino acid residues with charged side chains such as lysine, histidine, aspartic acid etc. and others with nonpolar side chains, such as valine, leucine, isoleucine etc. are quite common in epitopes. An epitope binds to the corresponding antibody or T cell receptor non-covalently through a combination of ionic bond hydrophobic forces and hydrogen bonding. Therefore, charged and hydrophobic residues are essential for antigen binding. A good "lock and key" fit between the conformation and charged profile of the epitope and the corresponding binding sites on B and T cells is important for ensuring effective and strong interaction between antigen and these lymphocytes.

Table 13.1 Size of antigenic determinants

Antigen	*Composition of antigenic determinant*	*Size* Å
Protein	5–7 amino acids residues	9–30
Nucleic acid	5 nucleotides	20
Polysaccharide	6 sugar residues	30
Hapten	one small molecule	5–15

13.3.1.3 Species of the animal

The immune response to the same antigen differs in different species and even strains of animals. For example, some guinea pig strains respond well to diphtheria toxoid while others are poor responders. This response is under the control of antigen specific immune response (Ir genes in the host).

13.3.1.4 Route of immunization

Mostly the routes of immunization are intravenous, subcutaneous, intraperitoneal, intramuscular, intradermal and oral. The route of immunization influences the primary site for antibody production, i.e. spleen or regional lymph nodes and it also effects the degree of immune response.

13.3.1.5 Antigen concentration

Concentration of antigen can also affect the immune response. The immune response can be produced only by a few of the antigen molecules, whereas sometimes even a large concentration of an antigen fails to elicit an immune response but induces a refractory state as immunological tolerance. The effect of antigen concentration varies depending upon the nature of the antigen which is an important factor in vaccine development.

13.3.1.6 Amino acid composition

Immunogenicity of proteins is enhanced by the presence of aromatic amino acids. It was first found that attachment of tyrosine residues to gelatin enhanced its immunogenicity in guinea pigs and rabbits whereas glutamic acid, alanine, proline, serine, lysine and cysteine had no significant effect. This increase in immunogenicity is due to the increased rigidity of the antigenic determinant which in fact is attributed to the rigid planar structure of the side chains of aromatic amino acids.

13.4 Haptens

Haptens are non-immunogenic chemical substances having molecular weight of less than 10 kDa. Haptens alone cannot induce an immune response. However, when coupled to a suitable carrier protein, they produce immune response specific to a hapten which thus, infact, acts as an antigenic determinant on the carrier molecule. The carrier is generally a large molecule and immunogenic. Normally diazonium salts are used to conjugate haptens to amino acids like lysine, histidine and tyrosine residues. The widely used coupling agent is toluene-2,4-diisocyanate. It is now known that antibodies are also produced against carrier molecules which may eventually contribute to the magnitude of immune response to the hapten. For example, animals immunised with 2,4-dinitrophenol (DNP) conjugated to bovine gammaglobulin (BGG) produce high level of anti-DNP antibodies when inoculated a second time with DNP-BGG but respond only weakly when inoculated with DNP coupled to KLH (keyhole limpet haemocyanin, a protein from *Megathura crenulata*), indicating that the carrier influences the intensity of immune response to that antigen. This effect is known as "carrier effect" and involves specific amino acid residues.

13.5 Adjuvants

Adjuvants are the substances which have the ability to enhance as well as prolong the immune response against any immunogen when simultaneously injected into a mammalian host. The commonly used adjuvants are Freund's complete and incomplete adjuvants. Freund's adjuvant having water in oil emulsion (containing lanolin or aracel A as emulsifying agent) is known as incomplete adjuvant. The Fruend's incomplete adjuvant containing killed *Mycobacteria* in addition, is called Fruend's complete adjuvant. The mechanism of adjuvant role in immunopotentiation is quite complex. A number of different mechanisms which have been ascribed to immunopotentiation include:

(*a*) *Depot effect*: It has been observed that adjuvants maintain a depot of the antigen with which they are injected and release it slowly to the immune system. Antigens injected without adjuvant disperse rapidly in the body, get catabolised and removed from the system before its detection by immune system and thus their degree of stimulation is comparatively poor. Adjuvants counteract the fast degradation of antigens and provide long lived reservoir of antigen. The most common adjuvants of this type used in man are aluminium compounds (phosphate and hydroxide) and Freund's incomplete adjuvant. Both types enhance antibody response probably by maintaining the antigen for a longer time.

(*b*) *Accumulation of immunoreactive cells at the site of inoculation:* Certain adjuvants have been noted to induce hypercellularity and enlargement of draining lymphnodes resulting in the accumulation of immunocompetent cells at the site of inoculation. Thus the adjuvants stimulate lymphocyte trapping and facilitate optimal exposure of antigen to immune system cells.

(*c*) *Activation of macrophages*: Macrophages form granuloma which provides site for interaction with B lymphocytes. Virtually all adjuvants activate macrophages. They are believed to act by improving the immunogenicity probably through an increase in amount of antigen on their surface and efficiency of its presentation to lymphocytes. They also enhance immune response by secreting lymphocyte activating factor (interleukin 1).

(*d*) *Specific effects on lymphocytes:* The immunopotentiating effects of mycobacterial components in complete Freund's adjuvant are well known. All types of lymphocytes are known to be stimulated by adjuvants.

13.6 Antibodies (Immunoglobulins)

Antibodies are heterogeneous group of globulins mainly found in the serum. Since the antibodies are proteins, they can also act as antigens when inoculated into a different mammalian species. Antibodies of all the classes are gammaglobulins and are collectively called immunoglobulins (Igs). There are five types of Igs. The classification of the Igs is based on the type of "heavy chain" present in the antibody. Thus, immunoglobulins containing heavy chains γ, μ, α, δ and ε are called immunoglobulin G (IgG), immunoglobulin M (IgM), immunoglobulin A (IgA), immunoglobulin D (IgD) and immunoglobulin E (IgE), respectively. Main characters of all the five classes of the immunoglobulins are presented in Table 13.2. In addition, several subclasses and inherited variants or allotypes are also known. When an individual is immunised against a single antigen, the specific antibodies are produced which usually consist of a mixture of IgG, IgA and IgM. The amount of each Ig produced depends on the nature of the antigen and the stage of immunity.

13.6.1 Structure of immunoglobulins

All classes of immunoglobulins have the same basic architecture which consists of two heavy (H_2) and two light (L_2) chains that are linked together by disulfide bridges to form 'Y' shaped structure (Fig. 13.2). Each L chain is bonded to an

Table 13.2 Important features of immunoglobulin isotypes

	IgG	*IgM*	*IgA*	*IgE*	*IgD*
Heavy chain	γ	μ	α	ε	δ
Light chain	k/λ	k/λ	k/λ	k/λ	k/λ
MW (kDa)	150	900	160	200	185
Percent of total immunoglobulins	80	6	13	0.002	0.2
Valency for antigen binding	2	5 or 10	2	2	?
Half life (days)	20–21	5	6	2.3–4	3
Complement fixation	+	+	+	–	–
Transport across placenta	+	–	–	–	–
Attachment to monocytes, polymorph-nuclear cells and NK cells	+	–	–	–	–
Presence in milk	+	0-traces	+	–	–
Presence in secretion	–	–	+	–	–
Approximate concent-ration (mg/ml)	12	1	1.8	0.0–0.00002	0.0–0.4

H chain by a disulfide bond and the H chains too are linked to each other by at least one disulfide bond. The structure of IgG, the major antibody in serum, is well established.

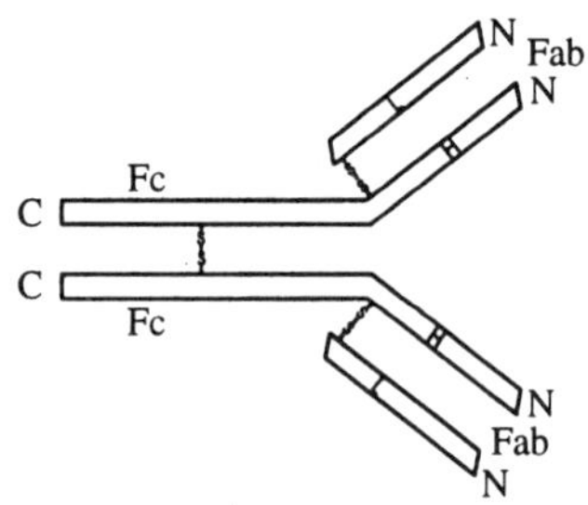

Fig. 13.2 Structure of immunoglobulin.

A fruitful approach to study such a large protein molecule is to split it into fragments that retain activity. In 1959, Rodney Porter showed that IgG can be cleaved into three active 50 kDa fragments by a limited proteolytic action of papain.

$$\underset{(150\text{ kDa})}{\text{IgG}} \xrightarrow{\text{Papain}} \underset{(50\text{ kDa each})}{\text{2Fab}} + \underset{(50\text{ kDa})}{\text{Fc}}$$

Papain cleaves H chain on the N terminal side of the disulfide bond that links L and H chains. Thus Fab (antigen binding fragment) consists of the entire L chain and amino terminal half of H chain whereas Fc (fraction crystallizable) consists of C-terminal halves of both H chains (Fig. 13.3). Each Fab contains combining site for hapten/antigen as does the whole molecule while the other fragment (Fc) does not bind antigen but it has other important biological activities. Fc mediates protective responses termed as effector functions such as complement fixation.

The structure of immunoglobulin has several unique features essential for their participation in immunological reactions. It has been noted that papain digestion of immunoglobulin molecule results in cleavage N-terminally to disulfide bridge between the heavy chains at the hinge region yielding two monovalent Fab fragments and a Fc fragment. On the other hand, pepsin digestion results in cleavage C-terminally to the disulfide bridge, resulting in divalent fragments referred to as F(ab′)$_2$, consisting of two Fab fragments joined by a disulfide bond and several Fc fragments (Fig. 13.3). The antigen binding capacity is located in the hypervariable region of light and heavy chains while the constant region of the heavy chain mediates complement fixation and other biological functions.

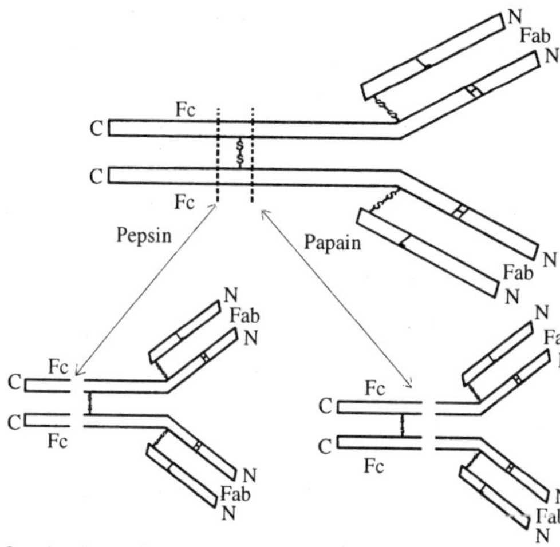

Fig. 13.3 Action sites and cleavage products after treatment of immunoglobulins with pepsin and papain.

13.6.2 Different types of immunoglobulins and their properties

The salient characteristics of main types of immunoglobulins are briefly described as follows:

13.6.2.1 IgM

It is called IgM as it contains μ type heavy chains. This antibody is synthesized during primary immune response and secreted into body secretions. IgM occurs both as monomers and pentamers. The latter is found in the serum while monomeric form is located on the surface of B lymphocytes and serves as an antigen receptor. The pentameric form is polymer of five monomers connected through their Fc (constant region) regions by disulphide bonds and a molecule of J (joining) chain. The normal valency of pentamer IgM is 10. Because of the high valency of the molecule, it forms many cross links between different molecules of the same antigen which enables it to neutralize the antigen more effectively and participate in the process of agglutination.

13.6.2.2 IgG

It is referred to as IgG because it contains γ type of heavy chains. Its molecular weight is 150 kDa and is the most abundant class of immunoglobulins and may be present as high as 8–16 g/L. Each molecule has two binding sites for the antigen. It diffuses into all tissue fluids and secretions except cerebrospinal fluid and is very effective in neutralizing soluble antigens like exotoxins, viruses and other proteins. This antibody is known to cross placental barriers from mother to foetus.

13.6.2.3 IgA

This class of antibody contains α type of heavy chains, hence is known as IgA. Predominantly it is a secretory type of antibody and hence its main function is to defend exposed surfaces of body particularly against bacteria and fungi. It is normally found in watery and mucous secretions of the body such as oral cavity, gut, nasal, lungs, urino-genital and eye secretions. Plasma cells found close to mucous membranes secrete dimers of this antibody which consists of two IgA monomers joined by a peptide of about 15 kDa called "J chain". There is another component associated with the dimer, called "secretory component". The IgA dimer has a valency of 4 and is thus more efficient in aggregating the antigen as compared to IgG but less as compared to IgM.

13.6.2.4 IgD

This type of antibody contains δ type of heavy chains and hence is designated as IgD. This antibody, alongwith surface IgM, has been found to be attached to B lymphocytes at certain stages of their development. Therefore, quite likely that this antibody is involved in antigen recognition process. It has also been hypothesized that it may be responsible for maintaining B cell tolerance to self antigens. It is highly susceptible to the attack by proteolytic enzymes and hence has a short half life in serum. This antibody is present in monomeric form and its structure is similar to IgG.

13.6.2.5 IgE

It is called IgE because it contains ε type of heavy chains. This type of antibody is normally bound on the surface of basophils and mast cells. Therefore, it is localised in the skin, lungs and mucous membranes of the body. IgE class of antibody is the most important in causing allergic reactions to non-infectious agents called "allergens" that include pollen, house dust, mites, food proteins and a variety of chemicals which may come in contact with the skin. It is also a monomer and has a valency of 2. This type of antibody is also known as reagenic antibody due to its role in allergic reactions. It is also similar to IgG in structure.

13.7 Affinity and Avidity

The binding force (a combination of ionic, hydrophobic and hydrogen bonding) between the epitope and the binding site on antibody is known as "affinity". The combination of antibody with the antigenic determinant or a hapten molecule is reversible and the complex may readily dissociate depending upon the strength of the binding. If the antigenic determinant and the antibody fit together strongly, the antibody is called as "high affinity" antibody. Analysis of serum for different antibodies indicates that it contains heterogeneous mixture of antibodies with different affinities.

The functional combining strength of an antibody molecule with its antigen is called "antibody avidity". It is related to both affinity of the reaction between the epitopes and paratopes and the valencies of the antibody and antigen. In most of the practical situations, the interaction of antiserum with antigens having multivalent antigenic determinants is quite complex and the term employed for multiple binding is called "antiserum avidity". The factors which contribute to avidity are complicated because a single molecule may have several immunodominant regions, thus creating several antigenic determinants. The multivalent nature of most antigens contribute to avidity in the antigen- antibody interaction.

13.8 Immune response

Both B and T cells are made up of clones of cells with receptor for a particular "epitope" (antibody or T cell receptor binding site on antigen) and each undergo clonal selection, expansion and memory cell formation when activated by appropriate signals including contact with a specific antigen. When an antigen is introduced in the body, antigen specific B lymphocytes and T lymphocytes are stimulated by the clonal selection process. It implies that out of millions of T and B lymphocytes present in the body, only those lymphocytes which are specific to the antigen will be stimulated to mount immune response. After specific stimulation of T and B cell clones by the antigen, proliferation of the clones takes place. This process is called "clonal amplification" or "clonal expansion". Stimulation of lymphocytes by an antigen leads to generation of memory B and T lymphocytes. The main function of memory cells is to quickly recognise the same antigen and elicit strong immune response in future encounter with the same antigen.

The common outcome of any antigen inoculation into an animal is generation of humoral and cellular immune responses. The humoral immune response is mediated by B cells in which the activated cells (plasma cells) produce antibodies specific to the antigen while cellular immune response is mediated by T cells. Since immunochemical techniques used are essentially based on antigen-antibody interaction the present discussion will be confined here with antibody response only. In response to the first inoculation of an animal with an antigen, primary antibody response is generated which is initially dominated by IgM antibody followed by IgG. The primary response generates memory B cells and hence when the same antigen is again inoculated into the animal, antibody response is much quicker and dominated by IgG antibodies. The secondary antibody response is also called anamnestic immune response which essentially is generated by the presence of memory B cells (Fig. 13.4).

13.9 Production of Antibodies

13.9.1 Polyclonal antibodies

All antisera are generally polyclonal because they contain antibodies against

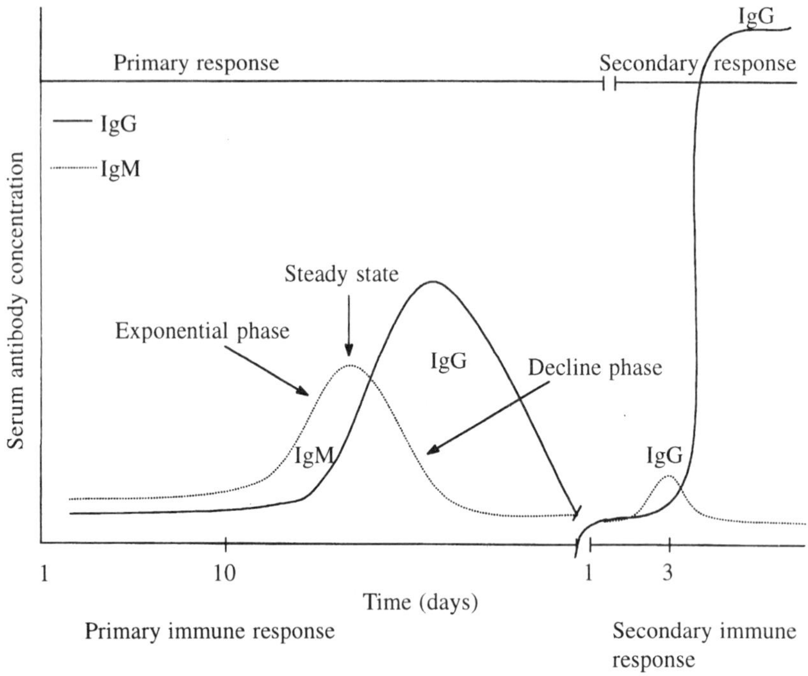

Fig. 13.4 Kinetics of antibody response.

more than one antigenic determinant and are being produced by different clones of plasma cells in the body of animals. Even very small molecules have several epitopes against which specific antibodies are generated. Antiserum can be prepared against an antigen or a hapten by inoculating an appropriate amount of antigen into a suitable animal or in man. The selection of an animal species is based on the amount of antiserum required. In principle, any animal can be used for antisera production. If large quantity of antiserum is required, large animals such as horse/donkey, sheep and goat are generally used. However, for a small requirement, laboratory animals such as rabbits, guinea pigs, rats, mice etc. are used. The outbred animals generally exhibit great variations in their immune response to a given antigen. Hence, 4-5 animals should be used for immunization and the animal giving highest titre should be selected for production of serum. In order to achieve high titre anti-serum, a number of factors should be kept in mind. These include:

(i) The animal used for immunization should be phylogenetically distant from the source of antigen used for antiserum production.

(ii) If the antiserum is to be raised against a hapten, the hapten molecule must be coupled to a carrier protein such as bovine serum albumin (BSA), tetanus toxoid, thyroglobulin or to RBC, bacteria etc.

(iii) Form and dose of antigen also affects the production of antiserum. Antigens can be of innumerable shapes and sizes as described earlier. Particulate antigens such as bacteria, yeast, normal mammalian cells, tumour cells, sub cellular components etc. are excellent immunogens and evoke very good immune response when inoculated without any adjuvant in a suitable animal. Soluble antigens such as proteins, carbohydrates, nucleic acids etc., produce poor immune response if inoculated directly without any adjuvant. To overcome this problem, the primary dose of the soluble antigens should be given with suitable adjuvant via a route which retains the antigen for sufficiently long time to activate the immune system cells.

(iv) The route of immunization also influences the immune response. The most commonly used routes for production of antiserum against soluble antigens are subcutaneous (s/c) and intramuscular (i/m) but some times intradermal (i/d) inoculations are also given. However, for production of antiserum against particulate antigens, intravenous (i/v) or intraperitoneal (i/p) routes are preferred.

(v) The dose of antigen against which antiserum is to be raised also affects the quality of serum. The dose of the antigen varies according to the type of antigen, the species of animal and the route of immunization. The protein antigens can be given in dose ranging from 10 μg to 5 mg/animal without inducing high or low dose tolerance. For raising antibodies against cell surface antigens, 1–100 million cells can be used per animal for immunization.

13.9.2 Monoclonal antibodies

Most of the protein antigens contain several antigenic determinants. When such antigens are introduced into animals and human beings, they produce antibodies against each of the antigenic determinants. Serum obtained from such animals will contain a mixture of antibodies with specificity to individual immuno-dominant domains. Such sera are called "polyclonal sera" because these are not only heterogenous with respect to antigenic specificity but also with respect to the heavy chain isotype, light chain type, allotype, variable regions sequence and idiotype. It was a long cherished dream of biomedical scientists to isolate a single antibody producing plasma cell and cultivate it *in vitro* to provide a continuous source of a homologous antibody produced by a single clone that would bind to a single antigenic determinant. Unfortunately till 1975, normal antibody producing B lymphocytes could not be propagated *in vitro* for an extended duration. Georges Kohler and Caesar Milstein overcame this difficulty by fusing antibody producing spleen B cells to a myeloma cell (cell line that originated from malignant B lymphocytes and can be easily propagated *in vitro*) in the presence of a fusogen (Sendai virus or polyethylene glycol). Such fused cells are called hybridomas. The hybridomas have the capability of antibody production of a normal spleen B cell and the *in vitro* growing capacity of the myeloma cells. The diagramatic representation of different basic steps involved in monoclonal antibody production is given in Fig. 13.5.

The biochemical strategy adopted for production of monoclonal antibodies in brief is as follows: Mammalian cells synthesize nucleotides both by *de novo* and/or salvage pathways. The *de novo* pathway requires ribose 5-phosphate, some amino acids, carbon dioxide and ammonia for the synthesis of purine and pyrimidine nucleotides. In addition, tetrahydrofolate coenzymes, which are regenerated from

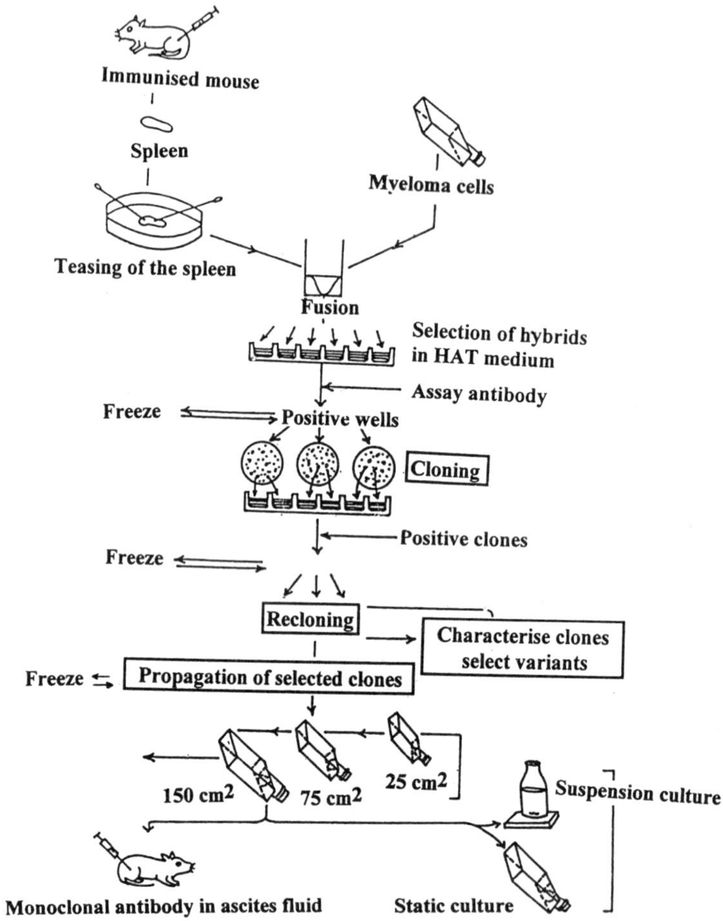

Fig. 13.5 Scheme showing different steps in the monoclonal antibody production.

dihydrofolate by dihydrofolate reductase, are also required in the *de novo* pathway. Aminopterin, an antagonist of folic acid, inhibits the dihydrofolate reductase and thereby blocks the *de novo* pathway. The salvage pathway, in which preformed purine and pyrimidine bases and their corresponding deoxyribonucleosides are recycled, depends on the enzymes hypoxanthine-guanine-phophoribosyl transferase (HGPRT) and thymidine kinase (TK). Thus, if a cell is grown in the medium containing hypoxanthine, aminopterin and thymidine (HAT medium), nucleic acid synthesis can proceed, provided both the enzymes are present. In the absence of one of the above enzymes, the cell is unable to produce nucleotides and hence dies in HAT medium. To select hybridoma, a myeloma cell line is used which lacks either HGPRT or TK. Only hybrids between myeloma cells and spleen cells can grow in the HAT medium since spleen cells contribute the HGPRT and TK enzymes and the myeloma cells possess the ability to grow in culture. Under these cultural conditions the spleen cells by themselves will die due to their inability to grow *in vitro* while myeloma cells will perish in HAT medium because they lack one of the enzymes of the salvage pathway. Myeloma cells deficient in HGPRT enzyme are produced by exposing the cells to a mutagenic agent followed by 3-azaguanine or 5-thioguanine selection. These chemicals interfere with nucleic acid synthesis mediated by HGPRT and the only cells lacking HGPRT enzyme can survive in presence of these compounds. Mutants of myeloma cells lacking TK can be identified in a similar way by selection with bromodeoxyuridine which is phosphorylated by TK enzyme and incorporated into DNA resulting in death of the cell.

13.10 Immunological Methods Based on Antigen-Antibody Reactions

13.10.1 Precipitation reaction based assays

Combination of molecules of soluble antigen with specific antibodies results into lattice formation and when this complex attains a certain size, it becomes insoluble and precipitate out in form of a visible band at the site of reaction. The phenomenon of precipitation is called precipitin reaction. Figure 13.6 depicts the qualitative precipitin reaction. When increasing concentration of the antigen are added to a series of tubes that contain a constant amount of the antibodies, varying quantity of precipitate is formed. If the amount of precipitate is plotted against the amount of antigen added, a precipitin curve as shown in Fig. 13.7 is obtained. There are three important regions in the curve: (a) the zone of antibody excess; (b) equivalence zone; and (c) zone of antigen excess. In the equivalence zone, the proportion of antigen to antibody is optimal for maximal precipitation while in the antibody excess or the antigen excess zone, the proportion of the reactants do not lead to efficient cross linking and formation of precipitate. Several tests have been developed based on the principle of precipitation.

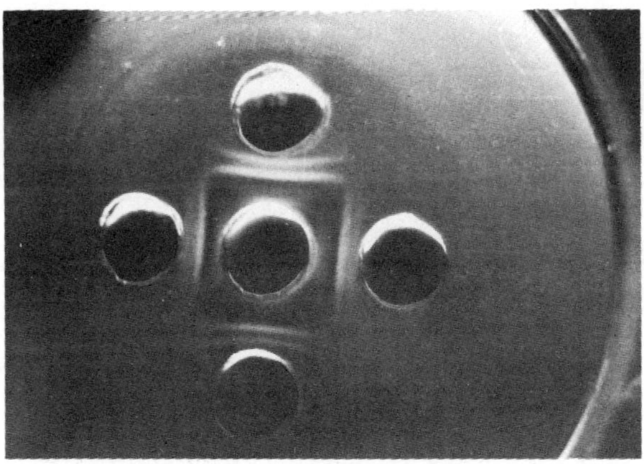

Fig. 13.6 Showing qualitative precipitation reaction in agar.

13.10.1.1 Tube mixture precipitation and ring test

In this test, when an optimal amount of soluble antigen is mixed with antibody, precipitation occurs in the tube, indicating the specific reaction. An excess of the

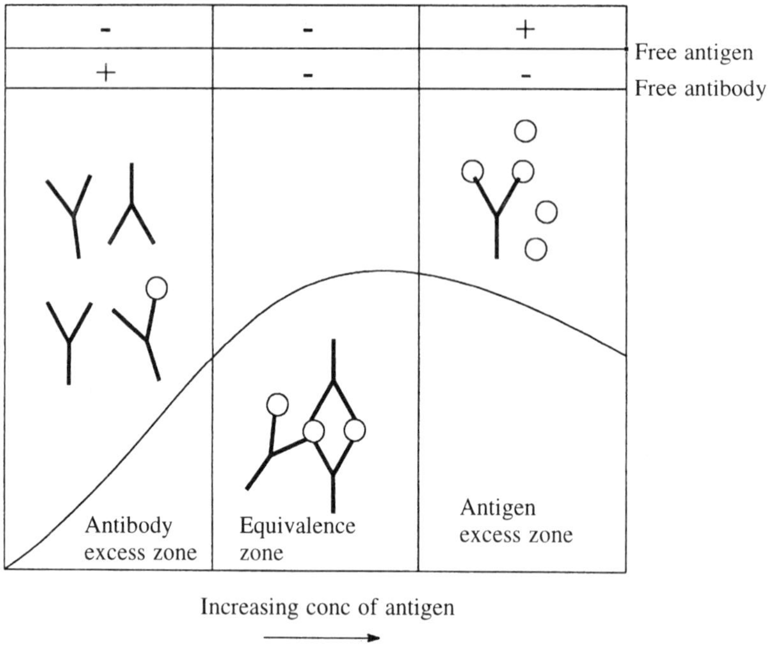

Fig. 13.7 Illustration of antigen-antibody reaction *in vitro* leading to precipitation.

antigen or antibody may inhibit the reaction. This effect is called "prozone phenomenon". An excess of either antigen or antibody inhibits the formation of lattice, which is responsible for precipitation due to steric hinderances. In a slight modification of this method, a ring is formed at the junction when the antigen is layered on the solution containing homologous antibody in a tube.

13.10.1.2 Agar gel precipitation test

This test is performed in agar medium in which wells are cut to fill the antigen and antibody solutions. Then the antigen and antibody are allowed to diffuse towards each other and a thin line of precipitation is formed at the junction of the two. Figure 13.8(a) indicates presence of antibodies against single antigen while Fig. 13.8(b) shows the presence of three types of antibodies and three types of antigens.

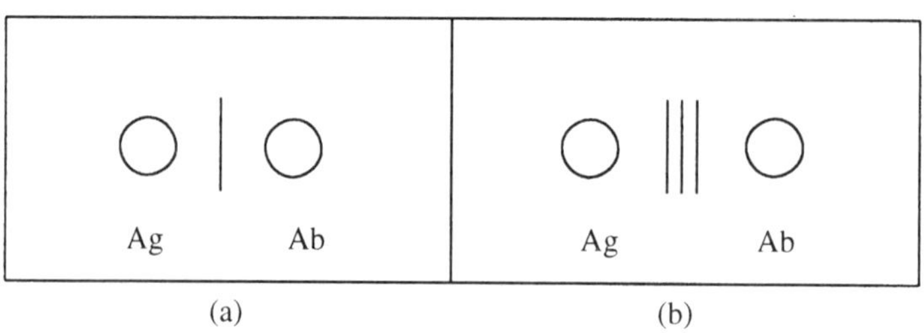

Fig. 13.8 Precipitation lines between single antigen and antiserum (a) and between mixture of multiple antigen and polyvalent antiserum (b).

If the lines of precipitation are fused to form a semi-ring like band, then the antigen and antibody are homologous (Fig. 13.9a). Alternatively when the lines originate at different angles and cross each other then the antigen and antibody are considered to be heterologous (Fig. 13.9b) while if there is formation of spur, the antigen and antibody are partially identical (Fig. 13.9c).

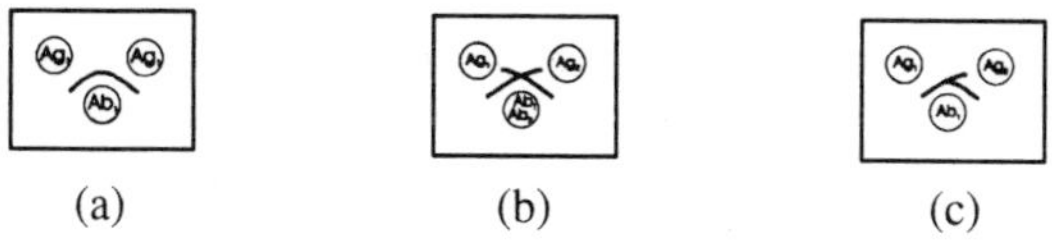

Fig. 13.9 Antigen-antibody reaction leading to precipitation lines. (a) Line of identity; (b) Line of no identity; (c) Line of partial identity.

13.10.1.3 Radial Immunodiffusion

This is similar to the agar gel precipitation test and the only difference is that the antigen solution is mixed in the gel and wells are cut in the same gel and filled with the varying amounts of antiserum. A precipitation ring is formed around the wells (Fig. 13.10). Diameter of the ring is directly proportionational to the concentration of the antibodies. This test is based on homologous antigen-antibody reaction. The test may also be performed in a reverse manner by mixing antibody solution in agar gel and serial dilutions of antigen in the wells.

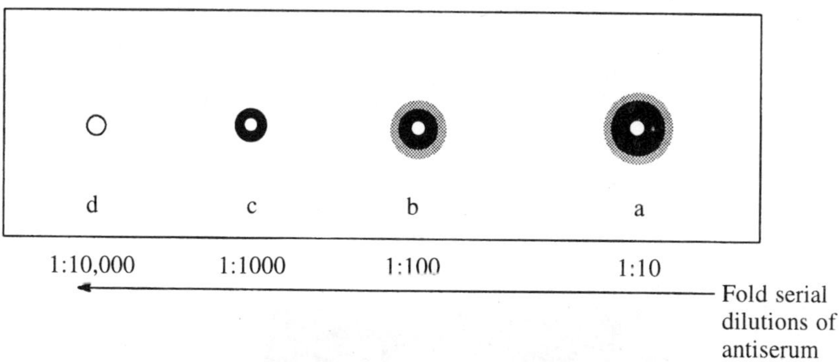

Fig. 13.10 Illustration of radial immunodiffusion test showing precipitation rings of different sizes around the wells filled with different dilutions of antibodies.

13.10.1.4 Immunoelectrophoresis

This technique combines gel diffusion precipitation and electrophoresis. A glass slide is coated with agar or agarose onto which a trough is cut in the centre and one well on each side equidistance from the trough. The wells are filled with samples of antigen. Electric current is passed to separate the components of antigen. After electrophoresis, the trough is filled with antiserum and diffusion is allowed to take place. Based on the composition of the antigen samples, lines of precipitation are formed (Fig. 13.11).

13.10.1.5 Counter current immunoelectrophoresis (CCIE)

This method is more sensitive and quicker as compared to simple agar gel

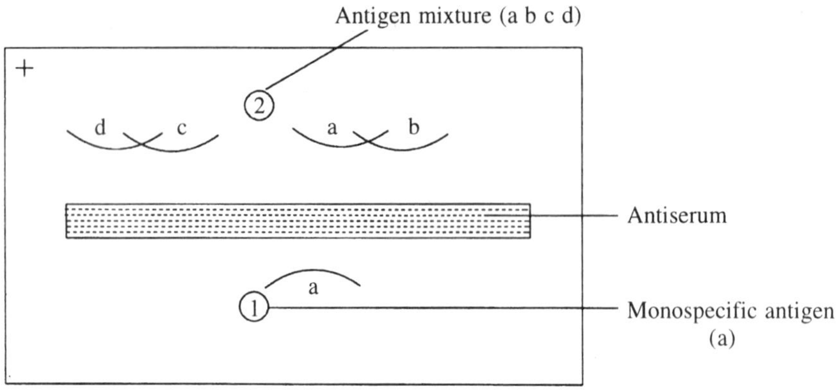

Fig. 13.11 Diagram showing types of precipitation lines obtained during immuno-electrophoresis. Well No. 1 contains monospecific antigen (a) whereas Well No. 2 contains a mixture of antigens a, b, c and d.

precipitation test. In this technique electric current is applied to move the antigen towards the antibody to form the precipitation band (Fig. 13.12). The CCIE makes use of endoosmotic flow, which during electrophoresis causes the slow moving immunoglobulins to be displaced toward the cathode, while antigen moves in opposite direction towards the anode. By cutting the wells at appropriate locations, the antigen and antibody can thus be made to converge, forming a precipitate between the two wells. The precipitation lines develop rapidly and usually reach maximum intensity within 30-90 min depending on concentration of antigen and antibody. The CCIE is different from the simple immuno-electrophoresis because of the fact that in CCIE, both antigen and antibodies are added in opposite wells before electrophoresis and the test is completed within 30-90 min while in immunoelectrophoresis antiserum is added in the trough after electrophoresis of the antigen and simple diffusion is allowed to develop precipitation lines.

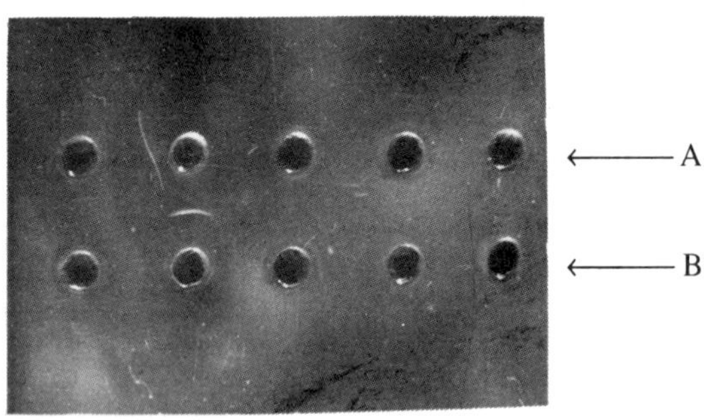

Fig. 13.12 Counter current immunoelectrophoresis. Wells of row A contain antisera while the antigen samples are added in the wells of row B.

Applications of precipitation based tests

Different variations of the precipitation test have been used for detection of antigens of bacteria, viruses and parasites. These tests are also suitable for detection of serum antibodies to identify antigens of a variety of infectious pathogens. For example, the gel diffusion test is used to establish presence of the diphtheria toxin and herpes virus in vesicle fluid. For rapid detection of bacterial antigens in clinical specimens such as sputum and cerebrospinal fluid, countercurrent immunoelectrophoresis test is frequently used.

Based on knowledge of specific reaction between antibody and antigen, a variety of methods have been developed. Conventionally, the source of antibody is usually the serum, and hence, the techniques involving antigen and antibody reaction are called serological methods. Serological methods have been very useful in identification and quantitative determination of enzymes from various sources and in examining their immunological relatedness. Theoretically all types of proteins, irrespective of source, can be identified by immunodiffusion assays if specific antibodies for them are available.

13.10.2 Agglutination based tests

When a particulate antigen combines with a specific antibody, it results in agglutination or formation of clumps of the reactants. These tests can be performed either on a slide or in a test tube.

13.10.2.1 Slide agglutination test

In this method, red blood cells or bacterial cells, which may be live or dead, are suspended in a drop of saline on a glass slide to which an equal amount of antiserum (*v/v*) is added. The contents are mixed by rocking the slide and the agglutination reaction is noted. For example, human blood group testing is done by this method. Standard antiserum against any bacteria or cells can be prepared in any suitable animal host using standard protocols for antiserum production.

13.10.2.2 Tube agglutination test

This is a quantitative test for determining antibody titre. A fixed amount of particulate antigen is taken in tubes to which serially diluted antisera are added and these are then incubated for an appropriate period of time. The highest dilution of the antiserum at which agglutination occurs is known as "titre" of that antiserum.

13.10.2.3 Latex particle agglutination test

In principle, this is similar to the indirect haemagglutination test. In place of erythrocytes, polystyrene latex particles are coated with either the antigen or the antibody. For example, latex agglutination test kits are commercially available for grouping Streptococci, detecting bacterial antigens in cerebrospinal fluid, hepatitis B surface antigens, rheumatoid factor etc. Advantages of this test over haemagglutination test include simplicity, rapidity and sensitivity. Moreover, latex particles are commercially available while RBCs used in haemagglutination test are to be collected from animals.

13.10.2.4 Coombs test

This test is useful when the incomplete antibodies which bind to the surface of cells fail to cause agglutination. This problem can be overcome by washing the cells coated with incomplete antibodies which are then resuspended in saline and allowed to react with specific antiglobulin serum. For example, in infections caused by *Salmonellae* and *Brucellae*, the antiglobulin test or Coombs test may yield positive result when ordinary agglutination tests are negative.

13.10.2.5 Indirect haemagglutination test

This is a more sensitive test and is used for detection of viral and bacterial antigens which are not easily adsorbed onto erythrocytes. To facilitate adsorption of such antigens, the erythrocytes are first treated with tannic acid. The cells derived after this treatment adsorb the antigens readily and are agglutinable by antiserum of particular antigen.

13.10.2.6 Haemagglutination-inhibition test

A variety of viruses are known to cause agglutination of RBC of different species of animals. The agglutination can be inhibited by using antiviral antibody. This test has been used for the identification of viruses and determination of titre of virus specific antibodies.

Application of agglutination based tests

Agglutination assays are used for the identification of species of *Salmonella* and *Shigella* and serological types of *Haemophilus, Neisseria* and *Streptococcus* by using specific diagnostic sera. These tests are also being employed to diagnose typhoid (Widal test), brucellosis, leptospirosis and rickettsial infections. The utility of agglutination assays is, however, mainly restricted to particulate antigens and these are not very effective for investigations with soluble antigens.

13.10.3 Complement fixation test

Complement system (a group of normal serum proteins) has ability to get activated in presence of antigen-antibody complexes and cause lysis of cells. The complement gets fixed and is not detectable in free form. Based on this property, complement fixation test has been developed. The test involves two stages: (a) antigen and antibody reaction (test system); (b) haemolysis (haemolytic system). In the test system, antigen and antibody are allowed to react in presence of limited amount of complement and then in haemolytic system it is determined whether complement has been fixed or not. Thus when antigen, antibody and complement are incubated and allowed to react, if the antigen encounters a specific antibody, they combine and the complement gets fixed but if the specific antibody is not present in the test sample, the reaction does not take place and the complement remains free. The presence or absence of specific antibody is tested by the haemolytic system using anti-sheep red blood cell serum sensitised sheep RBC. The sensitised RBC suspension is mixed with the test reaction mixture and incubated. In case the complement is fixed, it is not available for haemolysis, and the RBC are not lysed thereby indicating positive test. But if the antigen has not combined, the complement remains free and haemolysis of RBC takes place and the suspension clears with the release of haemoglobin and this is suggestive of negative result. Complement fixation test is quantitative as well as qualitative for detection of antibodies and /or antigen. It is used in the diagnosis of rickettsiae, chlamydiae, mycoplasmas, viruses etc.

13.10.4 Radioimmunoassay (RIA)

Radioimmunoassay (RIA) was developed by Berson and Yalow in 1959. The development of radioimmunoassay unleashed an enthusiastic application of this new assay particularly in the field of biomedical research and diagnosis of human and animal diseases. The technique enables measurement of picogram (10^{-12} g) of an antigen with relative ease, precision and rapidity by using radioactively labelled antigen. This method is highly sensitive and can thus be employed for accurate micro determination of specific molecules in unfractionated mixtures. For example, the sensitivity of RIA is 1000 times greater than that of gel diffusion, 500 times than that of cross over electrophoresis and 250 times than compliment fixation technique. This test for antibodies against influenza and mumps virus shows a close positive correlation with haemagglutination inhibition or neutralisation tests.

The radioimmunoassay involves reaction of a specific antibody with a mixture of radiolabelled and unlabelled antigen. Hence a kind of competition occurs between the labelled and unlabelled antigen to bind with the antibody. At the end of the reaction, unbound and antibody-bound antigen can be separated and radioactivity can be quantitatively measured by using suitable standards. In this way the extent of inhibition of the binding of radiolabelled antigen by the unlabelled antigen can be estimated. Being a very sensitive test it can be used for quantitation of a particular protein, such as an enzyme, immunoglobulin, viral protein, hormones, antibiotics, steroids, digoxin, morphine, drugs etc.

Reagents labelled with ^{125}I or ^{131}I are most commonly used in the RIA. Both isotopes emit high energy gamma radiation that can be detected by scintillation spectrophotometer equipped with a crystal detector. The half life of ^{131}I (8.05 days) is shorter than the half life of ^{125}I (60 days). Therefore, more frequent iodination of reagents is necessary when ^{131}I is used. Iodine atoms of these isotopes can efficiently be incorporated into protein molecules by several methods with a minimal loss of immunological specificity. Besides, β-emitters (^{14}C and ^{3}H) are also frequently used in RIA. Viruses are adequately labelled with ^{14}C or ^{3}H by reductive alkylation with ^{14}C-formaldehyde or ^{3}H-formaldehyde in presence of sodium borohydride. Although reagents labelled with ^{3}H and ^{14}C are advantageous on account of their long half lives, the determination of weak β-emitters requires use of liquid scintillation counters. Thus ^{3}H or ^{14}C containing specimens must be solubilized in a toluene based scintillation fluid. On the other hand materials labelled with γ-emitters can be counted directly in crystal detectors without further processing.

For performing RIA direct assays, an antibody specific to the molecule to be detected is needed. But for indirect assays an antibody against the species in which the primary antibody was raised is required. When a radiolabelled antigen is used it should possess a high degree of purity. However, in situations based on unlabelled antigen reacting with labelled antibody, a less pure antigen may be used but the antibody should be highly purified. A variety of versions of RIA have been developed which are described as follows.

13.10.4.1 Liquid phase RIA

The RIA is characterised as liquid if all immunological reactions occur between the materials in a liquid suspension. On the other hand, solid phase system is the one in which immunological reactants are bound to the surface of a solid material where all the interactions occur.

(i) Direct RIA

If the antigen is to be determined, the antibody is labelled with a radioisotope and incubated with different dilutions of the antigen-containing the sample preparation. Likewise if antibody is to be determined, the labelled antigen is incubated with various amounts of antibody-containing specimens. Typically, in a direct RIA for antibody, virus concentrations of 1 to 10 ng labelled with ^{125}I equivalent to approximately 10,000 cpm are reacted with serial dilutions of unlabelled antibody. The bound ^{125}I labelled virus is then separated from mixture by centrifugation or other physical means. End concentration of the antibody specimen is the dilution which precipitates 50% of the labelled antigen.

(ii) Indirect RIA

In this procedure, neither of the primary reactants is radiolabelled. However, the species specific antiglobulin to the primary antibody is radiolabelled. Antiglobulin can be prepared by injecting immunoglobulins of the species in which primary antibody has been prepared into another species. Thus the primary reaction is indirectly measured by a secondary immunological reaction. If viral antibody is to be measured, antibody-containing specimens are reacted with a standard quantity of the unlabelled antigen in a primary reaction. Similarly, if an antigens is to be determined a standard quantity of unlabelled antibody is used.

(iii) Competitive Inhibition RIA

In a model assay for an antigen, the specimen is preincubated with a standard concentration of the virus specific antibody. During this reaction antigen-antibody complex is formed in an amount which is directly proportional to concentration of antigen in the specimen. A standard amount of labelled antigen is then added to reaction mixture and it is re-incubated. Now the labelled antigen will react only with the remaining antibody sites not occupied by the unlabelled antigen of the specimen. The amount of labelled antigen bound in the reaction with antibody will be hence inversely proportional to the amount of antigen in the sample.

In most of the liquid phase RIA separation of the bound labelled reagents from the unbound is accomplished by centrifugation. Better sedimentation is assured when an antiglobulin binds in a secondary reaction with the primary antigen and antibody complex. Even ultrafiltration through cellulose acetate filter has been used to separate the immune complex from unbound components of the reaction mixture for subsequent counting of radioactivity. For an assay of rabies antibody, membrane filtration (0.45 μm Millipore filters) is employed to separate the antibody-antigen complex from the unbound reactants.

13.10.4.2 Solid phase RIA

Solid phase RIA involves attachment of the antigen or the antibody to an inert insoluble material and the immunological reactions then take place on the surface of this solid matrix. In direct RIA, labelled antibody can be used for quantitative estimation of an antigen. In this method, the antigen is attached to the solid phase which is then incubated with radiolabelled antibody preparation. For quantitation of antibody, the radiolabelled antigen is bound to a solid material and sample containing the antibody is added and sufficient time is given for formation of the complex. Similarly, in an indirect RIA, the antigen-antibody complex on solid phase can be quantified by a secondary immunological reaction in which a radiolabelled antiglobulin is allowed to bind to the primary antigen-antibody complex. The unbound radiolabelled indicator is removed by rinsing.

The solid supports used for this technique are various polymers in form of beads, plastic tubes and small discs. These are first activated by appropriate chemical treatments to produce protein binding sites on the polymer surface. A sandwich system of polystyrene tubes coated with rabbit anti-influenza globulin, influenza virus, serum specimen and finally with ^{125}I labelled anti-human globulin has been successfully developed for measuring antibodies to influenza virus. The titers obtained by RIA correlate well with titres obtained by a haemagglutination inhibition tests. Techniques have been described in which the viral antigens were prepared by infecting monolayer of host cells growing directly on cover slips or in small glass vials. The infected cell-cultures were histologically fixed on the coverslips in the preparation for subsequent immunological reactions employing RIA.

Applications of RIA

RIA has been used for a wide variety of purposes both for diagnosis, quantitative estimation of biomolecules such as proteins, hormones and synthetic drugs in human beings and animals. The RIA procedures have been adopted quite extensively used for estimation of different hormones in human beings as well as in animals.

13.10.5 Enzyme-linked immunosorbent assay (ELISA)

The RIA suffers from certain inherent drawbacks such as short half life of radiolabel, requirement of highly skilled technicians, expensive infrastructure, potential health hazards associated with routine use of radioactive material and the problem of disposal of radioactive waste. These drawbacks of RIA have been circumvented by development of even a more sensitive and versatile technique in which in place of radioactive compound, enzymes have been used as labels in immunoassays. The term "enzyme linked immunosorbent assay" (ELISA) was coined in 1971 by Engval and Perlaman for this enzyme based technique. Over the last two decades, application of enzyme immunoassays has expanded quite dramatically. The developments in automation of enzyme immunoassays and monoclonal antibody technology have enabled introduction of a variety of variations in the basic ELISA procedure to fulfil the requirements of different test systems. The application of monoclonal antibodies in ELISA protocols has enhanced specificity of the test. ELISA is presently one of the most commonly used tests for the detection of antigens and antibodies. There are a variety of modifications of this test and these are collectively known as enzyme immunoassays (EIA). The basic feature of this group of assays is that an appropriate enzyme is conjugated with the antibody. It is then incubated with the test sample or the test sample which has been previously treated with primary antibody against the antigen. The labelled antibody specifically combines with the antigen or the antigen-primary antibody complex. The unbound enzyme labelled antibody is removed by washing. The enzyme labelled immunoreactant is then conveniently detected and quantitatively estimated by assaying the enzyme associated with antigen-antibody complex.

The general experimental strategies employed in ELISA can primarily be classified into two categories: direct ELISA and indirect ELISA. In the direct ELISA, the labelling enzyme is conjugated either with antigen or an antibody specific for the antigen (Fig. 13.13) while in indirect ELISA, the enzyme is

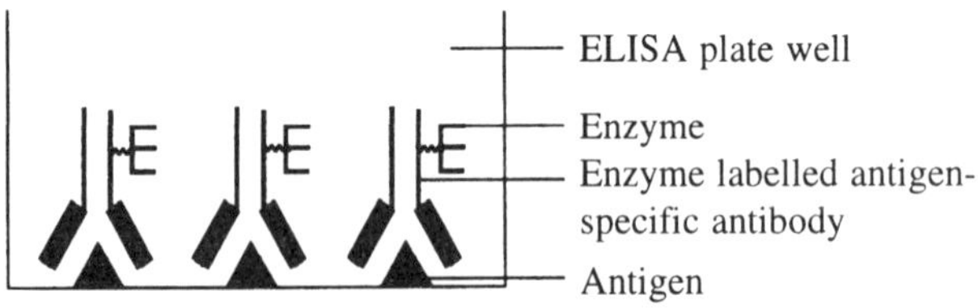

Fig. 13.13 Schematic representation of direct ELISA.

attached to an antibody raised against the primary antibody for the antigen. Antigen coated ELISA plates are incubated with primary antibody against the antigen and then plates are incubated with secondary antibody (developed against the immunoglobulin of the animal species in which primary antibody was raised) conjugated with the enzyme. The presence of antigen-antibody is detected by the binding of the enzyme conjugated anti-species antibody (secondary antibody) to the primary antibody bound to the antigen. Since antibody bound to the antigen is detected indirectly by the enzyme conjugated anti-species antibodies, the test is called indirect ELISA (Fig. 13.14). Indirect ELISA is more common because good quality of different anti-species conjugates are commercially available. Such conjugates can be used in different antigen antibody systems while direct labelling of the antibodies would restrict the use for that antigen only against which the antibody has been raised. The several variations of ELISA include, capture ELISA, sandwich ELISA, competitive ELISA, dot ELISA, ELOSA, immunoblotting and immunoperoxidase test.

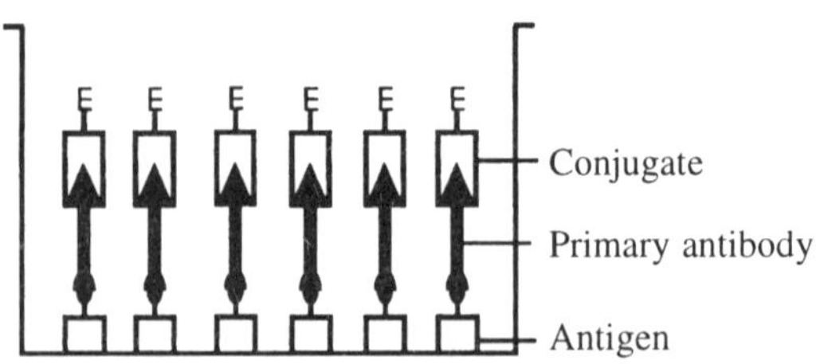

Fig. 13.14 Schematic representation of indirect ELISA.

A number of enzymes have been used for enzyme immunoassays. The selection of enzyme label is dependent on the type of enzyme immunoassay. In general, the enzyme should have following characteristics: catalyse a reaction which yields a coloured product that can be easily quantitated, it should be stable enough to allow conjugation, purification, and storage without appreciable loss of its activity, should possess a very high specific activity at physiological pH, should be readily available and be relatively inexpensive in a purified form and should not require expensive reagents such as substrates, cofactors or those needed for colour development. The enzymes which have been used successfully for ELISA include alkaline phosphatase (AP), horse radish peroxidase (HRP), β-galactosidase and glucose oxidase. The conjugate of β-galactosidase to rabbit IgG has proved to be the most sensitive label in ELISA so far. A number of reagents have been used for the conjugation of the enzyme with the antibodies. The principal reagents used for conjugation are carbodiimides, toluene diisocyochates, glutaraldehyde and sodium periodate. The activity of the above mentioned enzymes can be conveniently determined since either the product formed itself is coloured or yields a coloured compound upon reaction with appropriate reagents.

For the assays carried out in ELISA plates, a soluble coloured product of substrate is measured by monitoring absorbance using spectrophotometer. Many commercial ELISA readers based on spectrophotometric principle are now available which directly record the absorbance of samples in 96 well ELISA plates. Conversely, for the enzyme immunoassays carried out on nitrocellulose or diazotised paper, a soluble substrate that yields an insoluble coloured product at the site of reaction is used. These results can be evaluated by visual inspection and quantitated by reflection densitometry.

Use of solid carriers such as beads, discs, and tubes facilitate washing and separation steps. Antigen and antibodies have also been physically adsorbed onto plastic sufaces by non-covalent binding. Polystyrene, polyvinyl, polypropylene and polycarbonate coated disposable ELISA plates are commercially available which provide excellent adsorption of antigens and antibodies. Most proteins get adsorbed to plastic surfaces, probably as a result of hydrophobic interactions between nonpolar protein substructures and the nonpolar plastic matrix. The rate and extent of coating depends on the diffusion coefficient of the adsorbing molecule, the ratio of surface area to be coated to the volume of the coating solution, the concentration of adsorbing substance, temperature and the duration of adsorbing reaction. Polystyrene is the most widely used support in ELISA plates.

13.10.6 Enzyme-linked oligonucleotide sorbent assay (ELOSA)

ELOSA can be employed for detection of PCR amplified DNA using enzyme conjugated antibodies. The assay is carried out in ELISA plate and is suitable for testing a large number of amplified products. ELOSA is a sandwich solution hybridization assay procedure and is based on use of an internal 5′-biotinylated capture probe and a separate internal 5′-fluoresceinated detector probe. Following PCR, a small aliquot of the reaction mixture is denatured with an alkali and hybridized with internal biotinylated capture probe and a separate internal fluoresceinated detector probe. The hybridization mixture is then transferred in

ELISA plate well coated with streptavidin. Streptavidin is an avidin like protein isolated from Streptococcus and has a high affinity for biotin and binds to it very strongly. Thus the amplified DNA fragments or molecules with biotinylated 5′ ends get firmly captured on the plate. The immobilized hybrid is then reacted with an anti-fluorescein horse radish peroxidase conjugate which binds to fluoresceinated detector probe. The hybrid is then detected by addition of substrate (tetramethylbenzidine) which is converted to a coloured product and is read on a ELISA plate reader at 650 nm after 10–15 min incubation at 37°C. The basic ELOSA procedure is schematically represented in Fig. 13.15.

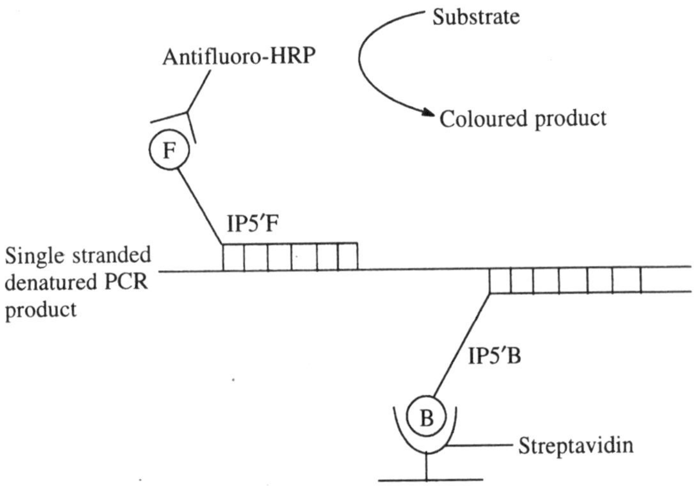

Fig. 13.15 Schematic representation of ELOSA. IP5′B and IP5′F denote internal 5′-biotinylated capture probe and internal 5′-fluoroscienated detector probe, respectively.

This method of PCR amplified DNA detection is very suitable for use in clinical and forensic laboratory tests because no radioisotopes are involved and the assay system is highly sensitive. A number of variations in respect of enzyme antibody conjugates and labels as detector probes have been developed.

13.10.7 Immunoblotting

Immunoblottings are immunoassays in which the antigens are immobilised on sheets of synthetic filter membranes and these firmly adhering antigens on the membrane are referred to as immunoblots. Immunoblotting involves transfer of electrophoretically separated proteins of the test sample from the gel to a solid support such as nitrocellulose membrane and subsequently probing the blots with specific antibodies. These antibodies would react specifically with antigenic epitopes on the target proteins and this facilitates in establishing the presence or absence of the antigen under investigation. Thus, immunoblotting combines resolution of SDS-polyacrylamide gel electrophoresis with specificity of immunochemical detection. It is therefore, very useful in determining the presence of antigen as well as relative molecular weight of polypeptide chain. It is also a powerful technique for ascertaining the presence and specificity of antibodies from polyclonal sera. A variety of highly specific and sensitive blotting procedures for proteins have been developed and these procedures, in one form or the other, are now routinely used in most biochemical laboratories. Immunoblotting is basically a qualitative test applicable to most protein samples. Such a technique which entails transfer of proteins to membrane followed by detection of a particular protein by use of antibodies is popularly known as "Western Blotting".

A typical immunoblotting technique is largely analogous to the indirect ELISA procedure and comprises of the following five steps:

(i) Proteins of the sample are immobilised on a membrane surface either by direct application as a spot or by transfer from an electrophoretic gel. Several types of synthetic membranes bind proteins tightly enough, while retaining antigenicity and accessibility to antibodies, and hence these membranes can be used as supports for immunoblots.

(ii) Unoccupied protein binding sites on the membrane are saturated with bovine serum albumin to prevent non-specific binding of antibodies to the membrane. This step is called either blocking or quenching.

(iii) The blot is probed for proteins of interest with specific primary antibody.

(iv) Secondary antibody, specific for the primary antibody type and conjugated to an appropriate enzyme, are then used to label primary antibodies. Staphylococcal protein A and Streptococcal protein G can also be used as secondary immunological probes.

(v) The label bearing immobilised proteins become visible by activity of the conjugated enzyme, which converts substrate into insoluble and detectable product at the sites where the antigenic protein is immobilised on the membrane.

13.11 Immunohistochemical Techniques

Immunocytochemistry has now become one of the indispensable and a versatile tools for studies on cytology and pathology. A variety of immunocytochemical assays have been developed which enables identification and localization of any intracellular antigen/molecule, against which antibodies can be produced, by either light or electron microscopy. Antigens are identified immunologically by using antibodies which are labelled with either fluorescent substances or compounds which have colour or possess very high electron density. However, due to the low resolving power, light microscope, has its own limitations and is not capable of detecting antigen sites very accurately. On the other hand, electron microscope offers better precision due to its much higher resolving capacity. The immunoprobes can readily be identified under electron microscope based on electron density or defined shape of the probe carrier. A wide range of probes are used for labelling of antibodies for a variety of cytochemical techniques. These probes can be broadly classified into 6 groups:

(i) Fluorescent compounds: Fluorochromes were the first to be used as labels for immunochemical localization of antigens and antibodies in tissue sections. Two most commonly used fluorescent dyes are fluorescein isothiocyanate (FITC) and tetramethyl rhodamine isothiocyanate (TRITC). FITC has emission wavelength at 525 nm and is detectable from its yellowish green fluorescence while TRITC upon excitation radiates reddish orange light of 620 nm.

(ii) ***Metals:*** Colloidal gold or silver and metalloproteins such as ferrocene, haemocyanin and ferritin are the most commonly used probes of this category. The colloidal gold is negatively charged and hydrophobic in nature. The charge helps in stabilising these particles by electrostatic repulsion. Antibodies can be conjugated with colloidal gold particles or carrier proteins. Colloidal gold particles of variable sizes can be prepared depending on the requirement of the immunocytochemical test. Both colloidal gold and silver are useful in light as well as electron microscopy. Based on the size of the particles, colloidal silver gives yellow colour while colloidal gold appears red to blue in transmitted light. Ferritin is a protein which contains 2000-3000 atoms of iron/molecule which gives it enough electron density to be visualised under electron microscope. Other iron containing compounds such as haemocyanin and ferrocene are also suitable for this purpose. Compounds of uranium, mercury and osmium are also being used to make electron dense antibodies. The procedure for conjugating ferritin with antibodies is similar to that employed for any other protein including enzymes.

(iii) ***Enzymes:*** The suitability of an enzyme as a probe depends on a number of factors. Firstly the enzyme should not act on the tissue/cell under investigation. Secondly the reaction product should be insoluble with distinct colour or have high electron density or get precipitated directly at the site of reaction. The commonly used enzymes as probes are acid phosphatase, alkaline phosphatase, horse radish peroxidase, glucose oxidase, tyrosinase etc. The enzyme is conjugated with monospecific antibodies using bifunctional reagents. It is important that the antigen binding sites of antibodies should remain free and accessible to interact with the antigen. The cellular site at which the conjugate binds, is revealed by an appropriate cytochemical method based on light or electron microscopic observations.

(iv) ***Radioisotopes:*** Iodine (^{125}I) is the most commonly used radioisotope probe. It is introduced into tyrosine residues of the antibody using chloramine T or lactoperoxidase. The lactoperoxidase gives better results in immunoradiography because it causes minimal damage to the protein as compared to chloramine T. Other radioisotopes such as tritium (^{3}H) and carbon (^{14}C) are also quite satisfactory for labelling of antibodies.

(v) ***Viruses and synthetic beads:*** Viruses can also serve as the labels in immunocytochemistry because they provide excellent results for location of surface antigen especially by scanning electron microscopy. When cells are exposed to the virus-antibody conjugate, the antibody along with the virus binds to the surface of antigen. The viruses, on account of their definite shapes, can be identified very easily by scanning electron microscopy. Similarly synthetic materials which

have definite shape and size are good substitutes for viruses as probes. For instance many synthetic beads, microspheres and such other materials which can be conjugated to carrier proteins and immunoglobulins can serve this role. Polystyrene latex spheres, co-polymer microspheres and silica spheres, among others, belong to this group of probes.

(vi) Carrier proteins: There are mainly four categories of carrier proteins viz. protein A, lectins, antibodies and a bungarotoxin. Probes are attached to carrier protein by physical binding at one particular site. The other site recognizes the specific antigen and thus probe can trace antigenic sites.

The immunocytochemical techniques employ labelled antibodies and antigens. The three important assay systems which have been widely adopted particularly for immunodiagnosis are: immunofluorescence, radioimmunoassays and enzyme based immunocytochemical assays.

13.11.1 Immunofluorescence assays

Most fluorescent compounds contain a ring structure whose electrons get excited and oscillate in resonance on exposure and absorption of light. A molecule can exist in its excited state for a very short period and soon returns to its more stable ground or native state. The average lifetime of the excited state is often less than 10^{-8} seconds for most of the fluorescent dyes. During reversion from the excited state back to the ground state the absorbed radiant energy is released in form of light of a longer wavelength with a very short time lapse between absorption and emission of light as fluorescence. There are two classical methods of immunofluorescence assays: direct and indirect methods. The indirect method is more commonly used. This test allows detection of antigens in smears and cryopreserved tissue sections. The specific antibody is coupled to a fluorescent compound such as FITC or TRITC. The site(s) of occurrence of the antigen in the specimen fluoresces after staining which can be localised under a fluorescent microscope.

In the direct method, an optimised fluorescent labelled antibody against the antigen is used to detect its presence and location in tissue/cells fixed on a slide. The staining with FITC conjugated antibody is done at 37° C for at least 30 min. The cytosmear/tissue specimens are then thoroughly washed to remove excess of the free labelled antibody. The specimens are then carefully blotted and mounted with buffered glycerol for examination under fluorescent microscope. Detection of fluorescence indicates presence of the antigen and the fluorescent zones denote its location within the cell or the tissue.

The indirect method involves incubation of smears/tissue slides with specific primary antibody solution followed by washing of the specimen and then again treated with secondary antibody (anti-species antibody) coupled to FITC. Following thorough washing, the slides are viewed under fluorescent microscope. It is an indirect method because a specific antigen-antibody reaction is visualised by addition of labelled antibody directed against the primary antibody. This method has an advantage of utilising a single labelled anti-antibody to detect many different antigen-antibody reactions occurring within different specimens in a given species.

Indirect assay is also relatively more specific as compared to the direct one. This test has been used for diagnosis of a variety of viral and bacterial infections in man and animals. It has been employed for rapid detection of parainfluenza viruses, influenza virus, poxviruses, herpes viruses, respiratory syncytial virus, adenoviruses and measles and *Legionella pneumophila*. In many viral infections immunofluorescent test can provide the required diagnostic information with precision within a few hours thus eliminating the need for traditional isolation procedures. Figure 13.16 shows fluorescing BHK 21 cells after indirect assay for ascertaining presence of bluetongue virus. The cells were firstly incubated with primary antibody against bluetongue virus group specific proteins followed by treatment with FITC labelled secondary antibody.

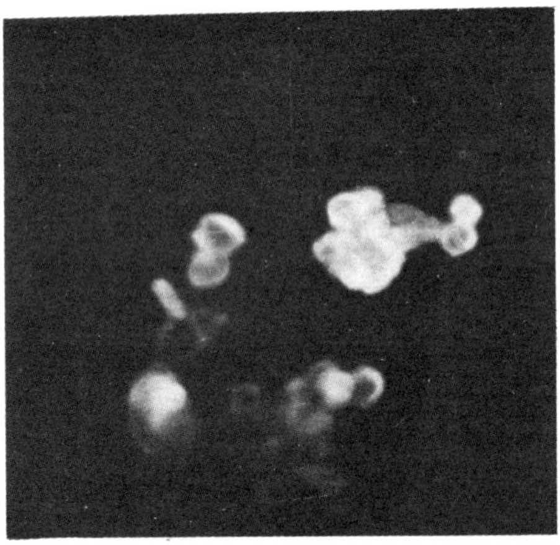

Fig. 13.16—Indirect immunofluorescence. BHK 21 cells showing fluorescence specific to bluetongue virus group specific protein.

13.11.2 Fluorescent activated cell sorter (FACS) and magnet activated cell sorter (MACS)

A very powerful method based on immunofluorescence approach called fluorescent activated cell sorting (FACS) has been developed. Through this technique, a particular type of cells can be separated using a specific cell surface marker antibody coupled to a fluorescent label. A cell suspension stained with the specific fluorescent antibody is passed through an apparatus that produces stream of small droplets each containing only one cell. These droplets are passed between a laser beam ultraviolet light and a detector responds to emitted fluorescence when the labeled cell is present in the droplet. The emitted signal is transmitted to an electrode that charges the droplet, leading to its deflection in an electromagnetic field. Thus, as all the emerging droplets while falling follow different paths, depending on whether or not they emit the signal and are collected and counted separately. Thus by sophisticated electronics, the fluorescein-stained and unstained cells may be counted as well as the intensity of fluorscence of each recorded.

A variation of this method has been developed which utilizes fluorescent antibody coupled to magnetic beads to separate the cell populations. The cells which bind to the fluorescent antibody get separated from unstained ones due to deviation in their paths under the influence of magnetic force. This technique is hence known as magnet activated cell sorting (MACS). Both FACS and MACS have thus facilitated isolation of very rare cells such as stem cells, lymphocytes, their subsets and a variety of other cell types for diagnostic as well as further investigations.

13.11.3 Immunoenzymatic cytochemical techniques

Immunoenzymatic techniques have been used extensively for the identification of cell surface markers, viral antigens, hormones and a variety of other antigens. These techniques utilise antibodies conjugated with an enzyme, generally peroxidase or alkaline phosphatase, as probe. After staining with this probe, the specimen is incubated in a mixture containing appropriate substrate for the conjugated enzyme which produces coloured product. The dual staining with the combination of alkaline phosphatase and peroxidase allows identification of two different antigens simultaneously. Immunohistochemical procedures utilising different antibodies made in the same species have been introduced in recent years. The double immunoenzymatic method is recommended for identifying with monoclonal antibodies (MAb) of two different tissue or cell antigens with non overlapping pattern of distribution. Since the range of available MAbs for cell surface antigens and infectious pathogens is rapidly increasing, the importance of double staining method is likely to expand. The use of immunoenzymatic technique for immunological localisation of antigen in cytosmears is a widely accepted technique especially because of its superiority in terms of the histopathological details over the immunofluorescence methods.

There are five main methods of immunoenzymatic assays that can be used to localise cellular/microbial antigens. These are: direct, indirect, peroxidase-anti-peroxidase (PAP), avidin-biotin and alkaline phosphatase-anti alkaline phosphatase (APAAP). Each method has certain advantages and disadvantages which must be evaluated for selection of the most effective procedure suited to the nature of investigation being undertaken. Each of these methods is described briefly.

13.11.3.1 Direct method

The simplest way to localise a certain antigen is by using an enzyme conjugated antibody directed specifically against it. This conjugated reagent is applied to the specimen and it forms a complex with the antigen. The immunological interaction is then detected by transferring the specimen to a buffer containing the substrate which is converted to a coloured end product and precipitates at the site and thus reveals the location of the antigen. Direct technique can be performed very quickly with a low probability of non specific reactions. The main drawback is that, for every antigen to be examined, a different conjugated antibody is required. The most common application of direct immunoperoxidase method is for the detection of immunoglobulin, complement and immune-complex deposit in kidney biopsies of patients with renal diseases.

13.11.3.2 Indirect method

In this technique the specimen is first treated with an unconjugated antibody which binds to the antigen. For perceiving this attachment, in the next step a peroxidase conjugated antibody specific to the primary antibody is used. On addition of the substrate the site of activity of the conjugated enzyme is localised. This method is more versatile than the direct method because a variety of primary antibodies raised in the same animal species can be used with one type of conjugated secondary antibody. The main use of indirect assay is to detect antigens of viral, bacterial and parasitic pathogens.

13.11.3.3 Peroxidase-anti-peroxidase (PAP) method

This procedure is based on use of three reagents: primary and secondary antibodies and PAP-complex consisting of the enzyme peroxidase and antibody against the enzyme (Fig. 13.17). The primary antibody is specific for the antigen and the secondary or "Link" antibody is capable of binding to both the primary antibody and the PAP complex. The link antibody is added in excess, so that only one of its Fab sites will bind to the primary antibody, leaving the other Fab site free to bind to the antibody in the PAP complex. The peroxidase enzyme activity is visualised with a substrate chromogen reaction. This method is very sensitive and specific which can be employed effectively in paraffin embedded sections.

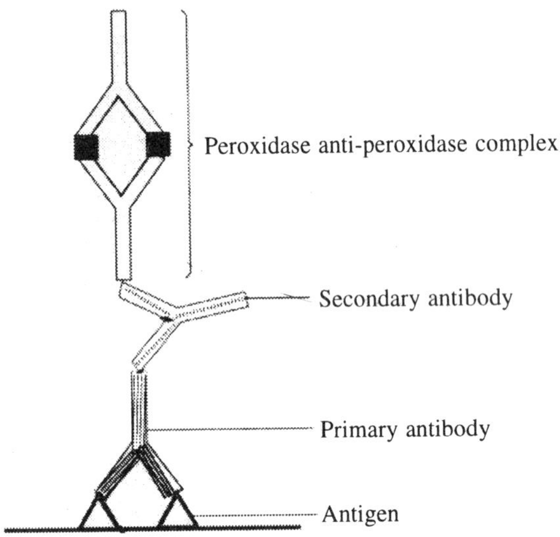

Fig. 13.17 Step-wise procedure of peroxidase-anti-peroxidase (PAP) method.

13.11.3.4 Alkaline phosphatase-anti alkaline phosphatase (APAAP) method

Its principle is analogous to that of PAP method, except that it utilises alkaline phosphatase and anti-alkaline phosphatase antibody as "unlabelled antibody bridge" instead of peroxidase anti-peroxidase. In contrast to PAP, this method can be used successfully on cytosmears having high endogenous peroxidase activity.

13.11.3.5 Avidin-biotin system

The biotin-avidin system is particularly attractive as a sandwich system used in conjunction with specific antibodies in immunohistochemistry. The technique exploits the capacity of avidin (egg white glycoprotein) to bind with four molecules of biotin which is conjugated with a specific antibody. An enzyme, ferritin, fluorochrome or other probes to which avidin is attached serves as a stable, high affinity second step reagent. These probes form a complex with the biotin bearing antibody due to rapid interaction of avidin with this compound. The main advantage of avidin as sandwich reagent is that it avoids the use of anti-immunoglobulin antibodies. The four binding sites together with its high affinity of avidin for biotin aid in accomplishing the sensitivity of the immunoassays. As illustrated in Fig. 13.18 this method utilises three basic reagents: a primary antibody specific for the antigen under investigation, the secondary antibody conjugated to biotin, capable of binding to the primary antibody and finally avidin coupled to peroxidase. The free sites of the avidin molecules allow binding to the biotin on the secondary antibody. The peroxidase enzyme, and therefore the presence as well as location of the original antigen, are probed with an appropriate chromogenic reaction. The most important application of this technique is to identify pituitary hormone in normal and neoplastic tissues, cell surface markers, viral proteins and lectin binding sites.

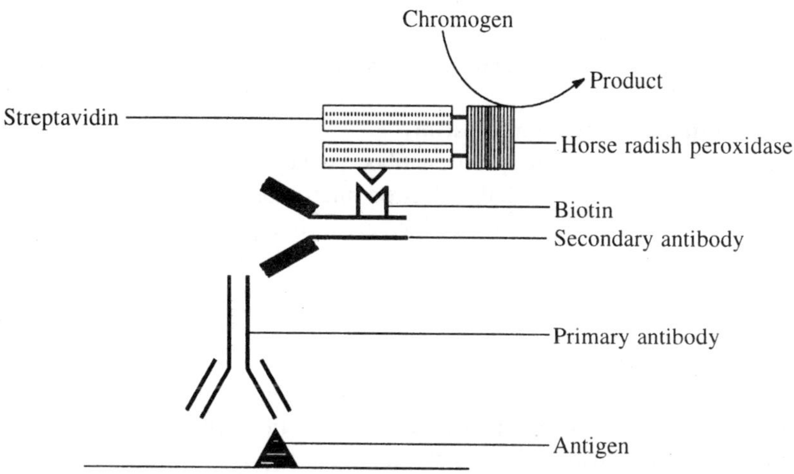

Fig. 13.18 Schematic representation of avidin-biotin system.

For example, the following basic steps will be involved for detection of types of T or B cells: Firstly allow the B/T cells to bind with monoclonal antibody (raised against the cell surface antigens) and then incubate them with biotinylated goat anti-mouse antibody. This is followed by their incubation with avidin conjugated with horse radish peroxidase. Finally the cells having specific cell surface antigens are visualised upon addition of 3,3′-diaminobenzidine (DAB).

13.11.4 Immunogold technique

Colloidal gold is frequently used as a label for localization of antigens in tissues by light as well as electron microscopy. The advantage of using immunogold probes is that a variety of proteins can be adsorbed on the colloidal gold particles and can thus be visualised under microscope. The colloidal gold particles can be prepared easily in different sizes which are highly stable for long time. The different size gold particles allow simultaneous localization of multiple antigen in the same tissue. The adsorption of proteins on colloidal gold is dependent on molecular weight, ionic forces, solubility, pH and electrical charge of the mixture.

The basic method of immunogold involves incubation of tissue sections with specific primary antibody against the antigen to be detected, followed by another incubation with colloidal gold labelled with second antibody raised against immunoglobulin of the species in which primary antibody was prepared or with protein A labelled with colloidal gold. The gold particles can be directly visualised under microscope while the colloidal gold can be localised under light microscope by silver enhancement method. The metallic gold catalyses the reduction of silver ions in the presence of a reducing substance thus creating a shield of silver around each gold particle that makes it visible under light microscope. The colloidal gold particles can readily be identified under electron microscope also. In this method, the tissue is fixed, dehydrated and embedded in resin. Ultrathin sections of the tissues are cut and mounted on electron microscope grids. The tissue mounted grids are incubated on a drop of primary antibody followed by another incubation with secondary antibody conjugated with colloidal gold or protein A conjugated with colloidal gold. After washing, the grids are stained with uranyl acetate in dark, washed, dried and viewed under transmission electron microscope.

13.11.5 Immuno-electron microscopy

When viruses are mixed with specific antibodies, they interact and form aggregates which are more easily seen under electron microscope. This test is particularly useful and suited for those viruses which are difficult to detect by conventional methods. For example, hepatitis A virus can readily be identified by immuno-electron microscopy.

13.12 Laboratory Exercises

In this Section a few selected experiments on immunochemical techniques such as production of antibodies and their applications in biochemical investigations as well as for diagnostic purposes are described.

13.12.1 EXPERIMENT: To produce polyclonal antiserum against bovine serum albumin

Principle

When a "foreign" substance such as a protein or other macromolecular substance is injected into an animal its immune system gets activated leading to synthesis of antibodies specific to the antigenic determinants (epitopes) present on the antigen and are released in the serum. Since different B lymphocytes recognise individual epitopes, the antibodies thus produced is a mixture of the ones having different specificities as these are produced by different clones of activated B cells. Hence the antiserum thus produced is referred to as polyclonal antiserum. To produce polyclonal antiserum against a particular antigen, a suitable animal is immunised with pure homologous sample of the substance. For example, antiserum against BSA can be raised by injecting it into rabbits. Normally antigen is mixed with appropriate adjuvant (see Section 13.5) to elicit a good immune response. A few days (usually 7–21 days) later a "booster" injection is given to the animal which considerably enhances antibody response. After few days, the blood is collected and allowed to clot. The clot is broken and the cells are removed by centrifugation and the supernatant which represents antibody containing blood serum is aseptically collected and kept at –20 to –70°C for long time storage and at 4°C if it has to be used within a short period.

Materials and Reagents

1. Syringes (2, 5 and 10 ml)
2. Needles (18, 21 and 25 G)
3. Glass tubes, pipettes, etc.
4. Plastic serum storage vials
5. Refrigerated centrifuge
6. White albino rabbits: Four rabbits of around one year age
7. Chloroform or anaesthetic ether
8. Fruend's complete and incomplete adjuvants. Available commercially
9. Bovine serum albumin (BSA): A highly purified preparation of BSA free from any contaminating proteins may either be prepared in laboratory or purchased from commercial sources
10. Normal saline solution: Normal saline solution comprises of NaCl and is prepared by dissolving 8.5 g NaCl in water and making its volume to 1 L with distilled water

Procedure

1. Suspend 10 mg/ml BSA antigen in normal saline solution in a clean and sterile beaker. To a 6 ml of BSA solution add an equal volume of complete Freund's adjuvant. Make emulsion with help of syringe by repeatedly drawing and ejecting the mixture through 18 G needle.
2. When a complete emulsion has been formed, inject 1 ml of emulsified BSA intramuscularly (I/M) into each rabbit.
3. Give booster injection (I/M) of the same dose of antigen but now in incomplete Freund's adjuvant after 4–6 weeks of the first injection.
4. Test bleed the animals from ear vein 5–7 days after the booster inoculation and collect the blood sample.
5. Separate the serum by allowing the blood to clot at room temperature. Remove any remaining cellular material from the serum by centrifuging it at 2000 rpm for 20 min at 4°C in refrigerated centrifuge.
6. If the serum contains desired level of antibody titre (see EXPERIMENT 13.12.2), bleed the animals by cardiac puncture. The bleeding of animals by this route allows collection of large volume of blood and thus higher quantities of polyclonal antiserum.
7. Store the antiserum at 4°C for short term use but at –20 to –70°C if it has to be preserved for long time. The antiserum can also be lyophilised and stored at 4°C for long time use.

8. To avoid contamination, either filter the antiserum through 0.45 μm membrane filter or add antimicrobial agents like merthiolate (0.01%) or sodium azide (0.1%).

13.12.2 EXPERIMENT: Titration of polyclonal antiserum by radial immunodiffusion assay

Principle

Titre value of the test sample can be determined by immunodiffusion assay (see Section 13.10.1.3). Homologous antigen is incorporated into agarose solution which is then allowed to form a gel. Circular wells of uniform size are cut into the gel to which test serum sample containing the antibody is added. On diffusion of the antibody, a circular precipitation zone is formed around the well whose area is proportional to the concentration of the antibody. In other wells, standard solutions containing known amount of the antibody are added. A plot of antibody concentration versus square of diameter of the precipitate ring gives straight line thus enabling preparation of standard curve. From the diameter of the ring formed around the well containing the test serum sample, concentration of antibody in it can be determined from the standard curve. If antiserum with known concentration of the antibody is not available, titre value of the test antiserum sample can be determined by serial dilution method as described in this experiment.

Materials and Reagents

1. Clean glass slide
2. Gel punch
3. Magnifying glass
4. BOD incubator
5. Needle (20 gauge)
6. Micropipettes
7. Pasteur pipettes
8. Petri dishes
9. Agarose
10. Bovine serum albumin (BSA)
11. Antiserum against BSA
12. Barbitone buffer (pH 7.8): Dissolve 1.5 g NaCl, 1.5 g KCl, 1.25 g sodium barbital, 1.0 g $MgSO_4$ and 100 mg NaN_3 in distilled water. Make the final volume to 500 ml with distilled water

Procedure

1. Take 2% (*w/v*) agarose in barbitone buffer and melt it by boiling.
2. Keep the melted agarose at 56°C in water bath.
3. Prepare several dilutions of BSA in barbitone buffer and keep it at 56°C water bath.
4. Mix equal volumes of melted agarose with diluted antigen solution and pour the mixture onto a glass slide or a Petri plate so that approximately

1 mm thick layer of agarose is formed. Prepare separate plate for each antigen dilution.
5. After solidification of agarose, punch 6 holes into agarose and remove the agarose from the punched holes by suction.
6. Seal bottom of the wells by adding melted agarose.
7. Fill each well by serially diluted anti-BSA antiserum.
8. Place the plates in humidified chambers and incubate at 25°C for 24 h in BOD incubator.
9. Observe the plates and record diameter of the precipitate ring around the wells.
10. The dilution of antiserum beyond which there is no visible ring around the well is the titre of the antiserum. However, the other way to find out the anti-BSA antiserum titre is by using standard curve obtained by using preparation containing known concentration of the antibody.

13.12.3 EXPERIMENT: Production of monoclonal antibody against purified protein

Principle

The antibodies obtained by direct injection of an antigen into an animal are derived from more than one clone of B lymphocytes and hence are polyclonal. Such mixtures of antibodies are relatively less specific in their immunoreactivity and the possibility that some of them might be cross-reacting with other antigenic compounds having similar or nearly identical epitopes can not be precluded. This limitation can be overcome to a large extent by use of monoclonal antibodies produced by hybridoma (see Section 13.9.2 for details). The monoclonal antibodies recognise only a particular epitope. Evidently these exhibit much greater stringent immunoreactivity and are thus highly specific. The hybrid cells, which are positive for producing the desired antibody, should be single cell cloned without any delay after the hybrids have been tested, because delay might result in the failure of cloning of hybrid cells. There are two conventional methods of single cell cloning viz. limiting dilution method which is common and the soft agar cloning technique. The hybridomas can be identified by observing growth in HAT (hypoxanthine, aminopterin and thymidine) medium. Unfused myeloma cells are unable to survive because they lack thymidine kinase or HGPRT enzymes whereas spleen cells perish due to their inherent inability to grow *in vitro* (see Section 13.9.2).

The four basic steps involved in the generation of hybridoma cell line producing a monoclonal antibody are given in the Fig. 13.5. These include: immunization of mice, fusion of myeloma cells with spleen cells, screening of hybrids and single cell cloning. Strictly inbred BALB/c (strain of mouse generated after brother-sister mating for more than 20 generations, these animals are almost identical in their genetic and immunological parameters) are used for the production of monoclonal antibodies. The spleen cells so obtained from the immunised mice are mixed with myeloma cells for fusion. This step is performed in the presence of polyethylene glycol as it facilitates fusion. A number of myeloma cell lines are commercially available but SP 2/0 myeloma cell line which is deficient in HGPRT enzyme is commonly used. The mixture is then grown on HAT medium to select hybridomas which are then examined for their capacity of producing antibody by monitoring the antibodies released in the medium by ELISA (see Section 13.10.5 for details) using peroxidase conjugated anti-mouse IgG. The hybridoma producing the desired antibodies are then single cell cloned. i.e. to obtain individual hybridoma line generated from a single hybrid cell. These are then cultured in static or suspension or injected into mice to produce ascites for obtaining the monoclonal antibodies in large amounts.

Materials and Reagents

1. Scissors and forceps
2. Sterilised needles
3. Sterilised syringes (1, 2, 5 and 10 ml)
4. Sterilised Petri dishes
5. Glass pipettes (1, 2, 5 and 10 ml)
6. High quality ELISA plates
7. 96 well high quality cell culture plates
8. Micropipettes (10, 100 and 1000 μl)
9. Multi-channel pipettes (40–200 μl)
10. Sterilised microtips
11. Plastic conical centrifuge tube (50 ml)
12. Water bath at 37°C
13. Refrigerated centrifuge
14. Ordinary incubator set at 37°C
15. Haemocytometer
16. Inverted microscope
17. CO_2 incubator set at 35–37°C and 5% CO_2 and 95% air mixture
18. ELISA reader
19. Immunised BALB/c mice
20. Myeloma cells
21. RPMI 1640 (Rose Park Mary Institute) medium and foetal calf serum (both commercially available)
22. HAT (hypoxanthine, aminopterin and thymidine) medium (commercially available)
23. HT (hypoxanthine and thymidine) medium (commercially available)
24. Hanks' balanced salt solution (commercially available)
25. Polyethylene glycol (PEG)
26. Feeder layer (mouse monocytes/spleen cells)
27. Antibiotics solution: It contains both antibacterial and antifungal antibiotics (commercially available)
28. Fruend's complete and incomplete adjuvants (Sigma Co., USA)
29. Antigen against which monoclonal antibodies are to be detected: The purified antigen may be prepared in the lab using appropriate techniques or procured from commercial sources
30. Anti-mouse IgG peroxidase conjugate (Sigma Co., USA) or from any other source

31. Substrates solution: Orthophenyl diamine (OPD). Dissolve 6 mg OPD in 10 ml of citrate buffer and add 10 ml of 30% H_2O_2.
32. H_2O_2 (30%)
33. Wash buffer: Phosphate buffered saline-Tween 20 (PBST, pH 7.4) and carbonate-bicarbonate buffer. Prepare the PBS and PBST as follows:
    (a) Phosphate buffered saline (PBS) pH 7.4: Prepare by 8 g NaCl, 0.20 g KCl, 1.45 g disodium hydrogen phosphate and 0.20 potassium dihydrogen phosphate in distilled water and make the final volume to 1 L with distilled water
    (b) Phosphate buffered saline-Tween 20 (PBST): Add Tween 20 to PBS so that its final concentration is 0.05% (*v*/*v*)
34. Trypan blue solution

Procedure

The procedure is described in four parts (A–D), each part dealing with particular aspect of the technique for production of monoclonal antibodies (MAb).

A. *Immunization of mice and preparation of spleen cells*

1. Immunise 2-3 months old strictly inbred BALB/c mice by injecting them sub-cutaneously at multiple sites with protein antigen against which the MAbs are to be produced (see EXPERIMENT 13.12.1 for details). Give a booster injection of the antigen after 21 days. A few days later, draw a drop of blood sample from the immunised mice and test (as described in part C of the *Procedure*) it for the presence of specific antibodies. After a month again give an intraperitoneal injection of the antigen solution (without Freund's adjuvant).
2. Sacrifice the immunised mice by cervical dislocation and swab their abdominal region with 70% ethanol. Cut open their abdomen with scissors and remove the spleen under sterile conditions.
3. Place the spleen in Petri dish containing 10 ml of RPMI-1640 supplemented with 2.5% foetal calf serum (FCS) and wash gently before transferring it to another Petri dish containing 10 ml of same medium and squeeze the cells with help of sterile needles.
4. Gently pour the cell suspension into 50 ml conical tube leaving behind any debris in the Petri plate. Fill the tube with RPMI-1640 medium supplemented with 2.5% FCS and centrifuge at 2000 rpm for 10 min. The pellet is resuspended in 10 ml fresh medium and centrifuged as earlier. Suspend the pellet in 10 ml of medium and count the cells.

B. *Fusion of spleen cells with myeloma cells and culturing of hybrids*

1. Culture the myeloma cells (SP 2/0) in RPMI 1640 medium supplemented with 10% FCS and harvest them at logarithmic growth phase by centrifugation at 4°C for 10 min at 2000 rpm.
2. Resuspend the pellet in 10 ml of 2.5% FCS-RPMI 1640 medium and count the cells.

3. Keep all the media and reagents required for fusion in an incubator or water-bath, maintained at 37°C. Carry out all the operations strictly under sterile conditions as followed for mammalian cell culture. Mix 10^8 spleen cells and 10^7 myeloma cells in 50 ml conical tube and add 50 ml of RPMI 1640 medium supplemented with 10% FCS. Centrifuge the cells for 10 min at 2000 rpm at room temperature and decant the supernatant completely to avoid dilution of PEG.
4. Place the centrifuge tube in a 200 ml beaker containing water at 37–40° C. Loosen and break the pellet by gently tapping the bottom of the tube. Slowly add 1 ml of 50% PEG, prewarmed to 37°C, to the pellet using 1 ml pipette, over a period of 1 min continually shaking the tube or stirring the cells with the pipette tip. Continue stirring of cells in 50% PEG for another 1 min and by then agglutination of the cells occurs.
5. Now add dropwise 10 ml of RPMI 1640 medium to the fusion mixture over a period of 10 min with a 10 ml pipette. Centrifuge the tubes at 2000 rpm at room temperature for 8–10 min. Discard the supernatant, loosen the pellet by gently tapping the bottom of tube and suspend in 50 ml of HAT medium. Distribute the fusion suspension in 96 well culture plates adding 200 μl of the suspension in each well. Incubate the plates at 37°C in CO_2 (5%) incubator.
6. After one day, remove about 100 μl of medium from each well without disturbing cells which get settled at bottom and add fresh 100 μl HAT medium. Repeat feeding with fresh HAT medium for next 2 or 3 days and after that once a week until vigorous growth of hybrids is obtained. The growth of hybrids can be judged under the inverted stage microscope and at this stage cultures become more yellow and are ready to be tested for antibody activity.

C. Screening of hybrids for antibodies production

ELISA is commonly used to detect the presence of required antibody in the spent medium. The details of the technique are given in Section 13.10.5.

1. Coat two rows of 8 wells each of the U-shaped ELISA plate modules with the antigen, against which monoclonal antibodies are being produced. Add 50 μl of carbonate-bicarbonate buffer containing appropriate amount of the antigen to each well of ELISA plate module and incubate at 37°C for 1 h or at 4°C overnight. Wash the wells 3 times with PBST.
2. Block the unsaturated sites by adding 50 μl of 3% BSA to each well and incubate at 37°C for 1 h. Rinse the wells with PBST twice to remove extra BSA. Take 50 μl of the spent medium from each well containing the hybrid clones to well of ELISA module. Preparation of MAb (if available) and mouse serum serve as controls. Incubate the modules for 1 h at 37°C. After incubation period wash the wells with PBST five times.
3. Add 50 μl of diluted anti-mouse horse radish peroxidase conjugate to each well and incubate the modules at 37°C for 45 min. Wash the wells again with PBST five times.
4. Add 50 μl of substrate (OPD) solution in each well and incubate the

modules for 10 min in dark at room temperature. Then stop the reaction with a few drops of 2 N H_2SO_4. The wells in which the positive colour reaction appears contain the antibodies.

5. After the positive clones have been selected, prepare the duplicates of the growing hybrid cultures for a week on hypoxanthine and thymidine (HT) medium. Amplify the cultures subsequently from 96 well plate to 24 well plate to 25 cm^2 flask and finally to 80 cm^2 flask. After a week, grow cultures in absence of HAT additives and adapt to lower concentration (10%) of FCS.

D. Single cell cloning by limiting dilution method

This is the conventional method of cloning and involves various steps described below:

1. A day before cloning, lay 9 plates/clone with feeder cells (mouse monocytes or spleen cells).
2. Pipette in and out the culture medium of a hybridoma repeatedly for proper mixing of the cells.
3. Pipette out 50 μl of aliquot of the cultured cells from the culture well and count the cell. Dilute the cells to achieve cell concentration of 3 cells, 1 cell and 0.3 cells per 100 μl volume. Prepare about 30 ml suspension of each cell concentration.
4. Pipette 100 μl of suspension in each well of first cell concentration suspension into three plates. Similarly, prepare three plates each for other two cell concentration suspensions. Secure the lid of the plates with cellophane tape and keep the plates in the CO_2 incubator (5% CO_2 and 95% air).
5. About 4–5 days after the cloning, remove 100 μl of the medium from each well of all the plates and add 100 μl HAT medium. Feed the cultures regularly with this medium. After about 12–14 days of the cloning, inspect the plates seeded with 0.3 cell/well cell concentration suspension under inverted microscope for growth of hybrids.
6. If hybrids have grown big enough to be screened, perform the screening test as described earlier (see part C of this Procedure). If about 33% wells give positive reaction in ELISA, it can be assumed that one clone is present in each of the wells of the plates. If greater per cent of wells show positive in ELISA then cell number is more than one cell per well. Clones seeded with higher cell concentrations are also partially cloned and could be used as backup in case the hybrids die and fail to grow when diluted to 0.3 cell/well cell concentration. Repeat the cloning, if the hybrid cells fail to grow. The cloning should be done 3 times to be reasonably sure about the viability capacity of single cell clones.

Ascites Production

For getting large amount of MAb, ascites is produced in BALB/c mice. The procedure to produce this is as follows:

1. Before the induction of ascites, inject the mice with 0.5 ml of pristane intraperitoneally (I/P). After 7–10 days of pristane treatment, suspend about 10^7 hybrid cells of the positive clone in 0.5 ml of medium and inject I/P into the mice. Observe the mice daily and the bulging of belly after about 5–10 days inoculation indicates the ascites production.
2. Withdraw the ascites fluid with the help of sterilized needle and syringe. Removal of ascitic fluid is necessary at appropriate time as ascites is metabolically very stressful for mice and they may die of this stress. Withdraw ascitic fluid after every 2–3 days till the mice succumb to stress.
3. Test the ascites fluid for specific monoclonal antibodies using ELISA as described earlier in part C of this Procedure.

Storage of hybridoma

1. Store the hybrid cultures in liquid nitrogen before cloning of the positive hybrids as well as after single cell cloning for future use. For freezing the hybrids, about 10^6–10^7 cells in logarithmic phase are pelleted by centrifugation at 2000 rpm at 4°C for 7–10 min.
2. Pour out the supernatant and resuspend the pellet in 1 ml of freezing medium (10% DMSO and 20% FCS-RPMI, *v*/*v*). Transfer the suspension to cryopreservation vials and place in insulated box and keep it at –70°C for atleast 5–6 h before finally transferring them to liquid nitrogen.

Thawing procedure

For reviving the hybridomas, take out the vials from liquid nitrogen cylinder and thaw the hybrids as quickly as possible in a 37°C water bath. After thawing is complete, transfer the cell suspension to 10 ml centrifuge tube and slowly add 10 ml of cold 10% FCS-RPMI and mix carefully. Centrifuge the cell suspension at 2000 rpm at 4°C for 7–10 min and resuspend in about 5 ml of fresh medium and transfer to a small tissue culture flask.

Applications

A large number of monoclonal antibodies are now commercially available against a number of antigenic determinants. The availability of hybridoma technology has revolutionary impact on biomedical research and development of more sensitive and reliable diagnostic tools against several infectious and non-infectious diseases of man and animals and also for more reliable immunological characterization of biomolecules.

13.12.4 EXPERIMENT: To detect nitrite reductase in leaf extracts by immunodiffusion test

Nitrite reductase is one of the enzymes of nitrate assimilation that catalyses the reduction of nitrite to ammonia. It is a nitrate inducible enzyme and its synthesis in leaves requires light. The objective of this experiment is to examine the presence of nitrite reductase in leaves of wheat seedlings grown in light in both nitrate free and nitrate containing media.

Principle

Presence of a particular enzyme in cell extract can reliably be established by immunodiffusion test. The test is based on the reaction of soluble antigen with homologous antibodies in agar/agarose gel leading to development of precipitin lines at the junction of the reaction. The test is simple and convenient to use for detection of biomolecules including enzymes in biological samples.

Materials and Reagents

1. Clean glass slide
2. Gel punch
3. Magnifying glass
4. BOD incubator
5. Needle (20 gauge)
6. Micropipettes
7. Pasteur pipettes
8. Petri dishes
9. Agarose
10. Purified nitrite reductase preparation
11. Antiserum against the enzyme to be tested: Prepare the antiserum using purified nitrite reductase from wheat leaves according to the procedure described in EXPERIMENT 13.12.1
12. Barbitone buffer: Prepare as in EXPERIMENT 13.12.2
13. Experimental material: About 8 days old wheat seedlings raised in a net house or illuminated growth chamber in well washed river sand and supplied daily with either distilled water or 50 mM potassium nitrate solution

Procedure

1. Preparation of sample extract: Take 1 g leaves of wheat seedling grown in distilled or 50 mM potassium nitrate solution. Macerate the leaves in chilled pestle and mortar in presence of 0.1 M phosphate buffer (pH 7.5) containing 1 mM cysteine. Centrifuge the homogenate at 4°C and 10, 000 × *g* for 20 min and use the supernatant as sample for detecting the presence of nitrite reductase.
2. Prepare 0.8–1% agarose solution in barbitone buffer by keeping it in boiling water bath.
3. Overlay about 4 ml of the molten agarose on each clean glass slide using glass pipette.
4. Allow the agarose to solidify and then punch wells using commercial gel punch with one central well and six peripheral wells equidistant from the central well.
5. Remove gel from the wells using needle and then seal bottom of the wells with small amount of molten agarose with help of Pasteur pipette.
6. Fill antiserum against nitrite reductase in the central well and the leaf extract of wheat seedlings in peripheral wells. Also add purified preparation of nitrite reductase in one of the peripheral wells which would serve as a reference.

7. Keep the slides in humidified glass Petri dishes and incubate at 25–30°C for 24 h and then keep them in refrigerator for 2–3 days and record the observations after 72 h.

Conclusion

The appearance of precipitation line in the space between the central well containing antiserum and peripheral sample well is considered as a positive indication for that particular antigen (nitrite reductase) in the extract.

13.12.5 EXPERIMENT: To detect antibodies in sera by dot immunobinding assay

Principle

Dot immunobinding assay (DIA), a modified form of ELISA is performed on a piece of nitrocellulose paper attached to a plastic dipstick. DIA has been used as the assay of choice for detection of antibodies against infectious agents since it is easy to perform, as well as sensitive and economical for application under field conditions. This test can be used both for detection of antibodies and antigens. This experiment describes monitoring of antibodies in experimental animals.

Materials and Reagents

1. Glass pipettes (1, 2 and 5 ml) 10 each
2. Sterilised capillaries 5
3. Beakers (500 and 50 ml) 3 each
4. Tubes (2 ml) 20 each
5. Micropipettes (100 and 1000 μl)
6. Micro tips 50
7. Ordinary incubator at 37°C.
8. Dipsticks and nitrocellulose membrane: Nitrocellulose membrane pieces (0.5 × 0.5 cm) of 0.45 μm pore size and bound to plastic strips (0.5 × 7.0 cm).
9. Antigen: Any antigen against which antibody is to be detected.
10. Serum samples: Select test sera, known to give positive and negative reaction.
11. Enzyme-antibody conjugate: Peroxidase conjugated with anti-species IgG.
12. Substrate solution: 0.5 mg of diaminobenzidine tetrahydrochloride is dissolved in per ml of PBS (reagent 14a) and add 30% H_2O_2 to it.
13. Blocking agent: Bovine serum albumin (3%) in PBS. Alternatively 4–5% aqueous solution of skimmed milk powder may be used.
14. Washing buffers:
    (a) Phosphate buffered saline (PBS), pH 7.4: Prepare by dissolving 8 g NaCl, 0.20 g KCl, 1.45 g disodium hydrogen phosphate and 0.20 g potassium dihydrogen phosphate in distilled water and make the final volume to 1 L with distilled water.
    (b) Phosphate buffered saline–Tween 20 (PBST): Add Tween 20 to PBS so that its final concentration is 0.05% (*v*/*v*).

Procedure

1. Deposit approximately 1 μl of the antigen preparation at centre of the piece of nitrocellulose membrane on each dipstick using sterilised microcapillaries.
2. Allow the antigen dots to dry at ambient temperature for 5–10 min.
3. Block the unoccupied binding sites on the membrane by immersing the membrane end of the dipsticks either overnight at 4°C or 2 h at 37°C in 3% bovine serum albumin or in 5% skimmed milk powder solution.
4. Rinse the dipsticks with PBST to remove excess of BSA or skimmed milk powder.
5. Dilute each of the test and reference serum to 1:20 in PBS in 5 ml vials.
6. Dip the dipsticks in the vials containing the diluted test and reference sera and incubate for 1 h at 37°C.
7. Thereafter, rinse the dipsticks thoroughly with PBST to remove unbound antibodies.
8. After washing, immerse the membrane end of the dipsticks in peroxidase conjugate (1:1000 in PBS) and incubate for 1 h at 37°C.
9. Rinse the dipsticks thoroughly and dip in diaminobenzidine tetra-hydrochloride (0.5 mg/ml in PBS + 1 μl of 30% hydrogen peroxide) substrate solution for 2–5 min.
10. Wait till brown colour develops at the site of the antigen dot (usually within 2–5 min) in the positive controls. Terminate the reaction by washing the dipsticks with PBS and dry each stick.

Conclusion

Appearance of brown coloured dot at the site of applied sample denotes positive reaction while failure of appearance of coloured dot indicates absence of the antibodies in sample.

13.12.6 EXPERIMENT: To Detect antibodies in sera by competitive ELISA (cELISA)

Principle

In this test, serial dilutions of the test sera are made and allowed to interact with the antigen. Then high titred monoclonal antibody prepared against the purified antigen is added. In the test serum of lower dilutions, the number of antibodies binding to antigen will be much larger and thus monoclonal antibodies would find fewer sites on the antigen to bind. Subsequently, upon addition of enzyme conjugated with anti-mouse IgG, a fewer molecules of this conjugate will bind and thus little colour would only develop. At higher dilution of the antibodies in the test sera get diluted and more sites on the antigen would become available for the monoclonal antibodies to bind and thus the colour developed would be more. Evidently the colour intensity obtained would be inversely proportional to the amount of antibodies in the test serum.

Materials and Reagents

1. ELISA plates.
2. Antigen.
3. High titre known serum specific to the antigen produced in any species other than mouse.
4. Low titre known serum specific to the antigen produced in any species other than mouse.
5. Antigen specific monoclonal antibody or standard polyclonal antiserum.
6. Anti-species immunoglobulin conjugate (anti-mouse)—commercially available.
7. Substrate solution: For preparation see serial No. 31 of materials and reagents of EXPERIMENT 13.12.3.
8. Carbonate-bicarbonate buffer (pH 9.6 coating buffer): Dissolve 0.15 g sodium chloride, 0.29 g sodium bicarbonate and 0.02 g sodium azide in water and make it to 1 L with distilled water.
9. Washing buffer, pH 7.4 (PBST): Prepare a solution comprising 8.0 g sodium chloride, 0.20 g potassium chloride, 1.45 g disodium hydrogen phosphate and 0.2 g potassium dihydrogen phophate in 1 L distilled water. Add Tween-20 to this solution so that its final concentration is 0.05% (*v*/*v*).
10. Skimmed milk powder: Prepare 5% (*w*/*v*) solution of skimmed milk powder in distilled water.
11. 1 M H_2SO_4.

Procedure

1. Coat each well of ELISA plates with 50 μl of diluted antigen. Dilute the antigen appropriately in coating buffer. Incubate on plate shaker for 1 h at 37°C or over night at 4°C.
2. Wash the plates thrice with washing buffer and tap the plates to remove the residual solution from the wells.
3. Add 50 μl diluted sera in triplicate to the antigen coated wells. Incubate on plate shaker for 1 h at 37°C.
4. Add 50 μl of diluted antigen specific monoclonal antibody (MAb) to each well. The MAb is diluted in PBST + 1% skimmed milk powder. Incubate on plate shaker for 30 min at 37°C. Wash plate thrice with washing buffer and tap out excessive liquid.
5. Add 50 μl of diluted antimouse IgG-conjugate to all wells. The conjugate is diluted in PBST + 1% skimmed milk powder. Incubate on plate shaker for 30 min at 37°C. Wash the plates three times and remove any remaining solution by tapping.
6. Add 50 μl of TMB substrate (as per the instructions in the kit) to each well. Incubate at 25°C for 10 min.
7. Stop the reaction with 50 μl of 1 M H_2SO_4.
8. Record absorbance of the formed colour in each well of the plates at 450 nm in ELISA reader.

9. Convert the recorded experimental values into % inhibition using the following formula:

   % Inhibition

$$= \frac{A_{450} \text{ non-reactor (negative-serum)} - A_{450} \text{ test serum}}{A_{250} \text{ non-reactor (negative serum)} - A_{450} \text{ high titre serum}} \times 100$$

The % inhibition values greater than 50 are considered as positive while less than 50 as negative for the antibodies in the test sera.

13.12.7 EXPERIMENT: Qualitative determination of antigens by immunoblotting technique

Principle

Immunoblotting involves transfer of electrophoretically separated proteins of the test sample from the SDS-PAGE gel to a solid support such as nitrocellulose membrane and subsequently probing western blots with specific antibodies. These antibodies would react specifically with antigenic epitopes on the target proteins and this facilitates in establishing the presence or absence of the antigen under investigation. Thus, immunoblotting combines resolution of SDS-polyacrylamide gel electrophoresis with specificity of immunochemical detection. It is, therefore, very useful in determining the presence of antigen as well as relative molecular weight of polypeptide chain. The proteins identified by immunoblotting are highly denatured. Hence conformational epitopes are lost leading to failure of binding of antibodies specific to conformational epitopes. Therefore, it is advisable to use polyclonal antiserum which contains antibodies to a variety of epitopes. This is the reason why many monoclonal antibodies fail to react in immunoblotting.

Materials and Reagents

1. PBS (phosphate butter saline, pH 7.4): For preparation see serial No. 14a of EXPERIMENT 13.12.5
2. Blocking solution (5% skimmed dry milk powder in PBS, pH 7.4)
3. Hyper immune serum against the antigen: For preparation see EXPERIMENT 13.12.1
4. Anti-species horse radish peroxidase (HRPO) conjugate (Sigma Co., USA)
5. Hydrogen peroxide 30%
6. DAB (3,3′-diaminobenzidine tetrahydrochloride)
7. Substrate solution: Dissolve 6 mg of DAB in 10 ml of 50 mM Tris (pH 7.6). Filter through Whatman No. 1 filter paper to remove precipitate. In the filtered substrate solution add 10 ml of 30% H_2O_2 just before use
8. Nitrocellulose membrane
9. Washing buffer (PBS, pH 7.4)
10. Bovine serum albumin (BSA)

Procedure

1. Blot the polypeptide pattern onto nitrocellulose membrane from the SDS-PAGE gel as described in Section 10.4.4.

2. Block non-specific binding sites on the nitrocellulose membrane by immersing overnight in blocking solution.
3. Remove the membrane from the blocking solution and wash twice for 5 min in PBS. Add appropriate dilution of hyper immune antiserum raised in guinea pig and diluted in PBS containing 3% (*w/v*) BSA. Incubate for at least 1 h at room temperature with agitation. Wash the membrane with 4 changes of PBS for 5 min each.
4. Add anti-guinea pig HRPO conjugate diluted in PBS containing 3% (*w/v*) BSA and incubate for 1 h at room temperature with agitation. Wash the membrane with 4 changes of PBS for 5 min each.
5. For development of immunoblot, incubate the membrane in appropriate amount of substrate solution to which 10 ml of 30% H_2O_2 is added and develop at room temperature with agitation until the sufficiently dark visible bands (brown colour) appear. Stop the reaction by rinsing with PBS.

Precautions

1. Use of appropriate amount of antibody and conjugate solutions and adequate shaking should be done to ensure proper antigen-antibody reaction.
2. The problems of high background and other artifactual stains on the immunoblot can be minimised by effectively blocking the binding sites on the membrane and adequate washing.
3. If background problem still persits, reduce the concentration of labelled antibody. For this, the concentration of labelled antibody may have to be optimized for best resolution without much background staining.
4. Sometimes uneven background may be seen in the immunoblots which may occur due to improper handling of membrane. Thus it is advisable to use gloves when manipulating the membrane with hands.

13.12.8 EXPERIMENT: To establish intracellular location of an enzyme by indirect immunofluorescence assay

Principle

In this method, an ultrathin section of the tissue is first treated with an antibody against the enzyme. Free antibody is removed by several washings and the section is then treated with a labelled antibody against the primary antibody. The labelling of the secondary antibody may be done by any of the commonly used methods such as by attachment of fluorescent dyes, electron dense materials as ferritin, gold, silver etc. (see Section 13.11). The section is then examined under fluorescent microscope (for antibody labelled with fluorescent dye) or electron microscope/ordinary light microscope when labelling is done with electron dense substances. This indirect method has an advantage of utilising a single labelled anti-antibody to detect many different antigen-antibody reactions occurring within a given species. In addition, the indirect assay is relatively more specific as compared to the direct one. In the following experiment described, FITC is used as fluorescent label.

Materials and Reagents

1. Microtome
2. BOD incubator
3. Fluorescent microscope
4. Dry ice
5. Glutaraldehyde
6. Tissue sections/cytosmears
7. Specific antibodies against the target enzyme
8. Anti-species IgG-FITC
9. Bovine serum albumin (BSA)
10. Phosphate buffer solution (PBS 0.01 M, pH 7.4)

Solution A: 0.2 M $NaH_2PO_4 \cdot H_2O$: Dissolve 2.78 g in 100 ml of distilled water.

Solution B: 0.2 M $Na_2HPO_4 \cdot 7H_2O$: Dissolve 5.37 g in 100 ml of distilled water.

To prepare 0.01 M PBS, take 9.5 ml of solution A and 40.5 ml of solution B and make a final volume of 1 L with distilled water. Autoclave and store at 4°C.

Procedure

1. Fix tissue sections/cells on microscopic glass slides with chilled 80 % acetone + 20% methanol for 10 minutes at 4°C. After fixing the tissues, rinse sections with PBS (0.01 M, pH 7.4) thoroughly.
2. Place a few drops of appropriately diluted primary antibody, against the enzyme to be detected on the tissue section and incubate in humidified chamber at 37°C for 30 min. Following incubation, wash the tissue sections thoroughly with PBS.
3. After the primary antibody reaction, incubate the tissue section in 3% BSA in PBS to avoid non-specific attachment of FITC conjugate.
4. Incubate the tissue sections at 37°C for 30 min in appropriate dilution of second antibody conjugated with FITC, which has been raised against the immunoglobulins of the species in which primary antibody was prepared.
5. Rinse the tissue section thoroughly in PBS and mount clean coverslips on the tissue sections using 90% glycerol-PBS buffer (*v/v*, 9ml glycerol + 1 ml PBS).
6. Examine the mounted tissue slide under fluorescent microscope and note the presence of fluorescence in the tissue sections which will indicate presence of the enzyme. If possible try to identify the fluorescence in sub-cellular components of the cell.

13.12.9 EXPERIMENT: To ascertain presence of bluetongue virus antigens in cells by immunoperoxidase assay

Principle

This technique is based on utilization of a secondary antibody conjugated with an enzyme, generally peroxidase; as probe and this antibody specifically interacts with the primary antibody for the antigen. Presence of labelled secondary antibody is then detected by incubating treated tissues with chromogenic substrate for peroxidase which produces a coloured product. This indirect method is more versatile than the direct method because a variety of primary antibodies made in the same animal species can be used with one type of conjugated secondary antibody. The main use of this indirect assay is to detect antigens such as viruses, bacteria and parasites by first treating the tissue with antibodies against these antigens.

Materials and Reagents

1. BOD incubator
2. Light mcroscope
3. Tissue sections/cytosmears/cell cultures
4. Group specific antibodies against bluetongue virus
5. Anti-species IgG-peroxidase conjugate
6. Diaminobenzidine (DAB)
7. Hydrogen peroxide (H_2O_2) 30%
8. Bovine serum albumin (BSA)
9. PBS 0.01 M, pH 7.4: Prepare as in EXPERIMENT 13.12.8
10. Quenching solution: Absolute methanol and 30% hydrogen peroxide (4:1 ratio)
11. Substrate solution: For preparation see Serial No. 12 of materials and reagents of EXPERIMENT 13.12.5

Procedure

1. Fix tissue sections/cells on microscopic glass slides with chilled acetone for 10 min at 4°C. Rinse the fixed tissue sections with PBS (0.01 M, pH 7.4) thoroughly.
2. Inactivate endogenous peroxidase by quenching the tissue sections with quenching solution (absolute methanol and 3% hydrogen peroxide in 4:1 ratio) for 20 min at 25°C. Wash the slides thoroughly with PBS.
3. Incubate the tissues sections with 3% BSA for 20 min at 25°C.
4. Add a few drops of appropriately diluted anti-bluetongue virus antibody on the tissue sections and incubate in humidified chamber at 25°C for 20 min. After incubation, rinse the tissue sections with PBS and then place them in PBS bath for 10 min.
5. Incubate the tissue sections at 25°C for 20 min in appropriate dilution of second antibody, which had been raised against the immunoglobulins of the species in which primary antibody was prepared and has been conjugated with horse radish peroxidase.
6. Rinse the tissue section thoroughly in PBS as in Step 4. Place the freshly prepared DAB substrate solution on tissue sections and incubate at 25°C for 5–10 min.
7. Gently wash the tissue slides thoroughly in PBS and observe them under a microscope. Development of brown colour will indicate presence of the antigen under investigation at the specific cellular site. The slides may be permanently mounted with Canada balsam and stored. To mount the slides, put one drop of Canada balsam on the tissue/cell smear and place a clean coverslip on top of it.

Suggested Further Reading

Bullock G.R. and Petrusze P. (eds) (1983). Techniques in Immunochemistry. Rd., Vol. 1, 2 and 3, Academic Press, New York.

Cambell A.M. (1984). Laboratory techniques in Biochemistry and Molecular Biology: Monoclonal antibodies. Vol. 13, Elsevier, Amsterdam.

Hudson L. and Hay F.C. (eds) (1989). Practical Immunology. Oxford Blackwell Scientific Publications, Oxford.

Johnston A. and Thorp R. (eds) (1987). Immunochemistry in Practice. Oxford Blackwell Scientific Publications, Oxford.

Tijssen P. (1985). Laboratory techniques in Biochemistry and Molecular Biology: Practice and Theory of Enzyme Immunoassays. Vol. 15, Elsevier, Amsterdam.

Weir D.H. (ed) (1986). Handbook of Experimental Immunology. Blackwell Scientific Publication, Oxford.

Williams C.A. and Chase M.C. (eds) (1967). Methods in Immunology and Immunochemistry. Vol. 1, Academic Press, New York.

14. Molecular Biology Techniques

14.1 Introduction

The rapid and fascinating advances in molecular biology have ushered in a new and exciting era with boundless opportunities of harnessing the already existing technology and also the expected developments in biological sciences, for further benefit of mankind. Remarkable success has already been achieved in the unravelling of genome organization in diverse groups of organisms, regulation and molecular mechanisms of gene expression, transferring of desirable genes from one organism to another and thus construct trangenics with a view to improvise the commercial performance of the organisms, and also alter very minutely the inherent genes through site directed mutagenesis to enhance biological efficacy of the gene products. All these remarkable accomplishments owe their success to development of highly innovative experimental techniques.

It is well established that the entire genetic information embodied in the chromosomes is not expressed at a constant rate at all the time during the life cycle of an organism. There are quantitative and qualitative changes in expression of various genes at different developmental stages, in different organs and in response to various chemical and physical stimuli. This well programmed and stringently controlled expression of genetic information is central and of immense importance in cell differentiation and modulation of metabolism to requirements of an organism under fluctuating internal and external environment. Identification of genes and regulation of their expression forms an integral and important aspect of cellular development and metabolic adaptation. Which genes are being expressed at a given time is usually established by estimating translational products *viz.* protein by Western blotting (see Section 10.3.3.4) or measurement of transcripts by Northern blotting (see Section 14.4). These techniques alongwith c-DNA cloning and Southern blotting also provide insights into the characterization and organization of gene and gene families, estimation of gene copy numbers and for determination of chromosomal location of structural genes. Polymerase Chain Reaction (PCR) is another revolutionary technique which amplifies DNA *in vitro* and has wide applications in diagnostics, fingerprinting, sequencing, mutagenesis and so on. Another group of techniques such as Restriction Fragment Length Polymorphism (RFLP) and Randomly Amplified Polymorphic DNA (RAPD) provide a way to directly follow chromosomal segments during recombination and are finding applications in construction of molecular maps and map-based cloning of genes. This chapter deals with some laboratory exercises on these techniques.

14.2 Isolation of Nucleic Acids from Biological Samples

Nucleic acids are present as nucleoprotein complexes in cells. The major problems encountered in isolation of pure and intact DNA or RNA molecules are: degradation of high molecular weight nucleic acids by mechanical damage or by hydrolytic action of nucleases, contamination of DNA preparations with RNA and *vice versa* and contamination of nucleic acids with proteins, polysaccharides and other high molecular weight compounds. Methods have been devised for isolation of nucleic acids from different sources taking adequate precautions to eliminate or minimize the above problems. The main steps involved in isolation of nucleic acids are:

(*a*) *Homogenization or disruption of cells:* The procedure employed for breakage of cells should be as gentle as possible. In case of plant material, this is generally done by grinding it in a pestle and mortar or a Waring blender at low temperature in buffers containing EDTA when isolating DNA for chelating Mg^{2+} ions which are required for DNase activity. For isolation of RNA, an inhibitor of RNase such as bentonite, diethyl-pyrocarbonate, placental RNase inhibitor etc. is included in the extraction buffer. Disintegration of bacterial cells can be achieved by treating them with cell coat hydrolysing enzyme, i.e. lysozyme in presence of a detergent.

(*b*) *Dissociation of nucleo-protein complexes:* The approach employed is such that the proteins either get dissociated or degraded while nucleic acids remain unaffected and intact. This is generally achieved by using detergents like SDS, phenol or broad spectrum proteolytic enzymes such as pronase or proteinase K. Alkaline pH and high concentration of salts improve efficiency of the process.

(*c*) *Removal of contaminating materials and precipitation of nucleic acids:* Proteins are removed by treatment of tissue homogenate with phenol or mixture of chloroform-isoamyl alcohol or phenol-chloroform. Upon centrifugation, the denatured proteins form a layer at the interface between aqueous and organic phases, lipids and other contaminants remain in the organic phase while nucleic acids are recovered in the aqueous phase from which they are precipitated with ethanol. For isolation of DNA, the contaminating RNA is removed by selective salt precipitation or treatment with DNase-free RNase. Conversely while isolating RNA, the preparation is incubated with DNase for eliminating DNA as an impurity.

(*d*) *General precautions while handling nucleic acids:*

1. All glassware/plasticware and solutions (except organic solvents) should be sterilized/autoclaved.
2. Gloves should be worn to avoid contamination of the experimental material and apparatus with nucleases which occur in fair abundance in skin exudates.
3. Phenol causes severe burns and phenol containing solutions should, therefore, be handled with care. Thoroughly rinse the burns with large volume of water. Do not use ethanol.
4. For work with RNA, rinse all glassware/plasticware with 1%

diethylpyrocarbonate solution to inactivate RNase and then autoclave them.

14.2.1 EXPERIMENT: Isolation of DNA from *E. coli* cells

Principle

This is a general method for isolation of chromosomal DNA from bacteria. Bacterial cells are lysed with lysozyme in presence of SDS, deproteinised with chloroform-isoamyl alcohol and finally DNA is precipitated with ethanol.

Materials and Reagents

1. Table centrifuge
2. Water bath
3. Spectrophotometer
4. *E. coli* cells: Grow the cells in liquid culture media following the standard microbiological techniques or use commercially available lyophilized cells
5. Extraction medium: 0.15 M NaCl containing 0.1 M Na_2EDTA
6. Lysozyme: Prepare solution containing 10 mg lysozyme per ml water
7. SDS: Make 25% (*w/v*) solution of SDS in water
8. 5 M sodium perchlorate
9. Chloroform : isoamyl alcohol (24 : 1)
10. 95% ethanol
11. Saline sodium citrate (1 ×): Prepare 0.015 M Tri-sodium citrate solution (pH 7.0) containing 0.15 M NaCl in it
12. 3 M sodium acetate buffer (pH 7.0)

Procedure

1. Suspend 2 to 3 g of wet packed bacterial cells or 0.5 g lyophilized cells in 25 ml saline-EDTA solution (Reagent No. 5) in a 125 ml Erlenmeyer flask.
2. Add 1 ml of lysozyme solution (Reagent No. 6) to the above suspension and incubate with occasional shaking at 37°C for 30 min.
3. To facilitate complete cell lysis, add 2 ml of SDS solution and heat the mixture in a water bath at 60°C for 10 min and then cool it to room temperature under tap water.
4. Add 7 ml 5 M sodium perchlorate so that its final conc in the above preparation is 1 M.
5. Mix well but gently and then add an equal volume (35 ml) of chloroform:isoamyl alcohol (24 : 1) and slowly shake in a tightly stoppered flask for 30 min at room temperature.
6. Centrifuge the above mixture at 10,000 × *g* for 5 min at room temperature. Three layers will be formed. Collect the upper aqueous phase in a beaker with the help of a wide mouth pipette.
7. Add two volumes of chilled 95% ethanol to this aqueous phase. Gently

stir the preparation with a glass rod and spool the thread-like DNA fibres on the glass rod. If DNA does not form fibrous thread like structures upon the addition of ethanol, centrifuge the mixture at 3000 rpm for 10 min. Dissolve the pellet in 10 ml of saline-EDTA solution (Reagent No. 5) and reprecipitate with 2 volumes of 95% ethanol.

8. Air dry the spooled DNA on the glass rod and dissolve it in a test tube containing 10 ml of saline citrate. Store the solution at 0–4°C.
9. Dilute 0.5 ml aliquot of DNA solution with 4.5 ml of saline-citrate. Measure the absorbance at 260 and 280 nm using saline-citrate as reference blank. Calculate the amount of DNA in the sample as described in EXPERIMENT 5.2.4.

Precautions

All precautions mentioned in Section 14.2(d) must be exercised.

14.2.2 EXPERIMENT: Isolation of DNA from wheat leaves

Principle

This is a rapid method for isolation of DNA from plants. Plant tissue is homogenised in a hot extraction buffer. Inclusion of sarkosyl facilitates dissolution of cell membranes, high concentrations of urea and NaCl help in dissociation of nucleo-protein complex. Dissociated proteins are denatured by phenol and are then removed by chloroform-isoamyl alcohol treatment. DNA in the aqueous phase is finally precipitated with chilled ethanol.

Materials and Reagents

1. Young wheat leaves
2. Liquid nitrogen or dry ice
3. Centrifuge
4. Water bath
5. DNA extraction buffer: This extraction media is prepared as follows:

Urea	42.0 g
5 M NaCl	7.0 ml
1 M Tris-HCl (pH 8.0)	5.0 ml
0.25 M EDTA	8.0 ml
10% Sarkosyl	20.0 ml
Phenol reagent (see Reagent 8)	5.0 ml

Make final volume to 100 ml with water

6. 20% SDS
7. Chloroform : isoamyl alcohol (24 : 1)
8. Phenol reagent: Take redistilled phenol and equiliberate it with 0.1 M Tris-HCI (pH 8.0) until pH of the phenolic phase is more than 7.8 (check with pH paper)
9. 3 M sodium acetate buffer (pH 7.0)
10. Tris-EDTA (TE) solution: To 1 ml of 1.0 M Tris-HCl (pH 8.0) add 0.4 ml

of 0.25 M EDTA (pH 7.0) and then make the final volume to 100 ml with distilled water

11. 95% ethanol
12. Isopropanol
13. RNase A stock solution: Dissolve 10 mg RNase A in 1 ml of sterile water or in 10 mM Tris-HCl (pH 7.5). Boil the solution for 10 min to inactivate DNase which might be present as an impurity and then let it cool. Store frozen in small aliquots

Procedure

1. Macerate 5–10 g of fresh wheat leaves in a mortar and pestle in liquid nitrogen or dry ice with a pinch of glass beads till a fine powder is formed. Do not allow the powder to thaw and transfer it to a capped polypropylene centrifuge tube containing 25 ml of extraction buffer which had been preincubated at 65°C. (Note: If liquid nitrogen or dry ice is not available, dried plant material may be used. Make a fine powder of the dried material and suspend in the extraction buffer at 65°C.)
2. Using a glass rod break any clumps. Add 0.75 ml of 20% SDS and mix the contents by inverting the capped centrifuge tubes. Incubate the mixture for 10 min at 65°C, inverting the tubes gently 2 or 3 times during this incubation step.
3. Add 15 ml chloroform : isoamyl alcohol (24 : 1) and mix by inversion 10–15 times to obtain an emulsion. Gently shake the emulsion for 15 min.
4. Centrifuge at 3000 rpm for 15 min at room temperature. Three layers will be formed. Transfer the top aqueous phase to another centrifuge tube with the help of a pipette.
5. Add an equal volume of cold isopropanol to the aqueous phase and invert the tube gently.
6. Spool the DNA threads on a glass rod and wash it by dipping the rod, into a beaker containing 70% ethanol. Dry the spooled DNA on the glass rod in air and then dissolve it in 4 ml of TE solution in a centrifuge tube. If DNA does not get dissolved, incubate it at 65°C for 1 h in a water bath. Centrifuge at 10,000 rpm for 10 min and transfer the supernatant to another centrifuge tube.
7. Add DNase free RNase to a final concentration of 10 μg/ml of the preparation and incubate for 5 min at room temperature. Precipitate DNA by adding 1/10th volume of 3 M sodium acetate and 2 volumes of 95% ethanol. Hook out DNA on a bent Pasteur pipette or a glass rod as before, wash it in 70% ethanol, air dry and dissolve in 1 ml TE solution in a test tube.
8. Dilute a portion of this DNA solution 20 times with TE and note absorbance at 260 and 280 nm in a spectrophotometer. Calculate the amount of DNA from A_{260} as described in EXPERIMENT 5.2.4.
9. The DNA solution can be stored at 0–4°C for use upto 1 month. For longer duration, layer a drop of chloroform on the top of DNA preparation and store at –70°C.

Precautions

Exercise all the precautions listed in Section 14.2(d).

14.2.3 EXPERIMENT: Isolation of total RNA from yeast

Principle

Yeast cells are homogenized with phenol which dissociates proteins from nucleic acids and also denatures the proteins. Upon centrifugation, DNA remains in the lower phenol phase, RNA is dispersed in the upper aqueous phase with the denatured proteins being at the interphase. RNA from the recovered aqueous phase is precipitated with ethanol and it is free of DNA but may contain polysaccharides as contaminants.

Materials and Reagents

1. Refrigerated centrifuge
2. Dried yeast
3. Phenol solution: Dissolve 900 g of phenol in 1 L of water
4. Potassium acetate: Dissolve 200 g potassium acetate in 1 L of water and adjust its pH to 5.0
5. Absolute ethanol
6. Diethyl ether

Procedure

1. Suspend 30 g of dried yeast in 120 ml of distilled water and incubate at 37°C for 15 min.
2. Add 160 ml of phenol solution. Stir the suspension for 30 min at room temperature and then centrifuge it at 3000 × *g* for 15 min in cold to facilitate the emulsion to separate out into different phases.
3. Carefully remove the upper aqueous phase and add potassium acetate solution to it to obtain a final concentration of 20 g of potassium acetate per L of the mixture.
4. Add two volumes of cold absolute ethanol to precipitate out RNA and leave it on ice for 1 h.
5. Centrifuge at 2000 × *g* for 5 min at 4°C. Discard the supernatant and wash the RNA pellet with about 10 ml of cold 70% ethanol, then absolute ethanol and finally with diethyl ether. Centrifuge at each washing step at 2000 × *g* for 5 min and collect the pellet. Air dry and weigh the pellet.
6. Calculate the amount of RNA per g of yeast cells. Generally RNA constitutes about 4% of the dry weight of yeast cells.

Precautions

Exercise all precautions mentioned in Section 14.2(d).

14.2.4 EXPERIMENT: Isolation of total RNA from plant tissue by SDS-phenol method

Principle

This is a general, relatively simple, economical and an all purpose procedure for isolation of RNA from leaves, shoots, roots and other plant tissues. SDS and phenol dissociate nucleoprotein complex and denature proteins. Further treatment with phenol : chloroform : isoamyl alcohol removes denatured proteins and RNA is recovered by precipitation with ethanol from the aqueous phase while DNA remains confined in the phenol phase.

Materials and Reagents

1. Refrigerated centrifuge.
2. Plant tissue such as leaves, embryos etc.
3. Phenol reagent: It contains the following components-

Redistilled phenol	500 g
8-Hydroxyquinoline	0.5 g
Water	150 ml
m-cresol (redistilled)	70 ml (optional)

4. SDS solution: It composes of the following reagents – 1% SDS, 0.15 M sodium acetate, 5 mM EDTA, 50 mM Tris-HCl (pH 9.0). Store the solution at room temperature.
5. Deproteinizing solution: Mix phenol reagent (Reagent No. 3) with an equal volume of chloroform and add isoamyl alcohol, an antifoaming agent so that its concentration is 4% (*v*/*v*).
6. Absolute ethanol.
7. Sodium acetate buffer: Prepare 0.15 M solution of sodium acetate and adjust its pH to 6.0 with acetic acid and then add SDS so that its concentration is 0.5% (*w*/*v*). Store this reagent at room temperature.

Procedure

1. Homogenise the plant tissue in a pestle and mortar in SDS solution at room temperature. The ratio of tissue weight to the SDS solution should be approximately 1:10.
2. Transfer the homogenate to a polypropylene centrifuge tube (do not use polycarbonate tubes). Add an equal volume of deproteinizing solution and mix for 5–10 min at room temperature.
3. Centrifuge the tubes at 10,000 × *g* for 10 min at 4°C. Three layers will be formed, with the upper aqueous phase containing RNA, the middle layer of denatured proteins and lower layer of the organic solvents.
4. Carefully remove the aqueous phase, taking care not to include the protein interphase and transfer it to a clean centrifuge tube. If the aqueous phase contains significant amount of proteins, re-extract RNA by repeating Steps 2 and 3.

5. Add about 2.5 volumes of chilled absolute ethanol to the aqueous phase, mix well and let it stand at – 20°C overnight or at – 70°C for 1 h.
6. Collect the precipitated nucleic acid by centrifugation at 10,000 × *g* for 10 min at 4°C. Discard the supernatant.
7. Air dry the pellet. Over drying should be avoided. Add a few ml of sodium acetate buffer to dissolve the pellet. Reprecipitate RNA from this solution by adding 2.5 volumes of ethanol and keeping it at –20°C overnight or at – 70°C for 1 h.
8. Centrifuge at 10,000 × *g* for 10 min at 4°C. Discard the supernatant and wash the pellet with 70% chilled ethanol.
9. The nucleic acid pellet can be stored in 70% ethanol in a freezer until use.

Precautions

1. Rinse glassware/plasticware with 1% diethylpyrocarbonate solution to inactivate RNase.
2. Exercise all precautions mentioned in Section 14.2(d).

14.2.5 EXPERIMENT: Isolation of mRNA by affinity chromatography

Principle

Messenger RNAs account for only 1–2% of total RNA in plant tissues. Eukaryotic mRNAs possess a poly (A) tail consisting of 3–200 adenylate residues at their 3′ ends and this characteristic structural property has been utilized for their separation from other species of RNA (tRNA and rRNA). Under high ionic strength, the poly (A) tail of mRNA binds with complementary base sequences such as poly U or oligo (dT) which are covalently attached to an appropriate affinity chromatographic media. Transfer RNA (tRNA) and ribosomal RNA (rRNA) which lack poly (A) tail do not interact with these sequences. After the unbound RNA molecules have been washed out, the bound mRNAs are released by lowering ionic strength of the elution buffer and are collected as they emerge out of affinity chromatography column.

Materials and Reagents

1. Spectrophotometer
2. Column for chromatography: A small column such as a Pasteur pipette or syringe is used for this purpose
3. Extracted RNA: Isolate total RNA as described in EXPERIMENT 14.2.3 or EXPERIMENT 14.2.4
4. Oligo (dT) cellulose
5. Binding buffer (pH 7.5): It is composed of 0.01 M Tris, 0.5 M NaCl, 0.001 M Na_2-EDTA. Add 10% SDS just before using this buffer and adjust the final pH of this solution to 7.5 with HCl
6. Elution buffer (pH 7.5): The elution buffer comprises of 0.01 M Tris, 0.001 M Na_2-EDTA and 10% SDS. Adjust pH of this solution to 7.5 with HCl

7. Ethanol
8. NaCl (solid)
9. TE buffer: To 1 ml of 1.0 M Tris-HCl (pH 8.0) add 0.4 ml of 0.25 M EDTA (pH 7.0) and make the final volume to 100 ml with distilled water

Procedure

1. Suspend about 500 mg of oligo (dT) cellulose in 20 ml of sterile binding buffer.
2. Prepare a column by packing the above affinity media into a Pasteur pipette or a syringe plugged with glass wool at the base. Equilibrate the column by passing 25–30 ml of the binding buffer through it.
3. Dissolve the isolated total RNA (for details see EXPERIMENT 14.2.4) in the binding buffer at a conc of approximately 1 mg/ml.
4. Load the RNA solution onto the column and recycle the initial effluent again through the column.
5. Wash the column exhaustively with the binding buffer till the absorbance at 260 nm of the column effluent reaches zero. The poly (A) containing mRNA is retained in the column while tRNA and rRNA are washed out during this step.
6. Now pass the elution buffer through the column. Collect 1 ml fractions each.
7. Measure the A_{260} of each fraction and pool the A_{260} absorbing fractions.
8. Add sodium chloride (solid) to the poly (A) containing RNA pooled fractions so that the final conc of NaCl in this preparation is 0.5 M.
9. Add 2 volumes of ice cold ethanol to it and keep at – 20°C overnight to precipitate mRNA.
10. Dissolve the precipitate in 1 ml TE buffer and estimate the amount of mRNA spectrophotometrically as described in EXPERIMENT 5.2.4.
11. After the bound mRNA has been eluted, the affinity material can be preserved for reuse at a later stage by passing 0.1 N NaOH through the column followed by washing with sufficient volume of buffer containing 0.1 % sodium azide and storing it at 0–4°C. Before use, the column must be thoroughly washed by passing 3–4 volumes of the elution buffer to remove azide.

Precautions
All the general precautions listed in Section 14.2(d) must be observed.

14.2.6 EXPERIMENT: Isolation of plasmid DNA by alkaline method

Principle
The bacterial cells are lysed by lysozyme and SDS at high pH and the lysate is then neutralized. The plasmid DNA undergoes renaturation under these conditions but the chromosomal DNA remains denatured and gets precipitated out in the form of a protein-DNA-SDS complex.

Materials and Reagents

1. Eppendorf centrifuge or microfuge
2. *E. coli* DH 5 α cells containing plasmid pUC 18 or any other strain possessing plasmid DNA
3. LB-Medium (Luria-Bertaini): Dissolve 10 g of Bacto-tryptone, 5 g of Yeast extract and 5 g NaCl in distilled water, ȧdjust the pH to 7.5 with NaOH and make final volume to 1 L with water. Autoclave it at 15 psi for 30 min
4. Solution A (pH 8.0): Prepare 25mM Tris-HCl(pH 8.0) containing 50mM glucose and 10 mM EDTA. Just prior to use, add lysozyme so that its concentration is 1% (*w*/*v*) in this solution.
5. Solution B: In 0.2 M NaOH solution add SDS to a final concentration of 1%
6. Solution C: Make 3 M potassium acetate (pH 4.8)
7. Solution D: It constitutes of 50 mM Tris-HCl (pH 8.0), containing 100 mM sodium acetate
8. Tris-EDTA (TE) buffer: 10 mM Tris-HCl (pH 8.0) containing 1 mM EDTA

Procedure

1. Grow *E. coli* DH 5 α cells overnight in LB medium at 37°C.
2. Transfer 1 ml of the above culture in an Eppendorf tube and collect the cell pellet after centrifugation at 5000 rpm for 5 min in a microfuge.
3. Resuspend the cells in 100 μl of solution A and place it on ice for 30 min.
4. Add 200 μl of freshly prepared Solution B and mix thoroughly by vortexing it briefly and keep on ice for 5 min. During this step the cells get lysed and the solution becomes viscous.
5. Add 150 μl of Solution C, vortex briefly and keep on ice for 1 h or in a freezer for 10 min.
6. Centrifuge for 5 min at 4000 rpm in microfuge or Eppendorf centrifuge at 4°C for 15 min.
7. Transfer the supernatant to a fresh Eppendorf tube and add two volumes of chilled ethanol. Leave it on ice for 10 min.
8. Again centrifuge it for 5 min at 5000 rpm at room temperature in an Eppendorf centrifuge/microfuge.
9. Discard the supernatant and dissolve the pellet in 100 μl of solution D. Add two volumes of ethanol, mix well and place it on ice for 10 min. Centrifuge at full speed for 5 min in Eppendorf centrifuge. Collect the pellet which contains DNA.
10. Air dry the DNA pellet and dissolve in 50 μl TE buffer.
11. Check the quantity and quality of DNA preparation as described in EXPERIMENT 5.2.4.

14.3 Fractionation of Nucleic Acids

Various classes of nucleic acids, i.e. DNA (single stranded or double stranded) and RNAs (rRNA, tRNA, mRNA etc.) can be separated by any of the following three methods:

(a) Density gradient centrifugation
(b) Column chromatography
(c) Polyacrylamide or agarose gel electrophoresis

(a) Density gradient centrifugation

(i) Cesium chloride (CsCl) density gradient centrifugation

In a density gradient of CsCl, which gets established during centrifugation, different molecules sediment to positions in the gradient which correspond with their buoyant densities. Various types and forms of nucleic acids differ in their buoyant density. Buoyant density of RNA is much higher (1.9 g/ml) than single stranded DNA (1.71 g/ml) which in turn is higher than that of double stranded DNA (1.70 g/ml) and hence these get separated during density gradient centrifugation. Buoyant density of double stranded DNA is in fact, a function of mole fraction of (G+C).

Plasmid DNA can be separated from bacterial chromosomal DNA on ethidium bromide-CsCl density gradient centrifugation. Circular plasmid DNA binds less of ethidium bromide and hence has higher buoyant density than linear chromosomal DNA which binds more ethidium bromide.

(ii) Sucrose density gradient centrifugation

This is the most commonly used method for fractionation of RNA molecules. In this case the separation occurs primarily on the basis of differences in size rather than equilibrium phenomenon of the macromolecules in the sample. Total cell RNA comprises of different RNA types including rRNA, tRNA and mRNA which vary in their size from approximately 100 bases in tRNA to more than 5000 bases in mRNA precursors and these can thus be resolved on a gradient based on differences in their sedimentation rates. Sucrose density gradient method can also be used for separation of polysomes of different sizes as well as obtaining large and small subunits of ribosomes.

(b) Column chromatography

Separation of nucleic acids can be achieved by chromatography on anion exchanger such as methylated albumin kieselguhr (MAK), gel permeation matrices like Sephadex G-50, G-100 etc. or adsorbants like hydroxylapatite. For theory and principles regarding different types of chromatographic techniques see Chapter 9.

(c) Polyacrylamide and agarose gel electophoresis

RNA and small DNA molecules (less than 1 kb) are generally separated on polyacrylamide gels while larger DNA molecules are fractionated on agarose gels of varying concentrations. Nucleic acid molecules migrate under the influence of electric field towards anode and their separation is achieved due to the differences in their size. Details of polyacrylamide and agarose gel electophoresis techniques are given in Chapter 10.

14.3.1 EXPERIMENT: Separation of different species of RNA by sucrose density gradient centrifugation

Principle

As explained above, the various forms of RNA in a cellular preparation get separated due to their dissimilar sedimentation rates during centrifugation through a sucrose density gradient because of differences in their molecular masses. Each type of RNA gets concentrated as a layer or zone in the density gradient. After centrifugation, the tubes are punctured at the bottom and a fixed volume of the dripping out droplets of the fluid is collected in each tube to obtain a number of fractions. A_{260} of each fraction is then recorded and the various species of RNA will be found to be recovered in the form of peaks each constituting of a number of successive fractions.

Materials and Reagents

1. Density gradient maker
2. Ultracentrifuge
3. Spectrophotometer
4. Total RNA: Extract total RNA from plant tissues according to the procedure given in EXPERIMENT 14.2.4. Adjust the conc of RNA so that the preparation contains 100 μg RNA/μl of TE buffer
5. Tris-EDTA (TE) buffer: Make 10 mM Tris-HCl (pH 8.0) containing 1 mM EDTA
6. Sucrose
7. Sodium acetate: Prepare a 3 M sodium acetate solution and adjust its pH to 6.0 with glacial acetic acid
8. Ethanol

Procedure

1. Prepare a gradient of 5–40% sucrose in TE buffer using a density gradient maker in a Beckman SW41 tube (or equivalent).
2. Prepare a solution containing 100 μg of RNA per μl in TE buffer and carefully layer 0.1 to 0.5 ml of it on top of the gradient.
3. Centrifuge at 4°C for 16 h at 80,000 × *g*.
4. Pierce bottom of the centrifuge tube with a hypodermic needle and collect 0.5 ml fractions each in different tubes.
5. Measure the absorbance at 260 nm, if necessary, after an appropriate dilution of each fraction.
6. Dilute the fractions containing RNA with an equal volume of water, add 0.1 volume 3 M sodium acetate (pH 6.0) and 2.5 volumes ethanol and keep at –20°C overnight for precipitating out RNA.
7. After centrifugation at 10,000 × *g* for 10 min at 4°C collect the pellet and wash it with 80% ethanol to remove traces of sucrose.
8. Add 1–2 ml ethanol to the RNA pellet and store at – 20°C or – 80°C.

Precautions

1. Layering of RNA solution on the top of sucrose gradient should be done as gently as possible.
2. Use sterile gloves while performing various operations.
3. Only properly sterilized glassware and plasticware should be used. The glassware/plasticware should be washed with 1% (*w*/*v*) diethylpyrocarbonate solution before being autoclaved.

14.3.2 EXPERIMENT: Separation of single stranded and double stranded DNA

Principle

At low ionic strength, double stranded DNA binds to hydroxylapatite (calcium phosphate gel) while single stranded DNA or RNA are washed out of the column. This differential behaviour is exploited for separation of double stranded DNA from single stranded DNA as well as from RNA.

Materials and Reagents

1. Boiling water bath
2. Small chromatographic column or disposable 3 ml plastic syringe may be used as a column
3. Eppendorf centrifuge
4. Hydroxylapatite slurry: Suspend hydroxylapatite powder (Bio-Gel HTP) in 0.01 M sodium phosphate buffer (pH 6.8)
5. 0.01 M sodium phosphate buffer (pH 6.8)
6. 0.12 M sodium phosphate buffer (pH 6.8)
7. 0.4 M sodium phosphate buffer (pH 6.8)
8. 2-Butanol
9. Total DNA: Use DNA sample prepared as described in EXPERIMENT 14.2.2. Precipitate DNA from the preparation with 2 volumes of chilled ethanol. Air dry the pellet and dissolve in appropriate volume of 0.01 M sodium phosphate buffer (pH 6.8) so as to give a solution containing 100–200 μg DNA/ml buffer. Boil the solution in a boiling water bath for 5 min and allow it to cool slowly at 25°C

Procedure

1. Plug the base of the column or syringe with glass wool and pour enough slurry of hydroxylapatite into it to obtain a column with a packed volume of about 1.0 ml.
2. Wash the column with several volumes of 0.01 M sodium phosphate buffer (pH 6.8).
3. Load the sample containing the nucleic acid onto the column and continue to pass about another 3 ml of this buffer through the column.
4. Drain the excess of buffer and add one bed volume of 0.12 M sodium phosphate buffer (pH 6.8) preheated to 60°C and collect 0.5 ml fractions each.

5. Now add one bed volume of 0.4 M sodium phosphate buffer (pH 6.8) preheated to 60°C.
6. Collect 0.5 ml fractions. Repeat Step 5 two times more. Let the eluate cool down to room temperature.
7. For extracting double stranded DNA, add an equal volume of 2-butanol to each fraction containing the nucleic acid as indicated by A_{260}. Mix by vortexing and centrifuge at 12,000 × *g* for 20 sec in an Eppendorf centrifuge at room temperature.
8. Discard the upper organic phase and repeat Step 7.
9. Again discard the organic phase and recover DNA from aqueous phase by precipitation with 2 volumes of ethanol at 0°C.

14.3.3 EXPERIMENT: Separation of RNA by anion-exchange column chromatography on MAK (Methylated Albumin Kieselguhr) column

Principle

Nucleic acids bind to an anion exchanger due to their negatively charged phosphate groups and these can then selectively be eluted by changing the ionic strength of the elution buffer. The method is commonly used for fractionation of different types of RNA. Kieselguhr acts as an inert support material for albumin. The negative charges on albumin are masked by methylation while the positive charges remain exposed and are thus available for binding nucleic acids which are negatively charged.

Materials and Reagents

1. Gradient maker
2. Spectrophotometer
3. Glass column
4. Kieselguhr
5. Methylated albumin: Prepare by suspending 5 g albumin in 500 ml absolute methyl alcohol and add 4.2 ml of 12 N HCl. The protein dissolves and eventually precipitates again. Allow the mixture to stand in the dark for 3 days or more with occasional shaking. Collect the precipitate in 250 ml centrifuge bottles, wash twice with methyl alcohol and twice with anhydrous ether in the centrifuge. Evaporate most of the ether in air and then *in vacuuo* over KOH till a dry powder is obtained and store it in dessiccator over KOH. Failure to remove promptly the residual acid lowers the basicity of the final product. The material is readily soluble in water and is stable both in solution and in dry form. Reproducible batches can easily be made. Make 1% solution of dry powder in water
6. NaCl solutions (0.1, 0.4 and 2 M)
7. Nucleic acid sample containing approx. 500 μg RNA in 1–2 ml of 0.4 M NaCl solution

Procedure

1. Suspend 20 g of kieselguhr in 100 ml of 0.1 M NaCl (pH 7.0) in a beaker and boil the suspension briefly to expel air. Cover the beaker, cool the solution and then add 5 ml of 1% (*w*/*v*) solution of methylated albumin and stir well. Wash the material thoroughly with 0.4 M NaCl and finally suspend it in 0.4 M NaCl. Label it as MAK suspension.
2. Boil and then cool a suspension of 6 g kieselguhr in 40 ml of 0.1 M NaCl. Allow it to cool and then pour it down the column whose base is plugged with glass wool. When kieselguhr settles down, add 10 ml of MAK suspension (prepared in Step 1) to form a second layer. The third layer in the column is obtained by adding a suspension of 1 g kieselguhr in 10 ml of 0.4 M NaCl. Wash the column with 0.4 M NaCl.
3. Load the nucleic acid sample on surface of the ion exchanger and elute the column with a 0.4 M to 2 M linear gradient of NaCl solution. Collect 5 ml fractions and record A_{260} of each fraction.
4. Pool fractions showing absorbance at 260 nm of the individual peaks and calculate the amount of nucleic acid in the pooled fractions as described in EXPERIMENT 5.2.4.

Note: The nucleic acids are eluted in the following order: rRNA, tRNA and finally mRNA.

14.4 Restriction Enzyme Digestion, Blotting and Hybridization of Nucleic Acids

Restriction enzyme digestion

This is done with the help of special class of enzymes called restriction endonucleases. There are three categories of restriction enzymes, namely, Type 1, Type II and Type III. Among these Type II enzymes cleave DNA at specific sites and are hence important tools in analysis and molecular cloning of DNA. Their recognition sites are generally 4–6 bp long palindromic sequences (sequences with two fold rotational symmetry) and cleavage often results in production of specific fragments with protruding overhangs that are complementary in nature, also called sticky or cohesive ends. For example, the restriction enzyme EcoR I recognises the following sequence and cleaves at sites shown by arrows to produce 5′ overhangs:

↓
5′——GAATTC——C′
3′——CTTAAG——5′
↑

↓

5′——G   +   5′AATTC——3′
3′——CTTAA 5′   G——5′

Some commonly used restriction enzymes alongwith their recognition sequences are listed in Table 14.1. In general, different restriction enzymes recognise different sequences. However, there are examples of enzymes isolated from different sources that cleave within the same target sequence. Such pairs of enzymes are called isoschizomers.

Table 14.1 Sequence specificity of some commonly used restriction endonucleases

Enzyme	*Source*	*Recognition sequence*
EcoR I	*Escherchia coli* Rr 13	G ↓A A T T C
Hind III	*Haemophilus influenzae* Rd	A ↓A G C T T
Bam HI	*Bacillus amyloliquefacience* H	G ↓G A T C C
Bgl II	*Bacillus globigi*	A↓ G A T C T
Mbo I	*Moraxella bovis* M	↓G A T C
Pst I	*Providencia stuartii*	C T C G A↓ G
Sau 3A	*Staphylococcus aureus* 3 A	↓G A T C
Taq I	*Thermus aquaticus*	T ↓C G A
Xba I	*Xanthomonas badrii*	T↓ C T A G A
Xho I	*Xanthomonas holicola*	C↓ T C G A C

Note: Arrow (↓) indicates the site of cleavage.

Activity of restriction enzymes is expressed in terms of units. One unit is defined as the enzyme required to digest 1 μg substrate (lambda or plasmid DNA) in 1 h under standard assay conditions. The assay conditions vary from enzyme to enzyme but generally they require $MgCl_2$ (10 mM), Tris–HCl buffer (10 mM, pH 7.8), NaCl (10–100 mM) and in most cases, an incubation temperature of 37°C. Commercially available restriction enzymes are supplied alongwith appropriate buffers and instructions for their assay conditions. Buffers for different restriction enzymes differ chiefly in the concentration of NaCl that they contain and are grouped as low (0–10 mM NaCl), medium (50 mM NaCl) and high (100 mM NaCl) ionic strength buffers.

Any alteration in the prescribed assay conditions such as change in pH or ionic strength of buffer and presence of glycerol (> 5%), often results in change in their recognition specificity and is called *star activity*. Restriction enzymes are stored as stocks (10X) in 50% glycerol at –20°C and thus have to be diluted in such a way that glycerol concentration in the assay mixture is less than 5%.

Nucleic acid blotting

Presence of a specific nucleic acid fragment in a digested sample is ascertained out by hybridization technique. This, however, involves an initial blotting of electrophoretically separated pieces of nucleic acids of the digested sample. The transfer of DNA fragments onto nitrocellulose paper or nylon membrane is called Southern blotting which has been named after its inventor Southern. The technique enables detection of very low concentrations of specific sequences within complex DNA populations and has wide applications such as identification of transformants, study of RFLPs and DNA fingerprinting. Hybridization efficiency of a labelled probe directly to DNA fragments embedded in a gel is rather low and such an approach, therefore, has poor sensitivity. Hence, after separation of restriction fragments of DNA by electrophoresis on agarose gels, they are transferred and fixed onto a membrane filter (blotting) and then hybridized with a radioactively labelled DNA or RNA probe to establish whether any of the blotted spots contains piece of the sequence under investigation. In a similar way, RNA molecules can be transferred to a nitrocellulose paper and the technique is known as Northern blotting by analogy to Southern blotting. This procedure has been used to detect specific mRNAs within complex RNA populations and hence is extremely valuable in studying the differential expression of specific genes in response to particular physiological or environmental stimulus.

DNA or RNA samples may also be applied directly as dots on the nylon membrane. The technique is called dot blotting and is often used for rapid and accurate measurement of abundance of specific RNA or DNA sequences.

A number of solid supports such as diazobenzyl-oxymethyl (DBM)-cellulose, o-aminophenylthioester (APT)-cellulose, nitrocellulose filter and nylon membranes (Hybond N, Hybond N^+, Genescreen etc.) are available. Nitrocellulose filter is generally recommended for transfer of RNA molecules while for transfer of DNA nylon membranes are commonly used.

Nucleic acid hybridization

The two strands of DNA are held together by hydrogen bonds between complementary bases. When the duplex molecule is subjected to conditions such as high temperature or alkali treatment, the two strands get separated resulting in denaturation of DNA. Under favourable conditions, renaturation or reannealing of the separated single strands into a duplex state can be restored. Reannealing of DNA is not limited to single strands of the same DNA molecule but can occur between DNA from different sources containing similar sequences and thus, has been used to study the degree of relatedness among organisms. The formation of DNA-RNA hybrids is also possible. The DNA sequence, which is used for examining hybridization with the fragments of digested nucleic acid is referred to as a probe.

Nucleic acid hybridization is a function of temperature, time, salt concentration, G+C content, probe length and hybridization volume. For DNA with a G+C content of 50%, temperature of 65°C is generally chosen which is approximately 20°C below the actual melting temperature (T_m) of DNA at standard salt concentration of 0.5 M NaCl. Addition of compounds such as formamide decreases the T_m of nucleic acids. So hybridization in presence of 50% formamide is usually done at 42°C. Hybridization volumes should be kept to a minimum to ensure maximum encounter rates between the probe and the bound DNA or RNA. Therefore, inert polymers like dextran sulfate are added since they have large hydrated volume and take up most of the space in the hybridization solution. At low ionic strength, hybridization of nucleic acids occurs slowly and can be improved by increasing the ionic strength. Prior to hybridization, the reactive empty sites on the membrane need to be blocked by non-specific DNA such as denatured salmon sperm DNA and gelatin to avoid background attachment of the probe to the membrane.

A wide variety of methods are available for labelling the probes among which nick translation and random hexamer labelling methods are most common. In nick translation, a nick is created in double stranded DNA by brief action of DNase 1. DNA polymerase I then replaces the strand starting at the point of nick with a new strand which gets radioactively labelled due to incorporation of ^{32}P-dNTP in the reaction mixture. In random hexamer labelling, synthetic hexanucleotides of random sequence are used as primers. The double stranded DNA is denatured by boiling and primers are annealed. Klenow enzyme (larger fragment of DNA polymerase I) then extends the primers and synthesizes a strand complementary to the template using four dNTPs one of which is labelled radioactively thus incorporating label into the probe.

Now-a-days non-radioactive labelling and detection systems are also available and have advantages over the radioactive ones in that they are safer to use, have much longer usable life and can thus be stored for much longer periods of upto 12 months. Some of the examples of non-radioactive systems include Biotin-labelling system, Digoxigenin-dUTP labelling and Electrochemiluminescence (ECL) system. The labelling and detection protocols for non-radioactive probes are supplied alongwith the kits which are available commercially. Detection of probe nucleic acid target hybrids can be accomplished by direct or indirect labelling methods. In the former case, label is attached directly to the nucleic acid by a covalent bond. For example, in ECL direct labelling system, DNA or RNA probes are labelled directly with the enzyme horse radish peroxidase. Detection is done based on generation of light in the form of enhanced chemiluminescence which is produced during oxidation of luminol (substrate) in a reaction catalyzed by the enzyme bound to the nucleic acid probe. Chemiluminescence thus generated is recorded by exposure for brief period (1h) on blue light sensitive autoradiography film. Indirect labelling employs a hapten (e.g. biotin-dUTP or digoxigenin-dUTP) attached to the nucleic acid probe. The hapten is detected using a labelled specific binding protein (e.g. avidin or antidigoxigenin) which gets bound to the labelled probe hybridized to the target DNA on one end and has an enzyme such as alkaline phosphatase or horse radish peroxidase bound at the other end, the later catalyses a reaction resulting in generation of coloured product or chemi-luminescence depending upon the substrate. A few laboratory exercises pertaining to these molecular techniques are as follows.

14.4.1 EXPERIMENT: Preparation of restriction enzyme digests of DNA samples

Principle

Restriction endonucleases (Type II) recognize specific sequences and cut double stranded DNA within their recognition sequence to produce fragments with protruding or blunt ends. Fragments thus produced get separated on agarose gel electrophoresis and are visualized after staining with ethidium bromide, a dye which intercalates between bases of DNA and gives orange coloured fluorescence under UV light.

Materials and Reagents

1. Microfuge
2. Horizontal slab gel electrophoresis apparatus
3. UV transilluminator
4. Water bath at 37°C
5. Eppendorf tubes (1.5 ml capacity) and rack
6. Adjustable micropipette (0–20 μl capacity)
7. DNA samples (1 μg/μl): Dissolve each of the following DNA samples in a suitable buffer (10 mM Tris-HCl, pH 8.0) at concentration of 1 μg DNA/μl.
   (i) Plant DNA: use plant DNA isolated as in EXPERIMENT 14.2.2
   (ii) Plasmid DNA: Plasmid preparation obtained in EXPERIMENT 14.2.6 may be used.
   (iii) λ DNA
8. Restriction enzymes: EcoRI and Hind III (use 2 units of enzyme/μg DNA to be digested). It is essential that the samples of restriction enzymes be taken out from –20°C freezer just before being used and these should immediately be kept on ice.
9. Restriction buffer (10 ×): As supplied by the supplier alongwith the enzyme.
10. Sterile water
11. Stop solution: Make a 20% solution of sucrose containing 10 mM sodium EDTA and 1% bromophenol blue
12. Ethidium bromide: Dissolve 10 mg/ml of ethidium bromide in distilled water

Procedure

1. Prepare reaction mixtures by adding the following components to different Eppendorf tubes using a micropipette:

Reagent	*Tube number*					
	1	*2*	*3*	*4*	*5*	*6*
	µl of the reagent					
DNA sample						
λ DNA	2	—	—	2	—	—
Plasmid DNA	—	2	—	—	2	—
Plant DNA	—	—	10	—	—	10
Restriction buffer (10 ×)	2	2	2	2	2	2
Sterile water	14	14	6	14	14	6
Eco RI	2	2	2	—	—	—
Hind III	—	—	—	2	2	2
Total volume (μl)	20	20	20	20	20	20

Note: The concentrations of restriction enzyme and buffer in the final volume of reaction mixture should be 1 ×. So, the reaction volumes should be adjusted accordingly.

2. Mix contents of each tube carefully and spin at full speed in a microfuge for 1 min.
3. Incubate the mixture at 37°C overnight for the restriction digestion of DNA samples to occur.
4. Stop the reaction by adding 5 μl of the stop solution to each tube.
5. Mix and spin down as before in Step 2.
6. Load the entire samples in separate wells on 0.8% agarose gel slab and perform electrophoresis at 80 V as described in EXPERIMENT 10.4.5 till the coloured bromophenol blue migrates upto 1 cm from the opposite edge of the gel.
7. Stain the gel with ethidium bromide and view under UV transilluminator. Presence of DNA on the gel will be indicated by orange coloured bands or smears.

Plant DNA digested with restriction enzyme generally appears as an elongated trailing smear on the gel whereas distinct bands are normally obtained for plasmid and λ DNA.

Precautions

1. Restriction enzymes should be taken out of –20°C freezer only at the time of use and should be immediately kept on ice.
2. Ethidium bromide is a powerful mutagen. Gloves should be worn while working with solutions containing this dye.
3. Follow all the general precautions for handling nucleic acids as mentioned in Section 14.2(d).

14.4.2 EXPERIMENT: Southern (capillary) blotting of DNA fragments from agarose gel

Principle

It is a technique for transferring DNA molecules from agarose gel to a solid support such as nitrocellulose filter or nylon membrane. Gel embedded DNA is apurinated in the presence of HCl followed by denaturation in alkali. Denaturation of DNA fragments prior to blotting is essential for dissociating the two polynucleotide chains which can then hybridize with the probe. Denatured DNA molecules move upwards by capillary action of buffer and on coming in contact with the nylon membrane they get bound to it. A brief UV treatment (2-3 min) or heating at 80°C for 2 h under vacuum facilitates firm binding of DNA to the nylon membrane.

Materials and Reagents

1. Agarose gel containing DNA fragments: As obtained in EXPERIMENT 14.4.1
2. Nylon membrane (Hybond N or Gene Screen Plus)
3. Whatman 3 MM paper
4. Glass tray (5 cm deep × 20 cm × 30 cm) and a glass plate (12.5 × 25 cm)

5. Rough filter papers
6. Parafilm or used X-ray film
7. 0.25 M HCl
8. Denaturating solution: Prepare 1.5 M NaCl solution containing 0.25 M NaOH
9. Neutralizing solution: Make 0.5 M Tris-HCl buffer containing 1.5 M NaCl and 1 mM EDTA. Adjust pH of this solution to 7.2
10. SSC solution (10 ×): It is made by preparing a solution of 1.5 M NaCl containing 0.15 M tri-sodium citrate. Adjust the pH to 7.0
11. SSC solution (2 ×): Dilute the above solution 5 times

Procedure

1. After electrophoresis, transfer the agarose gel to a glass tray containing sufficient HCl (0.25 M) to cover the gel and gently stir for 15 min at room temperature on a platform shaker. Unless otherwise stated, all the subsequent operations are also carried out at room temperature.
2. Pour off HCl and rinse the gel with distilled water.
3. Denature the DNA by adding sufficient volume of denaturating solution (Reagent 8). Gently stir for 30 min.
4. Decant the alkali and rinse with distilled water.
5. Neutralize the gel in neutralizing solution and gently stir for 30 min.
6. Remove the solution and wash the gel with 2 × SSC.
7. Make a bridge on a glass tray with glass plate and place 2 layers of Whatman 3 MM paper cut to proper size on top so that the ends of the paper hang down into the tray. Wet the paper with 10 × SSC and fill the tray with the same solution.
8. Place another piece of Whatman 3 MM paper, which is cut to the size of the gel and prewetted in 10 × SSC, on top of the bridge. Carefully transfer the gel on top of the wetted filter paper and remove any entrapped air bubbles by rolling a glass pipette gently over the gel.
9. Cut a piece of nylon membrane (Hybond N or Gene Screen Plus) of exactly the same size as that of the gel. Mark the membrane on the left hand bottom corner with a permanent ink marker. Wet the membrane in 2 × SSC and place it on top of the gel. Cover the membrane with 2 pieces of 3 MM Whatman filter paper cut to the same size and prewetted in 2 × SSC. Avoid entrapment of any air bubbles between the papers.
10. Seal all the sides of the 3 MM Whatman paper along edges of the gel with parafilm or strips of used X-ray films.
11. Keep a stack of dry rough filter papers cut to the same size on top of the wet filter paper.
12. Place a glass plate on top of stack of filter papers, keep a weight of about 0.5 kg (500 ml water in a bottle) over it and allow the transfer of DNA from gel to nylon membrane take place overnight.
13. Remove the weight, stack of rough filter paper and the Whatman filter papers from top of the membrane. Carefully remove the membrane and wash it in 2 × SSC.

14. Expose the membrane for not more than 2 min to UV light to achieve cross-linking and a steadfast binding of transferred DNA. (*Note:* Do not cross-link if 0.4 M NaOH has been used for denaturation. In this case, bake the membrane at 80°C for 2 h in a vacuum oven.)
15. Air dry the membrane briefly, keep between 2 sheets of Whatman 3 MM paper and store at 4°C in a refrigerator till use.

Precautions

1. Use gloves while handling gel and nylon membrane.
2. Take care that no air bubbles are entrapped between gel, membrane and Whatman papers.
3. Make sure that the buffer moves through the gel and the membrane. The layers of papers, placed on top should not touch the layers of filter paper placed beneath the gel.
4. Adopt all the general precautions used in handling nucleic acids as described in Section 14.2(d).

14.4.3 EXPERIMENT: Preparation of a radiolabelled probe by random primer method

Principle

The random primer method is one of the techniques employed for preparing a labelled DNA probe. The original DNA is first denatured by boiling to obtain single stranded DNA template. Synthetic hexanucleotide primers of random sequence are added and these anneal with the template. The Klenow enzyme then extends the primers and synthesizes a radioactively labelled complementary strand due to incorporation of α-^{32}P dCTP from the reaction mixture into the elongating chain.

Materials and Reagents

1. Boiling water bath
2. Water bath (37°C)
3. Eppendorf tubes (1.5 ml capacity)
4. Micropipettes (0–10 μl and 1000 μl)
5. DNA probe: Plasmid DNA containing plant DNA insert (5 ng/μl)
6. Random primer labelling kit: These kits are available commercially and usually contain the following components: Random primer mix (10 $\times$) constitutes of random primers in a buffer and mixture of dNTP (dATP, dGTP, dCTP and dTTP at 100 mM concentration each)
7. Klenow enzyme
8. λ DNA Hind III digest: It is available commercially or prepare as described in Experiment 14.4.1.
9. α-^{32}P d CTP (1 μCi/μl)
10. TE buffer (pH 8.0): Add 0.4 ml of 0.25 M EDTA solution to 1.0 ml of 1 M Tris-HCl buffer (pH 8.0). Make volume to 100 ml with distilled water
11. SDS (10%)

Procedure

1. Take 5 μl of probe DNA (25ng) in an Eppendorf tube with a hole poked in the cap, add 20 μl water and 8.0 μl TE buffer to it.
2. Place the Eppendorf tube in a boiling water bath in a floater for 5 min to denature DNA. After this immediately keep the tube on ice.
3. Add the following reagents, while the tube is on ice (or follow the instructions supplied by the manufacturer of the labelling kit): dNTP mix (6 μl), Random primer mix (5 μl), Klenow enzyme (1 μl) and α-^{32}P dCTP (5 μl).
4. Incubate the tubes at 37°C for 60 min.
5. Stop the reaction with 2.5 μl of 10% SDS.
6. Add 500 μl of TE buffer and keep in a boiling water bath for 5 min.
7. Cool in ice for 5 min.
8. Label λ DNA (Hind III digest) in a similar way.

Precautions

1. To keep probe DNA in denatured form, after boiling keep the Eppendorf tube immediately on ice without any undue delay.
2. Radioactive substances are hazardous to health. Hence necessary precautions such as use of gloves and protective clothing must be observed while handling these materials. Never handle radioactive material with bare hands.
3. Carry out radioactive work only behind radioactive shield or screen.
4. All glassware/plasticware used for work with radioisotopes should be clearly marked with a radiation symbol.
5. Always wash your hands and monitor your self before leaving radioactive area.
6. Dispose off radioactive wastes in approved containers provided for the purpose.
7. Use all the general precautions for handling nucleic acids as mentioned in Section 14.2(d).

14.4.4 EXPERIMENT: Detection of a specific fragment of plant DNA from restriction digest by hybridization with a labelled probe

Principle

Complementary strands of DNA hybridize with each other under suitable conditions of temperature and ionic strength. After blotting (see EXPERIMENT 14.4.2), the membrane is treated with radioactively labelled probe which hybridizes with the specific plant DNA fragment bearing sequence homology with the probe and the radioactive spot on the membrane is detected by autoradiography. Prior to hybridization, the membrane is prehybridized with salmon sperm DNA to block the non-specific binding sites where the probe may otherwise get bound spuriously.

Materials and Reagents

1. X-ray cassette, X-ray film, plastic bags and boxes.

2. Boiling water bath.
3. Shaking water bath at 65°C.
4. Ultrafreezer (–70°C).
5. Portable Geiger Müeller Counter.
6. HSB (5 ×): This mixture comprises of 3M NaCl, 100mM PIPES and 25 mM Na_2EDTA. It is prepared by dissolving 175.3 g of NaCl, 30.3 g of PIPES, 7.5 g of Na_2EDTA in distilled water and adjusting the pH to 6.8 with 4 M NaOH. Make the final volume to 1 L with distilled water.
7. Carrier DNA: To 0.5 g salmon sperm DNA, add 100 ml distilled water and autoclave. Store in refrigerator.
8. Denhardt's III: It contains 2 g gelatin, 2 g ficoll, 2 g polyvinylpyrrolidone, 10 g SDS and 5 g tetrasodium pyrophosphate. Dissolve these components in water and make the volume to 100 ml.
9. TES solution: Prepare 10 mM Tris-HCl containing 1 mM EDTA and 0.1% SDS.
10. 10 × SSC: It comprises of 1.5 M NaCl and 0.15 M trisodium citrate. Make 0.5 ×, 1 × and 2 × SSC solutions by diluting the above solution accordingly.
11. Prehybridization solution:

5 × HSB	8.0 ml
Denhardt's III	4.0 ml
Salmon sperm DNA	4.0 ml
(Freshly denatured by boiling for 5 min)	
Sterile distilled H_2O	4.0 ml

Mix well and incubate at 65°C for 5 min.

12. Hybridization buffer: Prepare by mixing the following:

5 × HSB	4.0 ml
Denhardt's III	2.0 ml
25% Dextran Sulfate	8.0 ml
Salmon sperm DNA	0.5 ml
(Freshly denatured by boiling for 5 min)	
Sterile distilled water	5.5 ml

13. Nylon membrane containing DNA fragments (see EXPERIMENT 14.4.2).
14. Labelled probe (as obtained in EXPERIMENT 14.4.3).

Procedure

The experimental procedure consists of two distinct phases:

(a) Hybridization during which blotted DNA on nylon membranes is incubated with labelled probe.
(b) Autoradiography for location of any radioactive zones on the membrane.

Various steps involved in performing hybridization and autoradiography are described as follows:

(a) Hybridization:

1. Place the nylon membrane containing the immobilized DNA fragments on it in a plastic bag, pour 20 ml pre-hybridization solution and seal the bag without leaving any entrapped air bubbles.
2. Incubate the bag at 65°C in a water bath with gentle shaking for at least 2 h.
3. Denature the probe by incubating in a boiling water bath for 5 min and then quickly chill it on ice.
4. Cut open the plastic bag and drain pre-hybridization solution. Add labelled probe (freshly denatured by boiling for 5 min) to 20 ml hybridization solution and pour the mixture into the plastic bag. Seal the bag taking utmost care that no air bubbles are entrapped.
5. Incubate at 65°C overnight with gentle shaking for hybridization to take place.
6. Next day, cut open the bag, remove the membrane with forceps and place it in a box with a lid. Dispose off the plastic bag and the solution in radioactive waste reservoirs.
7. Wash the membrane successively with 200 ml of the following buffers in a vigorously shaking water bath:

First wash	: 2 × SSC, 0.1% SDS 20 min at room temperature
Second wash	: 1 × SSC, 0.1% SDS 20 min at 65°C
Third wash	: 0.5 × SSC, 0.1% SDS 20 min at 65°C

(b) Autoradiography

8. For autoradiography, dry the washed membrane between folds of filter paper and then wrap in saran wrap. Measure the strength of radioactivity signal with a hand Geiger Müeller counter.
9. In a dark room (with red safety light) place the membrane in a film cassette with the DNA containing surface facing upwards and place a sheet of X-ray film on top of it. Carefully close the cassette and keep at –70°C for 2-3 days.
10. Develop the film in the dark room under red safety light for 3 min in X-ray developer, rinse briefly with deionised water and fix for 3 min in X-ray fixer. Rinse with deionised water for 5 min and let the film air dry.

Precautions

1. Follow all precautions for handling radioactive substances (see EXPERIMENT 14.4.3).

14.5 Polymerase Chain Reaction

Polymerase chain reaction (PCR), nicknamed as People's Choice Reaction, was devised by Karry Mullis in 1985 and is one of the most widely used basic molecular biology techniques due to its remarkable speed, specificity and sensitivity. Basically, it is an *in vitro* method for the enzymatic amplification of specific DNA sequences using two oligonucleotide primers that flank the region of interest in the target DNA. Amplification is done in a thermocycler and involves a repetitive series of cycles of template denaturation, primer annealing and the extension of annealed primers by DNA polymerase resulting in an exponential accumulation of a specific DNA fragment. Because the primer extended products synthesized in one cycle can also serve as a template in the next cycle, the number of target DNA copies approximately doubles in each cycle and results in a million-fold amplification in about 25 cycles. Thus this technique is particularly useful in production of probes in large quantities for molecular biology studies.

14.5.1 EXPERIMENT: To amplify DNA using PCR

Principle
As outlined earlier (Section 14.5).

Materials and Reagents

1. Thermocycler (PCR machine)
2. Eppendorf tubes (0.5 ml capacity)
3. Micropipettes
4. Template DNA (2 μg/10 μl)
5. 10 × PCR buffer: 100 mM Tris-HCl (pH 8.3) containing 500 mM KCl, 15 mM $MgCl_2$ and 0.1% gelatin
6. dNTPs mix: Mixture of dATP, dCTP, dTTP and dGTP each at a conc of 1.25 mM
7. Primer pair: Dissolve each primer strand in 5 ml of distilled H_2O. Primers are available commercially
8. Taq DNA polymerase
9. Mineral oil
10. Sterile distilled water

Procedure

1. In a sterile Eppendorf tube (0.5 ml capacity) mix the following stepwise: sterile distilled H_2O, 5.3 μl; 10 × PCR buffer, 10.0 μl; dNTP mix, 16.0 μl; Primer 1, 5.0 μl; Primer 2, 5.0 μl; template DNA, 10.0 μl.
2. Incubate the reaction mixture at high temperature (94°C) for 5 min to denature the DNA completely.
3. To this add 0.5 μl of Taq DNA polymerase and give a quick spin in a microfuge to collect the reaction mixture at the bottom.
4. Cover surface of the mixture with a drop of mineral oil to prevent evaporation of sample during repeated cycles of heating and cooling in thermal cycler.
5. Carry out the cyclic amplification in a thermal cycler by setting the following temperatures and time periods for each step of a cycle:

Denaturation	94°C	1 min
Annealing	35°C	2 min
Polymerization	72°C	3 min

Operate the thermal cycler for a total number of 30 cycles. Depending upon the template and primer pairs, these conditions can be varied.

6. After amplification, take out the PCR products carefully leaving behind the top mineral oil layer in the PCR tube. Store the products at –20°C.
7. Load a small fraction (5–10 μl) of PCR products on 1–2 % agarose gel alongwith 1 kb ladder DNA as a standard and run at 100 V for 60 min. Stain the bands with ethidium bromide and visualize under UV-transilluminator. (for details of agarose gel electrophoresis, see EXPERIMENT 10.4.6).

Precautions

1. Keep separate pipettes, solutions and other materials specifically for PCR work.
2. Use all general precautions used in handling nucleic acids mentioned earlier in Section 14.2(d).

14.6 DNA Cloning

DNA cloning refers to the techniques that are used to manipulate, move, recombine and propagate DNA. In other words, technique of inserting genes across species and genetic barriers is termed as gene cloning, genetic engineering or recombinant DNA technology. Development of this technique became possible in 1973 only after the discovery of enzymes that cut and join DNA molecules, i.e. restriction enzymes and DNA ligase. This opened up new vistas in our fundamental understanding of gene structure, function and expression. Besides this, gene cloning has great practical applications such as production in large quantities of proteins of therapeutic and commercial importance. Micro-organisms can be designed for special tasks. Similarly, plants and animals can be engineered with traits that are useful in agriculture and allied fields.

Basically gene cloning involves five main steps namely, (i) gene isolation; (ii) selection of vector; (iii) ligation of foreign DNA in vector; (iv) transformation or introduction of the recombinant DNA into host and (v) selection of the desired clone.

(i) Gene isolation

For isolation of foreign DNA or gene of interest, total genomic DNA is isolated from an organism and is subjected to restriction endonuclease digestion. DNA fragments thus produced are fractionated by agarose gel electrophoresis and specific DNA fragment coding for a particular gene product is identified by hybridization with a labelled probe. An alternate procedure is to prepare genomic library (which is a set of recombinant DNA clones that contain fragments corresponding to the entire genome of an organism) and then to select clones having specific DNA sequences by hybridization with a labelled probe. Another approach is to synthesize the desired gene if the amino acid sequence of the gene product is known and the size of the gene is small. In this case, based on the amino acid sequence of the gene product, the corresponding nucleotide sequence of the gene is predicted and it is chemically synthesized. If expression of mRNA of the gene product is known, the most common approach is to isolate mRNA corresponding to the gene product and synthesize a complementary DNA (cDNA) using reverse transcriptase enzyme. The single stranded cDNA thus produced is then converted to double stranded cDNA by Klenow enzyme and is used for cloning. Usually a cDNA library (a set of recombinant DNA clones that contain fragments corresponding to the total mRNA of the cell) is prepared which is then screened for presence of the desired clone by hybridization with a labelled nucleic acid or antibody probe.

(ii) Selection of vector

Vectors are the vehicles for moving and delivering genes from one organism to another. Depending upon the host, vectors are categorized into four types—bacterial, yeast, plant and animal vectors. Bacterial vectors are of three types namely plasmids, bacteriophages and cosmids. Yeast vectors are based on 2 μm plasmid present in *Saccharomyces cerevisiae*. Yeast artificial chromosomes (YACs) are another kind of vectors that are commonly used for cloning big DNA fragments in yeast. For cloning in plants, Ti plasmid of *Agrobacterium tumefaciens* has been most commonly exploited though some plant viruses such as cauliflower mosaic virus (CaMV) and gemini viruses have also been used as vectors. Animal vectors are all derived from animal viruses such as SV40, vaccinia virus, adeno virus, baculoviruses etc. Besides vectors, direct gene transfer methods are also used for cloning in plants and animals. These include microinjection, electroporation, biolistic gun, liposome mediated gene transfer etc. In the following section features of some commonly used bacterial vectors are discussed.

(*a*) *Plasmids*—These are circular DNA molecules that replicate independently of the host chromosome. A plasmid for being useful as a cloning vector should be relatively small and should replicate in a relaxed fashion so that multiple copies are obtained. It should contain one or more selectable markers to allow identification of transformants, should have unique sites for a large number of restriction endonucleases in a region that is non-essential for replication but in genes with a readily scorable phenotype. Many plasmid vectors suitable for cloning have been developed by modifying naturally occurring plasmids. The most widely used plasmid vector is pBR322. It is 4.3 Kb in size, has its own origin of replication, unique restriction enzyme sites present in two readily scorable marker genes, i.e. genes for resistance to tetracyclin and ampicillin and has relaxed control of replication. A foreign gene when cloned in either tetracyclin or ampicillin resistance genes, recombinants can easily be identified by insertional inactivation, i.e. loss of function of the marker gene. A foreign DNA of size less than 10 Kb can be cloned in plasmid vectors.

(*b*) *Bacteriophages*—Among bacteriophages, λ and M13 phages of *E. coli* have been extensively exploited as cloning vectors.

λ phage: Bacteriophage λ DNA is 48.5 Kb long linear double stranded molecule with single stranded complementary tails of 12 nucleotides length at both ends. Soon after entering the host bacterium, the DNA circularizes through pairing of cohesive ends and undergoes one of the two modes of DNA replication, lytic and lysogenic. During lytic growth, the circular DNA is replicated many folds in the cell, progeny bacteriophage particles are formed and matured and the cell finally lyses releasing many new virus particles. During lysogenic growth phage DNA gets integrated into the host DNA and is replicated and transmitted to progeny bacteria like other chromosomal genes.

The wild type DNA contains several sites for most of the commonly used restriction enzymes and is not suitable as a vector. The central 1/3rd portion of the phage DNA is not essential for lytic growth and can be replaced by foreign DNA. Two types of vectors have been constructed from λ phage. Firstly insertional vectors which have a single restriction enzyme site at which foreign DNA can be inserted. The second type are the replacement vectors which have a pair of restriction enzyme sites flanking a stuffer fragment which can be removed and replaced by foreign DNA. Phage derived vectors can accomodate more DNA (upto 20 Kb) compared to plasmids and are, therefore, commonly used for construction of genomic libraries. There is a limit to the size of DNA that can be packaged inside phage. It cannot accomodate DNA of size 5% more or 78% less than the normal size of the phage DNA. This is an advantage since after the removal of the central stuffer portion, the deleted genome of the non-recombinants is too short to be packaged while recombinants having foreign DNA get packaged thus providing a means of positive selection for recombinants. Phage vectors that carry a segment of *E. coli* DNA coding for β-galactosidase have also been developed. Such vectors form dark blue plaques when plated on *lac* Z hosts in the presence of chromogenic substrate X-gal (5-bromo, 4-chloro, 3-indolyl β-D-galactoside). Cloning in these vectors results in the replacement of *lac* Z gene with foreign DNA and transformants can be recognised from formation of colourless plaques by them when plated on *lac* hosts in presence of X-gal. Some of the examples of phage derived vectors include charon vectors, EMBL-series etc.

M13 phage: Bacteriophage M13 is a filamentous phage having a single stranded closed circular DNA of 6.5 Kb. The phage attaches to F-pili of *E. coli* and is able to infect only male bacterial cells (Hfr strain). Following infection, the phage DNA is converted into a double stranded replicative form (RF) which can be isolated and used as a vector for cloning. After 200–300 copies of RF DNA are synthesized per cell, M13 synthesis becomes asymmetric and large amounts of only one of the two strands, i.e. plus strand are produced which are packaged into mature phage particles. The phage particles are then extruded from the cells without causing cell lysis. Thus the particles contain ssDNA that is homologous to only one of the two complementary strands of cloned DNA and can, therefore, be used as template for DNA sequencing by Sanger's chain termination method.

All genes of bacteriophage M13 are essential for its replication except a small intergenic region of 507 nucleotides which can be used to insert foreign DNA. A series of vectors have been derived from M13 phage that carry a short segment of *E. coli* DNA corresponding to regulatory sequences and coding information for 146 N-terminal amino acids of β-galactosidase (*lac* Z) gene. The F' plasmid of host cell carries a defective β-galactosidase gene. The products of these defective genes complement each other to produce a functional β-galactosidase that results in production of blue colour when bacteria are grown on plates containing IPTG and X-gal. A number of useful cloning sites (polycloning sites- carry sites for a number of restriction enzymes) have been introduced into the *lac* Z region without any significant effect on ability of β-galactosidase peptide to carry out α-complementation. However, addition of extra DNA (foreign DNA) destroys α-complementation producing reccombinants that form white colourless plaques when grown in presence of IPTG and X-gal allowing screening of transformants. Two commonly used M13 derived vectors are M13 mp18 and M13 mp19 both of which contain a polycloning region having sites for 13 restriction enzymes but differ in the orientation of the polycloning region within the *lac* Z region. Using M13 mp 18 and M13 mp19 as dual vectors, it is possible to use a single primer (universal primer which has a sequence complementary to portion of vector) to determine the sequence of nucleotides.

Cosmids: These are the vectors which contain features of plasmids (such as origin of replication, unique restriction enzyme site and drug resistance character) and cos site from phage which allows the plasmid to be packaged like phage particles. DNA fragments upto 45 Kb length can be packaged in these vectors and thus cosmids are used in the construction of genomic libraries.

(iii) Ligation of foreign DNA to vector

The DNA to be cloned is usually obtained as a defined fragment using restriction enzymes and has cohesive ends which can be joined to a vector with the help of DNA ligase. Two types of DNA ligases known are: *E. coli* DNA ligase which requires NAD as cofactor and T_4 DNA ligase which requires ATP as cofactor. T_4 DNA ligase is generally used in recombinant DNA research and is available commercially. The enzyme catalyses the formation of a phosphodiester bond between 5′P and 3′OH ends of two DNA molecules. Both blunt as well as cohesive ends can be joined but ligation is more efficient with sticky or cohesive ends. Optimum temperature for the enzyme is 37°C but at this temperature hydrogen bonding between cohesive ends to be ligated is not stable and, therefore, ligation is usually carried out at 4–15°C. Before ligation vector molecule is dephosphorylated to avoid self annealing of the vector molecule on itself.

(iv) Transformation

The recombinant vector carrying foreign DNA is inserted into the bacterial cells such as *E. coli*. The bacterial cells are made competent by treatment with $CaCl_2$ which makes the cell walls permeable. This is followed by a brief heat shock treatment that allows the cells to take up any foreign DNA. *E. coli* cells from the exponential growth phase are most suitable for this purpose.

(v) Selection of desired transformants

A variety of methods are used to select desired transformants. Initially screening of transformants is done based on insertional inactivation of a marker gene, such as antibiotic resistance gene or gene coding for some enzyme (β-galactosidase), following insertion of a foreign DNA in the vector. This method helps in identifying cells that have taken up recombinant vector. Specific detection methods which involve hybridization with a labelled nucleic acid or antibody probe are required for determining whether desired gene is present in the clones and whether the gene is expressed and produces the required product. The cloned bacterial colonies or phage plaques to be tested are transferred by replica plating from a master plate to a nitrocellulose filter. The filter is treated with NaOH which lyses the cells/phages and denatures DNA so that the DNA is single stranded. The filter is then dried at 80°C for 2 h in vacuum oven to fix the DNA in place and subjected to hybridization under annealing conditions with a labelled probe for the gene of interest. Only those colonies/plaques containing the desired gene will bind the probe and will give a dark spot on autoradiography. The corresponding clones are then picked up from the master plate. The technique is called colony/plaque hybridization. The total number of clones that should be screened from a genomic or cDNA library is given by the following formula:

$$N = \frac{\ln(1 - P)}{\ln(1 - f)}$$

where N = Number of clones to be screened

P = Probability of obtaining a given sequence

f = fractional proportion of the genome in a single recombinant or fractional proportion of the total mRNA that is represented by a single type of rare mRNA. For example, to achieve a 99% probability (P = 0.99) of having a given DNA sequence represented in a library of 17 Kb fragments of a mammalian genome of size 3×10^9 bp

$$N = \frac{\ln(1 - 0.99)}{\ln\left(1 - \left[\frac{1.7 \times 10^4}{3 \times 10^9}\right]\right)}$$

So 8.1×10^5 clones need to be screened.

14.6.1 EXPERIMENT: Elution of plant DNA fragment from agarose gel

Principle

Freezing and thawing break agarose clumps releasing DNA which upon centrifugation appears in the aqueous phase and gets precipitated with two volumes of cold ethanol.

Materials and reagents

1. Agarose slab gel containing DNA fragments: Digest the plant DNA which was isolated in EXPERIMENT 14.2.2 with EcoRI and run the digested sample on agarose gel electrophoresis as described in EXPERIMENT 14.4.1. Also run in a separate lane a DNA Hind III digest as a molecular weight marker.
2. A knife or a razor blade.
3. Tris-saturated phenol: Take redistilled phenol and saturate it with 0.1 M Tris-HCl (pH 8.0).
4. Chloroform : *iso*amyl alcohol (24 : 1).
5. Tris-saturated butanol: Prepare in a similar way as for phenol (Reagent No. 3).
6. 3 M sodium acetate buffer (pH 5.5).
7. Absolute ethanol.
8. 70% ethanol (*v*/*v*).
9. TE Buffer: Prepare as in EXPERIMENT 14.2.6 (Reagent No. 8).

Procedure

1. Cut the agarose slab gel with a razor blade across the region corresponding to 0.5 Kb molecular weight as indicated by marker DNA.
2. Transfer the gel piece to a centrifuge tube and add Tris saturated phenol till the pieces dip in it.
3. Freeze at –70°C for minimum of 2 h.
4. Thaw out the frozen gel pieces gradually to 4–10°C by keeping the tube in refrigerator.
5. Centrifuge at 4°C at 10,000 rpm for 10 min.
6. Remove the aqueous layer and add equal volume of chloroform : *iso*amyl alcohol to it. Centrifuge as above (Step 5).
7. Collect the upper aqueous phase and add equal volume of Tris-saturated butanol to remove ethidium bromide. Centrifuge as above and collect the lower aqueous layer. (Note: The upper layer is of butanol.)
8. Add 3 M sodium acetate, pH 5.5 in such a way that its concentration in the solution is 0.25 M.
9. Precipitate DNA by adding two volumes of chilled absolute ethanol. Keep overnight at –20°C.
10. Centrifuge at 4°C at 10,000 rpm for 10 min. Wash the precipitates once with 70% ethanol, air dry and dissolve in suitable volume (10 μl) of TE buffer. Store at 4°C till further use.

14.6.2 EXPERIMENT: Restriction endonuclease digestion of vector M13 mp11 RF DNA

Principle

Vector (M13 mp 11 RF) is a double stranded circular DNA molecule which has unique restriction enzyme sites in it. Digestion of the vector DNA with EcoRI results in opening of the circular DNA at a unique EcoRI site in such a way that cohesive ends are produced which help in joining the vector molecule with foreign DNA.

Materials and Reagents

1. Vector M13 mp11 RF DNA: Use commercially available vector DNA.
2. EcoRI: Commercially available.
3. 10 × EcoRI assay buffer: As supplied by the manufacturer along with the restriction enzyme.
4. Sterile distilled water.
5. Chloroform : *iso*amyl alcohol (24 : 1).

Procedure

1. Make the following additions in an Eppendorf tube kept on ice:

(i) M13 mp11 RF DNA	4 μg
(ii) EcoRI	12 units
(iii) 10 × EcoRI buffer	3 μl
(iv) Make volume to 30 μl with sterile water	

2. Vortex or spin briefly (30 sec) in a microfuge at full speed to mix the contents thoroughly.
3. Incubate the mixture at 37°C overnight.
4. Add equal volume of chloroform : *iso*amyl alcohol (24 : 1) and spin in microfuge for 5 min at full speed. Collect the upper aqueous layer. Keep it at 4°C.
5. To check whether digestion is complete or not, load small aliquots each of digested DNA sample (5 μl) and undigested (control) DNA (1 μl) in separate lanes on 0.8% agarose gel and run electrophoresis at 80 volts for about 2 h. Stain the gel with ethidium bromide and visualize under UV-transilliuminator. (For agarose gel electrophoresis see Section 14.3(c) and EXPERIMENT 10.4.6).
6. Linear or digested DNA will move slowly compared to control DNA and will appear as a distinct band slightly above the position of control DNA indicating that digestion has occurred.

14.6.3 EXPERIMENT: Dephosphorylation of restriction enzyme digested vector M13 mp11 RF DNA

Principle

Alkaline phosphatase removes 5′ P from ends of DNA or RNA. Dephosphorylated vector DNA is thus prevented from self annealing prior to the insertion of foreign DNA fragments.

Materials and Reagents

1. Vector M13 mp11 RF DNA linearized with EcoRI (see EXPERIMENT 14.6.2).
2. Bacterial alkaline phosphatase (available commercially).
3. Tris-saturated phenol: Prepare as in EXPERIMENT 14.6.1, Reagent No. 3.
4. Chloroform: *iso*amyl alcohol (24 : 1).
5. Phenol-chloroform solution: Mix equal volume of Tris-saturated phenol and chloroform.

Procedure

1. Take 2–3 μg of linearized vector DNA (M13 mp 11 RF DNA cut with EcoRI) in an Eppendorf tube kept on ice.
2. Add bacterial alkaline phosphatase at a concentration 70 units/μg DNA to be dephosphorylated.
3. Incubate the reaction mixture at 65°C for 1 h.
4. Add equal volume of phenol-chloroform solution and spin in a microfuge at full speed for 5 min.
5. Collect the aqueous layer and add equal volume of chloroform : *iso*amyl alcohol to it. Spin as in Step 4.
6. Collect the aqueous layer and store at 4°C till further use.

14.6.4 EXPERIMENT: Construction of recombinant DNA

Principle

DNA fragments from two different sources obtained by digestion with the same restriction enzyme have same complementary or cohesive ends which anneal together due to hydrogen bonding between complementary bases and are joined covalently with the help of DNA ligase.

Materials and Reagents

1. Insert DNA: Plant DNA fragment having EcoRI ends as obtained in EXPERIMENT 14.6.2.
2. Dephosphorylated vector DNA (as obtained in EXPERIMENT 14.6.3).
3. T_4 DNA ligase (commercially available).
4. 10 × ligation buffer: Use the buffer supplied with the enzyme or prepare 0.5 M Tris-HCl, pH 7.6 containing 100 mM $MgCl_2$ and 100 mM dithiothreitol and label it as 10 × ligation buffer.
5. 10 mM ATP: This solution should be prepared fresh just before use.
6. EcoRI fragment: Use commercially available fragment as test DNA (control) which is already known to show ligation.
7. TE buffer: See EXPERIMENT 14.2.6, Reagent No. 8.

8. Sterile distilled water.
9. Phenol-chloroform solution: See EXPERIMENT 14.6.3. Reagent No. 5.
10. Chloroform:*iso*amyl alcohol (24 : 1).

Procedure

1. Make the following additions in Eppendorf tubes kept on ice:

Reagent	*Tube No.*		
	1 *Ligation reaction*	*2* *Control 1*	*3* *Control 2*
Dephosphorylated Vector DNA (ng)	100	100	100
Insert DNA (ng)	40	—	—
λ EcoRI fragment (ng)	—	—	40
10 × ligation buffer (μl)	50	50	50
10 mM ATP (μl)	50	50	50
DNA ligase (units)	2	2	2
Reaction volume (μl) (to be made with sterile distilled water)	500	500	500

2. Mix the contents by vortexing briefly and incubate at 15°C for 5–6 h. Now keep the tubes at 4°C for 24 h.
3. Check ligation by agarose gel electrophoresis. For this, take an aliquot (1 μl) each of ligated DNA and the controls in separate microfuge tubes. Add 9.0 μl TE buffer and 10 μl of gel loading buffer to each tube. Mix well and load on 0.7% agarose gel (for agarose gel electrophoresis see EXPERIMENT 10.4.6). Ligation will be indicated by increase in the size of vector DNA.
4. Add an equal volume of phenol:chloroform to ligated DNA and spin for 10 min. Collect the aqueous layer.
5. To the aqueous layer add an equal volume of chloroform : *iso*amyl alcohol (24 : 1) and spin at 12,000 rpm for 10 min. Collect upper aqueous layer and keep at 4°C till further use.

14.6.5 EXPERIMENT: Transformation of *E. coli* cells with recombinant DNA

Principle

Treatment of bacterial cells with ice-cold solution of $CaCl_2$ followed by a brief heat shock treatment induces a transient state of competence during which the bacteria are able to take up any foreign DNA.

Materials and Reagents

1. *E. coli* JM 103 cells.

2. LB medium

Bactotryptone	10 g
Yeast extract	5 g
NaCl	5 g
Distilled water	900 ml

Adjust pH to 7.5. Make volume to 1L and autoclave.
3. L-agar plates: Make LB medium as above. Add 15 g Bacto-agar to it and autoclave. Cool to 50°C and pour in the plates.
4. 2 × YT media

Tryptone	0.8 g
Yeast extract	0.5 g
NaCl	0.5 g

Add water to make 100 ml and adjust pH to 7.3.
5. Soft agar solution: To the 2 × YT media add 0.6 g Bacto-agar and autoclave.
6. M13 mp 11RF DNA.
7. Ligated DNA (from EXPERIMENT 14.6.4.).
8. 100 mM IPTG (Isopropylthiogalactoside): Prepare by dissolving 23.8 mg IPTG/ml water.
9. 2% X-gal (5-bromo, 4-chloro,3-indolyl-β-galactoside): Prepare by dissolving 25 mg X-gal in 1.25 ml of dimethyl formamide.

Procedure

(a) Preparation of competent cells: Frozen competent cells are availabl(commercially and may be used or prepared as follows:
1. Streak a loopful of the stock of *E. coli* JM 103 on a L-agar plate tc isolate single colonics.
2. Allow the cells to grow overnight at 37°C. (This plate may be used as a stock for single colonies for at least one month when stored at 4°C.)
3. Inoculate a single colony in 2 ml Y_T media (2 ×) and grow for 8–12 h at 37°C.
4. Dilute this fresh overnight culture 1:100 into YT media and incubate at 37°C, until a density of 0.2–0.3 O.D. at 660 nm is obtained.
5. Collect the cells by centrifugation at 8000 rpm for 10 min at 4°C.
6. To obtain exponentially growing JM 103 cells for later use, add an appropriate amount (about 100 fold) of fresh YT media (1 ×) to the remainder of the primary culture.
7. Resuspend pelleted cells in 1/2 of the growth volume in ice-cold 50 mM $CaCl_2$. Keep the suspension on ice for 20 min.
8. Centrifuge and collect the pellet again at 8000 rpm at 4°C for 10 min.
9. Resuspend the pellet in 1/10th of the growth volume in ice-cold 50 mM $CaCl_2$ solution. Label the tube as competent cells.

(b) Transformation

10. Take 40 ng each of M13 mp 11 RF DNA and ligated DNA (recombinant DNA) in separate microfuge tubes.
11. Add 300 μl of the competent cells to each tube and keep the tubes on ice for 40 min.
12. Heat shock the mixture at 42°C for 45 sec and put back on ice for 2 min.
13. Then directly add the following:

    10 μl of 100 mM IPTG
    50 μl of 2% X-gal
    50 μl of exponentially growing JM103 cells
    3 ml of soft agar at 45°C

    Mix and plate directly on YT agar plates. Allow the plates to solidify and then incubate at 37°C until plaques are seen. Turbid plaques will appear after 4 h and the colour reaction after another few hours.
14. After transformation, the recombinants will appear as white plaques while non-recombinants will be of blue coloured plaques.

14.6.6 EXPERIMENT: Isolation of recombinant M13 DNA from white plaques

Principle

Phage particles containing single stranded recombinant DNA are separated from bacterial cells by centrifugation. Polyethylene glycol forms coating around phage particles and allows the formation of pellet. The phage particles are lysed in the presence of protein denaturants such as phenol and chloroform releasing DNA which is then precipitated with two volumes of chilled ethanol.

Materials and Reagents

1. Plates containing white plaques (from EXPERIMENT 14.6.5.).
2. 2 × YT medium (see EXPERIMENT 14.6.5, Reagent No. 4).
3. TES solution: Prepare 20 mM Tris-HCl, pH 7.5 containing 10 mM NaCl and 0.1 mM EDTA.
4. 30% PEG (polyethylene glycol MW 6000) in 5 M NaCl.
5. Tris-saturated phenol (see EXPERIMENT 14.6.1, Reagent No. 3).
6. Chloroform : *iso*amylalcohol (24 : 1).
7. 3 M sodium acetate buffer, pH 5.8.
8. Absolute ethanol.
9. TE buffer (see EXPERIMENT 14.2.6, Reagent No. 8).
10. Tooth pick (autoclaved).
11. *E. coli* JM 103 cells (from EXPERIMENT 14.6.5).

Procedure

1. Transfer white plaques with a sterile tooth-pick into 1 ml of 2 × YT media. To increase the yield of DNA add 10 μl of fresh exponentially growing JM103 cells. Incubate at 37°C with shaking for 8 h.

2. Spin the cells at 12,000 rpm in microfuge for 5 min. Collect the supernatant. Re-spin the supernatant to ensure complete removal of host JM103 cells.
3. To the clear supernatant, add 80-100 μl of 30% PEG in 5 M NaCl. Mix well and keep on ice for 30 min.
4. Spin in microfuge at full speed for 5 min. Decant the supernatant. Remove all liquid with pipetman.
5. To the pellet add 20 μl of TES solution. Mix well by vortexing.
6. Add an equal volume of Tris-saturated phenol and spin at 12,000 rpm in microfuge for 5 min. Collect the upper aqueous phase.
7. Add equal volume of chloroform : *iso*amyl alcohol to the aqueous phase and spin as in Step 6. Collect the upper aqueous phase.
8. To the aqueous phase add 1/10th volume of 3 M sodium acetate buffer (pH 5.8). Add two volumes of chilled absolute ethanol. Keep at –20°C for 12 h or at –70°C for 1 h.
9. Spin in microfuge at full speed for 10 min. Layer the pellet with chilled absolute ethanol, do not mix.
10. Dry the pellet under vacuum for 5 min.
11. Dissolve in appropriate volume (20 μl) of TE buffer.

14.7 Nucleic Acid Sequence Determination

In fact, it is the sequence of nucleotides in DNA that directs the synthesis of RNA and proteins and also regulation of gene expression. Thus the knowledge about DNA sequence is basic for unravelling the structure, function and regulation of gene expression at the molecular level. The overall strategy for nucleic acid sequencing is analogous to that of sequencing of amino acids in proteins. It involves: (i) degradation and fractionation of the nucleic acid molecule to smaller fragments which can then be conveniently and fully sequenced, (ii) sequencing of individual fragments and (iii) ordering of the fragments by using a different degradation procedure that yields a set of nucleic acid fragments that overlap the cleavage points in the preceding degradation step.

Two most reliable techniques for DNA sequence determination are: (a) the chemical degradation and modification method of Maxam and Gilbert (1977) and (b) the enzymatic dideoxy chain termination method of Sanger et al. (1977). Sanger's method is widely used because the technique is simpler and amenable to automation. Both these methods are based on generation of specific sets of radioactively labelled fragments having one end common and differ with respect to a specific base at the other end. The fragments are then separated by high resolution polyacrylamide gel electrophoresis that allows separation of fragments differing from one another only by a single nucleotide and the sequence is then directly read off the autoradiogram of the sequencing gel.

(a) Chemical cleavage and modification method

This method can be used for sequencing both single stranded as well as double stranded DNA. The first step in chemical cleavage method is to radioactively label one end, usually the 5′ end with ^{32}P of the DNA to be sequenced. This DNA is then subjected to cleavage in a base specific manner to produce a set of radioactive fragments whose one end, which is labelled with ^{32}P, is common but the other end terminates with specific base along the polynucleotide chain. Four sets of reactions on separate aliquots of the nucleic acid samples are performed as described as follows:

(i) G cleavage: DNA is treated with dimethyl sulfate (DMS) which methylates the G residues at N 7 of the base thereby making the glycosidic bond of the methylated residue susceptible to hydrolysis. Subsequent treatment with piperidine cleaves the polynucleotide chain at a site just before the G residue.

(ii) A+G cleavage: In the second set, DNA is methylated at A and G bases using dimethyl sulfate. Under acidic conditions both A and G are released resulting in depurinated product. Piperidine treatment then breaks the strand at a point just before both A and G residues. The A residues are then identified by comparing the positions of the G and A+G cleavages.

(iii) C + T cleavage: In the third sample aliquot, DNA is made to react with hydrazine followed by piperidine treatment and this cleaves the DNA molecule before both its C and T residues.

(iv) C cleavage: If DNA is treated with hydrazine in presence of 1.5 M NaCl then only its C residues react appreciably and thus the piperidine cleavage occurs at C residues only. The comparison of C and C+T cleavage positions facilitates identification of the T residues.

In all the four reactions, the conditions are varied so that the strands are cleaved at an average of one randomly located position. These four sets of reaction mixtures are simultaneously electrophoresed in adjacent lanes on a long, thin denaturating polyacrylamide slab gel (0.1 mm thick × 40 cm long). The DNA fragments separate according to their size. After electrophoresis is over, the gel is subjected to autoradiography and sequence of the DNA can then directly be read from the autoradiogram in 5′–3′ direction (as bands from bottom to top).

(b) Dideoxy chain termination method

The method involves a primed synthesis of a complementary radioactive copy of a single stranded DNA template by a DNA polymerase. The base specific ends of the oligonucleotides thus synthesized are produced by incorporating 2′,3′-dideoxynucleoside triphosphates (2′3′ dd NTPs) into the newly synthesized DNA. 2′,3′ ddNTPs differ from the normal dNTPs in that they lack a hydroxyl residue at the 3′ position of deoxy-ribose. They can be incorporated by DNA polymerase (Klenow enzyme) into a growing DNA chain through their 5′-triphosphate groups. However, the absence of a 3′OH residue prevents formation of a phosphodiester bond with the succeeding dNTP. Further extension of the growing chain is, therefore, not possible. The reaction is performed in four different sets each containing a different dideoxynucleotide and this results in production of nested sets of fragments each terminating with dideoxynucleotide being used. The four reaction mixtures are run simultaneously in different lanes on denaturating polyacrylamide sequencing gels. The fragments get separated according to their size and sequence can then be read directly as bands from bottom to top of the autoradiogram. This will indicate the sequence of complementary strand and from this the sequence of the original (template) strand can be computed.

Since 1977, several advancements have occurred in Sanger's method such as development of sequencing vectors, use of non-radioactive fluorescent dideoxy analogues and PCR technology which have led to its automation.

14.7.1 EXPERIMENT: Determination of nucleotide sequence of DNA by dideoxy chain termination method

Principle

The method involves a primed synthesis of a complementary radioactive copy of a single stranded DNA template by a DNA polymerase (Klenow enzyme) in such a way that nested sets of radioactively labelled fragments having one common end but differing at their other end by a specific base are produced. The base specific ends are produced by incorporating 2′3′-dideoxynucleoside triphosphates (2′3′ dd NTPs) into the growing DNA chain, the ddNTPs lack hydroxyl group at 3′ end of deoxyribose and thus when these get incorporated the further elongation of the chain is stopped. The fragments thus synthesized are separated by high resolution polyacrylamide gel electrophoresis that allows fragments differing by a single nucleotide to be resolved and the sequence can then be directly read from the autoradiogram of the gel as dark bands.

Material and Reagents

1. DNA sequencing apparatus: It consists of the following:
   (a) Two glass plates (one notched and the other unnotched), 40×20 cm each.
   (b) Three spacers: Two 50 cm long and one 30 cm long, each 1 cm wide and 0.4 mm thick.
   (c) Shark's tooth comb.
   (d) Electrophoretic chamber.
   (e) Power pack 3000 volts capacity.
   (f) Gel loading pipette or syringe.
   (g) Clamps.
2. Gel dryer.
3. X-ray film.
4. Whatman filter paper.
5. Template DNA: M13 mp11 SS DNA containing plant DNA fragment obtained in EXPERIMENT 14.6.4.
6. TE buffer: See EXPERIMENT 14.2.6, Reagent No. 8.
7. Universal primer: Procure commercially available M13 sequencing primer. Dissolve in TE solution at concentration of 2 μg/ml.
8. $10 \times$ DNA polymerase reaction buffer: Prepare 70 mM Tris-HCl buffer (pH 7.5) containing 70 mM $MgCl_2$ and 500 mM NaCl.
9. ^{35}S dATP (8 μCi/µl, 650 Ci/mmol) or ^{32}P dATP (10 μCi/μl, > 400 Ci/m mol).

10. 0.1 M dithiothreitol (DTT).
11. Klenow enzyme (2 units/μl).
12. dNTP stock solutions: Prepare 0.5 mM solutions of each of dATP, dCTP dTTP and dGTP in TE buffer (pH 8.0). Store at –20°C.
13. dNTPs working solutions: Prepare working solutions of dNTPs by mixing the four 0.5 mM stock solutions of dNTPs in the ratios as shown:

dNTP (X°) working solution	*0.5 mM stock solutions of dNTPs (μl)*			
	dTTP	*dCTP*	*dGTP*	*dATP*
A°	20	20	20	—
C°	20	1	20	—
G°	20	20	1	—
T°	1	20	20	—

Note: Working solution of a dNTP contains lower concentration of a particular dNTP and higher concentrations of other two dNTPs. Working solution containing high concentration of dATP is not required as dATP is supplied in the reaction mixture as (^{35}S) dATP. Solutions may be stored at –20°C upto two weeks.

14. ddNTPs stock solutions: Prepare 10 mM solution of each of ddATP, ddCTP, ddTTP and ddGTP in TE buffer (pH 8.0). Store at –20°C.
15. ddNTPs (ddX) working solutions: Prepare by diluting 10 mM stock solutions of ddNTPs with distilled water as follows:

ddNTP (ddX) working solution	*10 mM ddNTP (μl)*	*H_2O (μl)*
ddA (1 mM)	1	9
ddC (0.35 mM)	1	27.6
ddG (0.7 mM)	1	14.5
ddT (2 mM)	1	5

These solutions can be stored at –20°C upto two weeks.

16. Formamide-dye mix (Gel loading solution). Prepare a mixture by taking:

Formamide	10 ml
Xylene cyanol FF	10 mg
Bromophenol blue	10 mg
0.5M EDTA (pH 8.0)	200 μl

Store at 4°C. Make fresh solution monthly.

17. Methanol-acetic acid solution: Prepare by mixing 100 ml methanol with 100 ml acetic acid and make volume to 1 L with distilled water.
18. X-ray developer.
19. X-ray fixer.
20. Acrylamide-bisacrylamide solution: Prepare a solution containing 38 g of acrylamide and 2 g of bis-acrylamide in 100 ml distilled water. Filter the

solution through nitrocellulose filter (0.45 μ pore size) and store in dark bottle at room temperature.

21. TBE buffer (10 ×): It comprises of

Tris	108 g
Boric acid	55 g
EDTA	9.3 g

Make volume 1 L with water and adjust pH to 8.3.

22. Urea.
23. TEMED as supplied.
24. 10% Ammonium persulfate: Dissolve 1g ammonium persulfate (APS) in 10 ml water.
25. Silconizing fluid.
26. Sealing tape.
27. Paraffin (liquid).

Procedure

(a) Sequencing reaction

1. Add the following to an Eppendorf tube to make a template-primer mixture:

Template DNA (700 ng)	7.0 μl
Primer DNA (5 ng)	2.0 μl
10 × polymerase reaction buffer, pH 7.5	1.0 μl
Paraffin (liquid)	15.0 μl

Keep in boiling water bath for 5 min. Switch off the water bath and allow the contents to cool slowly to room temperature.

2. Add the following components to another Eppendorf tube:

[^{35}S] d ATP	1.0 μl
0.1M DTT	1.0 μl
Klenow enzyme	2–3 units
Sterile distilled water	1.0 μl
Template primer mix (Prepared in Step 1)	10.0 μl

Mix the contents and label it as enzyme-template-primer mix.

3. Carry out the sequencing reaction in four different Eppendorf tubes labelled as G, A, T and C as follows:

	G	A	T	C
X°	G° 1 μl	A° 1 μl	T° 1 μl	C° 1 μl
ddX	ddG 1 μl	ddA 1 μl	ddT 1 μl	ddC 1 μl
Enzyme-template primer mix	3 μl	3 μl	3 μl	3 μl

4. Incubate the tubes at room temperature for 20 min.
5. Add 1 μl 0.5 mM dATP (chase) to each tube and allow the tubes to stand for another 20 min at room temperature.
6. Add 10 μl formamide-dye mix (gel loading solution) to all the tubes. Keep at –15°C till use.

(b) Electrophoresis

7. Wash the two plates (notched and unnotched or smooth) with detergent in hot water and rinse with deionized water. Dry the surface of the plates with tissue paper. Rinse with ethanol and again wipe dry the surface with tissue paper.
8. Put 1 ml of silane (5% solution of dichloro-dimethyl silane in chloroform) on a smooth plate, spread it evenly and keep at room temperature for 30 min. Rinse with ethanol and dry thoroughly with tissue paper (Caution: Ensure complete removal of all silane).
9. Place the spacers on the notched plate and align properly the smooth plate with the notched one and fix the tape at the bottom and on the sides (about 3–4 inches) of both the plates together. Clamp the plates.
10. Make 8% of denaturating polyacrylamide-bisacrylamide solution with 5 ml 10 × TBE buffer and 25 g urea. Keep in boiling water bath to dissolve urea and make the volume to 50 ml. Filter the solution through nitrocellulose paper (0.45 μ) and deareate.
11. Cool the solution and add 50 μl TEMED and 200 μl 10% ammonium persulfate.
12. Immediately transfer the solution between the two plates slowly and continuously using a big syringe, keeping the gel mould slightly slanting at an angle of 30°.
13. Insert the comb. Refill the acrylamide solution, if the gel retracts.
14. Allow the gel to polymerize for about 1 h.
15. Remove the tapes, clamp the plates to the electrophoretic unit.
16. Run the gel at 20–30 mA, ~ 2000 V for 30 min.
17. Heat samples at 90°C for 3 min and keep on ice. Load 1.5 μl of each sample on the gel in different lanes.
18. Run the gel at 20–30 mA, ~ 2000 V, till bromophenol blue runs out (90 min).

(c) Autoradiography

19. Take out the plates from the electrophoretic chamber and keep them on table top. Remove both spacers, insert a spatula at one corner and separate the plates carefully. Make sure that the gel is sticking evenly to the bottom plate.
20. Keep the gel in methanol-acetic acid solution for 10 min (this step is used when ^{35}S dATP is used as an isotope).
21. Transfer the gel onto a Whatman filter paper No. 3 and dry it at 80°C for about 1 h in a gel dryer.

22. Expose the dried gel to an X-ray film for 48–72 h at room temperature.
23. Develop the film for 3 min in X-ray developer in dark. Wash with water and keep for 2 min in fixer. Air dry and label the autoradiogram with date and name of the template. Mark each set of sequencing reactions clearly. Read the sequence from bottom to the top of the autoradiogram as dark bands, in different lanes. This will indicate the sequence of the complementary strand from 5′–3′ direction. From this sequence deduce the sequence of the template strand.

Further suggested reading

Boffey S. (1996). Molecular Biology Techniques. *In* Principles and Techniques of Practical Biochemistry, (Wilson K. and Walker J., eds.) 4th ed., Cambridge University Press.

Bryant J. (1993). Molecular Biology, *In* Methods in Plant Biochemistry (Dey P.M. and Harborne J.B., eds.), Vol. 10, Academic Press, New York.

Erlich H.A. (1989). PCR Technology—Principles and Applications for DNA Amplification, Stockton Press, New York.

Parish J.H. (1972). Principles and Practice of Experiments with Nucleic Acids, Longman Group Ltd., London.

Rickwood D. and Hames B.D. (1982). Gel Electrophoresis of Nucleic Acids, IRL Press, Oxford.

Sambrook J., Fritsch T. and Maniatis, J. (1989). Molecular Cloning—A laboratory Manual, 2nd edn., Cold Spring Harbour Laboratory Press, London.

Appendix

SI System

SI system (Systéme International d' Unités) is the internationally recommended and extensively adopted for measurement of physical quantity. The seven basic units for various physical quantities in SI system are:

Table 1 Standard units of SI system

	Physical quantity	*Unit*	*Symbol*
(i)	Standard of length	Metre	m
(ii)	Standard of mass	Kilogram	kg
(iii)	Standard of time	Second	s
(iv)	Standard of electric current	Ampere	A
(v)	Standard of temperature	Kelvin	K
(vi)	Standard of luminous intensity	Candela	cd
(vii)	Standard of amount of substance	Mole	mol

Very large or small quantities are expressed as multiples of the standard units. The various multiples and sub-multiples of units and their standard prefixes are given in Table 2.

Table 2 Multiples and sub-multiples of standard units

Prefix	*Symbol*	*Multiplication factor*
exa	E	10^{18}
peta	P	10^{15}
tera	T	10^{12}
giga	G	10^{9}
mega	M	10^{6}
kilo	k	10^{3}
hecto	h	10^{2}
deca	da	10^{1}
deci	d	10^{-1}
centi	c	10^{-2}
milli	m	10^{-3}
micro	μ	10^{-6}
nano	n	10^{-9}
pico	p	10^{-12}
femto	f	10^{-15}
atto	a	10^{-18}

Table 3 SI units of mass (weight)

Weight	*Symbol*	*Multiple of gram*
Megagram	M	10^{6}
Kilogram	k	10^{3}
Gram	g	1
Milligram	m	10^{-3}
Microgram	μ	10^{-6}
Nanogram	n	10^{-9}
Picogram	p	10^{-12}
Femtogram	f	10^{-15}

Table 4 Greek alphabets

Letter	*Name*	*Letter*	*Name*
α	Alpha	ν	Nu
β	Beta	ξ	Xi
γ	Gamma	o	Omicron
δ	Delta	π	Pi
ε	Epsilon	ρ	Rho
ζ	Zeta	σ*, ζ**	Sigma
η	Eta	τ	Tau
θ	Theta	υ	Upsilon
ι	Lota	φ	Phi
κ	Kappa	χ	Chi
μ	Mu	ω	Omega

*At the beginning and middle of a word.
**At the end of a word.

Table 5 Concentrations of some acids and bases

Acid or base	*Molecular weight*	*Specific gravity*	*Percent by weight*	*Normality N*	*ml/L for 1 N solution*
Acetic acid	60.1	1.05	99.5	17.6	56.9
Ammonium hydroxide	35.0	0.89	28.0	15.1	66.5
Formic acid	46.0	1.20	97.0	25.5	39.2
Hydrochloric acid	36.5	1.18	36.0	11.7	85.5
Nitric acid	63.0	1.42	69.5	15.6	64.0
Perchloric acid	100.5	1.69	70.0	11.6	85.7
Phosphoric acid	98.0	1.69	85.0	44.1	22.7
Sulphuric acid	98.08	1.84	96.0	35.9	28.4
Thioglycolic acid	92.1	1.26	80.0	10.9	91.3
Pyridine	79.1	0.98	100.0	12.4	80.6
2-Mercaptoethanol	78.13	1.14	100.0	14.6	68.5

Table 6 pK values and pH range of some commonly used biological buffers

Name	pKa_1	pKa_2	pKa_3	*Useful pH range*
Acetic acid	4.75	—	—	3.6–5.6
ADA (N-2-acetamidoiminodiacetic acid)	6.60	—	—	6.0–7.2
BIS-TRIS PROPANE	6.80	—	—	6.3–9.5
Bicine [N, N—*bis*(2-hydroxyethyl)glycine]	8.30	—	—	7.6–9.0
Boric acid	9.24	—	—	7.6–9.2
Citric acid	3.08	4.75	5.40	3.6–5.6
Glycine	2.35	9.78	—	
Glycine amide hydrochloride	8.20	—	—	7.4–8.8
Glycyl glycine	3.14	8.07	—	7.5–8.9
HEPES (N-2-hydorxyethylpiperazine-N′-2-ethano sulfonic acid)	7.55	8.00	—	6.8–8.2
Imidazole	7.07	—	—	6.2–7.8
MES [2-(N-morpholino)ethanesulfonic acid]	6.15	—	—	5.5–6.7
MOPS [3-(N-morpholino)propanesulfonic acid]	7.20	—	—	6.5–7.9
Phosphoric acid	1.96	7.12	12.32	5.8–7.8
TES (N-*tris*(hydroxymethyl)methyl-2-aminoethanesulfonic acid	7.50	—	—	6.8–8.2
TRICINE [N-*tris*(hydroxymethyl)methyl glycine]	8.15	—	—	7.4–8.8
TRIS [*tris*-(hydroxymethyl)aminomethane	8.30	—	—	7.2–9.0
Succinic acid	4.19	5.57	—	3.8–6.0

Table 7 Filters for colorimetry

Wavelength (nm)	*Colour of filter used*	*Colour observed*
400	Violet	Greenish yellow
425	Indigo blue	Yellow
450	Blue	Orange
490	Blue green	Red
510	Green	Purple
530	Yellow green	Violet
550	Yellow	Indigo blue
590	Orange	Blue
640	Red	Bluish green
730	Deep red	Green

Ranges of spectrophotometry

Range	*Wangelength (nm)*	*Radiation source*	*Cuvette*
UV	185–400	Deuterium or hydrogen lamp	Silica (Quartz)
Visible	400–760	Tungsten filament lamp	Borosilicate glass or plastic

Table 8 Ammonium sulphate fractionation of enzymes in sample preparations

Initial concentration of ammonium sulphate (% saturation)	*Final required concentration of ammonium sulphate (% saturation)*																
	10	20	25	30	33	35	40	45	50	55	60	65	70	75	80	90	100
	Grams solid ammonium sulphate to be added to 1 L of solution																
0	56	144	144	176	196	209	243	277	313	351	390	430	472	516	561	662	767
10		57	86	118	137	150	183	216	251	288	326	365	406	449	494	592	694
20			29	59	78	91	123	155	189	225	262	300	340	382	424	520	619
25				30	49	61	93	125	158	193	230	267	307	348	390	485	583
30					19	30	62	94	127	162	198	235	273	314	356	449	546
33						12	43	74	107	142	177	214	252	292	333	426	522
35							31	63	94	129	164	200	238	278	319	411	506
40								31	63	97	132	168	205	245	285	375	469
45									32	65	99	134	171	210	250	339	431
50										33	66	101	137	176	214	302	392
55											33	67	103	141	179	264	353
60												34	69	105	143	227	314
65													34	70	107	190	275
70														35	72	153	237
75															36	115	198
80																77	157
90																	79

Table 9 Extinction coefficient of some important biomolecules

Compound	λ_{max} *(nm)*	$A_m \times 10^3$*
Adenine nucleotides (AMP, ADP, ATP)	260	15.4
NAD(P)H	340	6.22
NAD(P)	260	17.8
CMP	270	9.0
UMP	260	7.8
GMP	252	13.7
FAD	445, 366	11.3 (at 445 nm)
Ascorbic acid	265	7.0
Pyridoxal 5-phosphate	388	4.9
Pyridoxamine	325	7.7
Coenzyme Q	275 (in 95% ethanol)	15.0
Vitamin A	325	52.4
Coenzyme A	260	16.8
Cytochrome *c*	550	30.2
2,6-Dichlorophenolindophenol (DCPIP)	600	20.6

*Molar extinction coefficient.

Table 10 Stains for electrophoresis

Stain	*Stain specificity*				*Comments*
	Glyco-protein	*Pro-tein*	*Nucleic-acids*	*Poly-saccharides*	
Acridine orange			√		RNA specific
Amido black 10B		√			
8-Anilinonaphthalene-1-sulphonic acid		√			Fluorescent stain
Commassie brilliant blue G-250	√	√			Commonly used stain
Commassie brilliant blue R-250	√	√			Sensitive, durable stain
Commassie brilliant violet R-150		√			Weaker background stain with IEF* gels compared with Commassie brillant blue
4′,6′-diimidino-2-phenylindol			√		Fluorescent stain
Ethidium bromide			√		Most commonly used stain
Fast green FCF		√			Quantitative determination of protein by colour intensity possible
Methylene blue			√		Specific for RNA
Methyl green			√		

(Contd)

Table 10 *(Contd)*

Stain	*Stain specificity*				*Comments*
	Glyco-protein	*Pro-tein*	*Nucleic-acid*	*Poly-saccharides*	
Panceau S		√			Particularly suitable for staining of albumins and globulins on cellulose acetate foil
Silver stain	√	√	√		
Stains all (E.Merck)	√	√	√	√	Universal stain
Sudan black		√			Special stain for lipoproteins

*IEF—isoelectric focussing.

Table 11 Characteristics of some commonly used ion-exchangers

Type	*Matrices*	*Functional group*	*Name of functional group*
Weak cation exchangers	Agarose Cellulose Dextran Polyacrylate	$—COO^-$ $—CH_2COO^-$	Carboxy Carboxymethyl
Strong cation exchangers	Cellulose Dextran Polystyrene	$—SO_3^-$ $—CH_2SO_3^-$ $—CH_2CH_2CH_2SO_3^-$	Sulpho Sulphomethyl Sulphopropyl
Weak anion exchanger	Agarose Cellulose Dextran Polystyrene	$—CH_2CH_2N^+H_3^-$ $—CH_2CH_2N^+H(CH_2CH_3)_2$	Aminoethyl Diethylaminoethyl
Strong anion exchangers	Cellulose Dextran Polystyrene	$—CH_2N^+(CH_3)_3$ $—CH_2CH_2N^+(CH_2CH_3)_3$ $—CH_2N^+(CH_3)_2$ CH_2CH_2OH $—CH_2CH_2N^+(CH_2CH_3)_2$ CH_2CHCH_3 OH	Trimethylaminomethyl Triethylaminoethyl Dimethyl-2hydroxyethyl aminoethyl Diethyl-2-hydroxypropyl aminoethyl

Table 12a Properties of gel filtration material

Gel type	*Fractionation range (mol. wt.)*		*Bed volume (ml/g dry material)*	*Dry bead diameter (μm)*
	Peptides and globular proteins	*Dextrans*		
Sephadex G—10	–700	–700	2–3	40–120
Sephadex G—5	–1500	–1500	2.5–3.5	40–120
Sephadex G—25 (Fine)	1000–5000	100–5000	4–6	20–80
Sephadex G—50 (Fine)	1500–30,000	500–10,000	9–11	20–80
Sephadex G—75	3,000–80,000	1000–50,000	12–15	40–120
Sephadex G—100	4,000–150,000	1000–100,000	15–20	40–120
Sephadex G—150	5,000–150,000	1000–150,000	20–30	40–120
Sephadex G—200 (Super fine)	5,000–250,000	1000–200,000		10–40
Sephacryl S—200	$5{,}000–2.5 \times 10^5$	$1 \times 10^3–8 \times 10^4$		40–105*
Sephacryl S—300	$1 \times 10^4–1.5 \times 10^6$	$1 \times 10^3–7.5 \times 10^5$		40–105*
Sepharose 2 B	$7 \times 10^4–40 \times 10^6$	$1 \times 10^5–20 \times 10^6$		60–200*
Sepharose 4 B	$6 \times 10^4–20 \times 10^6$	$3 \times 10^4–5 \times 10^4$		60–140*
Sepharose 6 B	$1 \times 10^4–4 \times 10^6$	$1 \times 10^4–1 \times 10^6$		45–165*

* Diameter of wet beads.

Table 12b Properties of gel filtration materials

Type	*Exclusion limit*	*Water regain (g/g gel)*	*Bed volume per g dry gel (ml)*	*Particle size (micron)*
Polyacrylamide types				
Bio–Gel P–2	1600	1.5	3.8	50–400
Bio–Gel P–4	3600	2.4	5.8	50–400
Bio–Gel P–6	4600	3.7	8.8	50–400
Bio–Gel P–10	10,000	4.5	12.4	50–400
Bio–Gel P–30	30,000	5.7	14.8	50–400
Bio–Gel P–60	60,000	7.2	19.0	50–400
Bio–Gel P–100	100,000	7.5	19.0	50–400
Bio–Gel P–150	150,000	9.2	24.0	50–400
Bio–Gel P–200	200,000	14.7	34.0	50–400
Bio–Gel P–300	300,000	18.0	40.0	50–400
Agarose types				
Bio–Gel A–0.5 m	500,000			50–400
Bio–Gel A–1.5 m	1.5×10^6			50–400
Bio–Gel A–5 m	5.0×10^6			50–400
Bio–Gel A–15 m	15×10^6			50–400
Bio–Gel A–50 m	50×10^6			50–400
Bio–Gel A–150 m	150×10^6			50–400

Table 13 Some characteristics of affinity chromatography matrices

Type of material	*Type of ligands attached*
Cyanogen bromide activated Sepharose 4B	All types of proteins and ligands containing primary amino group.
AH—Sepharose 4B	Has 6—C long spacer arm (1,6-diaminohexane) and has free amino group. Any ligand having carboxyl group.
CH—Sepharose 4B	Has 6—C long spacer arm (6—aminohexanoic acid) and has free carboxyl group. Any ligand having amino group can be attached.
Epoxy activated Sepharose 6B	Has a hydrophilic spacer and oxirane ring. Ligands having hydroxyl, amino or thiol groups can be attached.
Activated CH—Sepharose 4B	Activated form of CH Sepharose 4B formed by esterification of carboxylic group using N—hydroxysuccinimide compounds having free primary amino groups.
Con A Sepharose	Concanavalin A attached to Sepharose 4B. A group specific adsorbent for polysaccharides and glycoproteins.
Poly(U) Sepharose 4B	Used for isolation of mRNA having poly(A) sequences.
5′—AMP Sepharose 4B	For adsorption of kinases and dehydrogenases which require a cofactor having an adenylate residue.

Table 14 Important emission lines in flame photometeric spectra of some elements

Element		*Wavelength (nm)*	
Ag	328.1	338.3	
Ba	553.6	744.0*	873.0*
B	452.0*	548.0*	345.0*
Ca	422.7	554.0*	622.0*
Co	346.6***	353.0	387.4
Cr	360.5	427.5***	425.5
Cs	455.5	852.1	894.3
Cu	324.8	327.4	520.0*
Fe	373.7***	386.0***	385.6***
K	404.7**	766.5**	344.6**
Li	670.8	460.3	323.3
Mg	285.2	371.0*	383.0*
Mn	403.3***	543.3	279.5
Na	330.3**	589.3**	818.3**
Ni	341.5***	352.5***	385.8***
Pb	368.4	405.8	261.4**
Rb	420.2**	780.0	794.8
Sr	460.7	821.0*	407.8
Ti	377.6	535.0	276.8

*Band of the oxide.
**Dual line, the centre point between the two lines is given.
***Group of lines in the region of the stated wavelength.

Subject Index